Biosensors

Emerging Materials and Technologies
Series Editor
Boris I. Kharissov

MXene-Based Photocatalysts: Fabrication and Applications
Zuzeng Qin, Tongming Su, and Hongbing Ji

Advanced Electrochemical Materials in Energy Conversion and Storage
Junbo Hou

Emerging Technologies for Textile Coloration
Mohd Yusuf and Shahid Mohammad

Emerging Pollutant Treatment in Wastewater
S.K. Nataraj

Heterogeneous Catalysis in Organic Transformations
Varun Rawat, Anirban Das, Chandra Mohan Srivastava

**2D Monoelemental Materials (Xenes) and Related Technologies
Beyond Graphene**
Zongyu Huang, Xiang Qi, Jianxin Zhong

Atomic Force Microscopy for Energy Research
Cai Shen

**Self-Healing Cementitious Materials: Technologies, Evaluation Methods,
and Applications**
Ghasan Fahim Huseien, Iman Faridmehr, Mohammad Hajmohammadian Baghban

Thin Film Coatings: Properties, Deposition, and Applications
Fredrick Madaraka Mwema, Tien-Chien Jen, and Lin Zhu

Biosensors: Fundamentals, Emerging Technologies, and Applications
Sibel A. Ozkan, Bengi Uslu, and Mustafa Kemal Sezgintürk

**Error-Tolerant Biochemical Sample Preparation with Microfluidic
Lab-on-Chip**
Sudip Poddar and Bhargab B. Bhattacharya

Geopolymers as Sustainable Surface Concrete Repair Materials
Ghasan Fahim Huseien, Abdul Rahman Mohd Sam, and Mahmood Md. Tahir

For more information about this series, please visit:
www.routledge.com/Emerging-Materials-and-Technologies/book-series/CRCEMT

Biosensors
Fundamentals, Emerging Technologies, and Applications

Edited by
Sibel A. Ozkan, Bengi Uslu, and
Mustafa Kemal Sezgintürk

CRC Press
Taylor & Francis Group
Boca Raton   London

CRC Press is an imprint of the
Taylor & Francis Group, an **informa** business

First edition published 2023
by CRC Press
6000 Broken Sound Parkway NW, Suite 300, Boca Raton, FL 33487–2742

and by CRC Press
2 Park Square, Milton Park, Abingdon, Oxon, OX14 4RN

© 2023 Taylor & Francis Group, LLC

CRC Press is an imprint of Taylor & Francis Group, LLC

ISBN: 978-1-032-03865-0 (hbk)
ISBN: 978-1-032-03866-7 (pbk)
ISBN: 978-1-003-18943-5 (ebk)

DOI: 10.1201/9781003189435

Contents

PART I   Introduction to Biosensors

PART II   Transducers for Biosensors

PART III   Biorecognition Elements Used in Biosensors

PART IV   Applications of Biosensors

PART V   Challenges and Future Prospects

Preface

The story was started in 1956 by the invention of a glucose biosensor based on the electrochemical detection of oxygen, in which the critical role was attributed to Clark's oxygen electrode. Interestingly, before this invention, when Prof. Leland Clark tried to publish the results of his first bubble oxygenator, his manuscript was rejected by the journal's editor because it could not measure oxygen tension rising from the oxygenator. However, that did not stop him, motivating him to expose the oxygen electrode. Since his invention, the Clark electrode has been used in countless biosensing systems to detect numerous analytes associated with clinical, food, environmental, and other settings. Historically Clark's oxygen electrode was the milestone in developing novel biosensing concepts used for all scientific fields. Over the years, the fundamental sciences have improved the modern technological devices that have accelerated the developments of basic sciences. As a result of this progression, since the introduction of the first generation of glucose biosensors by Clark, new transduction systems and novel bioreceptors such as antibodies, whole cells, nucleic acids, aptamers, etc. have been introduced and have been manipulated for the detection of many different kinds of targets. However, portable, affordable, user-friendly, and integrated biosensing systems have been adequately used in people's services despite all these manipulations. Furthermore, there is an increasing need to integrate miniaturized electronics and biological/synthetic/engineered bioreceptors in a chip smaller than a pinhead.

The need for such portable diagnostic tools is vitally important for developing countries that do not possess modern laboratory facilities. Apart from these, personalization in medicine also requires rapid and portable biosensing concepts to be progressed by keeping up with technological progress. There is no doubt that nanomaterials have introduced new opportunities to develop biosensors that enable detection of trace amounts of target analytes economically, rapidly, and with only a small required sample volume. So, nanomaterials play a key role in any developmental processes that involve biosensors, biochips, and high throughput arrays. Moreover, the revolutionary advances in microfabrication technologies have allowed the development of lab-on-a-chip (LOC) and micro total analysis systems (μTAS), which are miniaturized and integrated biosensors in microfluidic-based chips. The integration of biosensors with microelectromechanical systems (MEMS) and nanomaterials has resulted in numerous miniaturized biosensing systems that could be performed to detect any targets coming from clinical, food industry, environmental, and other fields. Although there have been enormous advances in this field, there is still a need for convenient and affordable biosensing systems. This situation can be turned into an opportunity to make actual scientific progress.

This book aims to highlight the state-of-the-art field of biosensors with all aspects of the technology including nanomaterials applications, miniaturized and integrated systems, novel bioreceptors and immobilization strategies, and beyond. The prominent importance of the book is that it appeals to the general reader, from graduate/

postgraduate students to researchers and scientists in the field. In light of this statement, the topics in this book include:

Part I: Introduction to Biosensors
Part II: Transducers for Biosensors
Part III: Biorecognition Elements Used in Biosensors
Part IV: Applications of Biosensors
Part V: Challenges and Future Prospects

This book could not have finished successfully with many people's support. They have given their breadth of experiences to these book chapters step by step during the difficult COVID-19 pandemic duration. All book chapters are written by eminent scientists who are universally recognized in different branches of the biosensor field. Consequently, it should be mentioned that we appreciate all authors who made this book project possible by dedicating themselves to perfecting these chapters. We hope this book will be helpful to all people concerned.

We would like to thank our families for their unwavering support and love. In addition, we would like to express our gratitude to Gabrielle Vernachio, our project manager, and CRC Press Publishers, our publisher, for their assistance in publishing this book. Finally, we would like to express our sincere gratitude to leading authors who accepted our invitation to join us and dedicated their valuable time and efforts to guarantee the success of this book.

Sibel A. OZKAN
Bengi USLU
Mustafa Kemal SEZGINTÜRK

Editor Biographies

SIBEL A. OZKAN is currently working as Full Professor of Analytical Chemistry at Ankara University, Faculty of Pharmacy, where she has been active since 1986.

She is an active member of the European Chemical Society-DAC member on behalf of the Turkish Chemical Society. She is a member of the European Pharmacopoeia-EDQM-Chromatography Section. She is also a member of PortASAP: European Network for the Promotion of Portable, Affordable, and Simple Analytical Platform: Core Group of Cost Action CA 16215, Working Group 4.

She has been involved in several analytical chemistry projects related to Liquid Chromatography (LC) methods, separation techniques, chiral separation, drug analysis in dosage forms and biological samples, electrochemical biosensors, nanosensors, DNA biosensors, enzyme biosensors, biomarkers, environmental sensors, method development, and validation of drug assay.

She has published more than 360 original and review papers (indexed in ISI-WoS), and she is the editor of nine scientific books and about 50 book chapters. She has more than 8,000 citations of research papers with an h-index of 43. She has several international and national projects and awards.

She is the editor of the *Journal of Pharmaceutical and Biomedical Analysis* (SCI) and Regional Editor (Europe) of *Current Pharmaceutical Analysis* (SCI). She is also an editorial board member of *Talanta* (SCI), *Chromatographia* (SCI), *Biosensor & Bioelectronics X*, and other important journals.

BENGI USLU is currently a full professor in Ankara University Faculty of Pharmacy, Department of Analytical Chemistry, in Turkey. Her areas of interest are electrochemistry, biosensors, nanosensors, biotechnology, separation techniques, liquid chromatography, drug analysis, bioanalytical method development, redox mechanisms of drugs, spectrophotometry, and analytical method validation. Her postdoctoral research, sponsored by The Scientific and Technological Research Council of Turkey, was conducted with Prof. Emil Palecek at the Institute of Biophysics, Lab of Biophysical Chemistry and Molecular Oncology in the Czech Republic. She has authored or co-authored more than 180 peer-reviewed full papers and more than 200 oral or poster presentations in scientific conferences. She has more than 30 book chapters and took part in the translation of many books into Turkish. She has more than 3,500 citations of research papers with an h-index of 33. Prof. Dr. Bengi Uslu is the editor of the book *Electroanalytical Applications of Quantum Dot Based Biosensors*, which was published in Elsevier in 2021. She has completed 17 national/international projects and has 7 ongoing projects. She is an active member of the European Directorate for the Quality of Medicines (EDQM), 10A, and The International Council for Harmonization of Technical Requirements for Pharmaceuticals for Human Use (ICHQ14). She received the Ankara University Scientific Support Award in 2008, the Ankara University Scientific Award in 2014, and the Academy of Pharmacy Science Award of the Turkish Pharmaceutical Association in 2015.

MUSTAFA KEMAL SEZGINTÜRK studied chemistry at Ege University (Turkey) and was awarded a BSc degree in 1998 and an MSc degree in Biochemistry in 2002. Until completing his DSc of Biochemistry at Ege University in 2007, he served as a research assistant in Biochemistry. Sezgintürk's earlier research activities focused on electrochemical biosensors based on conventional solid electrodes to detect small metabolites, food additives, and toxins. His current research includes developing disposable electrode materials and novel immobilization strategies for electrochemical biosensing applications (especially for cancer biomarker diagnosis) and the development of new lateral flow assays. In these directions Mustafa Kemal Sezgintürk has over 130 publications (refereed journal papers, reviews, and book chapters) and two patents in the field of biosensors and lateral flow assays. He has a Google Scholar h-index of 31 (2021). He published the first extended and updated book on the field of commercial biosensors (*Commercial Biosensors and Their Applications: Clinical, Food, and Beyond*, Elsevier, 2020). He founded the National Congress on Biosensors in 2014. In 2016, Prof Sezgintürk was awarded one of the highest honors of Turkey, the Outstanding Young Scientist Awards by the Turkish Academy of Sciences (TÜBA).

Part I

Introduction to Biosensors

1 Overview of Biosensors
Definition, Principles, and Instrumentation

S. Irem Kaya[1,2], Didem Nur Unal[1],
Ahmet Cetinkaya[1], Bengi Uslu[1],
and Sibel A. Ozkan[1]

[1]Ankara University
Faculty of Pharmacy, Department of Analytical Chemistry
Ankara, Turkey

[2]University of Health Sciences
Gulhane Faculty of Pharmacy, Department of Analytical Chemistry
Ankara, Turkey

CONTENTS

1.1 INTRODUCTION

In the most general sense, a biosensor is described as an analytical device that consists of a bioreceptor (biological recognition element; enzyme, DNA, antibody, cell . . .) and a transducer (conversion of the information into a measurable signal;

DOI: 10.1201/9781003189435-2

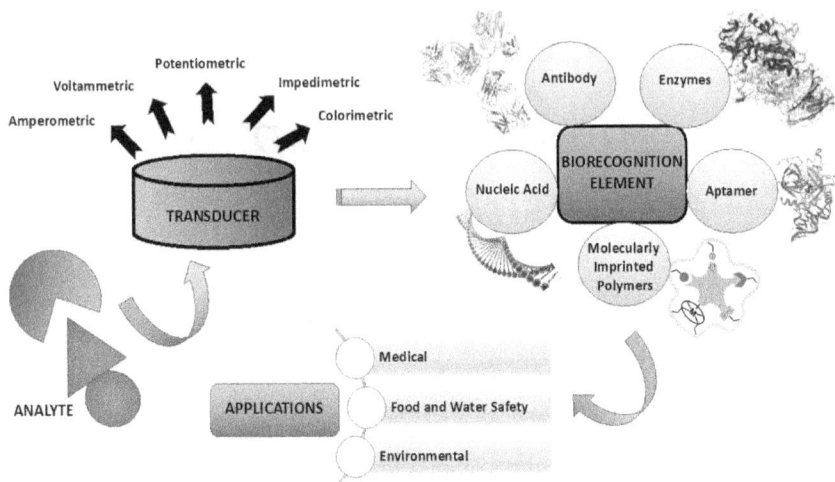

FIGURE 1.1 Schematic representation of the working principle of biosensors and applications in different fields.

optical electrochemical, magnetic . . .). Selectivity, sensitivity, stability, and biocompatibility are the most important parameters that describe a biosensor (1).

Since the first described as an enzyme biosensor in the literature in 1962 by L. Clark, biosensors have evolved and changed into efficient analytical tools. Thanks to their features such as miniaturization, affordability, rapidness, and portability, which cannot be achieved with conventional analytical techniques, biosensors are preferred over classic methods in a wide range of fields from environmental analysis to clinical analysis (2).

The working principle of biosensors is the detection with biorecognition element and then converting this into a measurable signal using the transducer (Figure 1.1). Although the basic principle of all biosensors is the same, various biorecognition elements and transducers can be selected according to the aimed analysis (3). The classification of biosensors based on bioreceptors and transducers will be explained in detail in the following sections.

One of the outstanding features of biosensors is that they have a great potential to be used as commercial products. When biosensors are evaluated in health care, their use for diagnosis, treatment, and disease monitoring is critical in clinical analysis. Additionally, developing devices that patients can apply themselves is a significant step in this topic. In addition to biomedical applications, biosensors are also utilized in environmental, food, and water safety analysis (4). On-site analysis of environmental resources for various pollutants requires advanced analytical tools. Therefore, biosensors are good options to overcome this issue.

The intensive work of researchers on biosensors combined with the developing technology has led to the formation of more advanced approaches for biosensors. Wearable biosensors are relatively novel applications that are still improving. They are able to offer some additional advantages, such as real-time analysis (2, 5).

This chapter gives an insight into biosensors by overviewing their classifications and applications.

1.2 CLASSIFICATION OF BIOSENSORS

1.2.1 BIORECOGNITION ELEMENT BASED

Analyte specificity is significant for a biosensor, and the purpose of the biorecognition element is to provide this specificity. The specificity that requires selective and robust affinity occurs between the bioanalyte and the biorecognition elements. Biorecognition elements can be classified differently in terms of affecting biosensor properties. Therefore, it is important to know the properties of the biorecognition element for a detailed analysis of biosensors (3). Biological recognition elements can be natural or synthetic structures. Structures derived through physiological interactions and biological processes are naturally occurring systems.

Conversely, structures developed to mimic physiological interactions are defined as synthetic structures. However, some methods emphasize both natural and synthetic recognition structures. This approach is known as the natural method, and supramolecular structures are called natural biorecognition elements. These structures were designed using natural subunits (3, 6–7). Figure 1.2 summarizes the advantages and disadvantages of biorecognition elements in biosensor studies.

1.2.1.1 Antibody

In biosensor development, the selection step of the biosensing element used for rapid and reliable analysis of the target analyte is important. Monoclonal antibodies (mAb) are the most prominent biorecognition element. Immunoglobulins (Ig) or antibodies are macromolecules (about 150 kDa) produced by the immune system. Immunoglobulins such as IgE, IgM, IgG are frequently used in biodetection fields. When the structure of an IgG antibody is examined, it consists of heavy and light protein chains. As shown in Figure 1.3, an asymmetric Y-shaped 3D conformational structure is formed by disulfide bonding of IgG, a heavy and a light protein chain (8–9).

The complexity of antibody binding to the sensor surface is a significant challenge in developing antibody-based biosensors. Although various methods of physical

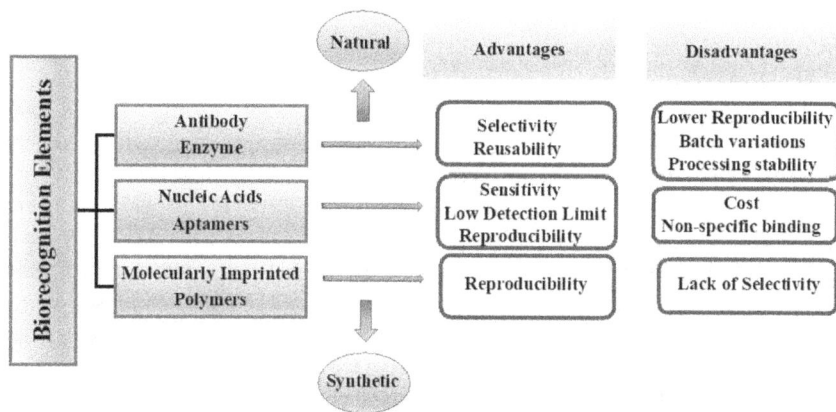

FIGURE 1.2 Schematic representation of the advantages and disadvantages of biorecognition elements.

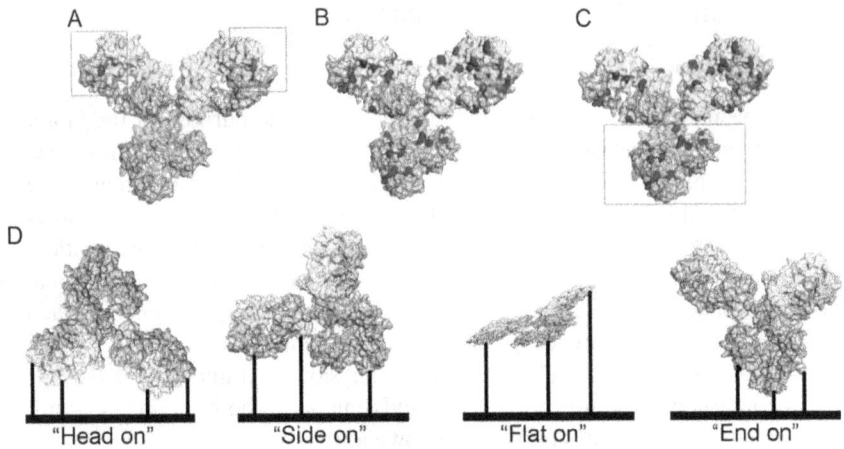

FIGURE 1.3   3D structure of IgG. (A) For protocol A, N-terminal groups are the only sites involved in the immobilization; (B) For protocol B, antibodies will be immobilized on the support surface through N-terminal groups and lysine residues; (C) For protocol C, the immobilization sites was limited on the Fc region of the antibody. The immobilization sites with the most possibility are indicated by the red boxes. (D) Four possible orientations when immobilizing antibodies onto solid supports. The light chains are represented in cyan, and the heavy chains are light blue. N-terminal amino groups are in brown, and polysaccharide moieties are presented in green. Lysine residues (blue color) are distributed on the whole area of the IgG surface, and histidine clusters (violet color) are located on the Fc fragment of the antibody. The figure was created with PyMol 3.3 (DeLano, USA) using the PDB ID: 1IGY. Reprinted from ref. (8) with permission from Elsevier.

adsorption, covalent bonding, and electrochemical polymerization have been suggested for the immobilization of the antibody, the appropriate behaviour of antibodies on the sensor surface in antibody-based biosensors differs for each method (10). Polysaccharide chains, sulfhydryl, and amino moieties on the antibody surface are promoter groups to conjugate Immunoglobulin G (IgG). Therefore, the biological activity of antibodies is provided at a high rate by using sulfhydryl and aldehyde groups (8). In addition, antibody production can be very costly as it involves process steps such as isolation and purification. Especially for discovering new antibodies, experimenting on animals is a disadvantage in terms of time and cost. Despite such recognized limitations of antibodies, they are still essential in antibody-based biosensor applications (11–12).

Rapid, sensitive, and reliable detection of SARS-CoV-2 (IgA)-specific immunoglobulin A (IgA) is important to prevent the spread of the SARS-CoV-2 (COVID-19) epidemic that has swept the world for the last two years and to detect the infection. Roda et al. (13) developed a lateral flow immunoassay (LFIA) immunosensor that includes both optical and chemiluminescence detection of IgA in serum and saliva. It is the beginning of the response to viral infection with the production of immunoglobulins secreted at a developing infection site in the body. Conventional enzyme-linked immunosorbent assay (ELISA) methods accurately measure serum IgA for COVID-19 infection. These studies have shown that serum IgA is produced in more significant quantities than IgM with time-dependent kinetics (13–15). Therefore, this study is based on IgA as an alternative and reliable biomarker for detecting COVID-19. Thanks to optical and chemiluminescence detection, they have enabled the highly sensitive, simple, and economical determination of IgA against SARS-CoV-2 with a simple smartphone-based immunosensor.

1.2.1.2 Enzymes

Enzymes capture the target bioanalyte through cavities within their 3D structure, and this interaction is converted into a measurable signal using amperometric or electrochemical methods. The enzyme-based biosensors create a specific recognition pattern using hydrogen bonding, electrostatic interaction, and non-covalent interactions in the binding cavities (6). This recognition pattern, which gives high specificity and accuracy to the biosensor, is extremely important for bioanalyte specificity (16–17). Enzymes embedded in surface structures allow the formation of diffusion pathways between biorecognition elements and electrochemical transducers (18). Enzyme-based biosensors are formed due to some changes (color change, mass change, thermal change, etc.) of the enzyme used as an analyte recognition unit. The resulting difference is converted into an electrical signal by the signal converter, such as electrochemical and optical. The concentration of the analyte is determined using these signals. Enzyme biosensors are used in a wide range of fields such as medicine, food, environment, and health. Enzyme biosensors are more preferred in the field of health and treatment due to their applications such as blood glucose sensor, cholesterol sensor, lactic acid sensor, and urine glucose sensor. In the food field, it is used as a biomarker for freshness testing as well as for the detection of components such as alcohols, ethers, cholesterol, and lipids. In addition, enzyme biosensors are used to detect chemical substances such as ammonia, nitric acid, and nitrite in soil and water quality. Enzymes are preferred for molecular recognition as antibodies cannot provide a high affinity for low molecular weight biotransformers (19).

When the latest technologies in enzyme-based biosensors are evaluated, it can be seen that nanozymes, nanoparticles (NPs), especially metal NPs (such as gold, platinum, molybdenum, and silver) provide an alternative to biological enzymes used in traditional enzyme-based biosensors (20–21). In a recent study, a new portable nanoenzyme-based biosensor was developed for monitoring fish freshness measuring hypoxanthine (HX), a product of nucleotide degradation in meat and fish. This work describes an integrated biosensing platform that exploits the ability of cerium nanoparticles (CeNPs) to react with hypoxanthine (HX) through surface oxidation and formation of charge-transfer complexes, producing a dark orange color on the surface of the NP due to the formation of cerium peroxide, whose density changes with HO. Concentration to impart selectivity for HX detection, CeNPs particles are immobilized with the xanthine oxidase (XOD) enzyme on silanized filter paper. As a result, sensitive, fast, efficient, and inexpensive detection of traced analytes has been achieved with this next-generation biosensor, which performs real sample analysis without sample treatment or the addition of exogenous reagents (22).

1.2.1.3 Aptamer

Aptamers, known as artificial antibodies, created from single-stranded deoxyribonucleic acid (DNA) or ribonucleic acid (RNA), are highly preferred in biosensor applications due to their ability to target small molecules, proteins, and more complex bioanalytes (6–7). High-affinity aptamers can be identified using a process known as the Systematic Evolution of Ligands by Exponential Enrichment (SELEX) (23–24). Aptamers, which are very advantageous compared to antibodies due to their easy synthesis step, high stability, easy modification, reusability, and low cost, have recently been preferred as bio-identity elements in biosensors (11, 23). Also, aptamers, smaller than antibodies, have a high surface density. This ensures that the desired response is high. Electrochemical impedance spectroscopy (EIS) is a widely used method to study surface changes in aptamer-based biosensors (24).

Liu et al. (25), using the SS2–55 aptamer array as the recognition molecule and lateral flow biosensor (LFB) technology, have developed a new type of biosensor that can detect isocarbophos for the first time, sensitive and fast. It shows that isocarbophos present in the real sample (selected some actual samples including green vegetables, tea leaves, and peaches) can bind with the SS2–55 aptamer in the aptamer-AuNP probe and lead to AuNPs exposure. The detection limit of the designed LFB is 20 µM, and the recovery rate in real samples was determined as 82.8%–93.6%. The designed LFB has good specificity, convenient operation, and does not need large-scale instruments (25).

1.2.1.4   Nucleic Acid

Nucleic acid biosensors, called genosensors, are widely used today because of their ease of application to complex samples, portability, and fast detection features. The specificity of the bioanalyte is determined by taking into account the complementary binding properties of DNA. The DNA fragment designed as a biological recognition element is immobilized on the sensor surface (26–27). As a result, a complementary pattern is obtained between the immobilized DNA fragment and the target analyte. Recently, locked nucleic acids (LNA) and peptide nucleic acids (PNA) are frequently used as nucleic acid recognition elements (28–29). In addition, biosensor technology is developing rapidly thanks to nanoparticles and developments in nanotechnology. Nanoparticles such as gold, silver, and copper are also used for nucleic acid immobilization. Apart from these, the immobilization of nucleic acids on the electrode surface is performed using carbon-based nanomaterials, and sensitivity, selectivity, and conductivity are facilitated (27).

Nucleic acid biosensors have long been used for food safety applications to monitor the presence of large numbers of pathogenic bacteria and the identification of genetically modified (GM) food using nucleic acid biomarkers such as DNA, mRNA, and microRNA (30–31). Recently, surface plasmon resonance (SPR) biosensors have been widely applied for the quantitative detection of nucleic acids due to their dynamic and sensitive detection ability for binding kinetics. Chen et al. (32) developed a new fiber-optic surface plasmon resonance (FOSPR) biosensor to detect the CaMV 35S promoter in GM foods. Dual signal amplification strategies based on gold nanoparticles reduction graphene oxide and CHA reaction were used to construct the FOSPR biosensor. Eighteen food samples were determined using the suggested method after DNA extraction without enzymes. This FOSPR biosensor system exhibits high sensitivity and specificity towards target ssDNA. In addition, CHA-based SPR biosensing methodology for detecting other substances via aptamer recognition can also be applied through the unique design of the trigger sequence. With this method, which can be completed without any enzyme, simpler, stable, and cost-effective systems can be developed. Therefore, the developed biosensing strategy has immense potential for food analysis and early clinical diagnosis (32).

1.2.1.5 Molecularly Imprinted Polymers

Molecular imprinting is a technique that creates artificial recognition sites for the bioanalytes. Non-specific interactions occur in the gaps created specifically to the molecular structure of the target molecule to obtain analyte specificity and selectivity. Components such as crosslinker, monomer, and initiator are required to prepare molecularly imprinted polymers (MIP) (33–34). Unlike natural biorecognition elements, synthetic-based biorecognition elements are generated specific to the target analyte. Also, the use of functional monomers helps the MIP to bind to the formed gaps (11). As a result, MIPs have advantages such as high stability, selectivity, short synthesis time, and low cost. Because of these properties, MIPs are the biorecognition elements that can replace natural antibodies (33).

In recent years, wearable sensors have been interesting devices to monitor the physical conditions of human health, thanks to their advantages such as ease of use, flexibility, and portability (35). Wearable tattoo biosensors for the analysis of human sweat have been popular recently. Zhang et al. (36) developed an Ag nanowires (AgNWs) modified MIP electrochemical biosensor for monitoring lactate in sweat (Figure 1.4). The MIPs-AgNWs sensor was applied with a screen-printed technique on a flexible substrate to monitor lactate in epidermal biofluids. They found the limit of detection (LOD) of 0.22 μM at concentrations ranging from 10^{-6} M to 0.1 M, with high sensitivity and selectivity for lactate. They applied the developed biosensor in the epidermis of volunteers to detect lactate in vivo in human sweat formed during exercise (36).

1.2.2 TRANSDUCER BASED

The design of a biosensor includes the selection of a suitable biorecognition element and a transducer. The transducer part of the biosensor is responsible for converting the biochemical signal into a measurable electrical signal (2, 4, 37–38). Transducers operate through different measurement principles. Therefore, it is possible to classify biosensors according to the used transducer. In order to acquire the most efficient results with a biosensor, it is important to bring together the most convenient biorecognition element and transducer for the target analyte and type of application (4). The main aim is to fabricate a highly sensitive biosensor, selective, rapid, easy-to-use, affordable, biocompatible, and stable (39).

Each type of transducer has its advantages and specific applications based on its measurement principle. Table 1.1 evaluates the transducer-based classification of the biosensors.

In their fluorescence-based biosensor development study, Chen et al. (45) focused on detecting DNA adenine methylation methyltransferase activity due to its clinical significance and gene regulation-related effects. This newly developed biosensor offers a rapid, simple, affordable, highly sensitive, and selective analysis using the advantages of the fluorescence method. They also applied this method to the anticancer drug 5-fluorouracil to evaluate the enzyme inhibition activity.

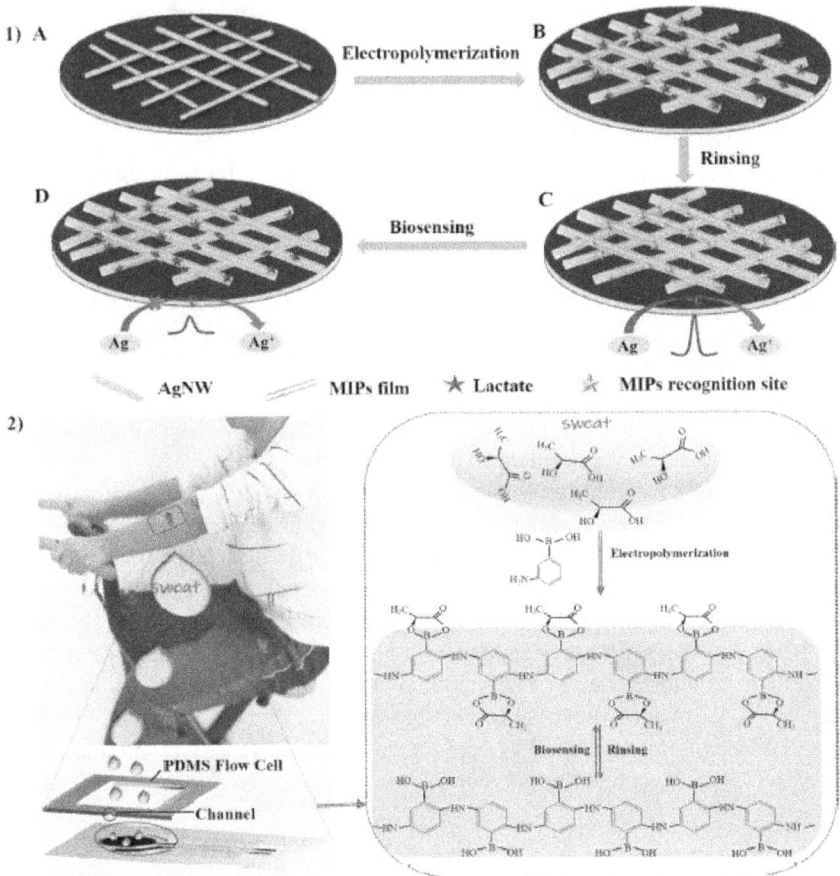

FIGURE 1.4 The fabrication steps of the MIPs-AgNWs electrochemical biosensor for the epidermal monitoring of lactate. Reprinted from ref. (36) with permission from Elsevier.

Haritha et al. (46) fabricated an amperometric biosensor based on cholesterol oxidase enzyme and various nanocomposites for cholesterol determination in their recent work. Cholesterol determination is vital for the clinical assessment and monitoring of several diseases. Electrochemical amperometric biosensors are widely used for the enzymatic detection of cholesterol due to their great sensitivity and selectivity. This present study offers a biosensor with a LOD value of 5 µM and 15 s response time.

For the electrochemical determination of the anticancer drug temozolomide whose mechanism of action is inhibition of DNA synthesis, a voltammetric biosensor based on Au nanoparticles and ds-DNA was developed by Jahandari et al. (47). Electrochemistry-based voltammetric biosensors are advantageous options for analysing a wide variety of biological compounds in complex media. They provide sensitivity, selectivity, portability, and short analysis time. This voltammetric biosensor successfully evaluated the intercalation of temozolomide. The biosensor was also

TABLE 1.1

Transducer-Based Classification of Biosensors

Transducer Type		Measurement Principle	Advantages	Applications	Reference
Optical	Fluorescence based	Measurement of fluorescence occurring after absorption of light or radiation by a molecule exposed to external light or radiation source	Convenience in labeling applications, Allowing real-time analysis	Detection of nucleic acids, antigens, antibodies, live cells	(3, 40)
	Chemilumine-scence based	Chemical reactions create the excited states of molecules, and measurements are based on the photon production rate related to light intensity	High sensitivity due to specific light-emitting reactions, wide calibration range	Diagnostics, detection of metal ions	(3, 40)
	Colorimetric	Measurement of color change due to changes in UV/visible absorption	Affordable, easy to use, short analysis time	Detection of microorganisms and pathogens, analysis of food samples	(3, 41–42)
Electro-chemical	Amperometric	Measurement of the current as a result of the electrochemical reaction occurred in the electrochemical cell in the presence of an applied potential	Easy to use, low-cost, improved sensitivity	Blood glucose analysis, cholesterol measurement and other enzymatic sensors	(1–2, 40, 42)
	Voltammetric	Measurement of the current in the presence of a varying potential	Allowing simultaneous analysis of different compounds, very high sensitivity, low-cost	Drug analysis, environmental analysis, food analysis	(2–3, 40)
	Potentiometric	Measurement of the potential changes after an electrochemical reaction during a constant current application	Evaluation of the ionic activity, simple application	pH measurement, analysis of biomolecules and ions	(1, 2, 40, 43–44)

(Continued)

TABLE 1.1

(Continued)

Transducer Type		Measurement Principle	Advantages	Applications	Reference
	Conductometric	Measurement of the changes in the conductivity after an electrochemical reaction	Cheapness, suitability for miniaturization, applicability without a reference electrode	Enzymatic analysis, immunosensor based applications	(1–3, 40, 42)
	Impedimetric	Measurement of the impedance (resistance of the electrical circuit against current changes during the potential application)	Label-free analysis, stability	Affinity-based analysis, non-invasive sensors	(3, 5, 40, 44)
Mass-sensitive	Piezoelectric	Measurement of the resonating frequency changes on piezoelectric materials after adsorption or desorption	Applicable for molecules with high molecular weight, interference-free	Immunosensor applications, detection of nucleic acids, whole cells	(42–44)
	Magnetoelastic	Exposure to magnetic fields change the magnetoelasticity of the materials and the emitted magnetic flux is measured	Wireless and real-time analysis, miniaturization, low-cost	Monitoring of blood coagulation, food sample analysis	(3, 44)
Electrical	Field Effect Transistor (FET) based	Changes in the surface charge effect of the conductance due to the application of a positive/negative potential	Simple application, affordability, high sensitivity, portability, small sample volume	Clinical analysis, blood pH measurement	(2, 3, 42)
Thermal		Measurement of the temperature changes influenced by the reaction between the analyte and bioreceptor	Rapid and label-free analysis, stability, good sensitivity	Environmental and food analysis	(3, 40, 44)

applied to capsule formulation and urine samples with good recovery results confirming the performance and applicability to real samples.

Piezoelectric biosensors use quartz crystal microbalances based on mass sensitivity, and they provide label-free DNA analysis for specific DNA sequences (48). They are highly advantageous with excellent specificity, rapid and real-time measurement. Utilizing piezoelectric biosensors for label-free DNA analysis in oncology, microbiology, and genetics is significant. A piezoelectric biosensor was designed for the identification and genotyping of human papillomavirus in this work (48). This sensor developed by Del'Atti et al. offers specificity and real sample application for clinical analysis.

Early diagnosis is the most vital and researched point for cancer treatment. FET biosensors, improved with nanomaterials as sensitive, miniature, easy-to-use, and label-free analytical tools, are utilized to detect various cancer biomarkers (49). For example, the prostate-specific antigen is a clinically valuable biomarker for cancer diagnosis. Therefore, Zhang et al. (49) designed a FET biosensor to determine prostate-specific antigen. It was also successfully applied for sensitive determination in human serum.

1.3 LATEST APPLICATIONS OF BIOSENSORS

Today, biosensors are essential in the medical, food, water safety, and environmental fields to detect various analytes. Since the first studies on biosensors, these devices that adapt to new technology and biosensor studies that allow faster and more reliable analyses are increasing day by day.

1.3.1 MEDICAL APPLICATIONS

Today, researchers in the medical field need inexpensive, practical, and safe tools to provide customized healthcare equipment to patients, perform early diagnosis, and treat various diseases (Figure 1.5). Thanks to the developing biosensor technologies, it is possible to reduce the costs of early diagnosis, diagnosis and treatment methods, and hospitalization in medicine. However, biosensors are competitors to modern medical devices thanks to their easy production processes, low cost, and fast response features.

Biomarkers for cancer are important indicators of tumor cell growth and are used to diagnose tumors. Ullah Khalid et al. (50) developed a lung-on-chip platform for lung cancer to evaluate Doxorubicin and docetaxel drug compounds in lung cancer using integrated biosensors.

Thanks to biosensor technology, fast and easy monitoring can be achieved in health. (35). Biosensors are also used to analyze interactions between drugs and biomolecules such as DNA, HSA, BSA (51–54). Examining these interactions contributes to human health by illuminating issues such as drug development studies and dosing studies. In vitro biosensors are analytical devices for diagnosis using blood, urine, or tumor tissues containing cancer markers, small molecules, enzymes, proteins, and living cancer cells. Mu et al. (55) developed a rapid, simple, and sensitive detection method for vancomycin detection that has been validated in vivo. The

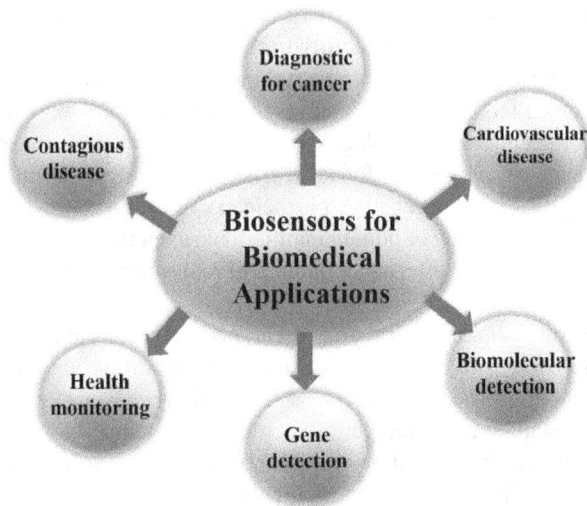

FIGURE 1.5   Biosensor applications of biomedical area.

developed way combines of vancomycin (Van) specific microdialysis sampling technique and fluorescent biosensor (Figure 1.6). They investigated the dosage and range of drug administration by comparing the pharmacokinetic profiles of Van in rabbits with normal and induced chronic renal failure (CRF). Accurate, rapid, and continuous detection of Van is important for physicians to adjust dosage and treatment plans in their real environment. They have demonstrated the feasibility of therapeutic drug monitoring in vivo thanks to the fluorometric biosensor they developed.

The detection of infectious diseases with biosensors has gained much more importance with the COVID-19 pandemic. Immunosensors developed by utilizing the antigen-antibody relationship and biosensors based on protein-DNA interaction are frequently used to detect infectious diseases. For example, Antipchik et al. (55) developed an electrochemical biosensor with a synthetic and antigen-specific binding peptide sequence to detect hepatitis C virus (HCV) surface antigen. Shariati et al. (56) developed an ultrasensitive FET biosensor to detect hepatitis B virus deoxyribonucleic acid (HBV DNA) using a label-free approach. The sensitivity-enhanced FET biosensor with ZnO-doped MoS_2 nanowires (NWs) measured DNA targets with a detection limit of 1 fM in a linear concentration range of 0.5 pM to 50 mM in as little as 25 s. The developed FET biosensor has been successfully applied to human serum samples for HBV detection.

1.3.2   Food and Water Safety

Biosensors are an important tool for public health because of their real-time response, relatively low cost, and applicability for evaluating and monitoring water and food. (57). Thanks to biosensors, simple, fast, and reliable measurements can be done to provide water quality monitoring (58). Using various biorecognition elements with

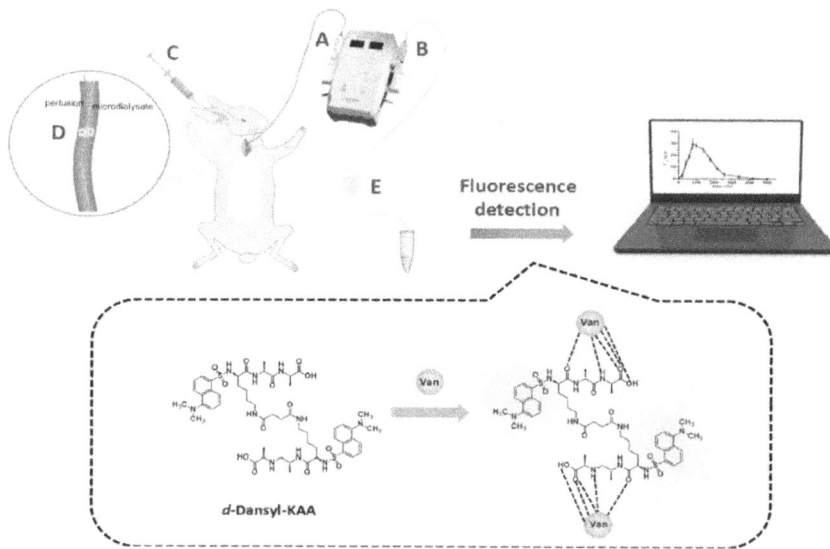

FIGURE 1.6 Schematic diagram of the vancomycin (Van) detection system. (A): perfusion fluid; (B): fluorescent probe (d-Dansyl-KAA); (C): vancomycin; (D): microdialysis probe; (E): three-way connector. Reprinted from ref. (55) with permission from Elsevier.

biosensors, biochemical oxygen demand (BOD), heavy metal, organic toxic substances, and even specific pollutants in the water source to be examined can be determined qualitatively or quantitatively by electrical, optical, thermometric, mass signals (Figure 1.7). Biorecognition elements are used in the early warning of microorganisms, including bacteria, microalgae, yeast, and fungi (59–60).

Various toxins such as Ochratoxin A, T-2 toxin, Aflatoxin derivatives can be analyzed with biosensors (61). Aflatoxin B1 (AfB1) has traditionally been detected with chromatographic, spectroscopic, and ELISA techniques, but these techniques are time-consuming and require costly equipment (62). In this respect, biosensors play a vital role in the efficient detection of AfB1 by minimizing the cost. Srivastava et al. created a highly sensitive rGO-based biodetection platform to detect AfB1 (63). Costa et al. (61) used a carboxyl-functionalized carbon nanotube modified gold electrode to produce an effective immunodetection platform to detect AfB1 from maize flour using cyclic voltammetry (CV) and EIS techniques. Nirbhaya et al. (62) prepared carbon nitride (g-C3N4) nanoplates with thionine-enhanced (Thn) chemical methods. Thn and anti-AfB1 and bovine serum albumin (BSA) molecules with the electrochemical biosensor platform in the linear range of 1 fM to 1 nM, the detection limit was found as 0.328 fM.

Currently, traditional detection methods for organophosphates include chromatography. The LFB has aroused intense interest in biological analysis and clinical diagnosis as an instant detection method due to its low cost, speed, and suitability for low-skilled personnel. However, LFBs using antibodies as biological recognition elements have some disadvantages such as high detection sensitivity and

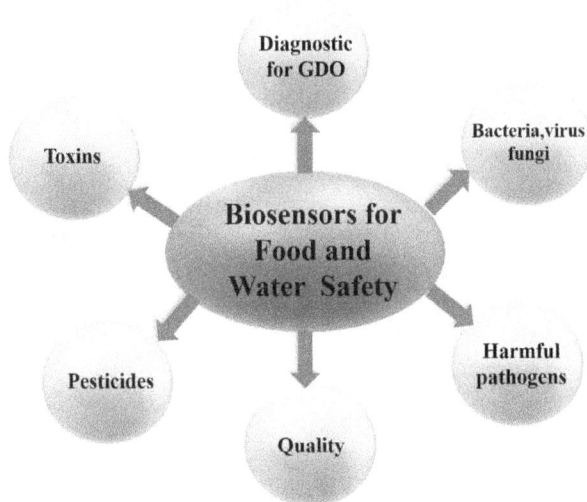

FIGURE 1.7   Biosensor applications of food and water safety.

selectivity, generally low stability, high production cost, and purification and modification of antibodies.

1.3.3   ENVIRONMENTAL APPLICATIONS

The release of environmental pollutants such as pesticides, toxins, toxic heavy metals, and pharmaceutical xenobiotics into the ecosystem is a global problem today (Figure 1.8). Therefore, rapid and reliable detection of these pollutants is critical to coping with environmental problems. Electrochemical nucleic acid biosensors and aptasensors, detect pollutants such as toxins, steroids, insecticides, heavy elements, and antibiotics from various ecological samples can be made with high sensitivity, selectivity, and less sample use (27, 64).

Pesticides-contaminated water and groundwater sources cause a great deal of environmental pollution, and it is very important to detect pesticides from these sources using small-volume sample samples. Khosropour et al. (65) developed a simple, sensitive, and selective electrochemical aptasensor to determine diazinon (DZN), an organophosphorus compound. It was modified on a nanocomposite glassy carbon electrode (GCE) prepared with vanadium disulfide quantum dots (VS_2QDs) and graphene nanoplatelets/carboxylated multi-walled carbon nanotubes (GNP/CMWCNTs). By following the oxidation of $[Fe(CN)_6]^{3-/4-}$ as redox probe with the modified electrode immobilized with DZN-binding aptamer (DZBA) via electrostatic interaction over a wide calibration range of 5.0×10^{-14}–1.0×10^{-8} M of DZN, 1.1×10^{-14} and 2.0×10^{-15} M were detected by the electrochemical method with LOD. Toxic heavy metals such as Hg^{+2}, Pb^{+2}, Cd, Cr, another environmental pollutant group, have critical importance in environmental protection. The analysis of these heavy metals with biosensors is a remarkable issue (58, 65–69). In addition to

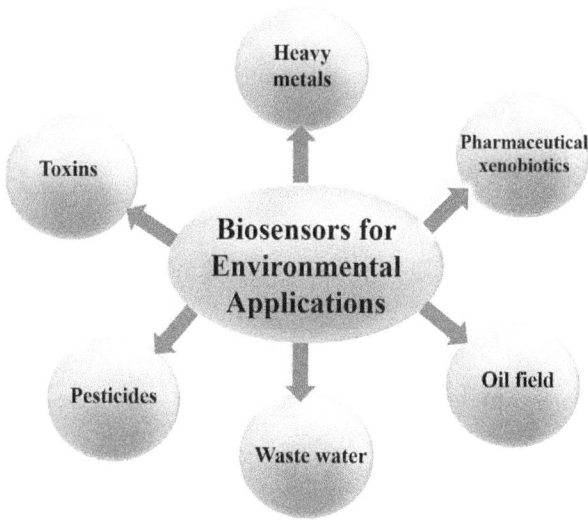

FIGURE 1.8   Biosensor applications of the environmental field.

electrochemical methods, biosensor studies are carried out to analyze heavy metals with the optical techniques in Table 1.1. Mei et al. (69) developed a cleavable PS-RNA-modified probe biosensor based on photochemical chemiluminescence to detect Hg^{+2}. The biosensor they created, detected Hg^{2+} in the linear range of 5 nM–5 µM with a detection limit of 1.4 nM. For the determination of mercury ions, they applied different lake water samples. A novel biosensor for the rapid and high-sensitive detection of mercury based on a cleavable phosphorothioate RNA probe.

There is also the use of fast, sensitive, and selective Surface-enhanced Raman Scattering (SERS)-based biosensors in environmental applications. SERS is a plasmonic analytical technique used to obtain the information of molecules adsorbed or attached to nanostructured metallic surfaces and provides the advantages of strong Raman scattering signals of these molecules. It is a sensitive, fast, and selective method to detect trace amounts of analytes in environmental applications (70).

In recent years, biosensors based on electroactive biofilm (EAB) have been developed for the next generation environmental detection of pollutants, including aromatic compounds. As can be seen in Figure 1.9, researchers analyze with EAB-based biosensors in various aquatic environments, including sewage treatment plants, waste from rivers, groundwater, and the oil industry. EAB-based sensors are shown in the figure to detect the pollution caused by industrial wastes with an early warning system thanks to biosensors (60).

1.4   CONCLUSION

Since the development of the first biosensor, biosensors consisting of various biorecognition elements and transducer parts have adapted to the developing technology. Due to the benefits of early diagnosis of various diseases that threaten human

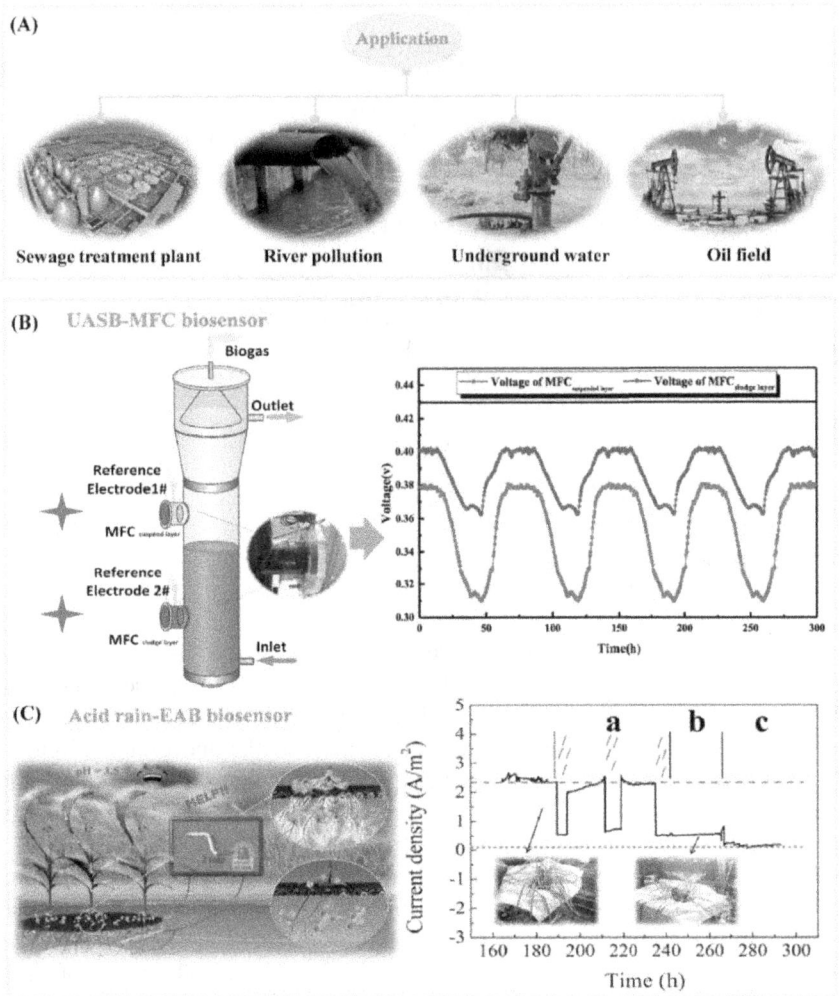

FIGURE 1.9 Applications of EAB-based biosensors in water quality early-warning. (A) EAB-based biosensor can serve different water environments. (B) An MFC biosensor in UASB reactor to monitor abnormal products. (C) EAB-based biosensor in the earth for acid rain detection. Reprinted from ref. (60) with permission from Elsevier.

health, it is highly needed in the health field today. It is important to develop sensitive and cost-effective diagnostic tools to detect diseases effectively and reliably. Biosensors are being developed with disease-specific biomarkers and different detection methods. In this way, various cancer, cardiovascular, and infectious diseases can be detected much more easily, quickly, and sensitively.

In recent years, the number of biomarkers specific to the substance to be determined has increased. With the rapid progress of science and technology, nucleic acids, aptamers, enzymes, and cells are being developed for their use in specific biosensor applications. With the increasing food and water safety problem with the increase in the world population, researchers are working on detecting toxins, genetically modified organisms (GMOs), and pathogens that adversely affect human health. Nanomaterials such as quantum dots, carbon-based nanomaterials, metal oxide nanoparticles, and nanozymes are developed and used in biosensor studies to develop nanotechnology. In this way, detecting elements that threaten food and water quality is fast, sensitive, and reliable. Biosensors can also detect the analytes with a small sample volume and appear reliable and practical devices in detecting environmental pests such as heavy metals, pesticides, and harmful organic compounds.

1.5 FUTURE PERSPECTIVES

With recent technological developments, studies involving biosensors have also begun to evolve into integration with new technology. In this context, it is possible to miniaturize biosensors, adapt them to chip technology, and develop fast and sensitive devices that measure low detection limits. In addition, practical and cost-effective biosensors are developed with recyclable paper-based sensors, and their application areas are increasing. For example, thanks to wearable biosensors, information about various health parameters will be obtained from patients with on-site detection without long analysis time. Furthermore, with the integration of biosensors with smart devices, compact, fast and new-generation devices will increase. Due to the increasing interest in biosensor development, new technological studies are carried out in this field every day.

ACKNOWLEDGMENTS

D.N.U. and A.C. thank the financial supports from Council of Higher Education 100/2000 (YOK) under the special 100/2000. A.C. also thanks the Scientific and Technological Research Council of Turkey (TUBITAK) under the BIDEB/2211-A PhD Scholarship Programme.

REFERENCES

1. Chakraborty M, Hashmi MSJ. 2017. An overview of biosensors and devices. *Reference Module in Materials Science and Materials Engineering*. 1–24
2. Bettazzi F, Marrazza G, Minunni M, Palchetti I, Scarano S. 2017. Biosensors and related bioanalytical tools. *Comprehensive Analytical Chemistry*. 77: 1–3
3. Karunakaran C, Rajkumar R, Bhargava K. 2015. Introduction to biosensors. In *Biosensors and Bioelectronics*, pp. 1–68. Elsevier
4. Lee YH, Mutharasan R. 2005. Biosensors. In *Sensor Technology Handbook*, pp. 161–180. Elsevier
5. Upasham S, Banga IK, Jagannath B, Paul A, Lin KC, et al. 2021. Electrochemical impedimetric biosensors, featuring the use of room temperature ionic liquids (RTILs): Special focus on non-faradaic sensing. *Biosensors and Bioelectronics*. 177(August 2020): 112940
6. Morales MA, Halpern JM. 2018. Guide to selecting a biorecognition element for biosensors. *Bioconjugate Chemistry*. 29(10): 3231–3239

7. Warriner K, Namvar A. 2011. Biosensors for foodborne pathogen detection. In *Comprehensive Biotechnology, Second Edition*. Vol. 4, pp. 659–674. Elsevier

8. Gao S, Rojas-Vega F, Rocha-Martin J, Guisán JM. 2021. Oriented immobilization of antibodies through different surface regions containing amino groups: Selective immobilization through the bottom of the Fc region. *International Journal of Biological Macromolecules*. 177: 19–28

9. Crivianu-Gaita V, Thompson M. 2016. Aptamers, antibody scFv, and antibody Fab' fragments: An overview and comparison of three of the most versatile biosensor biorecognition elements. *Biosensors and Bioelectronics*. 85: 32–45

10. Huy TQ, Hanh NTH, Van Chung P, Anh DD, Nga PT, Tuan MA. 2011. Characterization of immobilization methods of antiviral antibodies in serum for electrochemical biosensors. *Applied Surface Science*. 257(16): 7090–7095

11. Justino CIL, Duarte AC, Rocha-Santos TAP. 2016. Critical overview on the application of sensors and biosensors for clinical analysis. *TrAC—Trends Analytical Chemistry*. 85: 36–60

12. Miller E, Sikes HD. 2015. Addressing barriers to the development and adoption of rapid diagnostic tests in global health. *Nanobiomedicine*. 2: 1–21

13. Roda A, Cavalera S, Di Nardo F, Calabria D, Rosati S, et al. 2021. Dual lateral flow optical/chemiluminescence immunosensors for the rapid detection of salivary and serum IgA in patients with COVID-19 disease. *Biosensors and Bioelectronics*. 172: 112765

14. Dahlke C, Heidepriem J, Kobbe R, Santer R, Koch T, et al. 2020. Distinct early IgA profile may determine severity of COVID-19 symptoms: An immunological case series. *medRxiv*. doi:10.1101/2020.04.14.20059733

15. Yu H, Sun B, Fang Z, Zhao J, Liu X, et al. 2020. Distinct features of SARS-CoV-2-specific IgA response in COVID-19 patients. *European Respiratory Journal*. 56(2): 2001526

16. Gaudin V. 2017. Advances in biosensor development for the screening of antibiotic residues in food products of animal origin—a comprehensive review. *Biosensors and Bioelectronics*. 90: 363–377

17. Zhao WW, Xu JJ, Chen HY. 2017. Photoelectrochemical enzymatic biosensors. *Biosensors and Bioelectronics*. 92(November): 294–304

18. Zhou Y, Liu B, Yang R, Liu J. 2017. Filling in the gaps between nanozymes and enzymes: Challenges and opportunities. *Bioconjugate Chemistry*. 28(12): 2903–2909

19. Takeda K, Nakamura N. 2021. Biosensors: Enzyme sensors. In *Reference Module in Biomedical Sciences*, pp. 1–16. Elsevier

20. Ahangari H, Kurbanoglu S, Ehsani A, Uslu B. 2021. Latest trends for biogenic amines detection in foods: Enzymatic biosensors and nanozymes applications. *Trends in Food Science and Technology*. 112(March): 75–87

21. Huang Z, Zhang L, Cao P, Wang N, Lin M. 2021. Electrochemical sensing of dopamine using a Ni-based metal-organic framework modified electrode. *Ionics (Kiel)*. 27: 1339–1345

22. Mustafa F, Othman A, Andreescu S. 2021. Cerium oxide-based hypoxanthine biosensor for fish spoilage monitoring. *Sensors & Actuators, B: Chemical*. 332(September 2020): 129435

23. Nur Y, Gaffar S, Hartati YW, Subroto T. 2021. Applications of electrochemical biosensor of aptamers-based (APTASENSOR) for the detection of leukemia biomarker. *Sensing and Bio-Sensing Research*. 32: 100416

24. Bachour Junior B, Batistuti MR, Pereira AS, de Sousa Russo EM, Mulato M. 2021. Electrochemical aptasensor for NS1 detection: Towards a fast dengue biosensor. *Talanta*. 233(January): 122527

25. Liu B, Tang Y, Yang Y, Wu Y. 2021. Design an aptamer-based sensitive lateral flow biosensor for rapid determination of isocarbophos pesticide in foods. *Food Control.* 129(May): 108208

26. Li CZ, Karadeniz H, Canavar E, Erdem A. 2012. Electrochemical sensing of label free DNA hybridization related to breast cancer 1 gene at disposable sensor platforms modified with single walled carbon nanotubes. *Electrochimica Acta.* 82: 137–142

27. Hashem A, Hossain MAM, Marlinda AR, Mamun M Al, Simarani K, Johan MR. 2021. Nanomaterials based electrochemical nucleic acid biosensors for environmental monitoring: A review. *Applied Surface Science Advance.* 4(January): 100064

28. Ferapontova EE. 2018. DNA electrochemistry and electrochemical sensors for nucleic acids. *Annual Review of Analytical Chemistry.* 11: 197–218

29. Teengam P, Siangproh W, Tuantranont A, Henry CS, Vilaivan T, Chailapakul O. 2017. Electrochemical paper-based peptide nucleic acid biosensor for detecting human papillomavirus. *Analytica Chimica Acta.* 952: 32–40

30. Ahmed FE. 2002. Detection of genetically modified organisms in foods. *Trends in Biotechnology.* 20(5): 215–223

31. Vollenhofer S, Burg K, Schmidt J, Kroath H. 1999. Genetically modified organisms in food-screening and specific detection by polymerase chain reaction. *Journal of Agricultural and Food Chemistry.* 47(12): 5038–5043

32. Chen Z, Chengjun S, Zewei L, Kunping L, Xijian Y, et al. 2018. Fiber optic biosensor for detection of genetically modified food based on catalytic hairpin assembly reaction and nanocomposites assisted signal amplification. *Sensors & Actuators, B: Chemical.* 254: 956–965

33. Battaglia F, Baldoneschi V, Meucci V, Intorre L, Minunni M, Scarano S. 2021. Detection of canine and equine procalcitonin for sepsis diagnosis in veterinary clinic by the development of novel MIP-based SPR biosensors. *Talanta.* 230(January): 122347

34. Abbasy L, Mohammadzadeh A, Hasanzadeh M, Razmi N. 2020. Development of a reliable bioanalytical method based on prostate specific antigen trapping on the cavity of molecular imprinted polymer towards sensing of PSA using binding affinity of PSA-MIP receptor: A novel biosensor. *Journal of Pharmaceutical and Biomedical Analysis.* 188: 113447

35. Haleem A, Javaid M, Singh RP, Suman R, Rab S. 2021. Biosensors applications in medical field: A brief review. *Sensors International.* 2(May): 100100

36. Zhang Q, Jiang D, Xu C, Ge Y, Liu X, et al. 2020. Wearable electrochemical biosensor based on molecularly imprinted Ag nanowires for noninvasive monitoring lactate in human sweat. *Sensors & Actuators, B: Chemical.* 320(March): 128325

37. Kawamura A, Miyata T. 2016. Biosensors. *Biomaterials Nanoarchitectonics.* 157–176

38. Ekrami E, Pouresmaieli M, Shariati P, Mahmoudifard M. 2021. A review on designing biosensors for the detection of trace metals. *Applied Geochemistry.* 127(February): 104902

39. Huang X, Zhu Y, Kianfar E. 2021. Nano Biosensors: Properties, applications and Electrochemical Techniques. *Journal of Materials Research and Technology.* 12: 1649–1672

40. Malhotra BD, Ali MA. 2018. Nanomaterials in biosensors. *Nanomaterials for Biosensors.* 1–74

41. Mondal B, Ramlal S, Lavu PS, Bhavanashri N, Kingston J. 2018. Highly sensitive colorimetric biosensor for staphylococcal enterotoxin B by a label-free aptamer and gold nanoparticles. *Frontiers in Microbiology.* 9: 1–8

42. Parkhey P, Mohan SV. 2018. *Biosensing Applications of Microbial Fuel Cell: Approach Toward Miniaturization*, pp. 977–997. Elsevier

43. Sawant SN. 2017. Development of biosensors from biopolymer composites. In *Biopolymer Composites in Electronics*, pp. 353–383. Elsevier

44. Food US. 2021. Detection of foodborne organisms: In the perspective of biosensors. *Advance Food Analytical Tools.* 35–57

45. Chen L, Zhang Y, Xia Q, Luo F, Guo L, et al. 2020. Fluorescence biosensor for DNA methyltransferase activity and related inhibitor detection based on methylation-sensitive cleavage primer triggered hyperbranched rolling circle amplification. *Analytica Chimica Acta.* 1122: 1–8

46. Haritha VS, Kumar SRS, Rakhi RB. 2021. Proceedings amperometric cholesterol biosensor based on cholesterol oxidase and Pt-Au/MWNTs modified glassy carbon electrode. *Materials Today: Proceedings.* 50

47. Jahandari S, Taher MA, Karimi-Maleh H, Khodadadi A, Faghih-Mirzaei E. 2019. A powerful DNA-based voltammetric biosensor modified with Au nanoparticles, for the determination of Temodal; an electrochemical and docking investigation. *Journal of Electroanalytical Chemistry.* 840(January): 313–318

48. Dell'Atti D, Zavaglia M, Tombelli S, Bertacca G, Cavazzana AO, et al. 2007. Development of combined DNA-based piezoelectric biosensors for the simultaneous detection and genotyping of high risk Human Papilloma virus strains. *Clinica Chimica Acta.* 383(1–2): 140–146

49. Zhang Y, Feng D, Xu Y, Yin Z, Dou W, et al. 2021. DNA-based functionalization of two-dimensional MoS2 FET biosensor for ultrasensitive detection of PSA. *Applied Surface Science.* 548(October 2020): 149169

50. Khalid MAU, Kim YS, Ali M, Lee BG, Cho YJ, Choi KH. 2020. A lung cancer-on-chip platform with integrated biosensors for physiological monitoring and toxicity assessment. *Biochemical Engineering Journal.* 155(September 2019): 107469

51. Jalalvand AR, Ghobadi S, Goicoechea HC, Gu HW, Sanchooli E. 2018. Investigation of interactions of Comtan with human serum albumin by mathematically modeled voltammetric data: A study from bio-interaction to biosensing. *Bioelectrochemistry.* 123: 162–172

52. Eksin E, Senturk H, Zor E, Bingol H, Erdem A. 2020. Carbon quantum dot modified electrodes developed for electrochemical monitoring of Daunorubicin-DNA interaction. *Journal of Electroanalytical Chemistry.* 862: 114011

53. Kurbanoglu S, Dogan-Topal B, Hlavata L, Labuda J, Ozkan SA, Uslu B. 2015. Electrochemical investigation of an interaction of the antidepressant drug aripiprazole with original and damaged calf thymus dsDNA. *Electrochimica Acta.* 169: 233–240

54. Jalalvand AR, Ghobadi S, Akbari V, Goicoechea HC, Faramarzi E, Mahmoudi M. 2019. Mathematical modeling of interactions of cabergoline with human serum albumin for biosensing of human serum albumin. *Sensing and Bio-Sensing Research.* 25(June): 100297

55. Antipchik M, Korzhikova-Vlakh E, Polyakov D, Tarasenko I, Reut J, et al. 2021. An electrochemical biosensor for direct detection of hepatitis C virus. *Analytical Biochemistry.* 624(January): 114196

56. Shariati M, Vaezjalali M, Sadeghi M. 2021. Ultrasensitive and easily reproducible biosensor based on novel doped MoS2 nanowires field-effect transistor in label-free approach for detection of hepatitis B virus in blood serum. *Analytica Chimica Acta.* 1156: 338360

57. Kotsiri Z, Vidic J, Vantarakis A. 2022. Applications of biosensors for bacteria and virus detection in food and water—a systematic review. *Journal of Environmental Sciences.* 111: 367–379

58. Ejeian F, Etedali P, Mansouri-Tehrani HA, Soozanipour A, Low ZX, et al. 2018. Biosensors for wastewater monitoring: A review. *Biosensors and Bioelectronics.* 118(May): 66–79

59. Antonacci A, Scognamiglio V. 2020. Biotechnological advances in the design of algae-based biosensors. *Trends in Biotechnology.* 38(3): 334–347

60. Qi X, Wang S, Li T, Wang X, Jiang Y, et al. 2021. An electroactive biofilm-based biosensor for water safety: Pollutants detection and early-warning. *Biosensors and Bioelectronics*. 173(September 2020): 112822

61. Costa MP, Frías IAM, Andrade CAS, Oliveira MDL. 2017. Impedimetric immunoassay for aflatoxin B1 using a cysteine modified gold electrode with covalently immobilized carbon nanotubes. *Microchimica Acta*. 184(9): 3205–3213

62. Nirbhaya V, Chauhan D, Jain R, Chandra R, Kumar S. 2021. Nanostructured graphitic carbon nitride based ultrasensing electrochemical biosensor for food toxin detection. *Bioelectrochemistry*. 139: 107738

63. Srivastava S, Kumar V, Ali MA, Solanki PR, Srivastava A, et al. 2013. Electrophoretically deposited reduced graphene oxide platform for food toxin detection. *Nanoscale*. 5(7): 3043–3051

64. Congur G. 2021. Monitoring of glyphosate-DNA interaction and synergistic genotoxic effect of glyphosate and 2,4-dichlorophenoxyacetic acid using an electrochemical biosensor. *Environmental Pollution*. 271: 116360

65. Khosropour H, Rezaei B, Rezaei P, Ensafi AA. 2020. Ultrasensitive voltammetric and impedimetric aptasensor for diazinon pesticide detection by VS2 quantum dots-graphene nanoplatelets/carboxylated multiwalled carbon nanotubes as a new group nanocomposite for signal enrichment. *Analytica Chimica Acta*. 1111: 92–102

66. Tian C, Zhao L, Zhu J, Zhang S. 2021. Ultrasensitive detection of trace Hg2+ by SERS aptasensor based on dual recycling amplification in water environment. *Journal of Hazardous Materials*. 416: 126251

67. Sreekanth SP, Alodhayb A, Assaifan AK, Alzahrani KE, Muthuramamoorthy M, et al. 2021. Multi-walled carbon nanotube-based nanobiosensor for the detection of cadmium in water. *Environmental Research*. 197: 111148

68. Zou W, Tang Y, Zeng H, Wang C, Wu Y. 2021. Porous Co3O4 nanodisks as robust peroxidase mimetics in an ultrasensitive colorimetric sensor for the rapid detection of multiple heavy metal residues in environmental water samples. *Journal of Hazardous Materials*. 417: 125994

69. Yang Y, Li W, Liu J. 2021. Review of recent progress on DNA-based biosensors for Pb2+ detection. *Analytica Chimica Acta*. 1147: 124–143

70. Tamer U, Torul H, Dogan U, Eryilmaz M, Gumustas A, et al. 2021. SERS sensor applications in environmental analysis and biotechnology. In *Nanotechnology Applications in Health and Environmental Sciences*, ed. N Saglam, F Korkusuz, R Prasad, pp. 197–236. Springer

Part II

Transducers for Biosensors

2 Electrochemical Transducers for Biosensors

Ali A. Ensafi[1,2] and Parisa Nasr-Esfahani[1]
[1]Isfahan University of Technology
Department of Chemistry
Isfahan, Iran

[2]University of Arkansas
Department of Chemistry & Biochemistry
Fayetteville, AR, USA

CONTENTS

DOI: 10.1201/9781003189435-4

2.1 INTRODUCTION

Biosensors are one of the most striking areas in analytical chemistry. The glucose oxidase biosensor began research on biosensors in 1962 (1). So far, numerous studies have been done to develop biosensors for various analytes in different fields such as medicine, healthcare, the environment, and industry. A biosensor can be applied to monitor biological or non-biological samples. One of the main reasons for the increasing attention of biosensors is that they do not require sample preparation. On the other hand, disposable biosensors are very beneficial, such as measuring blood glucose for people with diabetes (2).

A biosensor is an analytical device that contains three main parts: a biological recognition element, a transducer or detector element, and a signal processor. The biological recognition element can be enzymes, antibodies, nucleic acids, proteins, and cells that react with the analyte selectively and reduce interferences from other species in a sample. The transducer transforms one signal into another one and, based on the transducer used, biosensors are divided into four types: electrochemical biosensors, mass-based biosensors, calorimetric biosensors, and optical biosensors. Finally, a signal processor gathers, amplifies, and shows the signal (2). As shown in Figure 2.1, different samples can be analyzed with biosensors, including cell cultures, human samples (blood, urine, salvia), food samples, and environmental samples (3).

Most biosensors utilize electrochemical methods as transducers because of their low price, simplicity of fabrication, portability, and ease of use. Electrochemical biosensors use electrochemical methods to produce measurable electrical signals, including current (amperometry), potential or charge (potentiometry), impedance (impedimetry), and conduction (conductometry) (4). If the current is measured at a constant potential, it is named amperometry, and when the current is followed at different potentials, it is called voltammetry. Electrochemical biosensors can easily be portable for environmental monitoring and clinical testing because they don't require intricate signalling elements, and electrochemical reactions deliver electronic signals

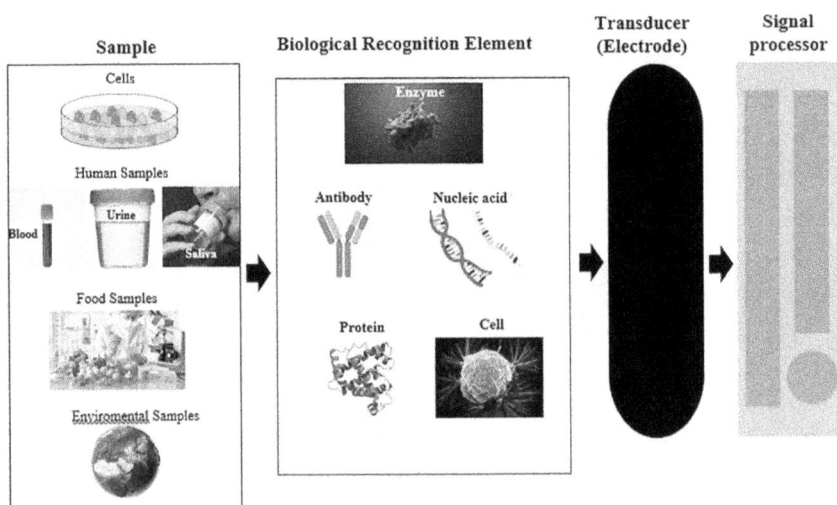

FIGURE 2.1   Different parts of an electrochemical biosensor.

(5). Actually, in electrochemical biosensors, the electroactive section of the biological recognition element acts as the main transduction element (6). Electrochemistry provides specific benefits for detection in biosensors because it is a surface technique. It needs a small sample volume and does not depend on the reaction volume (7). Moreover, electrochemical biosensors provide achievement too low detection limits because they usually do not need sample preparation. Also, electrochemical measurements can be done on turbid or colored samples because sample components, including particles, chromophores, and fluorophores, do not affect these measurements. At the same time, they are very effective in spectrophotometric detection (4).

Electrochemical biosensor studies are done using an electrochemical cell with two or three electrodes. Electrodes have a fundamental role in the performance of electrochemical biosensors because reactions are done near the surface of the electrode. Electrode features such as material, surface modification, geometry, and dimensions affect the electrochemical biosensor ability. There are three types of electrodes in a three-electrode system, as seen in Figure 2.2, including a reference, an auxiliary, and a working electrode. The reference electrode is used to apply a stable and known potential to the working electrode. Standard hydrogen electrode, calomel electrode (Hg_2Cl_2/Hg), and silver/silver chloride (Ag/AgCl) are used as reference electrodes. The calomel and silver/silver chloride electrodes are widely utilized more than the standard hydrogen electrode because hydrogen is explosive and difficult to prepare and maintain. Auxiliary electrodes are usually made of inert materials such as carbon, platinum, and gold. The auxiliary electrode potential is adjusted so that for every half-reaction that takes place on the working electrode, the opposite half-reaction is performed on the auxiliary electrode. Moreover, the surface of the auxiliary electrode is usually larger than that of the working electrode. The half-reaction at the auxiliary electrode should occur fast enough and not restrict the occurring half-reaction at the working electrode. The working electrode serves as the transducer element in biochemical or bioelectrochemical reactions. There are various working electrodes, including a gold electrode,

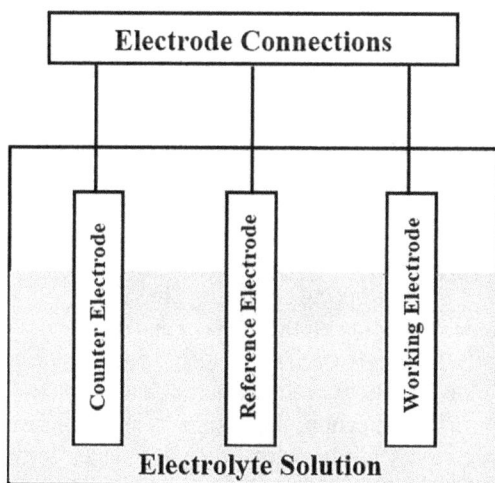

FIGURE 2.2 A schematic of a three-electrode electrochemical cell.

platinum electrode, silver electrode, carbon paste electrode, screen printed electrode, etc. In a two-electrode system, there are only the working and reference electrodes. If the current density is minimal, the reference electrode can transmit the charge. These two kinds of systems have been widely utilized for biosensing applications (8–9).

A significant step in fabricating electrochemical biosensors is the immobilization of the biological recognition elements on the electrode surface. The immobilization of biological recognition elements on the electrode surface can be performed using well-known immobilization techniques, including adsorption, microencapsulation, entrapment, covalent attachment, and cross-linking (10–11). Adsorption is the simplest way to immobilize recognition elements on the electrode surface. There are two types of adsorption methods: physical adsorption (physisorption) and chemical adsorption (chemisorption). Generally, physical adsorption is weaker than chemical adsorption. Therefore, the lifetime of an electrochemical biosensor prepared using the adsorption method is short, because the force linking the recognition element and the electrode surface in this method are van der Waals force and the hydrogen bond that are not very resistant and stable. In the microencapsulation method, biological recognition elements are trapped on the working electrode using an inert membrane trap such as cellulose acetate, chitosan, collagen, gluten aldehyde, Nafion, polyurethanes, etc.

Microencapsulation allows more concentrations of the biological recognition elements to be immobilized on the electrode surface. In the entrapment method, a solution of biological recognition elements in a polymer, including starch gels, nylon, polyaniline, or Nafion, is prepared and entrapped on the electrode surface. Covalent attachment immobilization prevents the biological recognition elements from releasing from the electrode surface when an electrochemical biosensor is used. It is important to note that this bonding should not cover the active site of the biological recognition element. Usually, OH, SH, NH_2, and CO_2H groups participate in this bonding. In the cross-linking method, bifunctional agents such as glutaraldehyde are utilized to bind the biological recognition element by covalent bonds. But in this method, a high ratio of enzyme activity is lost (4, 12).

Although there are different transducer types for biosensors, this chapter is limited to the electrochemical transducer, including amperometric, voltammetric, impedimetric, potentiometric, and conductometric transducers.

2.2   AMPEROMETRIC TRANSDUCERS

Commonly, the basis of amperometry is the continuous measurement of a current at a constant applied potential. The electrochemically active species cause the signal in the solution. It is proportional to the concentration of the species that oxidizes or reduces on the surface of the working electrode. Thus, the working electrode can act as an anode or cathode depending on the value of applied potential and the nature of the electroactive species. Amperometry is widely used in bioelectrochemical measurements because of its simplicity, wide dynamic range, and low detection limit (4). Actually, in amperometric detection, applying a constant potential minimizes the charging current, decreasing the background signal, which improves the detection limit. Furthermore, amperometric biosensors increase the sensor's selectivity for a particular analyte because the oxidation or reduction of the analyte occurs at a specific potential, which is one of its characteristics (12).

Establishing a fast electron transfer from the biological component to the electrode is necessary for developing sensitive and quick response amperometric biosensors. Therefore, the main goal in the design and architecture of amperometric biosensors is the fast electron transfer between the active site of the biological component and the electrode surface. Amperometric biosensors are divided into three classes (13):

- Mediatorless amperometric biosensors;
- Amperometric mediated biosensors;
- Amperometric biosensors based on direct electron transfer.

2.2.1 MEDIATORLESS AMPEROMETRIC BIOSENSORS

In each reaction, some species are produced, and some species are consumed. The concentration of these species can be determined by amperometric technique if they are electroactive. The enzymes used in these reactions are usually oxidases, dehydrogenases, or hydrolytic enzymes. As shown in Figure 2.3, the basis of the performance of the biosensors that use oxidases is oxygen absorption or generation of hydrogen peroxide during a biocatalytic reaction. Oxygen is monitored by O_2 reduction, and hydrogen peroxide is measured by H_2O_2 oxidation at the potential of -0.7V and +0.65V, respectively (13).

A common criticism of mediatorless amperometric biosensors is that the geometry of the electrode is very influential on the response, especially for sensor miniaturization (14). Also, the electrode must be pretreated to provide a reproducible surface. To solve this problem, the electrode surface is treated by chemical or heat method or cyclic voltammetry. These methods improve response reproducibility but do not have a long-term effect. A superior method to solve this problem is modifying the electrode surface chemically, using redox-polymers or conductive organic materials (13).

2.2.2 AMPEROMETRIC MEDIATED BIOSENSORS

In this group of amperometric biosensors, transducers are oxidizing agents used as mediators for electron transfer. Figure 2.4 demonstrates the operation basis of these

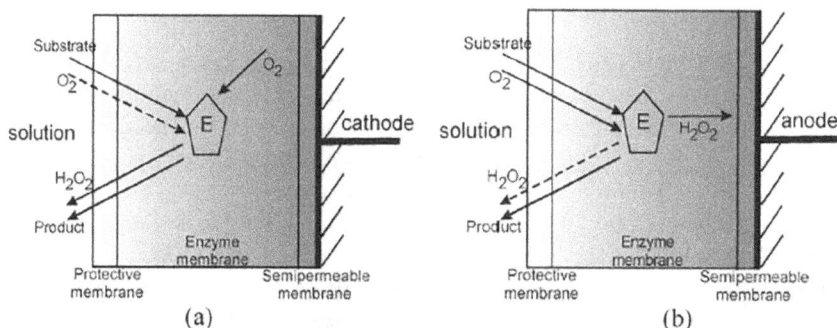

FIGURE 2.3 The principle of mediatorless amperometric biosensors base on (a) oxygen and (b) hydrogen peroxide detection (13).

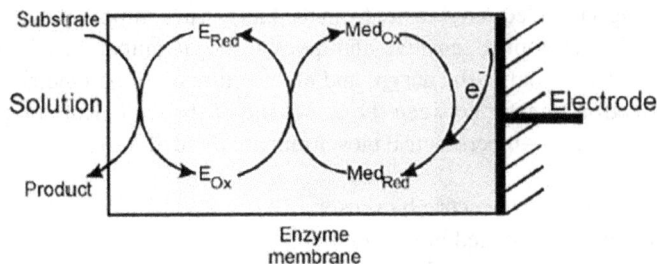

FIGURE 2.4   The principle of amperometric mediated biosensors (13).

biosensors. A mediator usually has a low molecular weight and transfers electrons between the enzyme and the working electrode. This decreases the potential required to perform the reaction and minimizes the effects of oxygen and other interferences. Ferrocene and ferricyanide are the most well-known and widely used mediators (13).

Mediators are immobilized on the electrode surface or added to the solution. Adding them to the measuring solution is a simple method but has some drawbacks because a number of mediators, such as organic colorants, are toxic, unstable, and pH-sensitive. Immobilization of mediators is the best and most technological method. But biosensors based on immobilized mediators have low stability, which has limited their development. For example, sometimes, mediator powder is mixed with carbon paste. Also, in some cases, the mediator is dropped directly on the electrode surface and absorbed. When the electrode is placed in the measuring solution, the mediator may be washed from the electrode surface.

Moreover, it may be oxidized or reduced by applying potential, and the amount of the mediator on the electrode surface reduces that decreases the output signal. For solving this problem, conductive polymers modified with intermediates are used. Another solution is using a graphite emulsion to introduce the mediator and enzyme into it. A proper mediator should have such features: the mediator should have fast electron transfer with the enzyme, the reduced mediator should not react with oxygen, and the mediator should not be sensitive to pH and toxicity (15).

2.2.3   AMPEROMETRIC BIOSENSORS BASED ON DIRECT ELECTRON TRANSFER

In this group of amperometric biosensors, electron transfer occurs directly and without any mediators between enzyme and electrode named bioelectrocatalysis. The rate constant of direct electron transfer between the enzyme and electrode depends on the distance between the enzyme active site and electrode surface and potential drop. Direct electron transfers between cytochrome c and gold electrode modified with 4,4′-bipyridine, TIO electrode, and mercury electrode has been studied by cyclic voltammetry. Direct electron transfers between the peroxidase and the electrode lead to hydrogen peroxide reduction. Thus, direct electron transfer between the enzyme and electrode has a significant role in developing amperometric biosensors. Amperometric biosensors based on direct electron transfer have substantial benefits, including high selectivity because of effective electric activation of enzymes, considerable reduction of interfering species signals, and high sensitivity because of large current density (13).

2.2.4 DIFFERENT APPLICATIONS OF AMPEROMETRIC BIOSENSORS

Today amperometric biosensors are widely utilized for healthcare applications, controlling industrial processes, and environmental monitoring. The design and instrumentation required to manufacture a biosensor depend on the field of application.

Vieira et al. synthesized vanadium dioxide (VO_2) film by hydrothermal treatment and immobilized glucose oxidase (GO_x) onto the film by covalent cross-linking to develop an amperometric biosensor detection of glucose. The effect of scan rate, pH, and glucose concentration on the VO_2/GO_x film was investigated by electrochemical studies. Under optimal conditions, the sensitivity of the VO_2/GO_x biosensor was obtained at 1.41 μA (mmol L^{-1}.cm^{-2})$^{-1}$ (16). Kumar et al. constructed an amperometric biosensor based on covalent immobilization of sarcosine oxidase nanoparticles (SO_x NPs) onto gold electrodes to determine prostate cancer. A linear relationship was observed from 0.1 to100 μM with a low detection limit of 0.01 μM for sarcosine. The biosensor was successfully used for measuring sarcosine levels in sera collected from persons who have prostate cancer (17). Azzouzi et al. developed an amperometric biosensor using screen-printed electrodes decorated with reduced graphene oxide-gold nanoparticles and L-lactate dehydrogenase to detect L-lactate tumor biomarkers. The biosensor displayed a linear range from 10 μM to 5 mM with a detection limit of 0.13 μM and a sensitivity of 154 μA/mM.cm^2. The performance of the proposed biosensor was investigated for L-lactate determination in an artificial serum (18). Vlamidis et al. developed two types of amperometric biosensors based on glassy carbon electrodes modified with graphene oxide and multi-walled carbon nanotubes using immobilization tyrosinase or laccase. The tyrosinase-based biosensor was used for the determination of polyphenols detection in fruit juices. The reliability of the biosensors was also studied by analyzing commercial fruit juices (19). Stasyuk et al. prepared two amperometric biosensors based on bimetallic PtRu nanoparticles coupled with alcohol oxidase and methylamine oxidase to detect ethanol and methylamine, respectively, in food analysis. The linear ranges for the ethanol and methylamine were 0.025 to 0.20 mM and 0.02 to 0.60 mM, with detection limits of 3 and 2.5 μM, respectively. The constructed biosensors were used for ethanol detection in wine and kefir and methylamine determination in fish tissue extracts (20). Conzuelo et al. reported an amperometric biosensor for the lactose determination. The biosensor was fabricated using coimmobilization of β-galactosidase, glucose oxidase, peroxidase, and the mediator tetrathiafulvalene via dialysis membrane on a gold electrode that was modified with 3-mercaptopropionic acid. The biosensor exhibited a linear response with the concentration of lactose in the range from 1.5×10^{-6} to 1.2×10^{-4} M with a detection limit of 4.6×10^{-7} M. The proposed biosensor was successfully employed for the determination of lactose in milk, butter, cheese, chocolate, margarine, mayonnaise, and yoghurt (21). Do and Lin prepared an amperometric biosensor based on immobilization of urease onto nanostructured polyaniline-Nafion/Au/Al_2O_3 for lead ion detection. The sensitivity of lead ion at the amperometric urease-based biosensor obtained 743.5 μA ppm^{-1} for the linear dynamic range of 0.1–1.0 ppm (22).

Zhang et al. constructed an amperometric biosensor via electrodeposition of gold nanoparticles on thiol graphene modified glassy carbon electrode to form a nano-composite film for CotA laccase immobilization. The bacterial laccase biosensor was used for the quantitative detection of hydroquinone. The linear range, detection limit, and sensitivity were obtained 1.6–409.6 µM, 0.3 µM, and 0.11 µA/mM, respectively. The feasibility of the CotA-based biosensor in the real sample was investigated by determining hydroquinone in the lake water (23).

2.3 VOLTAMMETRIC TRANSDUCERS

In voltammetric techniques, the voltage is scanned, and the current that is a result of analyte electrochemical reduction or oxidation at the working electrode is mea-sured. Actually, the voltage is applied between the working electrode and the refer-ence electrode, while the current is measured between the counter electrode and the working electrode. A voltammogram showed the obtained current *vs.* voltage. The current response is proportional to the analyte concentration and is usually a peak or a plateau (2). Depending on the type of applied voltage, there are various voltammet-ric methods, including polarography, cyclic voltammetry, linear sweep voltammetry, differential pulse voltammetry, square-wave voltammetry, stripping voltammetry, and AC voltammetry. In the following, we will discuss the most widely used of these methods in the field of electrochemical biosensors.

2.3.1 CYCLIC VOLTAMMETRY

The most famous form of voltammetry is cyclic voltammetry which gives a lot of information about the analyte's redox potential and electrochemical reaction rates. As shown in Figure 2.5a, the applied voltage to the working electrode is scanned linearly between two values from E_1 to E_2 and the direction of the scan is reversed at E_2 known as the switching potential. A critical factor in this technique is the scan rate defined as $(E_2-E_1)/(t_2-t_1)$. The choice of scan rate is crucial because there must be a suitable time to occur the electrochemical reaction. By increasing the voltage from V_1 toward the potential that the analyte reduces, the electrochemical reaction is performed, and the current also increases. As seen in Figure 2.5b, with increasing

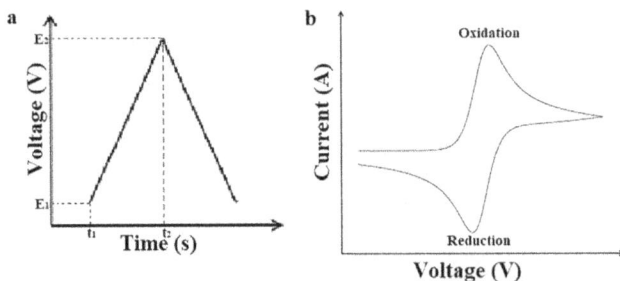

FIGURE 2.5 a) Voltage vs. time and b) typical voltammogram for cyclic voltammetry.

voltage toward V_2, the current decreases, and a peak is formed. As the voltage is reversed and scanned toward V_1, the oxidation reaction occurs, and the original analyte form is reproduced. As well, the reversibility of the electrochemical reaction can be specified with a reverse scan of potential at the desired scan rate. In the following, we examine a number of electrochemical biosensors that are based on cyclic voltammetry (24).

Kumar and D'Souza fabricated an electrochemical biosensor based on the screen-printed carbon electrodes modified with whole cells of recombinant Escherichia coli using glutaraldehyde as a sensor to detect methyl parathion. Before and after hydrolysis of methyl parathion, cyclic voltammograms were recorded. The detection range was obtained 2–80 μM, and the detection limit was 0.5 μM that estimated from signal to noise ratio (S/N = 3) in response to the blank sample. Also, the proposed biosensor was reused for 32 reactions, while 80% of its initial enzyme activity was maintained and stable for 22 days (25). Ahmadalinezhad and Chen reported a sensitive electrochemical biosensor for the detection of total cholesterol. For the biosensor fabrication, three enzymes, including cholesterol oxidase, cholesterol esterase, and horseradish peroxidase, were immobilized on nanoporous gold networks that were directly grown on a titanium substrate. Using cyclic voltammetry, the electrochemical behavior of the proposed biosensor was investigated, and results showed that it had high selectivity and high sensitivity. The detection limit of the biosensor was estimated at 0.5 mg dL^{-1}. The fabricated biosensor was used for cholesterol measurement in real food samples such as margarine, butter, and fish oil showed the biosensor has the potential to be used as a cholesterol detection tool in food (26).

2.3.2   DIFFERENTIAL PULSE VOLTAMMETRY

In differential pulse voltammetry, short pulses with a duration of 10–100 ms and an amplitude of 1–100 mV with a linearly increasing DC ramp are applied to the surface of the working electrode (Figure 2.6a). In this technique, the current is measured at two different times: before (I_1) and after the pulse is applied (I_2). The final signal is the difference between I_2 and I_1 (I_2-I_1), which is displayed as a function of applied potential (Figure 2.6b). A maximum in the differential pulse voltammogram is seen at the half-wave potential of the oxidation or reduction reaction (27). This technique has been used in numerous electrochemical biosensors, and we will mention some of them in this section.

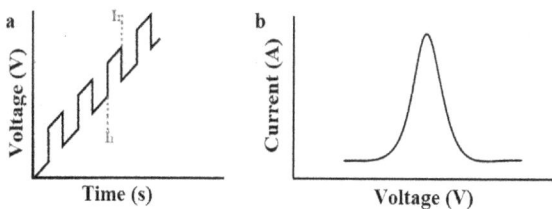

FIGURE 2.6   a) Voltage vs. time and b) typical voltammogram for differential pulse voltammetry.

Ensafi et al. developed an ultrasensitive electrochemical aptasensor based on Au electrode for mercury detection in water samples (Figure 2.7). In this study, zeolitic imidazolate framework -8-derived Ag@Au core-shell nanoparticles were synthesized and used as an aptasensor modifier. Actually, zeolitic imidazolate framework -8 as a substrate provided a large surface area to uniform distribution of Ag@Au core-shell nanoparticles. On the other hand, Ag@Au core-shell nanoparticles supported the covalent attachment of the aptamers on the surface of the Au electrode. Under optimal experimental conditions, using the differential pulse voltammetry technique, a wide linear range of 1.0×10^{-16}–1.0×10^{-12} M with a superior ultra-low detection limit of 1.8×10^{-17} M were obtained (28). Also, Ensafi's group reported a DNA biosensor to determine atropine sulfate, a cholinergic drug used as the medicine given before systemic anesthesia and as an anti-arrhythmic medicine. For biosensor fabrication, ds-DNA was immobilized on a pencil graphite electrode that was modified with titanium dioxide nanoparticles, multiwall carbon nanotubes, and poly-diallyl dimethylammonium chloride. Using differential pulse voltammetry under the optimal conditions, two linear ranges were obtained from 0.6 to 30.0 μmol L^{-1} and 30.0 to 600.0 μmol L^{-1} with a detection limit of 30.0 nmol L^{-1}. Finally, the applicability of the biosensor was investigated by measuring atropine sulfate in blood serum and urine samples (29).

Moreover, Ensafi et al. introduced an electrochemical biosensor based on naringin–DNA interaction for naringin detection. Naringin is a flavonoid compound that is found in different citrus fruits and has anti-carcinogenic, anti-inflammatory, free radical scavenging, and antioxidant effects. A pencil graphite electrode was modified with multiwall carbon nanotubes-poly diallyl dimethylammonium chloride decorated with ds-DNA. Under the optimized conditions, using differential pulse voltammetry, a linear relationship was observed between naringin concentration and the biosensor's response from 0.058 to 580.0 μg mL^{-1}, with a limit of detection of 0.010 μg mL^{-1}. Finally, this electrochemical biosensor was utilized for naringin determination in real samples (30).

2.3.3   SQUARE WAVE VOLTAMMETRY

In square wave voltammetry, to apply the potential in a square waveform, a square wave pulse is accompanied by a staircase wave (Figure 2.8). In this technique, τ is the time length of the pulse, E_{sw} is the pulse height of the square wave, and E_{sc} is the staircase increase for each step. At the end of the forward square-wave step, the first current (I_1) is measured that is related to the oxidation of R to O and is a positive current. At the end of the return square-wave pulse, the second current (I_2) is measured that is related to the reduction of O to R and is a negative current. Thus, the current signal is the difference between I_1 and I_2. The final signal is a higher signal than I_1 and I_2 because it is the difference between I_2 and I_1, which have the opposite sign (31). Square wave voltammetry has been widely utilized in developing electrochemical biosensors to detect diseases and environmental pollutants in recent years because of its excellent selectivity and high sensitivity (32).

Cardoso et al. developed a simple electrochemical biosensor to detect a cancer biomarker (miRNA-155) in breast cancer. At first, for the biosensor fabrication,

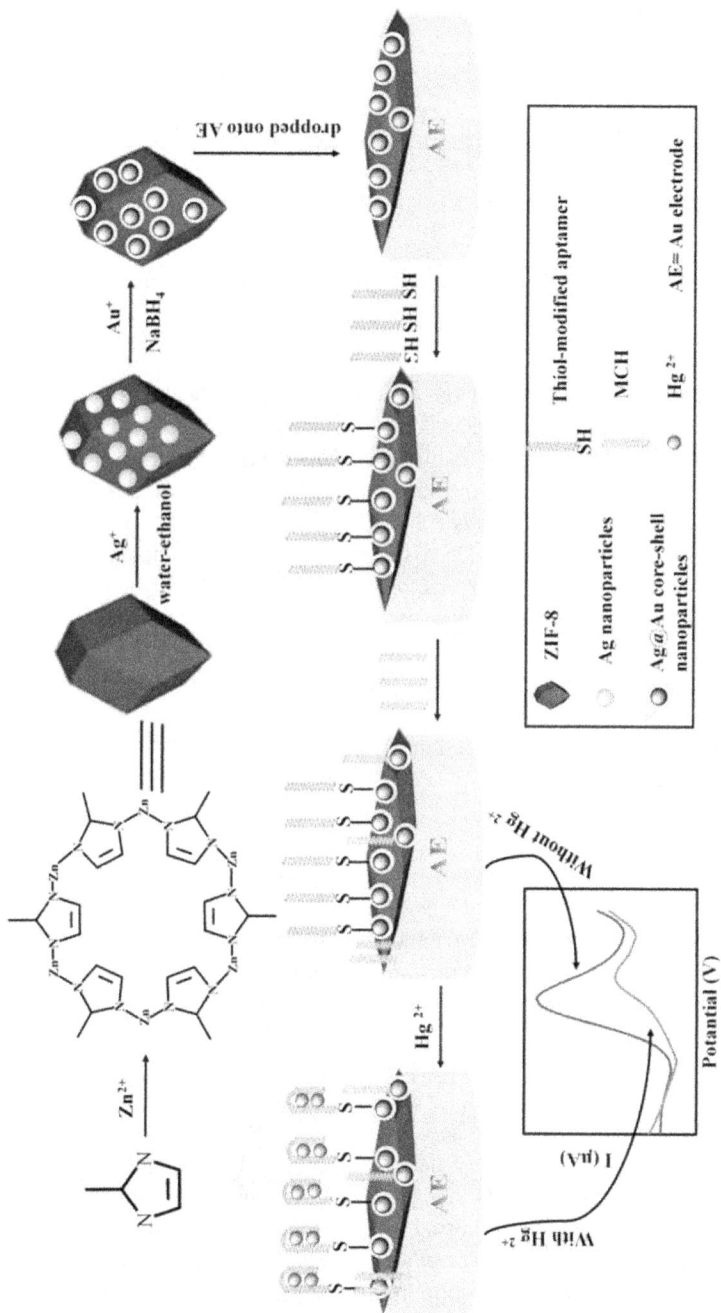

FIGURE 2.7 An ultrasensitive electrochemical aptasensor based on Au electrode for mercury detection in water samples (Ref. 28 with permission from Elsevier).

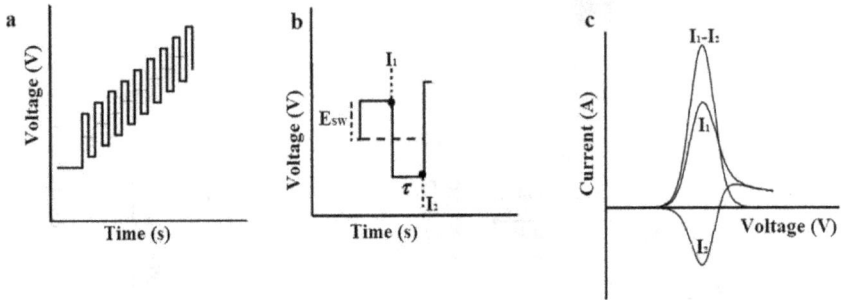

FIGURE 2.8 a) square-wave form, b) step potential, and c) typical voltammogram for square wave voltammetry.

anti-miRNA-155 was immobilized on an Au-screen printed electrode, and then, non-specific binding areas were blocked with mercaptosuccinic acid. Then, using square wave voltammetry, a linear relationship between assays current intensity and the logarithm of miRNA-155 concentration was obtained in the range of 1.0 fM and 10 nM, and the limit of detection was 2.8 aM. Finally, the electrochemical biosensor was successfully used for miRNA-155 assay in human serum samples (33). In addition, Fernandes's group reported an electrochemical biosensor for methomyl detection in vegetable extract samples. Methomyl is a very toxic and hazardous carbamate insecticide and can be absorbed through breathing, ingestion, and skin contact. This biosensor was fabricated using a carbon-ceramic electrode that was modified with laccase from Aspergillus oryzae. Under the optimum conditions, the calibration curve was obtained by square wave voltammetry in the range of 0.5 to 12.2 mM with a detection limit of 0.2 mM (34).

2.3.4 STRIPPING VOLTAMMETRY

Stripping voltammetry is one of the most important techniques in electroanalytical chemistry. It has two steps; at the first step, the analyte is accumulated on the surface of the working electrode at a constant potential that is called the preconcentration step; at the second step, the potential is scanned, and the preconcentrated analyte is stripped back into the solution that is called the stripping step. During the stripping step, the current in the voltammogram increases to a peak whose peak height or area is proportional to the analyte concentration, and the peak potential is a characteristic of the analyte. These are some of the important benefits of this technique, including extremely low detection limits (10^{-10}-10^{-12} mol L^{-1}) because of the enrichment of the analyte during the first step, ability to measure multiple elements simultaneously, and low costs for the purchase and management of instrumentation. These are the three types of stripping voltammetry techniques: anodic stripping voltammetry, cathodic stripping voltammetry, and adsorptive stripping voltammetry. In anodic stripping voltammetry, the analyte is electroplated on the working electrode during the preconcentration step and then oxidized from the electrode during the stripping step (Figure 2.9a). In cathodic stripping voltammetry, the analyte is electroplated on the working electrode at an oxidizing potential during the preconcentration step. Then, the oxidized species are stripped from the electrode by sweeping the potential negatively during the deposition step. This method is suitable for ionic species that have insoluble salts and deposit on the electrode during the deposition step. The difference between that adsorptive stripping voltammetry and the previous two methods is that in this method, the preconcentration step is not electrochemically and is performed by adsorption of the analyte on the working electrode surface (Figure 2.9b) (35). Stripping voltammetry has been used in numerous electrochemical biosensors.

Zhang et al. reported an electrochemical biosensor by anodic stripping voltammetry for Escherichia coli detection based on core-shell Cu@Au nanoparticles-labeled anti-Escherichia coli antibody. The principle of the proposed biosensor is shown in Figure 2.10. Core-shell Cu@Au nanoparticles-labeled anti-Escherichia coli antibody was captured on a polystyrene-modified ITO chip. Then, Cu@Au nanoparticles were converted to the ionic Au^{3+} and Cu^{2+} form by dissolving. Finally, the released Cu^{2+} ions were determined at glassy carbon/Nafion/Hg modified electrode by anodic stripping voltammetry. The stripping current response of Cu was proportional to the logarithmic of Escherichia coli concentration from 50 to 5.0×10^4 CFU/mL. The

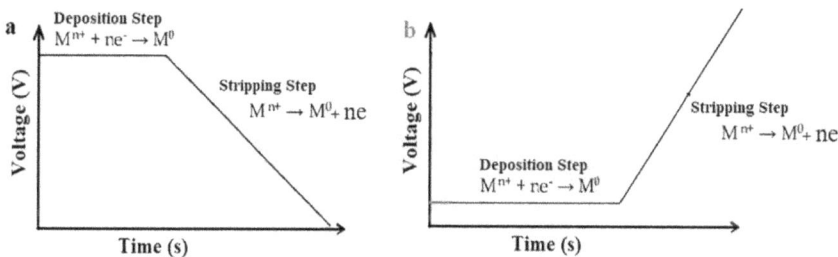

FIGURE 2.9 Steps in a) anodic and b) adsorptive stripping voltammetry.

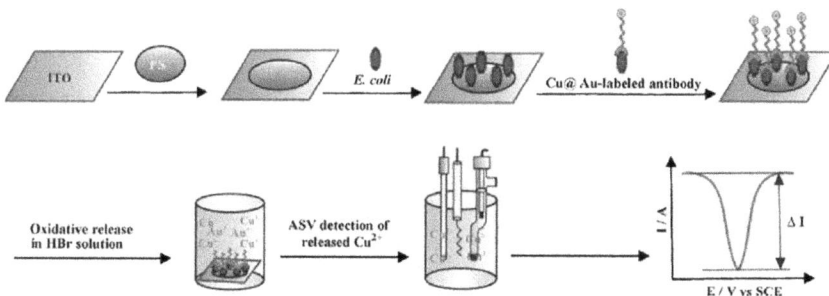

FIGURE 2.10 Preparation steps of the electrochemical biosensor based on Cu@Au nanoparticles as anti-Escherichia coli antibody labels (Ref. 36, with permission from Elsevier).

detection limit was obtained 30 CFU/mL. Also, the application of the electrochemical immunoassay was evaluated for Escherichia coli detection in surface water (36).

Dali et al. developed an electrochemical biosensor to detect lead (II) and cadmium (II) by differential pulse anodic stripping voltammetry. A glassy carbon electrode surface was chemically modified by a mixture of single-wall carbon nanotubes and biomass of fungal soil. The principle of lead and cadmium ions detection is illustrated in Figure 2.11. At first, complexes are formed between the immobilized biomass and the heavy metal ion (M^{2+}) at an open circuit. Then, the accumulated M^{2+} ions are reduced to M^0 at negative potential. Finally, M^0 is electrochemically stripped back into the solution by scanning toward positive potentials (stripping step). The stripping peak was proportional to the amount of M^{2+} in the accumulation solution. The detection limits were obtained 10^{-8}M and 10^{-7}M for Pb^{2+} and Cd^{2+}, respectively. The proposed biosensor was utilized for Pb^{2+} and Cd^{2+} detection in water samples (37).

2.4 IMPEDIMETRIC TRANSDUCERS

Electrochemical impedance spectroscopy (EIS) as a transducer for biosensors has recently attracted a lot of attention in the sensor community. Also, it has been widely utilized as an excellent technique for the study of fuel cells, batteries, and corrosion,

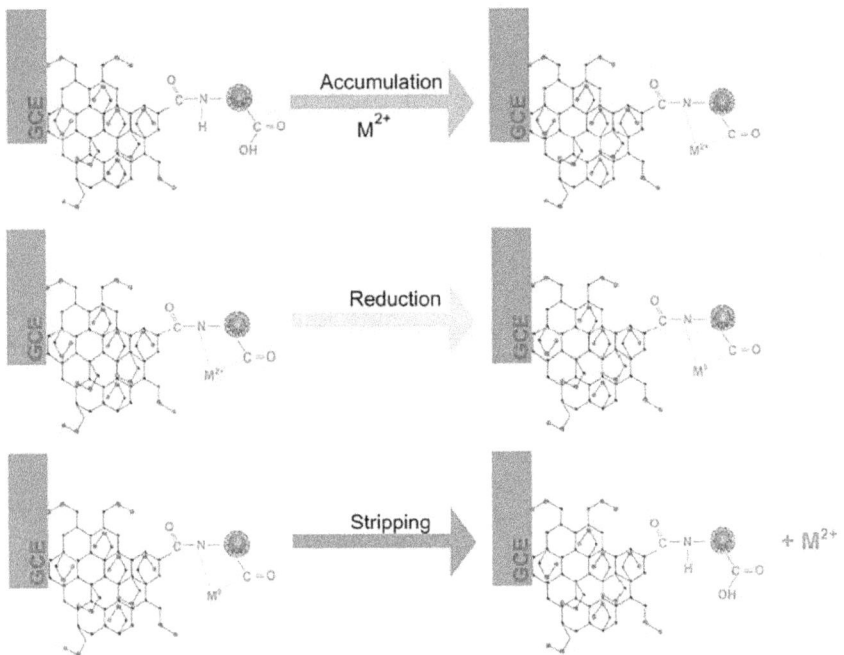

FIGURE 2.11 Schematic illustration of the detection principle for lead (II) and cadmium (II) (Ref. 37, with permission from Elsevier).

investigation conducting polymers, semiconductors, electrode kinetics, and most importantly, for characterization (38). EIS is a superb technique for determining the quantitative parameters of electrochemical systems because it is noninvasive, simple, fast, and inexpensive. If the conductivity, resistance, or capacity of the system changes during an electrochemical process, EIS can study it. Actually, EIS is a beneficial tool for developing electrochemical transducers for biosensors (2). Electron transfers that occur at the electrode surface involve mass transfer from the bulk solution to the electrode surface, adsorption of electroactive species, electrolyte resistance, and charge transfer at the electrode surface. These steps are demonstrated by resistance, capacitors, or constant phase elements in an electrical circuit (39). The Randles–Ershler electrical equivalent circuit model is the most famous model used in the EIS technique. This model includes parameters such as electrolyte resistance (R_s), charge-transfer resistance (R_{ct}), mass transfer resistance (R_{mt}), double-layer capacitance (C_{dl}), and Warburg impedance (W). R_s depends on the geometry of the electrochemical cell and solution conductivity. C_{dl} shows the electrostatic interaction between the electrode and the electrolyte, and parameters such as the electrolyte ionic strength, permittivity, and the electrode area affect C_{dl}. The charge transfer kinetics is related to R_{ct}. W and R_{ct} form the faradaic impedance (40).

The potential is scanned over a wide range of alternative current (AC) frequencies in the EIS technique. By applying a small sinusoidally varying potential, the current response can be measured. By changing the frequency of the applied potential, the complex impedance is obtained, which is the sum of real (Z_r) and virtual (Z_i) impedances. In the related equation (2.1), f is the frequency (2).

$$Z(j\omega) = \frac{U(j\omega)}{I(j\omega)} = Z_r(\omega) + j\,Z_i(\omega) \quad \omega = 2\pi f \qquad (2.1)$$

The Nyquist or Bode plots are used for displaying the impedance spectrum of an electrochemical system as a function of frequency (Figure 2.12). The Nyquist plots usually consist of a semicircle portion at high frequencies representing the charge transfer resistance and a linear portion at low frequencies corresponding to the diffusion process (30). The electron transfer resistance equals the diameter of the

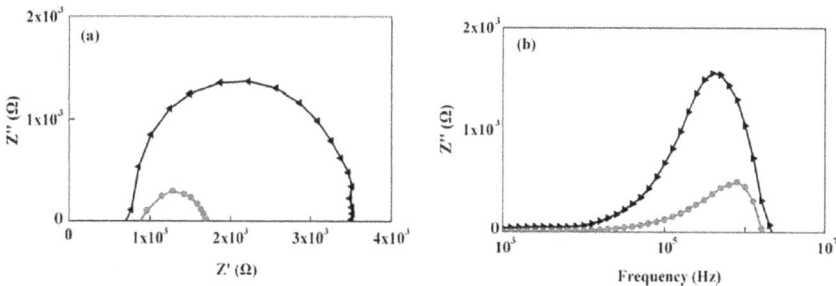

FIGURE 2.12   a): Nyquist plot and b): Bode plot.

semicircle. From this type of spectrum, information about the diffusional features and electron transfer kinetics is obtained. If the electron transfer process is very fast, the Nyquist plot consists only of the linear part. Still, if the electron transfer is prolonged, the spectrum consists of only the semicircle part (40). To form a Bode plot, the logarithm of the absolute value of impedance is plotted against the logarithm of the frequency.

Among benefits that can be considered for EIS, a couple of them are more significant. To begin with, it is a non-destructive technique due to the small amplitude perturbation from the steady-state. As well, the impedance can be measured in the presence (faradic impedance) or absence (non-faradic impedance) of a redox couple (41). However, the limitations cannot be ignored. There are three important principles for impedance measurements: linearity, stability, and causality. The operating procedures and the technical precision of the instrumentation affect the accuracy of EIS measurement (40).

2.4.1 APPLICATION OF IMPEDIMETRIC BIOSENSORS

As one type of transducer for electrochemical biosensors, the Impedance technique has attracted remarkable attention in different fields because of its sensitivity, rapidity, and portability (42–43). In this section, we express some examples of the use of these biosensors in various fields.

Detection of pathogenic bacteria in food has particular importance. The development of a sensitive and reliable method in this field is very significant so that impedimetric biosensors are widely used in bacteria detection. Joung et al. introduced an impedimetric biosensor based on a nanoporous membrane for bacterial pathogens detection in whole milk. In this study, an anodized alumina membrane was utilized for the fabrication of the impedimetric biosensor. Also, hyaluronic acid was used for surface modification to improve the immobilization of antibodies. Moreover, Escherichia coli O157:H7, one of the most detrimental food-borne pathogenic bacteria, was tested as a model. Finally, the prepared biosensor was applied for the Escherichia coli O157:H7 detection in whole milk samples with the detection limit of 83 cfu/ml (44).

Impedimetric biosensors are useful for rapid and precise quantitative detection of pathogenic bacteria to help rapid diagnosis in clinical applications. Asif Ahmed and coworkers developed an impedimetric biosensor for the detection of Streptococcus pyogenes in human saliva. Screen-printed gold electrodes were modified with polytyramine, and biotinylated antibodies were conjugated to amine groups of polytyramine via biotin-NeutrAvidin coupling (Figure 2.13). The sensor showed linear response from 100 cells/10 µL to 105 cells/10 µL) against S. pyogenes (45). Cervical cancer is the fourth most common type of deadly cancer in females.

Pradhan et al. fabricated a four electrode-based impedimetric biosensor (sensing electrode, working electrode, counter electrode, and the reference electrode) using photolithography techniques utilized to evaluate the cytotoxicity of tamoxifen on cervical cancer cell lines. The obtained results demonstrated that tamoxifen reduced the number of HeLa cells on the electrode surfaces. This research indicated that the EIS technique could be used as an efficient method to study the cytotoxicity of

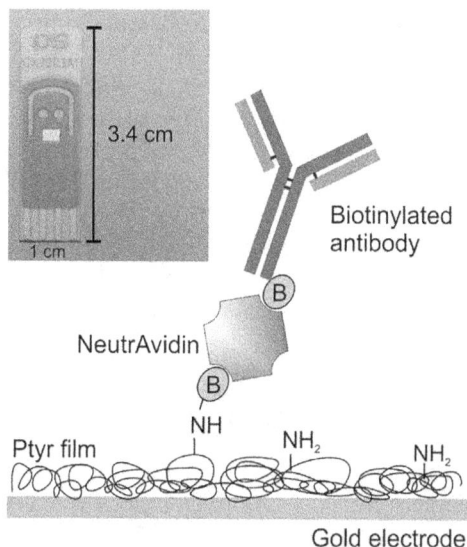

FIGURE 2.13 Schematic of an impedimetric biosensor to detect pathogenic bacteria streptococcus pyogenes (Ref. 61 with permission from American Chemical Society).

different novel drugs (46). One of the most common tumors in men is prostate cancer, and the well-known biomarker for its diagnosis is prostate-specific antigen (PSA). Diaz-Fernandez et al. prepared an impedimetric biosensor based on aptamer for the dual recognition of PSA. At first, two nanostructured gold electrodes are modified for biosensor fabrication using the immobilization of two various aptamers (PSAG-1, anti-PSA). Then, glycosylated-PSA and total PSA bind to the first and second biosensors, respectively. These bindings cause changes in the charge transfer resistance and determine the amount of glycosylated and total PSA in the sample. The dynamic range was from 0.26 to 62.5 ng/mL for the PSAG-1-based biosensor and between 0.64 and 62.5 ng/mL for the anti-PSA-based biosensor. This dual aptamer-based impedimetric biosensor is a promising method for the invasive diagnosis of prostate cancer (47). The presence of microbial pathogens in water can lead to some diseases such as typhoid, cholera, dysentery, and polio. For this reason, the development of rapid and low-cost point-of-need detection tools plays a significant role in preventing waterborne diseases.

Rengaraj's group reported an innovative impedimetric biosensor based on functionalized paper to recognize bacterial contamination in water. The biosensor was constructed using commercial hydrophobic paper and conductive carbon ink. The biorecognition element was Concanavalin A, because it can interact with mono- and oligo-saccharides on bacterial cells selectively. Results showed that the electron transfer resistance increased with increasing bacterial concentrations. A linear relationship between the charge transfer resistance and bacterial concentrations was observed from 10^3 to 10^6 CFU mL^{-1}, with a detection limit of 1.9×10^3 CFU mL^{-1} (48). One of the foodborne pathogens for humans and animals is the gram-negative bacterium Salmonella, a major factor in gastrointestinal infections found in milk, eggs, meat, and beverage.

Sheikhzadeh et al. developed a label-free aptamer-based impedimetric biosensor for Salmonella Typhimurium detection. Poly [pyrrole-co-3-carboxyl-pyrrole] copolymer was synthesized on a gold disk electrode. The presence of the carboxylic group in the poly [pyrrole-co-3- carboxyl-pyrrole] copolymer films allowed the covalent immobilization of the aptamer. The results showed that the relative charge transfer resistance increased with the logarithm of Salmonella Typhimurium concentration in the range of 10^2-10^8 CFU mL^{-1} and with a detection limit of 3 CFU mL^{-1}. Finally, the proposed aptasensor was used for Salmonella Typhimurium detection in apple juice samples (49).

2.5   POTENTIOMETRIC TRANSDUCERS

The potentiometric technique is based on measuring the potential difference between an indicator and a reference electrode, or two indicator electrodes in an electrochemical cell while the current is not significant (50–51). Actually, the potentiometric technique gives information about the ion activity in an electrochemical reaction and polymeric membrane. Ion-selective electrodes (ISEs) are the most common potentiometric sensors (52). The most well-known ion-selective electrodes are glass pH-electrode and ion-selective electrodes for ions such as Na^+, K^+, Ca^{2+}, and Cl- (4). Potentiometric sensors have advantages such as ease of use, small size, low cost, rapid response, and resistance to color and turbid interferences. The measurements using these sensors provide low detection limits in ranges between 10^{-8} to 10^{-11} M. On the other hand, potentiometric sensors are proper for low concentrations in small sample volumes (53).

In the potentiometric technique, the ion activity (or concentration) and the potential difference between the indicator and the reference electrodes are related to each other by the Nernst equation.

$$E_{cell} (emf) = E^0_{cell} - \frac{RT}{nF} \ln Q \tag{2.2}$$

In this equation, E_{cell} denotes the cell potential at zero current, which often is named electromotive force (emf), E^0_{cell} is the standard-state potential, R is the ideal gas constant, T the absolute temperature in Kelvin degree, n is the electron number that exchanges in the electrochemical reaction, F is the Faraday's constant, and Q shows the activity ratio of the ions at the anode to that at the cathode (2). This equation is established for a case where the membrane selectivity is infinite, and there is a constant or very low concentration of the interfering ions. Moreover, potential differences at diverse phase boundaries are negligible or constant (54). If the analyte ion concentration directly determines the Nernst equation, it is called direct potentiometry (12).

When a biological recognition element becomes part of a potentiometric sensor, a potentiometric biosensor is formed. Like any other biosensor, some points should be considered, such as analysis of substrate transfer to the biosensor surface, analyte diffusion to the biological layer, analyte reaction in the biological layer, and diffusion of products to the bulk solution. Like amperometric biosensors, the response of potentiometric biosensors is not an equilibrium response and only reaches a steady state or transition (54).

2.5.1 Application of Potentiometric Biosensors

Today, potentiometric biosensors have become powerful transducers for particular analytical purposes due to the introduction of novel biological recognition elements, including antibodies, aptamers, enzymes, and peptides. In this section, the applications of potentiometric biosensors have been discussed in different fields, such as detecting metal ions, small molecules, bacteria cells, and toxicities.

2.5.1.1 Metal Ion

Detection of traces of toxic heavy metals in air, soil, and water is critical because they can accumulate and store and cause irreparable damage to animals, plants, and humans. The inhibition of enzymatic activity as an approach for trace metal detection has attracted a lot of attention because of simplicity, rapidity, and some inhibitors' selective nature. Zeng et al. fabricated a potentiometric biosensor using a pH electrode for Hg^{2+} detection. The pH electrode was modified with chitosan–poly (vinyl alcohol) hydrogel film, and then urease immobilized onto the modified electrode. Enzyme-based biosensors are very sensitive for Hg^{2+} detection because a single inhibitor molecule can reduce enzyme activity noticeably due to the amplification effect. However, the inhibition of enzymatic reaction by Hg^{2+} is not specific and leads to poor selectivity. But in this study, the problem was solved using chitosan–poly (vinyl alcohol) hydrogel film. The results showed a good linear relationship between inhibition rate and the logarithm of Hg^{2+} concentration in the range from 0.002 µM to 2µM with a detection limit of 0.001 µM. The biosensor was successfully used for Hg^{2+} determination in industrial wastewater samples (31).

Ayenimo and Adeloju described a potentiometric biosensor to detect Cu^{2+}, Hg^{2+}, Cd^{2+}, and Pb^{2+} ions. The biosensor was constructed based on electropolymerization of polypyrrole and glucose oxidase on a platinum disc electrode. The results show that the inhibition by the metal ions is not only due to the -SH group but also due to additional binding sites of the enzyme. Linear concentration ranges were achieved 0.079–16 µM, 0.025–5 µM, 0.10–15 µM and 0.04–62 µM for Cu^{2+}, Hg^{2+}, Pb^{2+}, and Cd^{2+}, respectively. Also, detection limits were 0.079 µM, 0.025 µM, 0.024 µM and 0.044 µM for Cu^{2+}, Hg^{2+}, Pb^{2+}, and Cd^{2+}, respectively. The proposed biosensor was successfully applied to detect metal ions in tap water (55). Liao's group developed a microfluidic device with a potentiometric biosensor array for multiple detections of K^+ and Ca^{2+} using photolithography technology. This sensor included a pH indicator and potassium and calcium ion-selective microelectrodes. The pH indicator was a platinum microelectrode that was modified electrochemically with iridium oxide thin film. The K^+ ion-selective microelectrode was platinum coated with a silicone rubber-based ion-selective membrane with potassium (valinomycin) ionophore. Moreover, Ca^{2+} ion-selective microelectrode was platinum coated with silicone rubber-based ion-selective membrane with (ETH 1001) ionophore. The proposed sensor can measure the solution pH in the region between 2.00 to 10.00, and K^+ and Ca^{2+} concentration from 0.1 M to 10^{-6} M (56).

2.5.1.2 Small Molecules

Small molecule detection with high selectivity that ions do not disturb is difficult. Therefore, methods based on analyte-induced enzymatic reactions, inhibition of enzyme activities, antibodies, and aptamers are commonly used to identify small molecules. So far, many small molecules such as galactose, cholesterol, urea, and lactate have been detected using potentiometric biosensors. A disease in which the body cannot transform galactose to glucose is called galactosemia, a genetic disorder. The treatment for this disease is to avoid eating foods containing galactose. Therefore, it is very important to provide easy-to-use and straightforward tools for detecting galactose in nutrition and the blood of these patients outside of labs. Bouri et al. reported a disposable, low-cost, and simple paper-based potentiometric biosensor for galactose detection in whole blood.

Potentiometric measurements were performed using a standard two-electrode system. An Ag/AgCl standard electrode and a platinized paper were used as a working electrode and a reference electrode, respectively. Galactose oxidase was utilized as a biological recognition element in this biosensor. The proposed sensor presented a linear range from 0.3 to 31.6 mM and a detection limit of 0.25 mM for galactose detection. Also, sensor validation in whole blood samples was done with acceptable recovery (57).

Determining the cholesterol level in the human body is very important because increasing the cholesterol level could cause serious problems such as cardiovascular disease, diabetes, cancer, and dementia. Psychoyios's group fabricated a new potentiometric biosensor for cholesterol detection. The electrochemical cell included two electrodes, an Ag/AgCl reference electrode and a biosensor as a working electrode. For biosensor fabrication, cholesterol oxidase was incorporated into the lipid film and polymerization on the surface of zinc oxide nanowalls as a measuring electrode. The results presented a linear relationship between the potential and the logarithm of cholesterol concentration in the concentration range of 10^{-6} to 10^{-3} M. The fabricated biosensor was successfully used to determine cholesterol in blood serum samples (58).

Determination of urea level in blood is essential because abnormal urea levels in the blood indicate some diseases related to the liver and kidney. Ondes et al. reported a potentiometric biosensor for urea detection. An ammonium selective electrode was modified with a urease enzyme covalently immobilized onto poly (2-hydroxyethyl methacrylate-glycidyl methacrylate) nanoparticles, nano-carriers for urease enzyme. The potentiometric biosensor showed a linear range between 0.01 mM and 500 mM and a detection limit of 0.77 µM. Finally, the designed urea biosensor application was investigated for urea detection in artificial human serum (59).

Lactic acid determination has received much attention because its concentration is a significant factor for many applications in clinical diagnostics, food industry, beverages, and sports medicine. Lupu et al. developed a potentiometric biosensor based on lactate oxidase for lactate determination. First, the plasma polymerized acrylic acid layer is deposited on the Si_3N_4 surface for biosensor fabrication. Then lactate oxidase is immobilized by covalent bonding on the plasma polymerized acrylic acid layer. As a result, the biosensor exhibited a linear range of 10^{-5} M and a detection limit of 2×10^{-7} M for lactate (60).

2.5.1.3   Bacteria Cells

Detection of pathogens with a selective, specific, reliable, and rapid method is crucial because it plays a significant role in clinical diagnoses, food safety, and environmental monitoring. One of the common concerns is the effective prevention of infectious diseases caused by bacteria. Hernandez et al. reported a potentiometric biosensor based on graphene for the detection of Staphylococcus aureus. A glassy carbon rod was modified with graphene oxide as a transducer layer and aptamers as a sensing layer for biosensor fabrication. For potentiometric analysis, the electromotive force between the aptasensor as the working electrode and an Ag/AgCl electrode as a reference electrode was measured (61).

Silva's group reported a disposable paper-based potentiometric biosensor for Salmonella typhimurium detection that is a foodborne pathogen. The Salmonella monoclonal antibody was used as the biological recognition element. The electromotive force was measured between the paper-based potentiometric biosensor and a double junction commercial reference electrode. The results showed that the relationship between electromotive force difference and the logarithm of Salmonella typhimurium was linear in the concentration range from 12 to 12×10^3 cell mL^{-1}. Also, a limit of detection of 5 cells mL^{-1} was obtained. The proposed biosensor was used to detect Salmonella typhimurium cells in apple juice samples (62). Also, Lv et al. developed a magneto-controlled potentiometric biosensor for Staphylococcus aureus detection. The biosensor was constructed based on screen-printed electrodes and peptides-modified magnetic beads used as both recognition elements and transduction indicators. A magnetic field can control and modulate the response properties of the peptide on an ion-selective sensor and cause improved sensitivity and stability. This biosensor could detect Staphylococcus aureus with a linear concentration range of 1.0×10^2 to 1.0×10^6 CFU mL^{-1} and the detection limit was 10 CFU mL^{-1} (63). Ercole's group reported a biosensor to detect Escherichia coli that caused bacterial contamination in drinking water based on an immunoassay test. The biological recognition element was made by a rabbit polyclonal antibody obtained using an environmental strain of Escherichia coli. The designed biosensor was very sensitive and fast and could detect 10 cells per ml (64).

2.5.1.4   Toxicities

Developing low-cost and rapid methods for determining toxicities in the environment and industrial wastewaters is essential. Aflatoxins are natural fungal toxins produced by Aspergillus fungi. These fungi usually infect maize, wheat, barley, rice, tree nuts, and groundnuts easily. Among different types of aflatoxins, aflatoxin B1 is the most toxic and responsible for liver cancer in animals. Li et al. developed a potentiometric biosensor for aflatoxin B1 detection in food. Aflatoxin B1-bovine serum albumin conjugate was immobilized on a glassy carbon electrode, and gold nanoparticles were functionalized with polyclonal anti-aflatoxin B1 antibody. The basis of this sensor is that with adding target aflatoxin B1, competitive immunobinding occurs between the analyte and aflatoxin B1-bovine serum albumin for the labelled anti-aflatoxin B1 antibody on the gold nanoparticles. The measured output potential is linearly proportional to the logarithm of aflatoxin B1 concentration in the range of 0.1 to 5.0 $\mu g.kg^{-1}$, and the detection limit was obtained 87 $ng.kg^{-1}$. To assess the accuracy of the immunosensor, it was applied for the detection of aflatoxin B1 levels in peanut samples (65). The toxicity detection of pollutants in water has significant importance for ecosystems and human health. Zhang's group reported a potentiometric biosensor for toxicity detection in water. They used the ammonia-oxidizing bacterium Nitrosomonas europaea as a bioreceptor and a polymeric membrane ammonium-selective electrode as a transducer. The molecular recognition and transduction processes were performed individually using the ammonia-oxidizing bacterium cells were immobilized on polyethersulfone membranes. This biosensor acts based on the inhibition effects of toxicants on the activity of the ammonia-oxidizing bacterium, which is evaluated by measuring the ammonium consumption rates with the ammonium-selective membrane electrode. The developed sensor represented good sensitivity, simplicity, and speed for toxicity detection in water (66).

2.6   CONDUCTOMETRIC TRANSDUCERS

Conductometric technique is based on measuring the ability of an electrolyte solution or a nanowire to conduct an electrical current between electrodes. Conductometric equipment is a subset of impedimetric devices. Because conductometric transducers have received less attention than other transducers, the fundamental mechanisms have not been well studied (2). Conductometric transducers have significant advantages, including no need for a reference electrode, low-amplitude alternating voltage operation, and insensitivity to light. But, the conductometric techniques have some restrictions. Electrolyte solutions have significant background conductivity that can be modified by a variety of factors. This reduces the selectivity of this technique and limits its application. Because the ratio between the signal and noise level should not be lower than 2%, the concentrations of buffer and other materials added to the electrochemical cell are very effective and must be controlled. Moreover, in the presence of non-reacting ions, the sensitivity of the method is reduced. To measure low concentrations, electrolyte solutions with low ionic strength can be utilized until the signal/noise ratio is a suitable value (67–70).

Usually, the biological recognition element in conductometric biosensors is enzymes. When an enzymatic reaction occurs, the solution conductivity between two electrodes changes due to the change in ionic strength. Because the enzymes catalyze only specific reactions, in this case, selectivity is provided (71–73). In the following, different types of conductivity biosensors based on enzymes and their applications are examined.

2.6.1   GLUCOSE BIOSENSORS

Since glucose is an essential metabolite of living organisms and the most important carbon source in numerous microbial fermentation processes, glucose biosensors have received much attention among the developed biosensor. The basis of conductometric biosensors for glucose detection can be described using the following reactions (67):

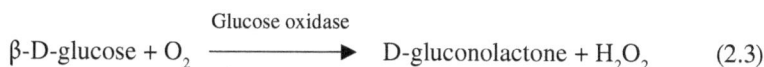

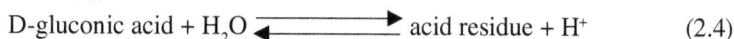

$$\beta\text{-D-glucose} + O_2 \xrightarrow{\text{Glucose oxidase}} \text{D-gluconolactone} + H_2O_2 \qquad (2.3)$$

$$\text{D-gluconic acid} + H_2O \rightleftarrows \text{acid residue} + H^+ \qquad (2.4)$$

Shulga et al. reported the first conductometric thin-film biosensor for glucose determination based on gold electrodes deposited on a ceramic substrate and modified with glucose oxidase. Different parameters that affect the biosensor's response, such as ionic strength and buffer capacity, were investigated. The detection limit for glucose determination obtained was 0.01 mM (60). Mahadeva and Kim described a conductometric biosensor for glucose detection using glucose oxidase immobilization on cellulose–tin oxide hybrid nanocomposite (Figure 2.14). The porous tin oxide layer was grown on the cellulose films via the liquid phase deposition technique. Glucose oxidase was immobilized into this nanocomposite via the physical absorption method. The proposed biosensor has a linear response in the range of 0.5–12 mM (48).

Li and his coworker presented an ion-sensitive conductometric glucose biosensor using layer-by-layer assembled single-walled carbon nanotube and glucose oxidase. The results showed that conductance is exponentially dependent on pH because of the

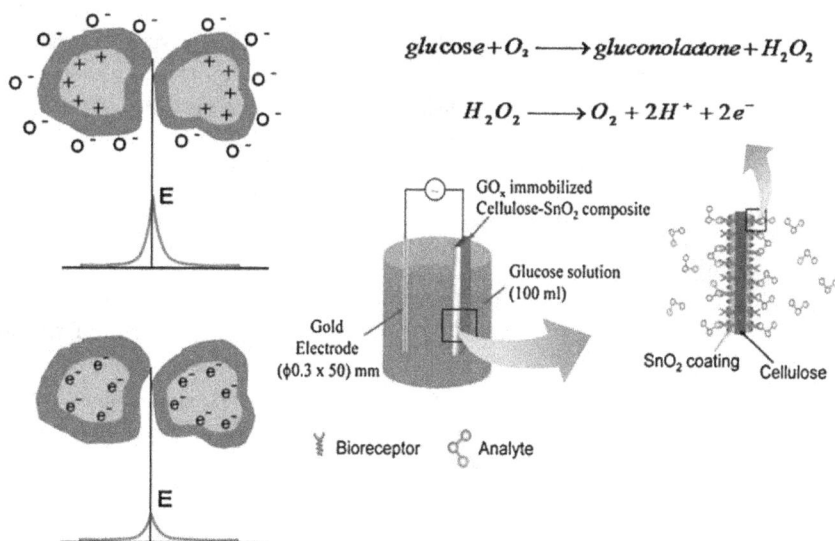

FIGURE 2.14 Schematic representation of a conductometric glucose biosensor based on immobilization of glucose oxidase on cellulose–tin oxide hybrid nanocomposite (Ref. 69, with permission from Elsevier).

protonation/deprotonation of carboxylic functional groups on the single-walled carbon nanotube. The overall sensitivity and the resolution of the proposed biosensor, at the bias voltage of 0.6 V, were 10.8 µA/mM and 1 pM, respectively. Dzyadevich et al. fabricated a glucose conductometric biosensor immobilization of glucose oxidase on the sensitive area of thin-film electrodes. Gold electrodes were constructed via vacuum deposition on a ceramic substrate. The enzymatically produced H^+ caused a decrease in the solution buffer capacity inside the enzymatic layer ions that made the desired signal (22).

2.6.2 Urea Biosensors

An important diagnostic test in clinical laboratories is the determination of urea in human blood. Abnormally high levels of urea in the blood indicate renal dysfunction. The enzymatic reaction that happens in urea biosensors is:

$$CH_4N_2O + 2H_2O + H^+ \xrightarrow{\text{Urease}} 2NH_4^+ + HCO_3^- \tag{2.5}$$

In this reaction, a proton is consumed, and ammonium and carbonate ions are produced, which change conductivity (21).

Lee et al. developed a conductometric urea biosensor using sol-gel-immobilized urease on a screen-printed interdigitated array. The screen-printed inter-digitated array electrode was an excellent conductometric transducer, and the conductance signal dominated the admittance signal. For urease immobilization, a conventional sol-gel process using tetramethyl orthosilicate as a precursor was used. The biosensor presented a linear dynamic range from 0.03 to 2.5 mM and a detection limit of 30 mM. The proposed sensor was used to determine urea in serum samples and the results were comparable with the results of blood urea nitrogen test kits (41). Saiapina's group developed a highly sensitive conductometric urea biosensor using a combination of ammonium-sieving and ion exchange properties of clinoptilolite and urease as a biological recognition element. The effect of pH, buffer capacity, and ionic strength on the biosensor's response were studied to optimize its performance. Under optimum conditions, the dynamic range and detection limit were found 0–64 mM and 10^{-6} M, respectively (57).

Velychko et al. reported a conductometric biosensor for urea detection. This biosensor was fabricated using adsorption urease on nanoporous particles of silicate. The designed biosensor has significant advantages, including the non-toxic compounds used in biosensor fabrication, simplicity, fast performance, and excellent reproducibility. The linear range and limit of detection were obtained 0.05–15 mM and 20 µM, respectively. This biosensor can be successfully utilized during renal dialysis for urea analysis (69).

2.6.3 Conductometric Biosensors for Phosphate Determination

Phosphate is an essential nutrient for plant growth, but increasing its concentration due to the use of fertilizers can cause eutrophication of lakes and rivers. In addition, the determination of phosphate levels in the body provides useful information about cells' energy, bone function, and different diseases. Also, the determination of phosphate is important for food quality control because it has a significant effect on human health (21). Zhang et al. developed a conductometric biosensor for phosphate detection by immobilization of maltose phosphorylase on a planar interdigitated electrode with the following reaction sequence:

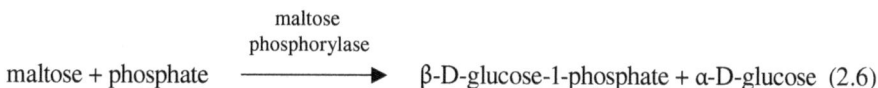

$$\text{maltose} + \text{phosphate} \xrightarrow{\begin{array}{c}\text{maltose}\\\text{phosphorylase}\end{array}} \beta\text{-D-glucose-1-phosphate} + \alpha\text{-D-glucose} \quad (2.6)$$

The principle of this biosensor is based on changes of conductance inside the maltose phosphorylase-based membrane. The biosensor provided two linear ranges from 1.0 μM to 20 μM and 20 μM to 400 μM with a detection limit of 1.0 μM. Finally, the performance of the proposed biosensor was investigated for the phosphate determination in real water samples (74). Upadhyay and Verma reported a novel and simple conductometric biosensor to determine the phosphate ions in an aqueous solution. At first, the glass surface was silanized with (3-Mercaptopropyl) trimethoxysilane and modified with cysteine functionalized silver nanoparticles for biosensor preparation. Then, the enzyme, alkaline phosphatase, was immobilized on the cysteine group of nanoparticles. This biosensor is based on the inhibition of immobilized alkaline phosphatase activity in the presence of phosphate ions. The developed biosensor showed a linear range from 0.5 to 5.0 mM with a detection limit of 50 mM. The proposed biosensor was finally utilized to determine phosphate levels in serum and tap water samples (67).

2.6.4 CONDUCTOMETRIC BIOSENSORS FOR PESTICIDES DETERMINATION

Every year, a wide variety of pesticides are used in agriculture. These toxins are degraded by microbial and photodecomposition and chemical hydrolysis and produce highly toxic products and contaminate air, water, and soil. The maximum level of pesticide residue is 0.5 mg/kg in vegetables and 0.1 mg/kg in rice. Therefore, the development of tests for toxicity assessment of environmental and agricultural samples is necessary (67). Tekaya et al. fabricated a conductometric biosensor for pesticide detection in water based on gold interdigitated electrodes modified with Arthrospira platensis cells. Arthrospira platensis cells were coated with poly (allylamine Hydrochloride)-gold nanoparticles through co-reticulation with bovine serum albumin by glutaraldehyde vapors cross-linking. Pesticides including paraoxon-methyl, parathion methyl, triazine, and diuron inhibited cholinesterase activity. Detection limits for paraoxon-methyl, parathion-methyl, triazine, and diuron were obtained 10^{-18} M, 10^{-20} M, 10^{-20} M, and 10^{-12} M, respectively (75).

Mulyasuryani and Prasetyawan proposed an enzyme biosensor for the organophosphate pesticides detection, including diazinon, malathion, chlorpyrifos, profenofos. A screen-printed carbon electrode was decorated with organophosphate hydrolase on a chitosan membrane for biosensor fabrication by cross-linking it with glutaraldehyde. The detection limits for diazinon, malathion, chlorpyrifos, and profenofos were 40 ppb, 30 ppb, 20 ppb, and 40 ppm, respectively (76).

2.7   CONCLUSION AND FUTURE PERSPECTIVES

Biosensors have always been of interest to researchers throughout history. Biosensors with electrochemical transducers play a significant role in pharmaceutical, clinical, environmental, and industrial applications because of their excellent sensitivity and selectivity. Their high sensitivity and selectivity are due to a combination of selective biochemical recognition with the high sensitivity of electrochemical detection. Although numerous electrochemical biosensors are still in the development and try-out steps, some of these biosensors have reached the stage of commercial use. They have been used in various fields thanks to current technological progress. In this chapter, the principles and the applications of electrochemical biosensors, including amperometric, voltammetric, impedimetric, potentiometric, and conductometric biosensors, were investigated. Perhaps the most significant limitation of these biosensors is the instability of the biological recognition element and the non-specific binding, which can be minimized with the help of some strategies.

Due to public concerns about human health, the rapid diagnosis of diseases and the detection of harmful substances to human health in the environment and food has particular importance. Despite the past and current extensive research in electrochemical biosensor development, there are still problems producing more reliable and improved devices. Therefore, more in-depth studies in the field of electrochemical biosensors are needed to develop intelligent and user-friendly biosensors.

REFERENCES

1. Fang A, Ng HT, Li SFY. 2003. A high-performance glucose biosensor based on monomolecular layer of glucose oxidase covalently immobilised on indium—tin oxide surface. *Biosensors and Bioelectronics*. 19(1): 43–49
2. Ding J, Qin W. 2020. Recent advances in potentiometric biosensors. *TrAC Trends in Analytical Chemistry*. 124: 115803
3. Grieshaber D, MacKenzie R, Vörös J, Reimhult E. 2008. Electrochemical biosensors-sensor principles and architectures. *Sensors*. 8(3): 1400–1458
4. Ronkainen NJ, Halsall HB, Heineman WR. 2010. Electrochemical biosensors. *Chemical Society Reviews*. 39(5): 1747–1763
5. Hernandez VG, Sosa HJE, Saldarriaga HS, Villalba RAM, Parra SR, Iqbal H. 2018. Electrochemical biosensors: A solution to pollution detection with reference to environmental contaminants. *Biosensors*. 8(2): 29
6. Do JS, Lin KH. 2016. Kinetics of urease inhibition-based amperometric biosensors for mercury and lead ions detection. *Journal of the Taiwan Institute of Chemical Engineers*. 63: 25–32
7. Ronkainen-Matsuno NJ, Thomas JH, Halsall HB, Heineman WR. 2002. Electrochemical immunoassay moving into the fast lane. *TrAC Trends in Analytical Chemistry*. 21(4): 213–225
8. Bard AJ, Faulkner LR. 2001. Fundamentals and applications. *Electrochemical Methods*. 2(482): 580–632
9. Bartlett PN. 2008. *Bioelectrochemistry: Fundamentals, Experimental Techniques and Applications*. John Wiley & Sons
10. Elgrishi N, Rountree KJ, McCarthy BD, Rountree ES, Eisenhart TT, Dempsey JL. 2018. A practical beginner's guide to cyclic voltammetry. *Journal of Chemical Education*. 95(2): 197–206

11. Koyun A, Ahlatcolu E, Koca Y, Kara S. 2012. Biosensors and their principles. *A Roadmap of Biomedical Engineers and Milestones*. 117–142
12. Eggins BR. 2002. *Chemical Sensors and Biosensors*, Second Edition. John Wiley & Sons
13. Dzyadevych SV, Arkhypova VN, Soldatkin AP, Elskaya AV, Martelet C, Jaffrezic-Renault N. 2008. Amperometric enzyme biosensors: Past, present and future. *ITBM-RBM*. 29(2–3): 171–180
14. Albareda SM, Merkoci A, Alegret S. 2000. Configurations used in the design of screen-printed enzymatic biosensors: A review. *Sensors and Actuators B: Chemical*. 69(1–2): 153–163
15. Rosen MI, Rishpon J. 1993. Novel approaches for the use of mediators in enzyme electrodes. *Biosensors and Bioelectronics*. 8(6): 315–323
16. Vieira NS, Souza FA, Rocha RCF, Cestarolli DT, Guerra EM. 2021. Development of amperometric biosensors using VO 2/GOx films for detection of glucose. *Materials Science in Semiconductor Processing*. 121: 105337–105342
17. Kumar P, Narwal V, Jaiwal R, Pundir CS. 2018. Construction and application of amperometric sarcosine biosensor based on SOxNPs/AuE for determination of prostate *Biosensors and Bioelectronics*. 120: 140–146
18. Azzouzi S, Rotariu L, Benito AM, Maser WK, Ali MB, Bala C. 2015. A novel amperometric biosensor based on gold nanoparticles anchored on reduced graphene oxide for sensitive detection of L-lactate tumor biomarker. *Biosensors and Bioelectronics*. 69: 280–286
19. Vlamidis Y, Gualandi I, Tonelli D. 2017. Amperometric biosensors based on reduced GO and MWCNTs composite for polyphenols detection in fruit juices. *Journal of Electroanalytical Chemistry*. 799: 285–292
20. Stasyuk N, Gayda G, Zakalskiy A, Zakalska O, Serkiz R, Gonchar M. 2019. Amperometric biosensors based on oxidases and PtRu nanoparticles as artificial peroxidase. *Food Chemistry*. 285: 213–220
21. Conzuelo F, Gamella M, Campuzano S, Ruiz MA, Reviejo AJ, Pingarron JM. 2010. An integrated amperometric biosensor for the determination of lactose in milk and dairy products. *Journal of Agricultural and Food Chemistry*. 58(12): 7141–7148
22. Do JS, Lin KH. 2016. Kinetics of urease inhibition-based amperometric biosensors for mercury and lead ions detection. *Journal of the Taiwan Institute of Chemical Engineers*. 63: 25–32
23. Zhang Y, Lv Z, Zhou J, Fang Y, Wu H, Xin F, Jiang M. 2020. Amperometric biosensors based on recombinant bacterial laccase CotA for hydroquinone determination. *Electroanalysis*. 32(1): 142–148
24. Mabbott GA. 1983. Cyclic voltammetry experiment. *Journal of Chemical Education*. 60(9): 697–702
25. Kumar J, Dsouza SF. 2011. Microbial biosensor for detection of methyl parathion using screen printed carbon electrode and cyclic voltammetry. *Biosensors and Bioelectronics*. 26(11): 4289–4293
26. Ahmadalinezhad A, Chen A. 2011. High-performance electrochemical biosensor for the detection of total cholesterol. *Biosensors and Bioelectronics*. 26(11): 4508–4513
27. Westbroek P, Priniotakis G, Kiekens P. 2005. *Analytical Electrochemistry in Textiles*. Elsevier
28. Salandari JN, Ensafi AA, Rezaei B. 2021. Ultra-sensitive electrochemical aptasensor based on zeolitic imidazolate framework-8 derived Ag/Au core-shell nanoparticles for mercury detection in water samples. *Sensors and Actuators B: Chemical*. 331: 129426
29. Ensafi AA, Nasr-Esfahani P, Heydari-Bafrooei E, Rezaei B. 2015. Determination of atropine sulfate using a novel sensitive DNA—biosensor based on its interaction on a modified pencil graphite electrode. *Talanta*. 131: 149–155

30. Ensafi AA, Nasr-Esfahani P, Rezaei B. 2018. Metronidazole determination with an extremely sensitive and selective electrochemical sensor based on graphene nanoplatelets and molecularly imprinted polymers on graphene quantum dots. *Sensors and Actuators B: Chemical.* 270: 192–199

31. Zeng X, Liu J, Zhang Z, Kong S. 2015. Sensitive and selective detection of mercury ions by potentiometric biosensor based on urease immobilized in chitosan—poly (vinyl alcohol) hydrogel film. *International Journal of Electrochemical Science.* 10: 8344–8352

32. Chen A, Shah B. 2013. Electrochemical sensing and biosensing based on square wave voltammetry. *Analytical Methods.* 5(9): 2158–2173

33. Cardoso AR, Moreira FT, Fernandes R, Sales MGF. 2016. Novel and simple electrochemical biosensor monitoring attomolar levels of miRNA-155 in breast cancer. *Biosensors and Bioelectronics.* 80: 621–630

34. Cadorin FS, Cruz VI, Barbosa AMJ, Souza FV. 2011. Methomyl detection by inhibition of laccase using a carbon ceramic biosensor. *Electroanalysis.* 23(7): 1623–1630

35. Abollino O, Malandrino M, Ruo Redda A, Valeria M, Berto S, La Gioia C, Giacomino A. 2018. Stripping voltammetry with solid electrodes for trace element determination in natural waters: Laboratory and field experiments. In *XVI Hungarian—Italian Symposium on Spectrochemistry*, p. 8. XVI Hungarian, Italian Symposium on Spectrochemistry

36. Zhang X, Geng P, Liu H, Teng Y, Liu Y, Wang Q, Jiang L. 2009. Development of an electrochemical immunoassay for rapid detection of E. coli using anodic stripping voltammetry based on Cu@ Au nanoparticles as antibody labels. *Biosensors and Bioelectronics.* 24(7): 2155–2159

37. Dali M, Zinoubi K, Chrouda A, Abderrahmane S, Cherrad S, Jaffrezic-Renault N. 2018. A biosensor based on fungal soil biomass for electrochemical detection of lead (II) and cadmium (II) by differential pulse anodic stripping voltammetry. *Journal of Electroanalytical Chemistry.* 813: 9–19

38. Guan JG, Miao YQ, Zhang QJ. 2004. Impedimetric biosensors. *Journal of Bioscience and Bioengineering.* 97(4): 219–226

39. Bahadır EB, Sezgintürk MK. 2016. A review on impedimetric biosensors. *Artificial Cells, Nanomedicine, and Biotechnology.* 44(1): 248–262

40. Yuan XZ, Song C, Wang H, Zhang J. 2010. *Electrochemical Impedance Spectroscopy in PEM Fuel Cells: Fundamentals and Applications*, pp. 193–262. Springer

41. Elshafey R, Tavares AC, Siaj M, Zourob M. 2013. Electrochemical impedance immunosensor based on gold nanoparticles—protein G for the detection of cancer marker epidermal growth factor receptor in human plasma and brain tissue. *Biosensors and Bioelectronics.* 50: 143–149

42. Liu Y, Gao P, Jiang X, Li L, Zhang J, Peng W. 2014. Percolation mechanism through trapping/de-trapping process at defect states for resistive switching devices with structure of Ag/SixCl– x/p-Si. *Journal of Applied Physics.* 116(6): 064505

43. Wang Y, Ye Z, Ying Y. 2012. New trends in impedimetric biosensors for the detection of foodborne pathogenic bacteria. *Sensors.* 12(3): 3449–3471

44. Joung CK, Kim HN, Lim MC, Jeon TJ, Kim HY, Kim YR. 2013. A nanoporous membrane-based impedimetric immunosensor for label-free detection of pathogenic bacteria in whole milk. *Biosensors and Bioelectronics.* 44: 210–215

45. Ahmed A, Rushworth JV, Wright JD, Millner PA. 2013. Novel impedimetric immunosensor for detection of pathogenic bacteria streptococcus pyogenes in human saliva. *Analytical Chemistry.* 85(24): 12118–12125

46. Pradhan R, Kalkal A, Jindal S, Packirisamy G, Manhas S. 2021. Four electrode-based impedimetric biosensors for evaluating cytotoxicity of tamoxifen on cervical cancer cells. *RSC Advances.* 11(2): 798–806

47. Diaz FA, Miranda CR, De LSAN, Lobo CMJ, Estrela P. 2021. Impedimetric aptamer-based glycan PSA score for discrimination of prostate cancer from other prostate diseases. *Biosensors and Bioelectronics*. 175: 112872

48. Rengaraj S, Cruz-Izquierdo A, Scott JL, Di Lorenzo M. 2018. Impedimetric paper-based biosensor for the detection of bacterial contamination in water. *Sensors and Actuators B: Chemical*. 265: 50–58

49. Sheikhzadeh E, CHamsaz M, Turner APF, Jager EWH, Beni V. 2016. Label-free impedimetric biosensor for Salmonella Typhimurium detection based on poly [pyrrole-co-3-carboxyl-pyrrole] copolymer supported aptamer. *Biosensors and Bioelectronics*. 80: 194–200

50. Malhotra A, Chaubey B. 2002. Mediated biosensors. *Biosensors and Bioelectronics*. 17: 441

51. Dorazio P. 2003. Biosensors in clinical chemistry. *Clinica Chimica Acta*. 334(1–2): 41–69

52. Bühlmann P, Chen LD. 2012. Ion-selective electrodes with ionophore-doped sensing membranes. In *Supramolecular Chemistry: From Molecules to Nanomaterials*. Wiley.

53. Eric B, Pretsch E. 2005. Potentiometric sensors for trace-level analysis. *Trends in Analytical Chemistry*. 24(3)

54. Thevenot DR, Toth K, Durst RA, Wilson GS. 2001. Electrochemical biosensors recommended definitions and classification. *Biosensors and Bioelectronics*. 16: 121–131

55. Ayenimo JG, Adeloju SB. 2015. Inhibitive potentiometric detection of trace metals with ultrathin polypyrrole glucose oxidase biosensor. *Talanta*. 137: 62–70

56. Liao WY, Weng CH, Lee GB, Chou TC. 2006. Development and characterization of an all-solid-state potentiometric biosensor array microfluidic device for multiple ion analysis. *Lab on a Chip*. 6(10): 1362–1368

57. Bouri M, Zuaznabar-Gardona JC, Novell M, Blondeau P, Andrade FJ. 2021. Paper-based potentiometric biosensor for monitoring galactose in whole blood. *Electroanalysis*. 33(1): 81–89

58. Psychoyios VN, Nikoleli GP, Tzamtzis N, Nikolelis DP, Psaroudakis N, Danielsson B, Willander M. 2013. Potentiometric cholesterol biosensor based on ZnO nanowalls and stabilized polymerized lipid film. *Electroanalysis*. 25(2): 367–372

59. Ondeş B, Akpınar F, Uygun M, Muti M, Uygun DA. 2021. High stability potentiometric urea biosensor based on enzyme attached nanoparticles. *Microchemical Journal*. 160: 105667

60. Lupu A, Valsesia A, Bretagnol F, Colpo P, Rossi F. 2007. Development of a potentiometric biosensor based on nanostructured surface for lactate determination. *Sensors and Actuators B: Chemical*. 127(2): 606–612

61. Hernandez R, Valles C, Benito AM, Maser WK, Rius FX, Riu J. 2014. Graphene-based potentiometric biosensor for the immediate detection of living bacteria. *Biosensors and Bioelectronics*. 54: 553–557

62. Silva NF, Almeida CM, Magalhaes JM, Goncalves MP, Freire C, Delerue-Matos C. 2019. Development of a disposable paper-based potentiometric immunosensor for real-time detection of a foodborne pathogen. *Biosensors and Bioelectronics*. 141: 111317

63. Lv E, Li Y, Ding J, Qin W. 2021. Magnetic-field-driven extraction of bioreceptors into polymeric membranes for label-free potentiometric biosensing. *Angewandte Chemie*. 133(5): 2641–2645

64. Ercole C, Del Gallo M, Pantalone M, Santucci S, Mosiello L, Laconi C, Lepidi A. 2002. A biosensor for Escherichia coli based on a potentiometric alternating biosensing (PAB) transducer. *Sensors and Actuators B: Chemical*. 83(1–3): 48–52

65. Li Q, Li S, Lu M, Lin Z, Tang D. 2016. Potentiometric competitive immunoassay for determination of aflatoxin B 1 in food by using antibody-labeled gold nanoparticles. *Microchimica Acta*. 183(10): 2815–2822
66. Zhang Q, Ding J, Kou L, Qin W. 2013. A potentiometric flow biosensor based on ammonia-oxidizing bacteria for the detection of toxicity in water. *Sensors*. 13(6): 6936–6945
67. Dzyadevych S, Jaffrezic-Renault N. 2014. Conductometric biosensors. In *Biological Identification*, pp. 153–193. Woodhead Publishing
68. Shulga AA, Soldatkin AP, Elskaya AV, Dzyadevich SV, Patskovsky SV, Strikha VI. 1994. Thin-film conductometric biosensors for glucose and urea determination. *Biosensors and Bioelectronics*. 9(3): 217–223
69. Mahadeva SK, Yun S, Kim J. 2011. Flexible humidity and temperature sensor based on —polypyrrole nanocomposite. *Sensors and Actuators A: Physical*. 165(2): 194–199
70. Dzyadevich SV, Arkhipova VN, Soldatkin AP, Anna V, Shulga AA. 1998. Glucose conductometric biosensor with potassium hexacyanoferrate (III) as an oxidizing agent. *Analytica Chimica Acta*. 374(1): 11–18
71. Lee WY, Kim SR, Kim TH, Lee KS, Shin MC, Park JK. 2000. Sol—gel-derived thick-film conductometric biosensor for urea determination in serum. *Analytica Chimica Acta*. 404(2): 195–203; Saiapina OY, Pyeshkova VM, Soldatkin OO, Melnik VG, Kurç BA, Walcarius A, Jaffrezic-Renault N. 2011. Conductometric enzyme biosensors based on natural zeolite clinoptilolite for urea determination. *Materials Science and Engineering*. 31(7): 1490–1497
72. Velychko TP, Soldatkin OO, Melnyk VG, Marchenko SV, Kirdeciler SK, Akata B, Dzyadevych SV. 2016. A novel conductometric urea biosensor with improved analytical characteristic based on recombinant urease adsorbed on nanoparticle of silicalite. *Nanoscale Research Letters*. 11(1): 1–6
73. Zhang Z, Jaffrezic RN, Bessueille F, Leonard D, Xia S, Wang X, Zhao J. 2008. Development of a conductometric phosphate biosensor based on tri-layer maltose phosphorylase composite films. *Analytica Chimica Acta*. 615(1): 73–79
74. Upadhyay LSB, Verma N. 2015. Alkaline phosphatase inhibition based conductometric biosensor for phosphate estimation in biological fluids. *Biosensors and Bioelectronics*. 68: 611–616
75. Tekaya N, Saiapina O, Ouada HB, Lagarde F, Ouada HB, Jaffrezic-Renault N. 2013. Ultra-sensitive conductometric detection of pesticides based on inhibition of esterase activity in Arthrospira platensis. *Environmental Pollution*. 178: 182–188
76. Mulyasuryani A, Prasetyawan S. 2015. Organophosphate hydrolase in conductometric biosensor for the detection of organophosphate pesticides. *Analytical Chemistry Insights*. 10:23–27

3 Optical Based Transducers for Biosensors

Hasan Ilhan[1], Sallahuddin Panhwar[2,3], Ismail Hakki Boyaci[4], and Ugur Tamer[3]
[1]Ordu University
Department of Chemistry, Faculty of Science
Ordu, Turkey

[2]Balochistan University of Engineering & Technology
Department of Civil Engineering
Balochistan, Pakistan

[3]Gazi University
Faculty of Pharmacy, Department of Analytical Chemistry
Ankara, Turkey

[4]Hacettepe University
Faculty of Engineering, Department of Food Engineering
Ankara, Turkey

CONTENTS

DOI: 10.1201/9781003189435-5

3.1   INTRODUCTION

A biosensor is a system that detects an analyte. It is typically composed of a biological recognition component (biological part) that serves as the primary transducer and a physicochemical detector component (non-biological part, signal amplification, and transduction) that serves as the signal conversion unit. Numerous biomolecules can serve as recognition elements, depending on the target analyte and application (1). The biological recognition elements can be nucleic acids such as DNA, aptamers, or proteins such as enzymes and antibodies, or even whole cells such as bacteria, neurons, or tissue slices. Similarly, due to their unique and sensitive electrical characteristics, a wide variety of nanomaterials can be employed as secondary transducers for efficient electron transport (2–5).

A wide range of biological and chemical compounds may be detected in real-time using optical biosensors, enabling direct, real-time, and label-free testing. It has the benefit of having high specificity, sensitivity, and tiny size as well as being relatively inexpensive. Many innovative concepts and several interdisciplinary techniques are employed in developing novel optical biosensors, such as micro/nanotechnologies, molecular biology, biotechnology, and chemistry. In the past decade, the rate of technological advancement in optical biosensors has risen dramatically. For the most part, research in optical biosensors has focused on the healthcare, environmental, and biotechnology industries. Biosensors have various uses in the medical, environmental, and biotechnology fields. Each field has distinct criteria for measuring an analyte's concentration. Each requirement influences the biosensor's output, how long it takes to run the test, the sample concentration required, how long it takes to reuse the biosensor, and the cleaning requirements of the system (6–7).

COVID-19 has increased the rate of viruses research by increasing testing and diagnosis. This year's fast global spread of COVID-19, caused by SARS-CoV-2, has underlined the need for more improved imaging and detection techniques (8). Optical biosensors are a viable alternative technique for viral detection because of their safety, simplicity of use, and cost-effectiveness. They also eliminate the requirement for nucleic acid amplification, which is a significant drawback of other methods (9). HIV, Ebola, norovirus, and influenza viruses have all been detected using fluorescence, surface plasmons, and colorimetry (8, 10). These methods have been used for nano-biosensors, allowing for targeted viral detection and single virus imaging, among other applications (11–12).

3.2   OPTICAL IMMUNOSENSORS

The electrochemical transducer is extensively used for immunosensing applications to receive a great deal of attention for optical transduction procedures. The optical immunosensor has exploited other methods for detecting analytes based on fluorescence, light absorbance, light polarization, and chemiluminescence with surface plasmon resonance

(SPR) as the most widely employed (13). However, the optical-based immunosensor observed a variation in phase polarization speed of input light frequency, corresponding to antigen-antibody multiple complexes (14). The main concept of multiple complexes is immersed in an optical sensor for detection based on analyte on the higher dielectric permittivity secured by the cell wall. All proteins and DNA and water affect these bio-molecules to reduce the reproduction of an electromagnetic field running through them. All molecules contain electrons and atomic nuclei in different orbital states because these molecules can interact with the electromagnetic field that passes through them. As the oscillating electromagnetic field by placing molecules with light transmission, electrons inside the molecules would vibrate due to the subjected force. Then the free electrons may polarize in the presence of a light magnetic field that generates a polar-ization current that was causing the movement of electrons, where it moves slower even though a biomolecule than in free space (15). Usually, optical immunosensors employ light either from a laser or white-hot light bulb, allowing them to observe an alteration in the attributes of the light passed or reflected from the sensor. Measurement was con-ducted by casting light upon the sensor at different angles on a single sensor plane. This type of sensor changes light characteristics to measure all the light sources externally, making it energy-efficient, especially in low illumination powder to generate signals (16). Several kinds of transducer components are used to generate optical changes, such as surface plasmon resonance (SPR) interferometry, grating couplers resonant mirror, reflectometric interference spectroscopy, and a total internal reflection fluorescence (TIRF) (17). However, the optical immunosensor was operated using two detection schemes, including label-free detection and fluorescence-based detection. On the other hand, the label-free detection method is a relatively simple technique that requires basic steps due the bioreceptor is not labelled and used in its native form (18–19). However, both approaches are commonly applied in optical sensors to yield dynamic information on the interaction among the biological molecules instead of a label-free process.

3.3   SURFACE PLASMON RESONANCE (SPR)

The core functions of SPR technology are to establish an affinity and kinetic rela-tionship between the biomolecules and receptor-ligands interactions to determine the nucleic acid hybridization. The SPR immunosensor contains these parts, including a light source, prism, transduction surface (generally gold film), biomolecule (antigen or antibody), flow system, and detector (20). A surface plasmon is a comprehensive charge density wave that occurs with two interfaces as gold in the air where the dielectric amounts of these two media have opposite signs. However, in SPR immunosensors, an antibody is immobilized on the surface of thin metal films as typically gold where the polarized light has emitted from the back surface over a prism, and a targeted ligand is introduced. The metal film has reflected with the light as acting as a mirror to strengthen the reflected light that can be assessed and quantified. When an immo-bilized antibody is bound to its targeted analyte, then it shifts to the SPR angle for an observation, which depends on the concentration of the target (21–22). SPR bio-sensor has obvious merits of being rapid, and it is relatively suitable as compared to other approaches. For example, in the detection of antibiotic-resistant bacteria (ARB), microbial growth has stopped employing quick biological readouts (23). Since signal outputs were a resonance angle and refractive index values, labelling with radioactive or fluorescent labels is optional. However, that technology makes the identification of small molecules such as toxins highly feasible and food allergens (24). While the tags are not essential in an SPR immunosensor, utilized tags could nevertheless amplify the signals and ultimately result in a sensitive detection, for example, with labelling antibodies with gold nanoparticles (AuNPs) in sandwich-type immunoassay between aptamers and antibodies (25). Therefore, the optical-based biosensor's main drawbacks are expensive instrument setup and immobilization of biomolecules as material losses during the immobilization of biomolecules process on the solid substrate; contamination due to the leakage of chemicals biomolecules was employed in a biosensor as leakage out of the biosensor. Another is the sterilization problem since biomolecules may become denaturized if non-sterile probes are used (26). Presently, surface plasmon resonance technology for the observation of biomolecular interaction in real-time samples has been commercialized by several companies. However, few problems have been reported about SPR sensors. The surface chemistry has been restricted to a noble metal surface, particularly to gold. The signal produced as the refractive index has been significantly affected by the physical environment (27).

3.4   DIRECT IMMUNOSENSOR

A direct immunosensor comprises a physical component as prism and light source. A physical part contains a transduction surface, chemical substance or biomolecule, flow system, and detector. Commonly, these systems have an optical reader, biorecognition element, and sample handling system with four main sub-systems: light source, detector module, data processing unit, and coupling optics. The biological component that the biomolecules were immobilized on the surface of the gold film acts as a molecular probe. In the physical part, the infrared light generated a resonance with surface plasmon on the film and minimized the reflected intensity obtained at a particular angle. The reaction of targeted molecules with the immobilized probe affects minimum angle changes and optical properties. The reflected light has converted into an electronic signal with a linear array detector, and the position of the minimum intensity was employed to determine the level of biochemical interaction. Such types of the test have been completed within ten minutes. However, in immunology research, the surface plasmon resonance plays an essential role in detecting targeted pathogen bacteria using an antibody, proteins, virus, and other types of bacteria (28). It may offer various advantages with a high-throughput tool for a diversity application compared with other transduction techniques.

Furthermore, a direct optical immunosensor produces a continuous real-time response to biomolecular interaction at the interface to deliver a rapid, immediate evaluation of the analyte. After the detection of the targeted analyte, the surface of an active sensor has recovered by repetitive use of the sample sensor chip (Figure 3.1). It has been achieved by employing regeneration eluent while carefully monitoring the process of reactivation. Moreover, the automated flow system can use the same sensor chip to perform consecutive and serial analyses of various sample solutions, including no control analytes (29).

3.4.1   DIRECT ASSAY

Indirect detection of medium-to-large analytes usually provokes a sufficient sensor response with the limit of detection (LOD) in the range of ng/mL or sub-ng/mL. In

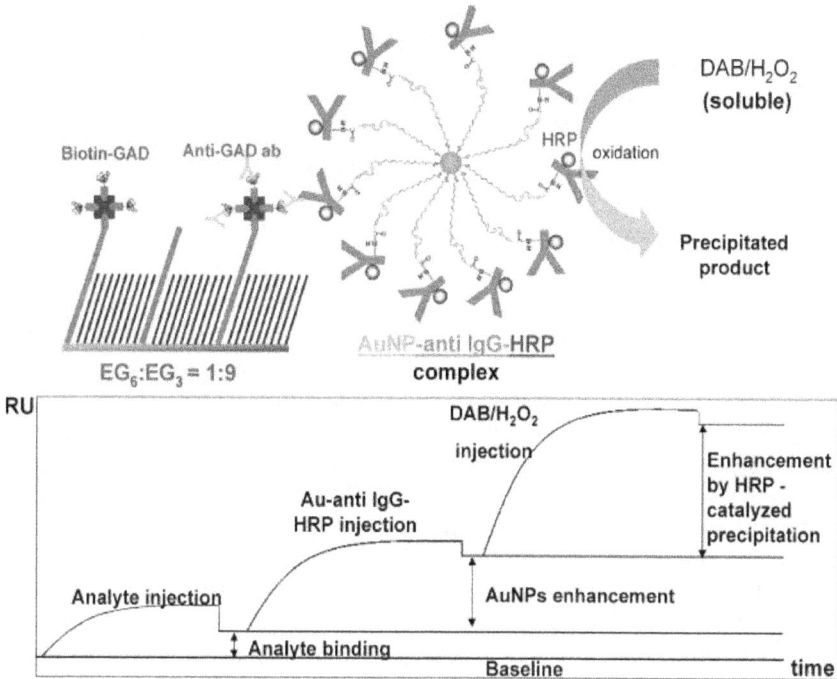

FIGURE 3.1 SPR immunosensor for anti-GAD Ab detection. Reprinted from ref. (29) with permission.

the direct assay, the detection element has mainly been used for antibody immobilization on a metal layer. First, the sample is inoculated on the biosensor surface, then incubated to establish a baseline. After that, the sensing and reference channel flow of the cell was loaded with the buffer solution. However, chemical or biological analyte samples have been attached with the biorecognition elements—the result changes in the mass and resonance angle to correlate with the concentration of the analyte.

Direct assay immunosensor has considerably improved along with the BIAcore™ technology, and it was first introduced in 1991. This technology system has been used in many types of molecules as elements for biomolecular recognition. At the same time, antibodies have been commonly employed for microbial identification. However, the antibody modification process is laborious and costly. Therefore, other molecules such as antigens, antibody fragments, peptides, engineered affibodies, and molecular imprinted polymers (facilitating the attachment of labelled proteins complete site-specific non-covalent interaction of a capture molecule and tag) are used in the biosensors (30).

3.4.2 SANDWICH ASSAYS

The sandwich assay is employed to enhance the lowest detection limit of direct surface plasmon resonance (SPR) by improving detection sensitivity. This type of sensor was successfully utilized to detect complex matrices. In a sandwich assay, each analyte molecule "Sandwich type" between the primary bio-recognition element is immobilized on the sensor's surface and then injected into a secondary biorecognition element. The biotin-avidin interaction has been widely employed as a molecular recognition for the immobilization of biomolecules on the sensor's surface. Therefore, this approach is used in a small number of secondary detection reagents. Furthermore, by modifying avidin or streptavidin, an almost unlimited amount of analyte molecules can be easily captured, immobilized, recovered, and detected in the avidin-biotin detection system (31). First, metal-organic frameworks (MOFs) nanoparticles were modified with the surface activators such as, N-(3-dimethyl aminopropyl) N-ethyl carbodiimide hydrochloride and N-Hydroxy succinimide sodium salt EDC/NHS solutions for 45 minutes at the shaker. After the activators process, then nanoparticles were magnetically separated and washed one time with a buffer solution (pH 7.4). The surface of particles was covered with avidin by incubating nanoparticles in avidin for 40 m for the covalent bond interaction between the carboxylic groups and avidin. Then, nanoparticles were collected with magnate and washed two times with PBS buffer. To avoid nonspecific interaction, 1% (v/v) of ethanolamine was employed to block the unfilled spaces and activate the groups after treating the surface of particles with an avidin solution. Then again, biotin-labelled *E. coli* antibody was mixed in buffer solution and interacted with nanoparticles. The surface of nanoparticles has been developed with a biotin-avidin conjugated antibody. After all, the washing process was done with a buffer solution to remove the un-conjugated biotin-labelled antibodies for the modified nanoparticles. Then, conjugated nanoparticles were magnetically collected and added to the new phosphate buffer for further experiments (32–33) (Figure 3.2).

3.4.3 Enzyme-Based Biosensors

Enzymes have often been used as biocatalysts that increase the speed of a biological process. Enzyme-based biosensors use a catalytic reaction and a binding potential for

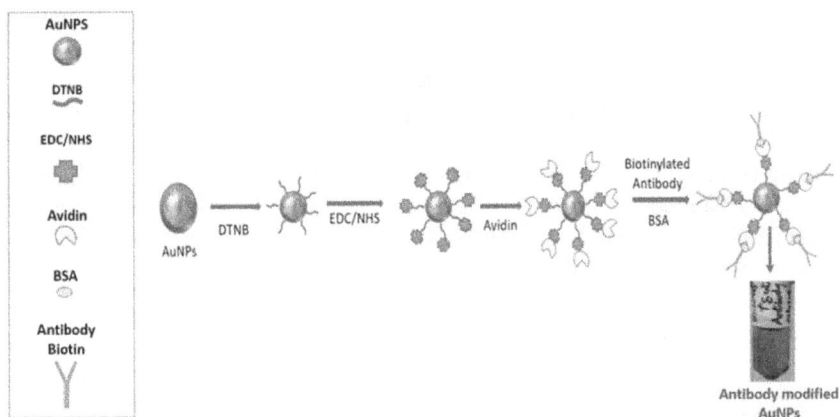

FIGURE 3.2 Schematic representation of a modification of organic metal frameworks using a monolayer of DTNB. Reprinted from ref. (32) with permission.

the target analyte detection to implement their working principles (34). In order to evaluate the intensity of a signal, researchers test the various possible factors affecting it: (i) Enzymes turn metabolites into active forms, so measuring the rate of the catalytic transformation of the analyte allows you to estimate the concentration of the enzyme, (ii) An enzyme inhibited or activated by the analyte, so the amount of the enzymatic product is tracked to see how much has been reduced, and additionally, changes in enzyme properties may be tracked (35–38). It is possible to create different biosensors using enzymes as a source of selectivity because of the significant presence of enzyme-based biosensors. On the other hand, the enzyme structure is susceptible, making it costly and difficult to increase its sensitivity, stability, and adaptability, among other properties (38). The glucose and urea biosensors are the most often used enzyme-based biosensors. Kidney disorders frequently stay undetected because of the poor techniques of screening available. The use of biosensors may be a very effective method of early detection and managing renal disease. Srivastava and colleagues (39) present single and dual fluorophore ratiometric biosensors based on alginate microspheres for measuring pH and urea in urine samples. These polymeric fluorophores and enzyme-based biosensors were created using a simple air-driven atomization technique. Polyelectrolytes coupled to reference fluorophores were used to build ratiometric biosensors. In biochemical assays, biosensors were shown to allow the measurement of samples in the pathological range of pH ranging from 4 to 8 and having urea concentrations from 0 to 50 mM. The pH and urea detection accuracy of these biosensors was tested on urine samples. As a result, research groups think FITC-Dextran and FITC-Dextran/RuBpy-based pH and urea biosensors have tremendous potential for use as point-of-care devices for early diagnosis and monitoring of kidney disorders. An inexpensive and easy technique is offered to the market, enabling a non-invasive approach to detect uric acid in human biological samples. Current research (40) describes a novel fluorescence biosensor capable of detecting urease and horseradish peroxidase using CdS quantum dots encapsulated in a sol-gel matrix. The QDs are employed as a fluorescence indicator to reveal the system's fluorescence signal caused by uricase/HRP enzymatic reaction with uric acid. Allantoin, CO_2, and H_2O_2 are the hybrid QDs-uricase/HRP upon the addition of uric acid. The H_2O_2 generated may extinguish QDs fluorescence which is proportional to uric acid concentration. The developed microplate biosensor has demonstrated its advantage, as it enables the development of a comprehensive system for 96 samples per assay that take 20 minutes to complete. In the concentration range of 60–2000 M, the detection limit of uric acid was 50 M. The designed sensor effectively detected uric acid in human urine with findings similar to test kits.

Park and Cho described (41) a carbon dots (CD's) and Rhodamine 6G (Rrh6G) ratiometrician fluorescent glucose biosensor is designed as both an aqueous and a solid-state poly(acrylic acid) cross-linking solution. Fluorescence quenching due to glucose oxidase's bienzymatic interaction (GOx) and horseradic Peroxidase (HRP) with glucose was responsible for a ratiometric fluoresce color shift. The bienzymatic interaction with glucose inhibited the blue fluorescence emission of CDs produced by solvothermal technique with citric acid and ethylene diamine, while Rh6G fluorescence was inert to glucose. As the glucose content rose, a ratiometric fluorescence color shift from blue to green was detected. Achievable ranges for 0.1 to 500 μM with a LOD of 0.04 μM in a CD/Rh6G/GOx/HRP aqueous solution are useful for quantifying glucose in human blood and are compatible with serum from humans. The immobilization of CD/rh6G/GOx/HRP in the hydrogel film, produced with UV curing of an acrylic acid combination with poly(ethylene glycol) (70:30, w/w) has produced a solid-state biosensor film. The hydrogel film had comparable fluorescence ratiometric color change, sensitivity (linear range of 0.5–500 µM with a LOD of 0.08 µM), and selectivity when compared to the CD/Rh6G/GOx/HRP aqueous solution. Aqueous solution instability was overcome by the solid-state glucose biosensor film, which was intrinsically stable (owing to aggregation, enzyme denaturation, etc.). Compared to turn on/off biosensors, ratiometric biosensors can detect glucose with the naked eye. Thus, this technique increases the range of possibilities for CDs in biosensors, making detection easier and more practical. Sales' team (42) is reporting the first study of a glucose biosensor that has been constructed using a nanocellulose (NC) coated cellulose paper substrate utilizing drop-deposition technique. 2,2,6,6-Tetramethylpiperidine-N-oxyl radical (TEMPO), sodium hypochlorite, and potassium bromide were used to oxidize MCC samples to carboxylated NC. It was utilized in several approaches to describe the characterization details: TEM, FTIR, and conductometric titration. The main alcohol groups were preferentially oxidized into carboxyl groups in all samples when sodium hypochlorite was introduced dropwise and the reaction was kept at pH 10. This method was developed in order to provide a colored detecting system for glucose. GOx was integrated onto a carboxyl NC/cellulose substrate to create a sensor system. The enzyme generated hydrogen peroxide from glucose, which reacted with HRP and ABTS to form a blue product. The test strip was calibrated in various glucose concentrations. The colors were then analyzed using image analysis tools. This varied from 1.5 to 13.0 mM.

3.4.4 WHOLE CELL-BASED BIOSENSORS

The whole cell-based biosensor is one of the most sustainable biosensors, which can immediately detect and transform the results into a digital electric signal through sensors or transducers via live cells (43–46). Biology and electronics have been connected by the fact that they both rely on engineering. Cell-based biosensors are devices that integrate live cells with sensors or transducers to monitor cellular physiological parameters, analyze pharmacological effects, test for environmental toxicity, and perform other tasks (47–49). The cell-based biosensors have a wide range of detection capabilities, whereas molecules-based technologies lack the capability to examine a broad spectrum of molecules. Cell-based biosensors have the benefits of quick and sensitive in situ analysis as well as sensing and detecting analytes (50–52). Molecular sensor arrays are naturally encapsulated in cells. Enzymes, receptors, and ion channels may react to their respective analytes through a natural cellular process.

In comparison to molecular biosensors, it is anticipated that cell-based biosensors would react best to bioactive analytes. The development of cell-based biosensors serves to aid in the research of analyte effects because of this. Nonetheless, cell-based biosensors have specific inherent weaknesses. Common challenges in optimizing cell-based biosensors include improving stability, selectivity, and cell lifetime. As a result of the development of biotechniques such as nanotechnology, microfluidics, and high-content screening, it has been anticipated that cellular mimicking and sensing will be utilized shortly. Cell-based biosensors, such as the ones described at the beginning of this section, have found widespread use in fields such as cellular physiological analysis, pharmaceutical evaluation, environmental monitoring, and medical diagnosis because of their apparent advantages, including long-term nonin-vasive recording, quick response time, and label-free experimentation (50, 53–55). Choo and coworkers (56) created a novel SERS-based test platform that can con-currently identify scrub typhus and murine typhus, the most prevalent acute fever illnesses in South Korea. The related O. tsutsugamushi IgG/IgM and R. typhi IgG/IgM biomarkers were quantitatively evaluated utilizing a SERS-based microarray chip arrayed in line with 5×5 mm^2 gold nanopopcorn substrates. The LODs for this technique were considerably lower than those for the ELISA method. In addition, the LOD values for the four titers were lower than the cut-off values for diagnosis, sug-gesting that this assay platform may be utilized for the clinical diagnosis of the two most prevalent typhus illnesses in South Korea. The RSD values for O. tsutsugamushi IgG/IgM and R. typhi IgG/IgM show excellent repeatability. Furthermore, the asso-ciated Raman mapping images show excellent selectivity for four distinct IgG/IgM target antibody titers. Thus, this research shows that the SERS-based typhus immu-noassay technique utilizing gold nanopopcorn substrates may be used for the simul-taneous diagnosis of scrub typhus and murine typhus. When the array sensor devices have coupled with portable Raman spectrophotometers, which are commercially accessible globally, it is feasible to diagnose the illnesses on-site without sending patient blood samples to a hospital.

Liu and coworkers (57) MicroRNA (miRNA) is becoming a significant role in early clinical illness diagnosis, therefore developing sensitive and quantitative miRNA detection techniques is critical. A label-free SERS assay with DSN signal amplification was presented here for easy ultrasensitive and quantitative miRNA-21 analysis. First, magnetic beads loaded with capture DNA were used to hybridize miRNA-21. In the process of DSN cycling, DNA-RNA heteroduplexes were broken, yielding tiny nucleotide fragments in the supernatant and the miRNA-21 in the solu-tion. Another DNA strand was created, and another DSN cycle was initiated. Because magnetic beads produced large amounts of capture DNA, the miRNA-21 signal was transferred and amplified by SERS signals from total phosphate backbones, which are rich in nucleotides. SERS signals were also generated using iodide-modified Ag nanoparticles (AgINPs). The suggested technique detected miRNA-21 with a linear range of 0.33 fM to 3.3 pM and a detection limit of 42 aM. This method also demon-strated excellent base separation and monitored miRNA-21 expression in several cancer cell lines and human serum.

Wang's team (58) proposed a new study that has shown that having a deficiency of dopamine in the retina may cause myopia, which has expected to impair over half of the world's population's vision by the year 2050. To get a full knowledge of the molecular process that underlies the development of myopia, a method for obtaining dopamine in the retina with great spatial accuracy is needed. While the retina has few detection methods for dopamine, it is tough to specifically detect dopa-mine in the retina. Researchers used surface-enhanced Raman scattering (SERS) for imaging in cells and retinal tissues to detect dopamine with great spatial accuracy. N-butylboronic acid-2 mercaptoethylamine and 3,3'-dithiodipropionic acid di(N hydroxysuccinimide ester) were the best molecules for specific interaction dopamine on the surface of gold nanoparticles. The aggregation of gold nanoparticles that forms plasmonic hot spots provides a significant rise in the Raman signal of dopamine, thereby proving the presence of dopamine. The SERS-synthesized nanoprobes have been assessed in live cells and retinal tissues in form-deprivation (FD) myopic guinea pigs. The nanoprobes are now being examined in living mice that have FD myopia to better determine localized dopamine levels. The study findings show that two weeks of FD therapy causes a decrease in dopamine levels in mouse retinas, which may explain the onset of myopia. Myopia's localized dopamine level will be improved via our method. Other essential chemicals in the biological samples may also be detected using the imaging platform. Matteini and coworkers (59) claimed that the use of this new technique to realize SERS active substrates using low-cost and stable soda-lime glass microrods is described. To make the microrods, the fiber was cut just 1 cm long, with the cut end held in an ion-exchange chamber containing sodium hydroxide until the fibers shrunk to a radius of less than 30 microns. Because of the excellent SERS characteristics of silver, silver nitrate was chosen as an ion source. To develop the surface layer of silver nanoparticles and prevent any additional chemical etching, the ion-exchange and thermal annealing post-process parameters were adjusted. For this assay, labelled molecular beacons (MBs) were immobilized on the tested substrates to measure DNA. These studies show that the DNA target has been successfully immobilized on the silver nanoparticles, and both SERS and fluorescence techniques show its location. With these findings, the way is cleared for the creation of inexpensive and reliable hybrid fibers that integrate SERS and fluorescence methods in the same optical device.

3.4.5 NANOPARTICLES AS SIGNAL TRANSDUCERS FOR OPTICAL BIOSENSOR

Nanomaterials are three-dimensional materials having at least one dimension (0.1–100 nm) or are made of them as fundamental units (60). Due to their unique chemical and optical characteristics and high surface-to-volume ratios, the use of new nanoparticles (NPs) can accelerate the development of "label-free" detection and biosensors with increased sensitivities, reaction times, and mobility (61–62).

The nanomaterial used in secondary transducers works as an interface, detecting the physical change caused by biochemical interactions and converting it to a quantifiable output signal. Thus, by combining a susceptible nanomaterial-based substrate with biomolecules, high-performance biosensors may be created. The two components of biosensors detect the signal generated by a change in response and generate electrical, thermal, or optical signals that may be digitally translated for further processing (63). The other two critical criteria for the sensor's functioning are its selectivity and sensitivity (64–65). The sensitivity of a biosensor is defined as the magnitude of the sensor signal change in response to a change in the analyte concentration. The selectivity of a biosensor is determined by the biological receptor used. Thus, when nanomaterial-based sensors are coupled to various biological recognition components, this selectivity may be altered.

Numerous nanomaterials, including magnetic nanoparticles (MNP), gold nanoparticles, polymer nanomaterials, and carbon-based nanomaterials, are being used in biosensors due to their unique physical, chemical, mechanical, magnetic, and optical properties, and these nanomaterials have significantly increased detection sensitivity and specificity (66). Nanomaterials, in particular, offer enhanced surface area, electrical conductivity and connection, chemical accessibility, and biocompatibility (67). Thus, while designing an effective nanobiosensor, one must consider the sort of nanomaterial to employ, since this element might significantly affect the biosensor's effectiveness. In general, the majority of nanomaterials have been produced in one of two ways: One approach is referred to as the "bottom-up" approach, which involves the self-assembly of small structures into larger structures; the other is referred to as the "top-down" approach, which involves the reduction of large systems to nanoscale dimensions to create multifunctional nanoscale structures (63).

3.4.5.1 Magnetic Nanoparticle-Based Optical Biosensors

Magnetic nanoparticles (MNPs) serve a critical role in the biosensor's function and quality, and owing to the presence of the magnetic field, the sample concentration is increased as well as the sample purity (68–69). Fe_3O_4 nanoparticles are engaging in target surface immobilization because of their oxidative stability, compatibility in aqueous environments, and their nontoxicity and suitable groups at the superficial surface. Additionally, unlike conventional purification procedures, the immobilized biomolecules on magnetic nanoparticles may be collected or dispersed in solution using an external magnetic field, allowing for fast sample preparation. Zengin and colleagues (70) proposed that adsorption kinetics, high adsorption capacity, good selectivity, stability, and reusability were all observed in the molecularly-imprinted magnetic nanoparticles (MIP@Fe_3O_4 NPs). In addition, a silicone covered by Ag NPs surface was created and utilized as SERS substrate to evaluate malachite green (MG) using SERS. The SERS active surface has been developed, showing excellent homogeneity, stability, recyclability and superficial purification characteristics to extend the SERS analytical application. Appropriate regeneration rates and related standard deviations from the SERS sensing system effectively determined MG in tap water with lower LL (1.50 pm for tap water and 1:62 pm for carp samples). Compared to previously published molecularly impressed polymer based techniques, the suggested approach exhibits excellent sensitivity. This can be attributed to several advantages of the proposed method: (i) the presence on the nanoparticles of highly selective thin polymer layers ($\alpha = 3,86$); (ii) the high magnetic nanoparticle surface that enhances the transportation of mass; (iii) the high sensitivity of the SERS compared to other spectroscopic, chromatographic and electrochemical methods. The findings show that efficient detection and separation characteristics with the synergistic impact of MIPs and SERS were achieved via the technique in Figure 3.3. In addition, new technologies are believed to be developed for the creation of MIPs via the SI-RTCP. Tamer et al. (71) described a new analytical method, SERS, which provides a reliable detection capability in trace quantities of analytes. Tamer and coworkers used magnetic and Raman-labeled nanoparticles to create a fast and straightforward SERS technique in this research. In short, the crucially important component of this process is the melamine-linked amino groups and the nanoparticle's carboxyl groups interacting via covalent bonding. This technique may be used to find low amounts of melamine. These

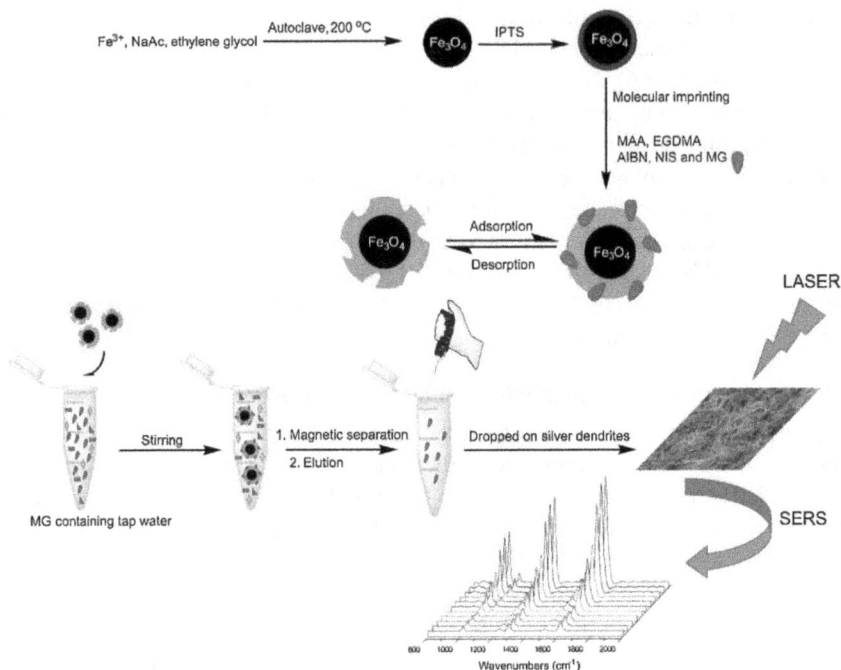

FIGURE 3.3 MNPs and SERS based optical biosensor platform. Reprinted from ref. (70) with permission.

findings show that the suggested technique is appropriate for the real-sample detection of melamine.

The approach is thought to be more user-friendly when compared to chromatography-based processes. In industry, rapid detection of melamine is critical. Trace quantities of melamine may be detected in milk using this technique in about 15 minutes, and a LOD of 0.39 mg L^{-1} is obtained, which is acceptable for monitoring melamine in milk. To screen for melamine in food quickly, use the technique with portable Raman equipment in an industrial environment. Zengin and colleagues (72) studied for the tau protein sandwich assay developed in their research, which is fast and ultra-sensitive because it employs homogeneous sandwich assays (where multiple components are bound at one time) combined with monoclonal antitau-functionalized hybrid magnetic nanoparticles as the recognition component and polyclonal anti-tau-immobilized gold nanoparticles as the SERS component. To bioconjugate with the monoclonal antitau protein, poly(2-hydroxyethyl methacrylate) (RAFT) was used to coat the magnetic silica particles. Then, the mixture was treated with monoclonal antitau protein, which binds to tau, followed by collection using a simple magnet. It was prepared using the SERS substrate, which consisted of polyclonal antitau and 5,5-dithiobis(2-dinitrobenzoic acid) on gold nanoparticles after being separated from the tau sample matrix. In the region of 25 fM to 500 nM, the relationship between the tau concentration and SERS signal was linear. The sandwich test has a LOD of less than 25 fM. Moreover, the sandwich assay was also assessed for bovine serum albumin and immunoglobulin G to investigate the tau specificity.

The excellent electrical conductivity and outstanding biocompatibility of MNPs have made them a popular choice for biosensors (73). Although MNPs are effective at detecting bacteria, Researchers have trouble seeing low quantities of target analyte in real samples; therefore, MNPs must include additional materials for surface modification (74).

3.4.5.2   Carbon Nanoparticle-Based Optical Biosensors

Carbon nanoparticles are among the most extensively discussed, researched, and used synthetic nanomaterials today (75). Carbon nanoparticles (CNPs) have exact mechanical, electrical, thermal, and optical characteristics that make them especially useful in fields where extreme accuracy is required, such as electronics, materials science, and biosensor technology (76). Carbon dots and carbon nanotubes (CNTs) are a type of tubular material mainly composed of carbon.

Carbon dots (CDs) are attractive because of their simple production, cheap raw ingredients, and sustainability (77). Most CDs are made of spherical nanoparticles with diameters between 10 and 20 nm (78–79). CDs conjugated with vancomycin were used as part of a method that Zhong and colleagues had previously proposed (80) to detect gram-positive bacteria. Vancomycin was applied to the CDs in this study. The modified CDs exhibited a solid attraction for gram-positive bacteria in the presence of $S.$ $aureus$ because vancomycin interacted with the cell wall ligand-receptor and caused the CDs to preferentially adhere to gram-positive bacteria. A detection range of 3.18×10^{5}–1.59×10^{8} cfu mL^{-1} was seen using the technique described, and a detection limit of 9.40×10^{4} cfu mL^{-1} was found. Yang and coworkers (81) described in that paper a straightforward and effective way to identify DA and 4-NP, utilizing nitrogen-ferric-doped carbon dots (N/Fe-CDs). Quantum yield of 49.52% was obtained when the N/FeCDs were produced utilizing the microwave digestion method. The N/Fe-CDs showed exceptional selectivity in distinguishing between dopamine and 4-nitrophenol. This is based on quenched fluorescence signals from N/FeCDs and DA, and on 4-NP, which has a quenched fluorescence signal between N/FeCDs and DA, both of which show linear response between DA concentration and detection limits of 0.07 or 0.02, respectively.

Furthermore, this method has already been used in practice for chemical sensing, and effectively so for human blood and urine samples and lake water and tap water. Thus, the fluorescent probe described here allows for more accurate detection in biological and environmental applications. The benefits of CDs in biosensors include high quantum yield, light stability, and low toxicity; however, mixing CDs with other nanomaterials must be done with caution since the resultant material could become more hazardous (82–83).

Carbon nanotubes (CNTs) are tubular substances, usually made of carbon, resembling cylindrical or nanotubular features. The Rai team (84) presents a simple, cheap method for synthesizing homogenous and extremely nano-crystalline plasmonic silver nanowire (AgNW) waveguides using polyol. The SEM and TEM methods were used to describe the morphology of the synthesized Ag-NWs, and it shows that the nanowires form a homogeneous AgNW with an average diameter of 400–450 nm and 15–25 μm in length. The enhancement factor of SERS combined with Ag-NW is at least 103 and 102, respectively, for the typical Raman signals (G-peak and radial breathing modes, respectively). It is proven using the numerical approach of finite difference time domain (FDTD) that the enhancement increases in value and is maximized when the SWNTs are oriented at 45° to the Ag-NWs. This technique produces high-quality SERS substrates that are suitable for distant laser stimulation, and it is effective, repeatable, and simple to use. Graphene, which looks like a honeycomb lattice with hundreds of parallel, overlapping hexagonal rings, is a carbon nanomaterial consisting of carbon atoms with a sp^2 hybrid orbital (85). Due to its enormous comparative area and semimetallic stability, graphene is very useful in optical biosensors (86–87). The peroxidase-like activity of $ZnFe_2O_4$ was exploited to construct a colorimetric biosensor out of $ZnFe_2O$ using rGO as the substrate. Conjugated with an aptamer specific to S. Typhimurium, $ZnFe_2O_4$/rGO was used in that experiment. This method was devised to detect Salmonella Typhimurium with a linear range of 11 to 1.10×10^5 colony-forming units per milliliter (cfu mL^{-1}) and a LOD of 11 cfu mL^{-1} in a buffer solution (88).

3.4.5.3 Metal Nanoparticle-Based Optical Biosensors

Due to their many and diverse characteristics, metal oxide-based nanomaterials have been used extensively in a number of different areas, such as electrochemistry, magnetism, catalysis, and sensors during the past decade. Scientists are employed as efficient electrocatalyst for detecting other analytes in the fields of biology and biomedicine because of their high electrocatalytic activity, cheap cost, and high organic capture ability. These oxide-based materials have a wide range of applications. Copper oxide (CuO), nickel oxide (NiO), iron oxide (Fe_2O_3), cobalt oxide (Co_3O_4), manganese oxide (MnO_2), zinc oxide (ZnO), titanium oxide (TiO_2), tin oxide (SnO_2), cadmium oxide (CdO), molybdenum oxide (MoO_3), and cerium oxide (CeO_2) are the metal oxide nanoparticles that are most frequently used (89). There has been a lot of interest in gold (AuNP) and silver nanoparticles (AgNPs) because of their simple synthesis and great safety/performance (90–91).

Due to good biocompatibility, high conductivity, and enhanced catalytic activity, silver nanoparticles (Ag NPs) have attracted considerable study attention in biological applications (92–94). Ag NPs were used to produce localized surface plasmon resonance (LSPR) and lossy-mode resonance (LMR) in a susceptible optical fiber sensor by the team of researchers led by Rivero et al. (95). A quick reaction time (476 ms and 447 ms for rising and fall, respectively) accompanied by a broad dynamic range (42.4 nm for RH variations between 25% and 70%) and high sensitivity (0.943 nm per RH percent). This monitor is ideal for keeping tabs on the breathing patterns of human subjects. Fe^{2+}, H_2O_2, and glucose detection were established by Mehdinia et al. (96) based on the Fenton reaction and AgNPs were biosynthesized. Fe^{2+} had a low detection limit of 0.54 μM, H_2O_2 had a detection limit of 0.032 μM, and glucose had a detection limit of 0.29 μM.

Akcan and coworkers (97) focused on researching to detect low concentrations of illicit drugs. The effect of Ag nanoparticles on Raman signal intensity of illegal drugs such as heroin and morphine, and AuNRs. SERS and its metabolites have been investigated by Inscore et al. (98), who found that optimal results may be obtained using AuNRs to show SERS spectra of MM. To further improve the overall reliability of the analysis, Akcan's team avoided uncontrolled aggregation of the material in the study by using an angled AuNRs-coated surface. Table 3.1 displays comparative information about Raman-active surfaces and supporting remarks.

Overall, the research shows that it is possible to identify and quantify the substances of heroin, morphine, and metabolites in saliva samples using a SERS-based method. The study's findings suggest that the addition of gold nanoparticles with the capacity to regulate liquid motion has a synergistic impact on the rise in both band numbers and intensities. As a result, authors propose a technique for detecting and quantifying heroin in biological fluids, namely saliva, that is quick, accurate, and cost-effective to implement.

The improvement in assay time and LOD due to the use of gold nanoparticles in a colorimetric bacterial assay increases the viability of traditional testing techniques for colorimetric bacterial detection (99). In an experiment, Nguyen's team (100) focused on the approach described here that uses biodegradable polysaccharides to achieve in situ production of gold nanoparticles (AuNPs). It is possible to prepare a novel composite of lactose/alginate (Lac/Alg) via an ionotropic gelation process that uses an in situ gold salt as a catalyst. In FTIR analysis, lactose has used as a reducing agent, which can be seen in the characteristic absorption peak. Analysis methods have verified the crystalline structure of AuNPs with a mean size of 10 nm. The nanocomposite powder can degrade 4-nitrophenol, methyl orange, rhodamine 6 G, and rhodamine B with excellent catalytic efficiency. AuNPs@Lac/Alg was an efficient probe for highly selective detection of Fe^{3+} ions because of its high dispersion

TABLE 3.1
Comparison of Nanomaterials for Detecting Illicit Drugs in SERS Findings (97)

Surface	SERS findings and related comment
AgNS	· Second least background noise · Second most efficient results · More identifiable peaks with higher intensities compared to AuNP/Nanocellulose, TLC and aniline
AuNRs	· Third least background noise · The most efficient results · More identifiable peaks in higher intensities compared to all
AuNPs/nanocellulose	· The least background noise · The intensity of main band (627 s-1) was significantly lower compared to AgNRs and AuNRs · Lower number of identifiable bands with extremely low intensities
TLC	· Useless in SERS-based detection and quantification of MM*
AuNPs/polyaniline	· The most background noise · The least number of identifiable peaks · Useless in SERS-based detection and quantification of MM*

* MM: Morphine monohydrate

solubility of AuNPs. The response has monitored when Fe^{3+} ions get together with nanocomposite in the presence of the composite. In a linear range of 2.0–80.0 μM, the LOD value was determined to be 0.8 μM.

Tang and colleagues (101) proposed a solution of the chitosan-stabilized gold nanoparticles that have been created to make it easier to detect uric acid using an H_2O_2 catalyst. According to theory, when stabilized gold nanoparticles are produced by chemical reduction, Tang's team found a peroxidase-like activity. These newly developed chitosan-stabilized gold nanoparticles (i.e., a catalyst) are able to catalyze H_2O_2, thereby producing OH radicals. In the presence of radical OH radicals, 3,3′,5,5′-tetramethylbenzidine is oxidized, resulting in a noticeable color shift (from colorless to blue). This innovative, sensitive, and easy-to-use technique is applicable for detecting uric acid using gold nanoparticles coated with chitosan. With a linear range from 0.1 to 30 μM, and a limit of detection (S/N ratio of 3) as low as 0.04 μM, this method is excellent work.

Additionally, human serum and urine samples have been effectively examined by optical detection. In human serum and urine, uric acid levels were about 96.1–103.1% and 95.2–97.7%, respectively. This investigation shows that visual colorimetry may serve as an alternative to daily monitoring and clinical care.

3.4.6 NANOBIOSENSORS BASED ANTIBODIES

Antibodies are proteins that are generated by the body when it is triggered by antigens to defend themselves. They are Y-shaped proteins produced by plasmocytes and are utilized by the immune system to detect and destroy foreign substances such as bacteria and viruses introduced into the body. An antibody is widely employed in detecting the biological target because antibodies are particular to antigens on the surface of a specific target and therefore are very effective (102). Antibodies should be extensively researched and studied to verify that they have a strong binding capability and affinity. The features of the biosensor should be considered while evaluating an antibody-antigen interaction. There are many benefits of using antibody-based biosensors, such as high sensitivity, specificity, noninvasive capabilities, and direct recognition ability (103). Recent research by Pal et al. (104) has shown that antibody-functionalized zinc doped magnetic nanoparticles (Zn-doped Fe_3O_4) were used to trap Salmonella bacteria from dairy products. The bacteria concentration was then quantified using a handheld luminometer. In this case, it has calculated that the test's detection limit was 10 cfu mL^{-1} in milk contaminated with Salmonella. Using the specific Zn-doped Fe_3O_4 nanoclusters, it is possible to identify different bacterial infections such as E. coli simply by adjusting the antibodies.

Pingli's team (105) proposed a colorimetric enzyme-linked immunosorbent test (ELISA) based on antibody coated gold nanoparticles to investigate ractopamine. The ELISA is based on an indirect method that competes with another sample. H_2O_2 oxidizes gold (III) ions in the presence of Ractopamine to produce red-colored metallic nanoparticles. If the sample does not include ractopamine, the AuNPs will form a purple-blue solution. At 560 nm, the absorbance is constant at levels between 2 and 512 ng/mL, and the detection limit is as low as 0.35 ng/mL in urine. In the presence of other β-agonists and antibiotics, ractopamine may also be visually identified.

The acquired findings match the results of LC-MS/MS, as shown by the analysis of sheep urine. In this ELISA technique, ractopamine detection is quick, simple, and cheap, and it is ideal for screening samples of animals for ractopamine. Vibrio infection, which is passed through drinking water on a regular basis, is a deadly danger. Guo's group (106) attached prominent gold vesicles to tiny gold nanowires (AuNWs) to create a SERS dual-mode immunosensor for colorimetric measurements. In order to improve the sensitivity for Vibrio parahemolyticus, the detection antibody was modified to the AuNWs, which were subsequently coated with silver staining, which amplified the detection. A LOD of 10 cfu mL^{-1} was observed for the colorimetric technique based on the biosensor used earlier.

Antibodies may be used to detect a specific analyte when modified for a particular purpose during analysis. Different labels in experimental tests, such as enzymes, biotin, fluorophores, and radioactive compounds, have been extensively used to enhance the detection signal. These labels are covalently linked to the antibody, resulting in high sensitivity and specificity, and thus creating an excellent detecting probe for the target antigen. On the other hand, antibodies have some disadvantages from drawbacks that restrict their potential uses, including production difficulties, a lack of batch-to-batch consistency, long production periods and high production costs, and the inability to distinguish between live and dead bacterial cells (107–108).

3.4.7 Nanobiosensors Based on Aptamer

Aptamers are single-stranded oligonucleotides of about 15 to 30 bases in length that bind to particular analytes such as bacteria, proteins, toxins, and hormones with a high degree of sensitivity and specificity. SELEX is an efficient library screening method in which several generations of oligonucleotides are progressively purified and amplified from an extensive random synthetic oligo-nucleic acid library. The analyte is recognized not by sequence but by form (109). Aptamers offer two distinct benefits as compared to antibodies. First, their nucleic acid characteristics allow for them to undergo nucleic acid replication and target-induced structural change. Second, aptamers have higher specificity and affinity, smaller size, cheaper cost, and greater thermality (110–112).

Tamer and coworkers (113) described a novel technique for detecting staphylococcal enterotoxin B (SEB), in which core–shell-structural iron–gold magnetic nanoparticles and a gold nanorod SERS probe is used in solution. Magnetic gold nanorod particles were used as SEB target scavengers, which were functionalized with a peptide ligand (aptamer). In order to perform the sandwich assay, the SEB molecules were removed from the matrix (Figure 3.4). It was found that the SEB: peptide–nanoparticle complex binding constant was 8.0×10^7 M^{-1}. Within a range of 2.5 fM to 3.2 nM, the connection between the SEB concentration and SERS signal was found to be linear. The homogeneous assay's limit of detection was set at 224 aM (about 2697 SEB molecules/20 μL sample volume). Additionally, gold-coated surfaces were also employed as capture substrates, and the two techniques were put to the test. The proposed technique was tested to see whether it was capable of identifying SEB in milk, blood, and urine that had been contaminated.

Heterogeneous sandwich immunoassay

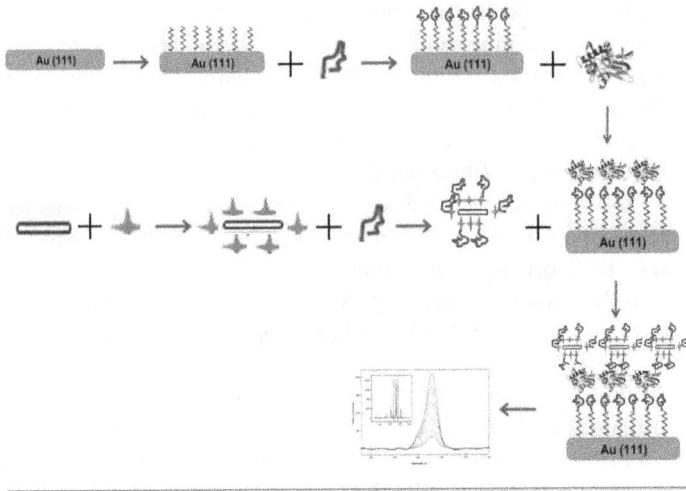

Homogeneous sandwich immunoassay

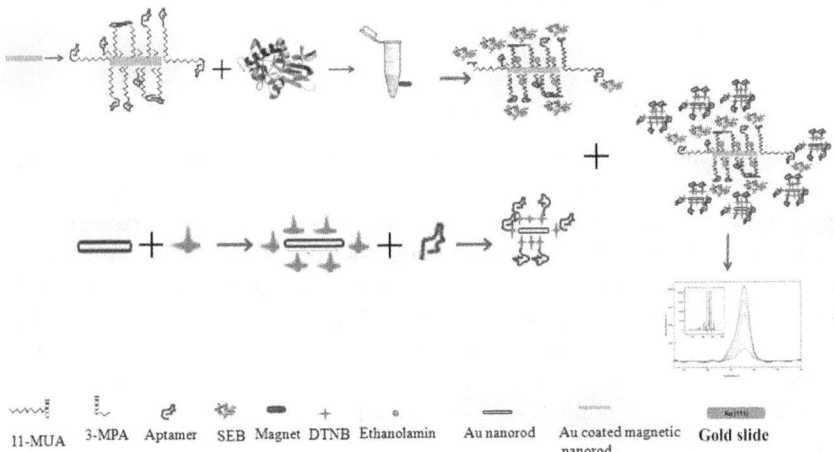

～～～	⌇	⌀	✿	▬	+	∘	▭	▭	▬
11-MUA	3-MPA	Aptamer	SEB	Magnet	DTNB	Ethanolamin	Au nanorod	Au coated magnetic nanorod	Gold slide

FIGURE 3.4 An aptamer-based heterogeneous and homogeneous sandwich immunoassay system with graphical depiction. Reprinted from ref. (113) with permission.

3.4.8 NANOBIOSENSORS BASED ON BACTERIOPHAGE

Recently, the use of bacteriophages (phages) in biosensor applications has been growing, and is now considered an attractive alternative to antibodies (114–117). Phages are viruses that exclusively infect the bacteria that are their host (118–119). They are less costly, quicker to manufacture, and have a longer shelf life than antibodies. Most bacteria may be detected by using phage-encoded biosensors since phages are bioreceptors that target bacterial cells. Nevertheless, the vast majority of phages infect just one species or a small number of closely related strains. A phage is capable of precisely detecting any existing bacterium since bacteria have a unique DNA signature. Some phages are resistant to high temperatures, such as at 76°C for three days (120), and some do not exhibit activity after being exposed to organic solvents (121–122). It is also possible to produce huge numbers of phages by causing a bacterial infection. Because of these characteristics, phages are well suited as biosensors for the fast detection of bacteria (114–117).

Watanabe and coworkers' (123) phage-immobilized SiO2@ AuNP core-shell nanoparticles were used in dark-field microscopy imaging as a plasmon scattering probe. *S. aureus* was targeted by the S13′ phage, which showed that it could identify and attach to the target bacterium. Meanwhile, SiO2@AuNP core−shell nanoparticles were used to increase the intensity of the target bacteria's light scattering. The detection limit for *S. aureus* was 1×10^4 cfu mL^{-1}, and the quantification of target bacteria across a broad range of concentrations was achieved within 15–20 minutes following the addition of the probes to the bacterial solution. By altering the phages used, the current technique may be used to detect any bacteria selectively, sensitively, and quickly. In our study, bifunctional plasmonic-magnetic nanoparticles are being used to detect specific bacteria at concentrations less than 1×10^4 cfu mL^{-1}.

Ilhan and colleagues (124) proposed a new method using phage-modified MOF nanoparticles to determine bacteria in the solution. The Salmonella biotinylated antibody binding method was used for the DTNB-labeled gold nanoparticles after conjugation with F5−4 Salmonella bacteriophage. DTNB has a disulfide structure and was adsorbed to the surface of AuNP as an adlayer. The disulfide linkages in DTNB were broken during this chemisorption process, causing the production of 2-nitro-5-thiobenzoate (TNB). TNB (diethanolamine benzoate) was utilized instead of DTNB (diethanolamine benzoate). For an efficient sandwich complex in solution, the spherical magnetic MOF bacteriophage (Salmonella *Enteritidis*) and an AuNP-TNB-Ab have been incubated to obtain the sandwich complex for 30 minutes at room temperature (Figure 3.5). The nanoparticles were removed from the sandwich complex and put into solution. 5 μL of suspension was placed on the TLC paper, and the complexes were redispersed in 50 μL of PBS buffer (pH 7.4). Measurements of TNB signals with SERS were performed at least five times, each time for 30 seconds, in three separate studies. Salmonella *Enteritidis*: (10^1-10^7 cfu mL^{-1}) is measured using SERS of sandwich immunoassay.

3.4.9 Novel Corona Virus-19, Challenges, and Opportunities

The novel coronavirus disease 2019 (COVID-19) has been affected by severe respiratory severe syndrome coronavirus (SARS-CoV-2). One of the most symptoms is high fever, diarrhea, headache, throat ache, fatigue, and loss of taste and smell (126). The clinical diversity of this disease has become challenging for rapid diagnosis at a clinical level. COVID-19 is a large family of viruses characterized by their sharp viral mirid bug that has been responsible for several outbreaks such as middle east respiratory syndrome (MERS) and SARS. Therefore, the COVID-19 virus has been caused by the SARS-CoV-2 coronavirus. The virus was first reported in Wuhan City of China in December 2019 and has spread to more than 187 countries globally (127–128). However, scientists are still enthusiastically working to characterize the

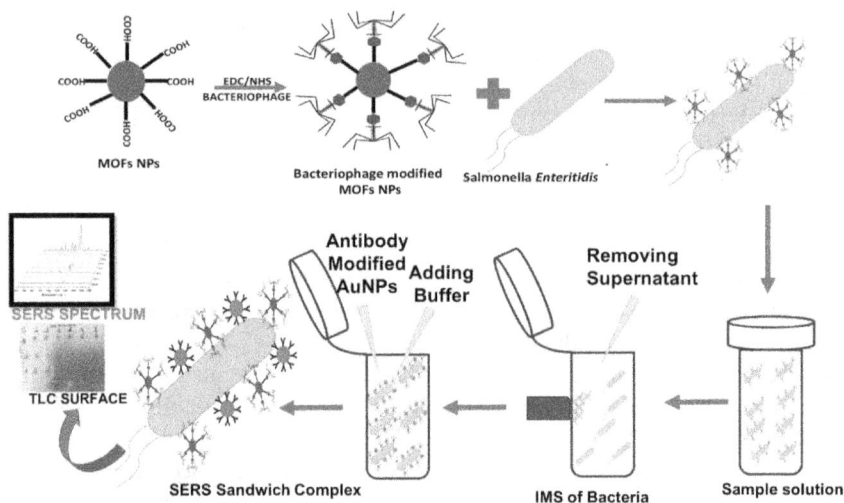

FIGURE 3.5 SERS measurement study conducted by magnetic-MOF and bacteriophages. Reprinted from ref. (124) with permission. Chen's team (125) developed a new method for identifying microbes by combining our phage-bacteria interactions in nature. 47–49 The RBP of a foreign phage was shown on an M13 scaffold, resulting in chimeric phages that bind various host bacteria. Additionally, the phages were chemically modified to bind with AuNPs, creating a chemical link between the target bacteria and the AuNPs, resulting in a noticeable change in SPR absorbance due to aggregate formation. *Pseudomonas Aeruginosa* (human pathogen) and *Vibrio cholerae* (plant pathogen) were detected as well as two strains of *Escherichia coli*. 100 cells may be detected with this method without any significant cross-reactivity between the Gram-negative bacterial species that were tested. The test takes less than an hour and works well with seawater and human serum. Gold nanoparticles are used to develop a method for determining the specific species of bacteria through a complex evolutionary process.

virus and its epidemiology in humans to understand the disease transmission and clinical diagnosis (129). There are two alternative methods for pandemic viral infection diagnostic that have been possible and practised.

It has comprised a serological investigation to analyze the biomarker levels such as immunoglobin G (IgG) and immunoglobin M (IgM). Whereas the second is involved in the direct determination of the virus itself and utilized with cellular proteins. The concentration limits the earlier immunosensing approach and generally hosts an antibody. But, in the severe case of a disease such as COVID-19, it may serve as a helpful pre-screening test. The second method is involved in the direct detection of the virus by its cell surface proteins. However, the detection of the new virus such as SARS-CoV-2 for the identification of unique antigen as the preliminary investigation revealed that the new SARS-CoV-2 has four main structural proteins, including the S (spike), E (envelope), M (membrane), and N (nucleocapsid) protein (130). Previous studies stated that MERS-CoV and SARS-CoV primarily utilize the receptor-binding domain of S protein and N protein as antigens to develop the detection assays. Commonly used receptors for these are antibodies (131), aptamers, and nanobodies (132).

3.4.10 OPTICAL BIOSENSORS FOR COVID-19

The increasing importance of novel optical biosensors can compete with the conventional nucleic acid test sensitivity and selectivity. Nanomaterials-based optical biosensors presented an ideal alternative due to their progress in designing and constructing methods such as high sensitivity and selectivity and good reproducibility. A biosensor has been used for the COVID-19 pandemic. Its future viral epidemics could be readily available and affordable, ideally employing either an at-home test or a little volume sample preparation. Optical biosensor offers alternate and additional methods for the testing and the potential as generated rapid results compared to test currently used for COVID-19 testing.

Moreover, it has the ability for rapid virus detection as compared to other techniques such as PCR (qRT-PCR), RT-PCR, and single-photon emission computed tomography (SPECT), including computed tomography (CT) and SERS. However, novel viral epidemics will occur in the future. It is essential to have rapid and reliable sensing technology to detect and reduce the spread of COVID-19 (133–134). The reverse-transcription polymerase chain reaction is linked with immunofluorescence to detect viruses and pathogens (135). The rapid worldwide spread of the pandemic COVID-19 provokes the SARS-CoV-2. RT-PCR is a gold standard for detecting SARS-CoV-2, and it is a multi-step technique that involves sample purification and fluorescence detection (136). RT-PCR technique is laborious, requires a trained operator, and has limited accessibility in resource-limited settings (137).

Moreover, surface plasmons, fluorescence, and colorimetry techniques have been used earlier to identify influenza viruses, Ebola, HIV, norovirus, and others (138). These techniques have been applied for targeted virus detection and virus imaging (139). Optical biosensors offer an alternating method for pandemic virus detection due to their cost-effective technology and safe and straightforward use (140–141). Additionally, an Optical biosensor has been used as a point-of-care (POC) diagnostic tool for virus detection, with no need for sample preparation and trained personal or expensive analysis apparatus and low test manufacturing cost (142). But merely a handful of optical biosensors are currently available on the market for virus detection and mostly consist of lab-on-a-chip (LOC) techniques added to nucleic acids for fluorescent analysis (143). Enhancing optical imaging efficiency in a particular single virus can be applied to track and monitor virus repetition and cell interaction to develop treatment options more quickly and efficiently. Continuing research to use these imaging techniques to identify COVID-19 has already begun. But further work is necessary to bring these technologies from benchtop to market (144–145).

3.4.10.1 Principle of Lab-on-a-Chip Optical Biosensors

Several advantages of lab-on-chip devices are primarily relevant to the current pandemic COVID-19. Lab-on-chip devices are vigorous, low-cost, rapid, and sensitive to provide results at the point of care (146). In the perspective of COVID-19, these advantages would help and support the efforts to increase testing access (147). To monitor the real-time sample of biomolecules such as proteins, biosensors with ultra-high sensitivity are required to detect the targets. However, the optical transducer-based biosensor is widely researched, commercialized, and utilized in clinics.

A label-free optical biosensor is a sensitive and durable point-of-care testing (POCT) device that are essential for pandemic SARS-CoV-2 because it can be easily operated at the point-of-risk without any special training (148).

On top of that, the label-free optical biosensor has displayed a solid and credible potential to grow in biomedical and healthcare fields. It provided a shortened, accurate analytical tool to promote the mass-scale screening on a broad range of samples over different parameters (149). The optical biosensing technique works in other physical transduction principles such as plasmonic, resonators, and interferometers to monitor many viruses with reasonable accuracy in different studies. The lab-on-a-chip platform is suitable for the examiner of SARS-CoV-2 spike protein detection and other infectious viruses such as influenza and middle east respiratory syndrome (150).

3.4.10.2 A Fiber-Optic Biosensor-Based Rapid Diagnosis of COVID-19

The fiber-optic biosensor platform has diagnosed COVID-19 by developing two bioanalytical approaches, label-free and labelled bioassays. First, set the label-free bioassay based on the biosensor matrix to immobilize the gold nanoparticles (Au-NPs) on the U-bent fiber-optic probe (150), followed by covalent conjugation of an anti-N protein monoclonal antibody via a suitable thiol-PEG-NHS linked with chemistry. The antibody immobilized probe was treated with bovine serum albumin (BSA) solution to prevent non-specific interactions. Then, the biofunctionalized investigation was used for the diagnosis of COVID-19 by introducing the patient's saliva sample. The binding of SARS-CoV-2 offered a drop in the light intensity, and results were obtained within 15 min (151). Second, the labelled bioassays strategy in Figure 3.6 was developed based on sandwich immunoassay and used capture and detector antibodies labelled with Au-NPs (152–153). Initially, the biosensor matrix needed to immobilize with anti-N protein monoclonal capture antibody on the U-bent fiber-optic probe, followed by treatment with a BSA solution to reduce the

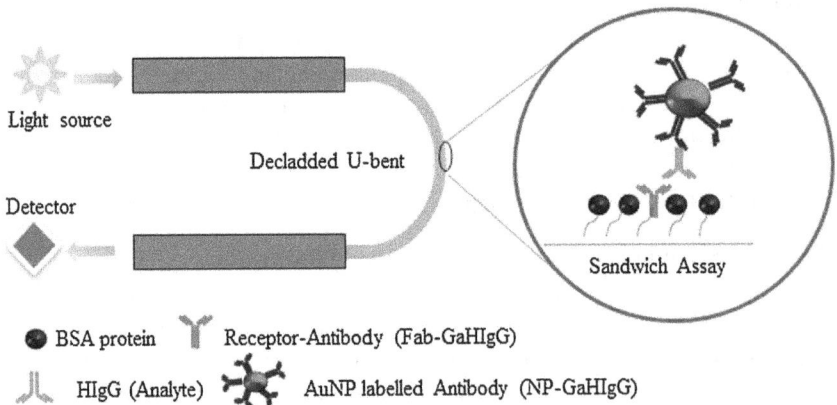

FIGURE 3.6 The detection of Human IgG in U-bent fiber optic biosensor. Reprinted from ref. (152) with permission.

non-specific interactions. Therefore, the samples were obtained from a virus-infected person and then mixed anti-N protein monoclonal capture antibody to conjugated Au-NPs within 5 min followed by introducing the sandwich complex to the sensing region of the bent probe. Then the quick signal response was generated in intensity count or absorbance change, leading to saturation within 10–15 min. However, the viral loaded small sample as 10^6 particles/mL caused infection and was detected using the label-free sensor. Therefore, the label-free sensor is desirable due to the one-step response without any reagents but low specificity due to high non-specific binding. Hence, both approaches have been evaluated and compared to choose one for clinical testing (154).

3.5 CONCLUSION

This chapter has summarized recent advances for the many kinds and processes of optical biosensors that use receptors (enzymes, antibodies, whole-cell detection, and aptamers), transducers (SERS, SPR, Fluorescence, gravimetric, and colorimetric), and nanomaterials as the basis for their function (gold-NPs, AgNPs, MNPs). Optical biosensors have a wide range of applications in the areas of medicine and biomedical science, toxicology and ecotoxicology, food safety, and drug delivery. In recent years, Tamer and colleagues have seen a significant increase in the use of nanomaterials in biosensors, which has resulted in fast development in biosensing technology. The focus was on molecular recognition elements and nanomaterials, comparing nanostructured optical biosensors to traditional methods. SARS-CoV-2 detection and treatment techniques are being developed by scientific groups. Future virus outbreak optical biosensors should be easily accessible, inexpensive, and need minimum sample processing. Optical biosensors provide alternative testing techniques that can produce findings considerably faster than the existing COVID-19 testing procedures. Optical biosensors effectively detect and support major breakthroughs in clinical diagnostics, drug development, food industry management, and environmental monitoring.

REFERENCES

1. Seok H, Park TH. 2011. Integration of biomolecules and nanomaterials: Towards highly selective and sensitive biosensors. *Biotechnology Journal*. 6(11): 1310–1316
2. Cui Y, Duan X, Hu J, Lieber CM. 2000. Doping and electrical transport in silicon nanowires. *The Journal of Physical Chemistry B*. 104(22): 5213–5216
3. Elfström N, Juhasz R, Sychugov I, Engfeldt T, Karlström AE, Linnros J. 2007. Surface charge sensitivity of silicon nanowires: Size dependence. *Nano Letters*. 7(9): 2608–2612
4. Yogeswaran U, Chen SM. 2008. A review on the electrochemical sensors and biosensors composed of nanowires as sensing material. *Sensors*. 8(1): 290–313
5. Lee SH, Sung JH, Park TH. 2012. Nanomaterial-based biosensor as an emerging tool for biomedical applications. *Annals of Biomedical Engineering*. 40(6): 1384–1397
6. Du H, Li Z, Wang Y, Yang Q, Wu W. 2020. Nanomaterial-based optical biosensors for the detection of foodborne bacteria. *Food Reviews International*. 1–30
7. Naresh V, Lee N. 2021. A review on biosensors and recent development of nanostructured materials-enabled biosensors. *Sensors*. 21(4): 1109

8. Maddali H, Miles CE, Kohn J, Ocarroll DM. 2021. Optical biosensors for virus detection: Prospects for SARS-CoV-2/COVID-19. *ChemBioChem.* 22(7): 1176

9. Arranz A, Ripoll J. 2015. Advances in optical imaging for pharmacological studies. *Frontiers in Pharmacology.* 6: 189

10. Saylan Y, Erdem Ö, Ünal S, Denizli A. 2019. An alternative medical diagnosis method: Biosensors for virus detection. *Biosensors.* 9(2): 65

11. Huang YF, Zhuo GY, Chou CY, Lin CH, Chang W, Hsieh CL. 2017. Coherent brightfield microscopy provides the spatiotemporal resolution to study early stage viral infection in live cells. *ACS Nano.* 11(3): 2575–2585

12. Ueki H, Wang IH, Zhao D, Gunzer M, Kawaoka Y. 2020. Multicolor two-photon imaging of in vivo cellular pathophysiology upon influenza virus infection using the two-photon impress. *Nature Protocols.* 15(3): 1041–1065

13. Monroe D. 1990. Amperometric immunoassay. *Critical Reviews in Clinical Laboratory Sciences.* 28(1): 1–18

14. Moina C, Ybarra G. 2012. Fundamentals and applications of immunosensors. *Advances in Immunoassay Technology.* 65–80

15. Zhang H, Ma L, Li P, Zheng J. 2016. A novel electrochemical immunosensor based on nonenzymatic Ag@ Au-Fe3O4 nanoelectrocatalyst for protein biomarker detection. *Biosensors and Bioelectronics.* 85: 343–350

16. Cooper MA. 2009. *Label-Free Biosensors: Techniques and Applications.* Cambridge University Press

17. Abdolrahim M, Rabiee M, Alhosseini SN, Tahriri M, Yazdanpanah S, Tayebi L. 2015. Development of optical biosensor technologies for cardiac troponin recognition. *Analytical Biochemistry.* 485: 1–10

18. Alves RC, Barroso MF, Gonzalez MB, Oliveira MBP, Delerue MC. 2016. New trends in food allergens detection: Toward biosensing strategies. *Critical Reviews in Food Science and Nutrition.* 56(14): 2304–2319

19. Mejard R, Griesser HJ, Thierry B. 2014. Optical biosensing for label-free cellular studies. *TrAC Trends in Analytical Chemistry.* 53: 178–186

20. He F, Qin X, Bu L, Fu Y, Tan Y, Chen C, Yao S. 2017. Study on the bioelectrochemistry of a horseradish peroxidase-gold nanoclusters bionanocomposite. *Journal of Electroanalytical Chemistry.* 792: 39–45

21. Zhang Y, Yuan XL, Lyu FL, Wang XC, Jiang XJ, Cao MH, Zhang Q. 2020. Facile one-step synthesis of PdPb nanochains for high-performance electrocatalytic ethanol oxidation. *Rare Metals.* 39: 792–799

22. Ritzefeld M, Sewald N. 2012. Real-time analysis of specific protein-DNA interactions with surface plasmon resonance. *Journal of Amino Acids.* doi:10.1155/20121816032

23. Homola J. 2003. Present and future of surface plasmon resonance biosensors. *Analytical and Bioanalytical Chemistry.* 377(3): 528–539

24. McWhirter A, Wahlström L. 2008. The benefits and scope of surface plasmon resonance-based biosensors in food analysis. *Handbook of Surface Plasmon Resonance.* 333–352

25. Myszka DG, Rich RL. 2000. Implementing surface plasmon resonance biosensors in drug discovery. *Pharmaceutical Science & Technology Today.* 3(9): 310–317

26. Lim SA, Yoshikawa H, Tamiya E, Yasin HM, Ahmed MU. 2014. A highly sensitive gold nanoparticle bioprobe based electrochemical immunosensor using screen printed graphene biochip. *RSC Advances.* 4(102): 58460–58466

27. Dey D, Goswami T. 2011. Optical biosensors: A revolution towards quantum nanoscale electronics device fabrication. *Journal of Biomedicine and Biotechnology.* 348218

28. Jain Y, Rana C, Goyal A, Sharma N, Verma ML, Jana AK. 2010. Biosensors, types and applications. In *BEATS 2010: Proceedings of the 2010 International Conference on Biomedical Engineering and Assistive Technologies, India,* pp. 1–6. BEATS

29. Green RJ, Frazier RA, Shakesheff KM, Davies MC, Roberts CJ, Tendler SJ. 2000. Surface plasmon resonance analysis of dynamic biological interactions with biomaterials. *Biomaterials.* 21(18): 1823–1835

30. Cao C, Sim SJ. 2007. Signal enhancement of surface plasmon resonance immunoassay using enzyme precipitation-functionalized gold nanoparticles: A femto molar level measurement of anti-glutamic acid decarboxylase antibody. *Biosensors and Bioelectronics.* 22(9–10): 1874–1880

31. Lee D, Hwang J, Seo Y, Gilad AA, Choi J. 2018. Optical immunosensors for the efficient detection of target biomolecules. *Biotechnology & Bioprocess Engineering.* 23(2)

32. Lakshmipriya T, Gopinath SC, Tang TH. 2016. Biotin-streptavidin competition mediates sensitive detection of biomolecules in enzyme linked immunosorbent assay. *PLoS One.* 11(3): e0151153

33. Panhwar S, Ilhan H, Hassan SS, Zengin A, Boyacı IH, Tamer U. 2020. Dual responsive disposable electrode for the enumeration of escherichia coli in whole blood. *Electroanalysis.* 32(10): 2244–2252

34. Panhwar S, Aftab A, Keerio H. A, Sarmadivaleh M, Tamer U. 2021. A novel approach for real-time enumeration of escherichia coli ATCC 47076 in water through high multi-functional engineered nano-dispersible electrode. *Journal of The Electrochemical Society.* 168(3): 037514

35. Morrison DW, Dokmeci MR, Demirci U, Khademhosseini A. 2008. Clinical applications of micro-and nanoscale biosensors. *Biomedical Nanostructures.* 1: 433–458

36. Justino CI, Freitas AC, Pereira R, Duarte AC, Santos TAR. 2015. Recent developments in recognition elements for chemical sensors and biosensors. *TrAC Trends in Analytical Chemistry.* 68: 2–17

37. Lim SA, Ahmed MU, Perumal V, Hasim U. 2016. Introduction to food biosensors. Advances in biosensors: Principle, architecture and applications. *Journal of Applied Biomedicine.* 2: 1–15

38. Liu H, Ge J, Ma E, Yang L. 2019. Advanced biomaterials for biosensor and theranostics. In *Biomaterials in Translational Medicine*, pp. 213–255. Academic Press

39. Chaudhari R, Joshi A, Srivastava R. 2017. pH and urea estimation in urine samples using single fluorophore and ratiometric fluorescent biosensors. *Scientific Reports.* 7(1): 1–9

40. Azmi NE, Rashid AHA, Abdullah J, Yusof NA, Sidek H. 2018. Fluorescence biosensor based on encapsulated quantum dots/enzymes/sol-gel for non-invasive detection of uric acid. *Journal of Luminescence.* 202: 309–315

41. Cho MJ, Park SY. 2019. Carbon-dot-based ratiometric fluorescence glucose biosensor. *Sensors and Actuators B: Chemical.* 282: 719–729

42. Neubauerova K, Carneiro MC, Rodrigues LR, Moreira FT, Sales MGF. 2020. Nanocellulose-based biosensor for colorimetric detection of glucose. *Sensing and Bio-Sensing Research.* 29: 100368

43. Gopal KV. 2003. Neurotoxic effects of mercury on auditory cortex networks growing on microelectrode arrays: A preliminary analysis. *Neurotoxicology and Teratology.* 25(1): 69–76

44. Fromherz P. 2003. Semiconductor chips with ion channels, nerve cells and brain. *Physica E: Low-Dimensional Systems and Nanostructures.* 16(1): 24–34

45. Maher M, Pine J, Wright J, Tai YC. 1999. The neurochip: A new multielectrode device for stimulating and recording from cultured neurons. *Journal of Neuroscience Methods.* 87(1): 45–56

46. Neher E. 2001. Molecular biology meets microelectronics. *Nature Biotechnology.* 19(2): 114–114

47. Thomas JCA, Springer PA, Loeb GE, Berwald NY, Okun LM. 1972. A miniature microelectrode array to monitor the bioelectric activity of cultured cells. *Experimental Cell Research.* 74(1): 61–66

48. Keefer EW, Gramowski A, Stenger DA, Pancrazio JJ, Gross GW. 2001. Characterization of acute neurotoxic effects of trimethylolpropane phosphate via neuronal network biosensors. *Biosensors and Bioelectronics*. 16(7–8): 513–525

49. Ecken H, Ingebrandt S, Krause M, Richter D, Hara M, Offenhäusser A. 2003.64-Channel extended gate electrode arrays for extracellular signal recording. *Electrochimica Acta*. 48(20–22): 3355–3362

50. Bousse L. 1996. Whole cell biosensors. *Sensors and Actuators B: Chemical*. 34(1–3): 270–275

51. Fromherz P, Offenhausser A, Vetter T, Weis J. 1991. A neuron-silicon junction: A Retzius cell of the leech on an insulated-gate field-effect transistor. *Science*. 252(5010): 1290–1293

52. Rudolph AS, Reasor J. 2001. Cell and tissue based technologies for environmental detection and medical diagnostics. *Biosensors and Bioelectronics*. 16: 429–431

53. Paddle BM. 1996. Biosensors for chemical and biological agents of defence interest. *Biosensors and Bioelectronics*. 11(11): 1079–1113

54. Rogers KR. 1995. Biosensors for environmental applications. *Biosensors and Bioelectronics*. 10(6–7): 533–541

55. Wilkins E, Atanasov P. 1996. Glucose monitoring: State of the art and future possibilities. *Medical Engineering & Physics*. 18(4): 273–288

56. Das A, Kim K, Park SG, Choi N, Choo J. 2021. SERS-based serodiagnosis of acute febrile diseases using plasmonic nanopopcorn microarray platforms. *Biosensors and Bioelectronics*. 192: 113525

57. Yao Y, Zhang H, Tian T, Liu Y, Zhu R, Ji J, Liu B. 2021. Iodide-modified Ag nanoparticles coupled with DSN-Assisted cycling amplification for label-free and ultrasensitive SERS detection of MicroRNA-21. *Talanta*. 235: 122728

58. Ren X, Zhang Q, Yang J, Zhang X, Zhang X, Zhang Y, Wang Y. 2021. Dopamine imaging in living cells and retina by surface-enhanced raman scattering based on functionalized gold nanoparticles. *Analytical Chemistry*. 93(31): 10841–10849

59. Berneschi S, Dandrea C, Baldini F, Banchelli M, Deangelis M, Pelli S, Matteini P. 2021. Ion-exchanged glass microrods as hybrid SERS/fluorescence substrates for molecular beacon-based DNA detection. *Analytical and Bioanalytical Chemistry*. 413(24): 6171–6182

60. Umesha S, Manukumar HM. 2018. Advanced molecular diagnostic techniques for detection of food-borne pathogens: Current applications and future challenges. *Critical Reviews in Food Science and Nutrition*. 58(1): 84–104

61. Gilmartin N, O'Kennedy R. 2012. Nanobiotechnologies for the detection and reduction of pathogens. *Enzyme and Microbial Technology*. 50(2): 87–95

62. Koedrith P, Thasiphu T, Tuitemwong K, Boonprasert R, Tuitemwong P. 2014. Recent advances in potential nanoparticles and nanotechnology for sensing food-borne pathogens and their toxins in foods and crops: Current technologies and limitations. *Sensors and Materials*. 26(10): 711–736

63. Yogeswaran U, Chen SM. 2008. A review on the electrochemical sensors and biosensors composed of nanowires as sensing material. *Sensors*. 8(1): 290–313

64. Thevenot DR, Toth K, Durst RA, Wilson GS. 1999. Electrochemical biosensors: Recommended definitions and classification. *Pure and Applied Chemistry*. 71(12): 2333–2348

65. Turner AP. 2000. Biosensors-sense and sensitivity. *Science*. 290(5495): 1315–1317

66. Zhang X, Guo Q, Cui D. 2009. Recent advances in nanotechnology applied to biosensors. *Sensors*. 9(2): 1033–1053

67. Suni II. 2008. Impedance methods for electrochemical sensors using nanomaterials. *TrAC Trends in Analytical Chemistry*. 27(7): 604–611

68. Sobczak KA, Venkatesan J, AlAnezi AA, Walczyk D, Farooqi A, Malina D, Tyliszczak B. 2016. Magnetic nanomaterials and sensors for biological detection. *Nanomedicine: Nanotechnology, Biology and Medicine.* 12(8): 2459–2473

69. Wang Q, Yang Q, Wu W. 2020. Progress on structured biosensors for monitoring aflatoxin B1 from biofilms: A review. *Frontiers in Microbiology.* 11: 408

70. Ekmen E, Bilici M, Turan E, Tamer U, Zengin A. 2020. Surface molecularly-imprinted magnetic nanoparticles coupled with SERS sensing platform for selective detection of malachite green. *Sensors and Actuators B: Chemical.* 325: 128787

71. Yazgan NN, Boyacı İH, Topcu A, Tamer U. 2012. Detection of melamine in milk by surface-enhanced Raman spectroscopy coupled with magnetic and Raman-labeled nanoparticles. *Analytical and Bioanalytical Chemistry.* 403(7): 2009–2017

72. Zengin A, Tamer U, Caykara T. 2013. A SERS-based sandwich assay for ultrasensitive and selective detection of Alzheimer's tau protein. *Biomacromolecules.* 14(9): 3001–3009

73. Cardoso VF, Francesko A, Ribeiro C, Banobre LM, Martins P, Lanceros MS. 2018. Advances in magnetic nanoparticles for biomedical applications. *Advanced Healthcare Materials.* 7(5): 1700845

74. Bohara RA, Pawar SH. 2015. Innovative developments in bacterial detection with magnetic nanoparticles. *Applied Biochemistry and Biotechnology.* 176(4): 1044–1058

75. Huang X, Liu Y, Yung B, Xiong Y, Chen X. 2017. Nanotechnology-enhanced no-wash biosensors for in vitro diagnostics of cancer. *ACS Nano.* 11(6): 5238–5292

76. Baptista FR, Belhout SA, Giordani S, Quinn SJ. 2015. Recent developments in carbon nanomaterial sensors. *Chemical Society Reviews.* 44(13): 4433–4453

77. Dhenadhayalan N, Lin KC, Saleh TA. 2020. Recent advances in functionalized carbon dots toward the design of efficient materials for sensing and catalysis applications. *Small.* 16(1): 1905767

78. Xu X, Ray R, Gu Y, Ploehn HJ, Gearheart L, Raker K, Scrivens WA. 2004. Electrophoretic analysis and purification of fluorescent single-walled carbon nanotube fragments. *Journal of the American Chemical Society.* 126(40): 12736–12737

79. Roy P, Chen PC, Periasamy AP, Chen YN, Chang HT. 2015. Photoluminescent carbon nanodots: Synthesis, physicochemical properties and analytical applications. *Materials Today.* 18(8): 447–458

80. Zhong D, Zhuo Y, Feng Y, Yang X. 2015. Employing carbon dots modified with vancomycin for assaying Gram-positive bacteria like Staphylococcus aureus. *Biosensors and Bioelectronics.* 74: 546–553

81. Mathivanan D, Mohan A, Yang Y. 2021. Facile fabrication of nitrogen—ferric-doped carbon dots for highly sensitive and selective detection of dopamine and 4-nitrophenol. *Journal of Materials Science: Materials in Electronics.* 32(7): 9005–9017

82. Nekoueian K, Amiri M, Sillanpaa M, Marken F, Boukherroub R, Szunerits S. 2019. Carbon-based quantum particles: An electroanalytical and biomedical perspective. *Chemical Society Reviews.* 48(15): 4281–4316

83. Kang Z, Lee ST. 2019. Carbon dots: Advances in nanocarbon applications. *Nanoscale.* 11(41): 19214–19224

84. Das TK, Goel R, Awasthi V, Singh T, Shukla V, Kumar A, Rai P. 2021. Surface Enhanced Raman scattering from single-walled carbon nanotube decorated on Ag nanowires. *Plasmonics.* 1–10

85. Geim AK, Novoselov KS. 2010. The rise of graphene. *Nanoscience and Technology: A Collection of Reviews from Nature Journals.* 11–19

86. Suvarnaphaet P, Pechprasarn S. 2017. Graphene-based materials for biosensors: A review. *Sensors.* 17(10): 2161

87. Wang Q, Yang Q, Wu W. 2020. Graphene-based steganographic aptasensor for information computing and monitoring toxins of biofilm in food. *Frontiers in Microbiology*. 10: 3139

88. Luan Q, Gan N, Cao Y, Li T. 2017. Mimicking an enzyme-based colorimetric aptasensor for antibiotic residue detection in milk combining magnetic loop-DNA probes and CHA-assisted target recycling amplification. *Journal of Agricultural and Food Chemistry*. 65(28): 5731–5740

89. Shi X, Gu W, Li B, Chen N, Zhao K, Xian Y. 2014. Enzymatic biosensors based on the use of metal oxide nanoparticles. *Microchimica Acta*. 181(1–2): 1–22

90. Wu W, Yu C, Chen J, Yang Q. 2020. Fluorometric detection of copper ions using click chemistry and the target-induced conjunction of split DNAzyme fragments. *International Journal of Environmental Analytical Chemistry*. 100(3): 324–332

91. Panhwar S, Hassan SS, Mahar RB, Canlier A, Arain M. 2018. Synthesis of l-cysteine capped silver nanoparticles in acidic media at room temperature and detailed characterization. *Journal of Inorganic and Organometallic Polymers and Materials*. 28(3): 863–870

92. El DR, Georges M, Azzazy HM. 2012. Silver nanostructures: Properties, synthesis, and biosensor applications. In *Functional Nanoparticles for Bioanalysis, Nanomedicine, and Bioelectronic Devices*, volume 1, pp. 359–404. American Chemical Society

93. Loiseau A, Asila V, Boitel AG, Lam M, Salmain M, Boujday S. 2019. Silver-based plasmonic nanoparticles for and their use in biosensing. *Biosensors*. 9(2): 78

94. Malekzad H, Zangabad PS, Mirshekari H, Karimi M, Hamblin MR. 2017. Noble metal nanoparticles in biosensors: Recent studies and applications. *Nanotechnology Reviews*. 6(3): 301–329

95. Rivero PJ, Urrutia A, Goicoechea J, Arregui FJ. 2012. Optical fiber humidity sensors based on localized surface plasmon resonance (LSPR) and lossy-mode resonance (LMR) in overlays loaded with silver nanoparticles. *Sensors and Actuators B: Chemical*. 173: 244–249

96. Basiri S, Mehdinia A, Jabbari A. 2018. A sensitive triple colorimetric sensor based on plasmonic response quenching of green synthesized silver nanoparticles for determination of Fe^{2+}, hydrogen peroxide, and glucose. *Colloids and Surfaces A: Physicochemical and Engineering Aspects*. 545: 138–146

97. Akcan R, Yildirim MŞ, Ilhan H, Güven B, Tamer U, Sağlam N. 2020. Surface enhanced Raman spectroscopy as a novel tool for rapid quantification of heroin and metabolites in saliva. *Turkish Journal of Medical Sciences*. 50(5): 1470–1479

98. Inscore F, Shende C, Sengupta A, Huang H, Farquharson S. 2011. Detection of drugs of abuse in saliva by surface-enhanced Raman spectroscopy (SERS). *Applied Spectroscopy*. 65(9): 1004–1008

99. Verma MS, Rogowski JL, Jones L, Gu FX. 2015. Colorimetric biosensing of pathogens using gold nanoparticles. *Biotechnology Advances*. 33(6): 666–680; Ho TTT, Dang CH, Huynh TKC, Hoang TKD, Nguyen TD. 2021. In situ synthesis of gold nanoparticles on novel nanocomposite lactose/alginate: Recyclable catalysis and colorimetric detection of Fe (III). *Carbohydrate Polymers*. 251: 116998

100. Li F, He T, Wu S, Peng Z, Qiu P, Tang X. 2021. Visual and colorimetric detection of uric acid in human serum and urine using chitosan stabilized gold nanoparticles. *Microchemical Journal*. 164: 105987

101. Liu P, Han L, Wang F, Petrenko V A, Liu A. 2016. Gold nanoprobe functionalized with specific fusion protein selection from phage display and its application in rapid, selective and sensitive colorimetric biosensing of Staphylococcus aureus. *Biosensors and Bioelectronics*. 82: 195–203

102. Justino CI, Freitas AC, Pereira R, Duarte AC, Santos TAR. 2015. Recent developments in recognition elements for chemical sensors and biosensors. *TrAC Trends in Analytical Chemistry*. 68: 2–17

103. Pal M, Lee S, Kwon D, Hwang J, Lee H, Hwang S, Jeon S. 2017. Direct immobilization of antibodies on Zn-doped Fe3O4 nanoclusters for detection of pathogenic bacteria. *Analytica Chimica Acta*. 952: 81–87

104. Han S, Zhou T, Yin B, He P. 2018. Gold nanoparticle-based colorimetric ELISA for quantification of ractopamine. *Microchimica Acta*. 185(4): 1–8

105. Guo Z, Jia Y, Song X, Lu J, Lu X, Liu B, Chen T. 2018. Giant gold nanowire vesicle-based colorimetric and SERS dual-mode immunosensor for ultrasensitive detection of vibrio parahemolyticus. *Analytical Chemistry*. 90(10): 6124–6130

106. Liu J, Cao Z, Lu Y. 2009. Functional nucleic acid sensors. *Chemical Reviews*. 109(5): 1948–1998

107. Feng C, Dai S, Wang L. 2014. Optical aptasensors for quantitative detection of small biomolecules: A review. *Biosensors and Bioelectronics*. 59: 64–74

108. Pan P, Huang YW, Oshima K, Yearsley M, Zhang J, Arnold M, Wang LS. 2019. The immunomodulatory potential of natural compounds in tumor-bearing mice and humans. *Critical Reviews in Food Science and Nutrition*. 59(6): 992–1007

109. Kim YS, Raston NHA, Gu MB. 2016. Aptamer-based nanobiosensors. *Biosensors and Bioelectronics*. 76: 2–19

110. Zeng Z, Zhang P, Zhao N, Sheehan AM, Tung CH, Chang CC, Zu Y. 2010. Using oligonucleotide aptamer probes for immunostaining of formalin-fixed and paraffin-embedded tissues. *Modern Pathology*. 23(12): 1553–1558

111. Dong H, Chen H, Jiang J, Zhang H, Cai C, Shen Q. 2018. Highly sensitive electrochemical detection of tumor exosomes based on aptamer recognition-induced multi-DNA release and cyclic enzymatic amplification. *Analytical Chemistry*. 90(7): 4507–4513

112. Temur E, Zengin A, Boyacı IH, Dudak FC, Torul H, Tamer U. 2012. Attomole sensitivity of staphylococcal enterotoxin B detection using an aptamer-modified surface-enhanced Raman scattering probe. *Analytical Chemistry*. 84(24): 10600–10606

113. Richter L, Janczuk RM, Niedzioka JJ, Paczesny J, Hołyst R. 2018. Recent advances in bacteriophage-based methods for bacteria detection. *Drug Discovery Today*. 23(2): 448–455.

114. Van der Merwe RG, Van HPD, Warren RM, Sampson SL, Pittius NG. 2014. Phage-based detection of bacterial pathogens. *Analyst*. 139(11): 2617–2626

115. Tawil N, Sacher E, Mandeville R, Meunier M. 2014. Bacteriophages: Biosensing tools for multi-drug resistant pathogens. *Analyst*. 139(6): 1224–1236

116. Anany H, Chou Y, Cucic S, Derda R, Evoy S, Griffiths MW. 2017. From bits and pieces to whole phage to nanomachines: Pathogen detection using bacteriophages. *Annual Review of Food Science and Technology*. 8: 305–329

117. Salmond GP, Fineran PC. 2015. A century of the phage: Past, present and future. *Nature Reviews Microbiology*. 13(12): 777–786

118. Tyler JS, Livny J, Friedman DI. 2005. Lambdoid phages and Shiga toxin. *Phages: Their Role in Bacterial Pathogenesis and Biotechnology*. 129–164

119. Brigati JR, Petrenko VA. 2005. Thermostability of landscape phage probes. *Analytical and Bioanalytical Chemistry*. 382(6): 1346–1350

120. Royston E, Lee SY, Culver JN, Harris MT. 2006. Characterization of silica-coated tobacco mosaic virus. *Journal of Colloid and Interface Science*. 298(2): 706–712

121. Bruckman MA, Kaur G, Lee LA, Xie F, Sepulveda J, Breitenkamp R, Wang Q. 2008. Surface modification of tobacco mosaic virus with "click" chemistry. *ChemBioChem*. 9(4): 519–523

122. Imai M, Mine K, Tomonari H, Uchiyama J, Matuzaki S, Niko Y, Watanabe S. 2019. Dark-field microscopic detection of bacteria using bacteriophage-immobilized SiO2@ AuNP core—shell nanoparticles. *Analytical Chemistry*. 91(19): 12352–12357

123. Ilhan H, Tayyarcan EK, Caglayan MG, Boyaci İH, Saglam N, Tamer U. 2021. Replacement of antibodies with bacteriophages in lateral flow assay of Salmonella Enteritidis. *Biosensors and Bioelectronics*. 113383

124. Peng H, Chen IA. 2018. Rapid colorimetric detection of bacterial species through the capture of gold nanoparticles by chimeric phages. *ACS Nano*. 13(2): 1244–1252

125. Yang MM, Wang J, Dong L, Teng Y, Liu P, Fan JJ, Yu XH. 2017. Lack of association of C3 gene with uveitis: Additional insights into the genetic profile of uveitis regarding complement pathway genes. *Scientific Reports*. 7(1): 1–8

126. Lan J, Ge J, Yu J, Shan S, Zhou H, Fan S, Wang X. 2020. Structure of the SARS-CoV-2 spike receptor-binding domain bound to the ACE2 receptor. *Nature*. 581(7807): 215–220

127. Matsuyama S, Nao N, Shirato K, Kawase M, Saito S, Takayama I, Takeda M. 2020. Enhanced isolation of SARS-CoV-2 by TMPRSS2-expressing cells. *Proceedings of the National Academy of Sciences*. 117(13): 7001–7003

128. Forster P, Forster L, Renfrew C, Forster M. 2020. Phylogenetic network analysis of SARS-CoV-2 genomes. *Proceedings of the National Academy of Sciences*. 117(17): 9241–9243

129. Nag P, Sadani K, Mukherji S. 2020. Optical fiber sensors for rapid screening of COVID-19. *Transactions of the Indian National Academy of Engineering*. 5(2): 233–236

130. Li Z, Yi Y, Luo X, Xiong N, Liu Y, Li S, Ye F. 2020. Development and clinical application of a rapid IgM-IgG combined antibody test for SARS-CoV-2 infection diagnosis. *Journal of Medical Virology*. 92(9): 1518–1524

131. Zhao G, He L, Sun S, Qiu H, Tai W, Chen J, Zhou Y. 2018. A novel nanobody targeting Middle East respiratory syndrome coronavirus (MERS-CoV) receptor-binding domain has potent cross-neutralizing activity and protective efficacy against MERS-CoV. *Journal of Virology*. 92(18): e00837–e00838

132. Maddali H, Miles CE, Kohn J, O'Carroll DM. 2021. Optical biosensors for virus detection: Prospects for SARS-CoV-2/COVID-19. *ChemBioChem*. 22(7): 1176

133. Ruuskanen O, Lahti E, Jennings LC, Murdoch DR. 2011. Viral pneumonia. *The Lancet*. 377(9773): 1264–1275

134. Bai H, Lu H, Fu X, Zhang E, Lv F, Liu L, Wang S. 2018. Supramolecular strategy based on conjugated polymers for discrimination of virus and pathogens. *Biomacromolecules*. 19(6): 2117–2122

135. World Health Organization. 2020. *Laboratory Testing for Coronavirus Disease (COVID-19) in Suspected Human Cases: Interim Guidance, 19 March 2020*. World Health Organization

136. Nguyen T, Duong BD, Wolff A. 2020. 2019 novel coronavirus disease (COVID-19): Paving the road for rapid detection and point-of-care diagnostics. *Micromachines*. 11(3): 306

137. Saylan Y, Erdem Ö, Ünal S, Denizli A. 2019. An alternative medical diagnosis method: Biosensors for virus detection. *Biosensors*. 9(2): 65

138. Ueki H, Wang IH, Zhao D, Gunzer M, Kawaoka Y. 2020. Multicolor two-photon imaging of in vivo cellular pathophysiology upon influenza virus infection using the two-photon IMPRESS. *Nature Protocols*. 15(3): 1041–1065

139. Fang Y, Zhang H, Xie J, Lin M, Ying L, Pang P, Ji W. 2020. Sensitivity of chest CT for COVID-19: Comparison to RT-PCR. *Radiology*. 296(2): e115–e117

140. Arranz A, Ripoll J. 2015. Advances in optical imaging for pharmacological studies. *Frontiers in Pharmacology*. 6: 189

141. Pashchenko O, Shelby T, Banerjee T, Santra S. 2018. A comparison of optical, electrochemical, magnetic, and colorimetric point-of-care biosensors for infectious disease diagnosis. *ACS Infectious Diseases*. 4(8): 1162–1178
142. Zhuang J, Yin J, Lv S, Wang B, Mu Y. 2020. Advanced "lab-on-a-chip" to detect viruses—current challenges and future perspectives. *Biosensors and Bioelectronics*. 163: 112291
143. Alafeef M, Dighe K, Moitra P, Pan D. 2020. Rapid, ultrasensitive, and quantitative detection of SARS-CoV-2 using antisense oligonucleotides directed electrochemical biosensor chip. *ACS Nano*. 14(12): 17028–17045
144. Maddali H, Miles CE, Kohn J, O'Carroll DM. 2021. Optical biosensors for virus detection: Prospects for SARS-CoV-2/COVID-19. *ChemBioChem*. 22(7): 1176
145. Tadimety A, Zhang Y, Kready KM, Palinski TJ, Tsongalis GJ, Zhang JX. 2019. Design of peptide nucleic acid probes on plasmonic gold nanorods for detection of circulating tumor DNA point mutations. *Biosensors and Bioelectronics*. 130: 236–244
146. Seo G, Lee G, Kim MJ, Baek SH, Choi M, Ku KB, Kim SI. 2020. Rapid detection of COVID-19 causative virus (SARS-CoV-2) in human nasopharyngeal swab specimens using field-effect transistor-based biosensor. *ACS Nano*. 14(4): 5135–5142
147. Nguyen T, Duong BD, Wolff A. 2020.2019 novel coronavirus disease (COVID-19): Paving the road for rapid detection and point-of-care diagnostics. *Micromachines*. 11(3): 306
148. Fernandez GA, Grajales GD, Ramirez JC, Lechuga LM. 2016. Last advances in silicon-based optical biosensors. *Sensors*. 16(3): 285
149. Nguyen HH, Park J, Kang S, Kim M. 2015. Surface plasmon resonance: A versatile technique for biosensor applications. *Sensors*. 15(5): 10481–10510
150. Satija J, Karunakaran B, Mukherji S. 2014. A dendrimer matrix for performance enhancement of evanescent wave absorption-based fiber-optic biosensors. *RSC Advances*. 4(31): 15841–15848
151. Sai VVR, Kundu T, Mukherji S. 2009. Novel U-bent fiber optic probe for localized surface plasmon resonance based biosensor. *Biosensors and Bioelectronics*. 24(9): 2804–2809
152. Ramakrishna B, Sai VVR. 2016. Evanescent wave absorbance based U-bent fiber probe for immunobiosensor with gold nanoparticle labels. *Sensors and Actuators B: Chemical*. 226: 184–190
153. Hassan SS, Panhwar S, Nafady A, Al-Enizi AM, Sherazi STH, Kalhoro MS, Talpur MY. 2017. Fabrication of highly sensitive and selective electrochemical sensors for detection of paracetamol by using piroxicam stabilized gold nanoparticles. *Journal of the Electrochemical Society*. 164(9): B427
154. Murugan D, Bhatia H, Sai VVR, Satija J. 2020. P-FAB: A fiber-optic biosensor device for rapid detection of COVID-19. *Transactions of the Indian National Academy of Engineering*. 5(2): 211–215

4 Mass-Sensitive Based Biosensors

Erdoğan Ozgür[1], Yeşeren Saylan[2],
Semra Akgönüllü[2], and Adil Denizli[2]
[1]Hacettepe University
Advanced Technologies Application and Research Center
Ankara, Turkey
[2]Hacettepe University
Department of Chemistry
Ankara, Turkey

CONTENTS

4.1   INTRODUCTION

The term "biosensing" refers to the detection of biomolecules utilizing an analytical tool such as a biosensor integrating a sensing part (bioreceptor) combined with a suitable physicochemical transducer that converts the binding of an analyte to its bioreceptor into a quantifiable output signal (1–4). Scientists across many disciplines (e.g. material science, biotechnology, chemistry, electronics, physics, biology, and nanotechnology) struggle to improve biosensing technologies through the development of new binding approaches, new materials, and instrumentation (5–6). The improvements in biosensing could be attended in two fundamental ways such as the transduction and biorecognition mechanisms. Biosensors widely used in the fields of clinical diagnosis, environmental monitoring, and food safety (7–11), can be categorized as electrochemical, optical, thermal, acoustic, mechanical, magnetic, etc. depending on the signal transduction mechanism (Figure 4.1) (3, 12–13).

Among them, acoustic wave biosensors have been widely utilized for the detection of mass, viscosity, conductivity, and density (1). Acoustic wave biosensors are mass-sensitive sensors that exploit both piezoelectric and inverse piezoelectric effects. Their transduction mechanism relies on the interconversion and acoustic waves and electrical energies detection (14–16). The piezoelectricity phenomenon was discovered by Jacques and Pierre Curie and has been employed to support the

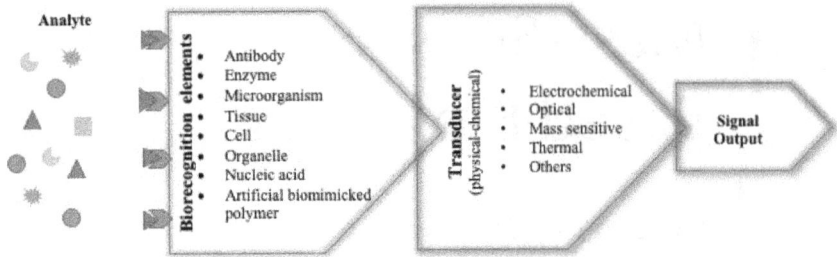

FIGURE 4.1 Schematic illustration of a biosensor assembly.

spread of transverse acoustic waves through a bulk of crystalline (piezoelectric) i.e. quartz crystal, lithium tantalite (LiNbO$_3$), and lithium niobate (LiTaO$_3$). The piezo-electric effect has been used most broadly both to receive and to generate acoustic waves. The piezoelectric effect is defined as the generation of an electrical charge at the opposite sides of the piezoelectric materials by applying mechanical stress (pull-ing, pressure, or torsion), which causes a deforming in the crystal lattice of materials. Besides, the piezoelectric effect is a reversible process. The converse piezoelectric effect also occurs by applying an alternative voltage to the piezoelectric materials (17–20). Acoustic wave biosensors can be classified into two categories according to their acoustic wave propagation modes: Surface acoustic wave (SAW) biosensors and bulk acoustic wave (BAW) biosensors. The acoustic wave spreads as guided and/or unguided throughout the only surface of the substrate in SAW biosensors, and in BAW biosensors the acoustic wave spreads without directing a piezoelectric substrate volume (12, 21). Also, acoustic waves can be divided into the polarization and direc-tion of propagation of the vibrating particles or the displacement of particles relative to the piezoelectric substrate surface (horizontal or vertical) (1). BAW biosensors contain longitudinal BAW (L-BAW), shear BAW (S-BAW), thickness-shear mode-quartz crystal microbalance (TSM-QCM), thickness extensional mode (TEM), film bulk acoustic wave (FBAR), lateral field excitation (LFE), Lamb waves, and acoustic plate wave (APW). SAW biosensors also contain Rayleigh SAW (R-SAW), Sezawa mode waves, shear-horizontal SAWs (SH-SAWs), Love wave, pseudo-surface acous-tic waves (PSAW), Leaky SAW (LSAW), or Lamb waves (Figure 4.2) (22–31).

The influences on the wave propagation or perturbations in crystal lattices of piezoelectric materials benefit to follow the binding events by sensing the velocity and/or energy of acoustic waves in any type of acoustic wave biosensors (32–33). Acoustic wave mass-sensing techniques respond to any changes in the frequency (f) or phase (Ph) change-related directly with the interfacial mass amount accumulated on the surface of the resonator, while dissipation (D) or amplitude (A) is correlated with the acoustic wave energy (33–36). The viscosity of viscoelastic changes occur-ring at the biosensor/liquid interface are associated with the dissipation/amplitude changes (37). The plot of shifts in bandwidth/-the resonance frequency (ΔΓ/-Δf) ratio vs the shift of frequency are used to characterize the adsorbed particle size (38–39). The factors of resonance frequency (f), energy dissipation (D), and bandwidth (Γ) are related via equation $D = 2\Gamma/f_n$ where n represents overtones (40). Besides, con-formation changes of biomolecules can be determined by using the findings of the

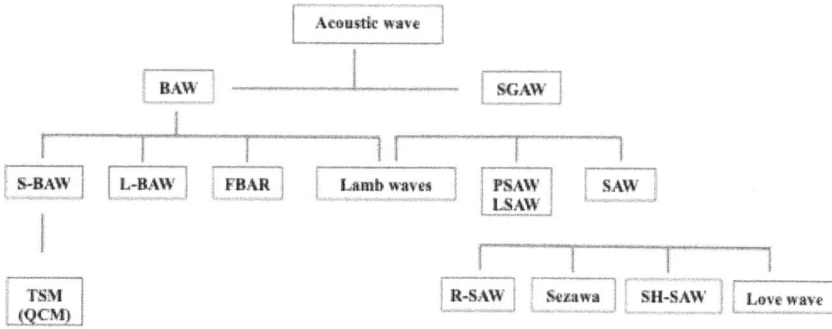

FIGURE 4.2 Classification of different wave modes.

viscoelastic properties of the adsorbed biomolecules layer (41–42). Various biological recognition elements (e.g., enzymes) (43), DNA and RNA probes (44), antibodies (45), aptamers (46), cell components (47), and protein/peptides (48), artificial polymeric layers/structures mimicking their natural analogues (49–51), nanomaterials such as nanoparticles (52), nanotubes (53), nanofibers (54), nanorods (55), nanowires (56) etc., can be employed as recognition unit of acoustic wave sensors. Hereby, the diversity of receptors enables the determination of a broad range of molecules such as nucleic acids, drugs, food additives, environmental toxins, proteins, and cells, etc. in both liquid and gaseous phases (32).

4.2 ACOUSTIC WAVE BIOSENSORS

4.2.1 Bulk Acoustic Wave Biosensors

Quartz crystal microbalances (QCM) emerges as bulk thickness-shear-mode acoustic wave mass-sensing BAW devices respond to any changes in the resonant frequency (mostly between 5–20 MHz) (57–60) associated with the interfacial mass accumulated on the surface of a piezoelectric material, frequently a quartz crystal (61). As aforementioned, the mode of oscillation relies on the angle of the crystal cut (generally, AT-cut (a 35° 10' angle from the Z-axis) crystals), thickness, and surface shape of the resonator. AT-cut quartz crystals have high-frequency stability of $\Delta f/f \sim 10^{-8}$ between 0–50 °C, so used in many electronic devices (62).

In 1959, Sauerbray reported as the shift of frequency of an oscillating crystal resonator has a linear relationship with mass accumulated on the surface of the crystal resonator (35), so measured the weight of thin film using the following equation:

$$\Delta f_m = -\left(f_o^2 / F_q \rho_q \right) m_s = -\left(f_o^2 / F_q \rho_q \right) \left(\Delta m_s / A_{el} \right) \tag{4.1}$$

where f_0 is basic resonance frequency of quartz resonator, F_q frequency constant of the crystal ($F_q = f_0 \cdot d_q$), d_q thickness, ρ_q mass density, m_s mass per area (absolute mass, Δm), and A_{el} electrode area of crystal resonator. However, **Equation 4.1** is applied to the quartz-air interface, and solid layers accumulated on the quartz resonator (62). In the 1980s, the QCM technique could be applied in the liquid phase by including mass shift related to liquid density-viscosity changes. QCM is also sensitive to viscosity and density changes as well as biomolecules (e.g. protein, enzyme, nucleic acid, and microorganism) detection as depicted in **Equation 4.2**.

$$\Delta f = - f_o^{3/2} \left(\rho_l \eta_l / \pi \rho_q \mu_q \right)^{1/2} \qquad (4.2)$$

where ρ_l is the density of the liquid, η_l the viscosity of the liquid, ρ_q the quartz density and μ_q the shear modulus (63). The total frequency shifts can be determined in two-layer systems (rigid mass and viscous liquid layers) by using **Equation 4.3**.

$$\Delta f = \Delta f_m + \Delta f_l = - f_o^2 \left(\left(\Delta m_s / F_q \rho_q A_{el} \right) + \left(\rho_l \eta_l / f_o \pi \rho_q \mu_q \right)^{1/2} \right) \qquad (4.3)$$

QCM based biosensors have emerged as efficient tools to detect and quantify disease-related biomarkers related to various types of cancers, chronic and infectious diseases (1, 13, 64–66). Atay et al. improved QCM biosensor for human breast cancer cells (MDA-MB-231) detection. Poly(2-hydroxyethyl methacrylate) nanoparticles were accumulated on a commercial QCM resonator functionalized transferrin attachment via carbodiimide activation (67). A highly sensitive and reusable chitosan-based QCM biosensor was generated to recognize MCF-7 cells over-expressed folic acid receptors using chitosan functionalized with folic acid. The MCF-7 cell was determined online in a broad range (4.5×10^2–1.01×10^5 cells/mL). The limit of detection was found to be 430 cells/mL (68). An aptamer-modified QCM biosensor was designed for the detection of leukemia cells via using silver-enhanced gold nanoparticles modified aminophenyl boronic acid. Hereby, leukemia cells were recognized by the QCM resonator surface in

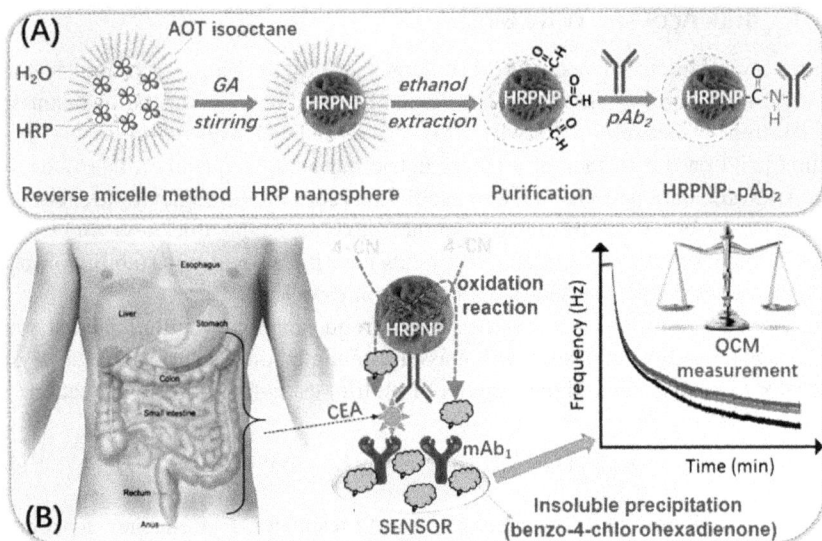

FIGURE 4.3 Scheme of QCM biosensor for CEA detection: (A) Nanospheres preparation with the reverse micelle method and polyclonal anti-CEA detection and (B) monoclonal anti-CEA detection with QCM biosensor. Reprinted with permission from Reference (70).

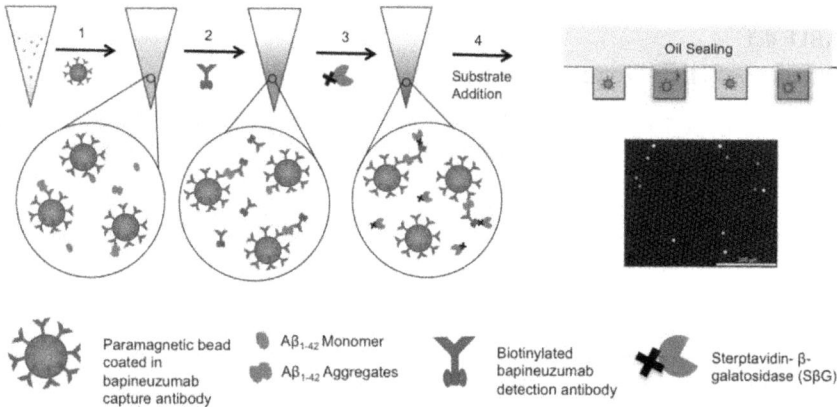

FIGURE 4.4 Scheme of the Aβ1–42 analysis. Reprinted with permission from Reference (71). In addition, numerous QCM-based biosensors were proposed to detect various biomarkers (e.g. immunoglobulin E (72), a-fetoprotein (73), cancer antigen 15.3 (74), parathyroid hormone-related peptide (75), anti-cyclic citrullinated peptide (76), human serum albumin (77), insulin (78), cortisol (79), urinary proteins (diabetes nephropathy) (80), cardiac troponin T (81) for clinical diagnosis. QCM-based biosensors have been reported also for infectious diseases (e.g. influenza, hepatitis B, dengue, malaria, human immunodeficiency virus infection, and tuberculosis) (66) using different biorecognition elements such as DNA probe (82), peptide nucleic acid probe (83), polyclonal or monoclonal antibodies (84–85), DNA aptamer (86), artificial imprinted polymeric structure (87), etc. High sensitivity on the surface accumulated weight in nanogram per cm^2 provides low detection limits like ELISA (88–89), hereby, the combination of QCM with specific biorecognition elements makes QCM an important promising candidate to be used in the early clinical diagnosis of diseases and also treatment stages for diseases.

the 2×10^3–1×10^5 cells/mL range with an 1160 cells/mL limit of detection (69). A feasible QCM biosensor was fabricated to detect low-abundance carcinoembryonic antigen (CEA) using horseradish peroxidase (HRP) nanospheres functionalized with polyclonal anti-CEA (Figure 4.3). Enzyme-labelled QCM based biosensors showed long storage stability and high specificity with a low limit of detection value (7.8 pg/mL) (70).

Monoclonal antibody (bapineuzumab)-based QCM biosensor through the detection of amyloid β peptides (Aβ) accumulate in a brain was also proposed for primary diagnosis of Alzheimer's (71). As shown in Figure 4.4, bapineuzumab was the first bonded to carboxylated beads. Antibody-coated beads interacted with Aβ1–42 molecules as shown in reaction 1, then biotinylated antibodies were incubated to modify the captured molecules in reaction 2, and complexes were tagged with streptavidin-β-galactosidase in reaction 3. At the end of the experiment, the beads and enzyme-substrate are loaded into wells (reaction 4) and arrays are imaged for detection in the wells.

FBARs known as thin-film bulk acoustic resonators are another type of BAW sensors that share a related structure and working idea as QCMs, though, their operating resonance frequency (in the GHz range) (Table 4.1) is much higher due to the reduced size and thickness of the piezoelectric layer (most commonly of aluminium nitride (AlN) or zinc oxide (ZnO) films) sandwiched between two conducting layers

TABLE 4.1
Overview of High-Frequency Acoustic Biosensors

	Quartz Crystal Microbalance (QCM)	Surface Acoustic Wave (SAW)	Film Bulk Acoustic Resonator (FBAR)
Acoustic Wave Type	Bulk	Surface	Bulk
Operating Frequency	5–200 MHz	100–1500 MHz	1.0–4.5 GHz
Common Piezoelectric Materials	Quartz	Quartz, $LiNbO_3$, $LiTaO_3$	AlN, ZnO
Working Medium	Liquids and gases	Gases	Liquids and gases
Sensitivity	Low	Intermediate	High

(90–94). The FBAR biosensors have been broadly utilized to detect disease-related biomarkers. They have been used to detect immunoglobulin E (95), acute myocardial infarction biomarker cardiac troponin I (96), heart-type fatty acid-binding protein (97), tumor biomarker MUC1 (98), carcinoembryonic antigen (99), human prostate-specific antigen (100), and alpha-fetoprotein (101).

4.2.2 SURFACE ACOUSTIC WAVE (SAW) BIOSENSORS

SAW biosensors generally utilize interdigital transducers (IDTs) to produce and sense acoustic waves on the piezoelectric substrate surface. Thus, the acoustic energy is intensely localized at the biosensor surface in the variety of the acoustic wavelength, independent of the whole substrate thickness (102–104). In particular, SAW devices allowing tunable operation frequencies from MHz to GHz have been used in various applications as resonators and filters in electronics and also communications in biosensors (105–110). Higher frequencies compared to QCM enable to detection of any variations on the surface much sensitive including accumulation, conductivity, and viscosity. The first SAW biosensors were developed based on Rayleigh-type SAWs (the most general mode used for gas sensing) in 1979 for organic gas detection (111) and this attempt was successfully configured to operate in contact with liquids by employing horizontal shear SAW (SH-SAW). SH-SAW biosensors are mainly used for the detection of liquids because of their ability to provide real-time detection. Nowadays, SAW biosensors are used to detect disease-related biomarkers related to various types of cancers and chronic and infectious diseases (Table 4.2).

As an example of SAW biosensor, Jandas et al. prepared an LW-SAW biosensor for carcinoembryonic antigen (CEA) detection (52). As demonstrated in Figure 4.5, a microfluidic chamber was prepared with three channels inlet system, contact points to the SAW tool, and analyzer. They performed CEA solutions of different concentrations (0.1–80 ng/mL) to explore the sensing capability of the biosensor and validated the feasibility of the biosensor employing clinical serum samples.

As another example, researchers fabricated and investigated the sensing features of LW-SAW biosensors for ammonia and ethylene detection using gold nanoparticle-modified and non-modified polypyrrole (120). The responses of the LW-SAW biosensors to different amounts (2, 5, and 10 ppm) of ethylene and ammonia showed that the change of frequency depended on the proportional increase to the gas

TABLE 4.2

Examples of SAW Biosensors

Target	Biosensor Type	Limit of Detection	Ref.
Antihemoglobin and antimyoglobin antibodies	SAW	0.32 mg/mL and 0.35 mg/mL	112
Interleukin-6	LW-SAW	-	113
Coxsakie B4 and Sin Nombre Virus	SH-SAW	-	114
Hepatitis B virus	LW-SAW	10 pg/mL	115
HIV-1 and HIV-2	SAW	12 TCID50s for HIV-1 and 87 TCID50s for HIV-2	116
Ebola Zaire virus	SAW	1.9×10^4 PFU/mL	117
Legionella and Escherichia coli	SAW	down to 10^6 and 10^5 cells/mL for Legionella and Escherichia coli	118
Colorectal cancer	LW-SAW	0.084 ng/mL	52
Uric acid	LW-SAW	5 mM	119
Ammonia and ethylene	LW-SAW	0.067 and 0.087 ppm	120

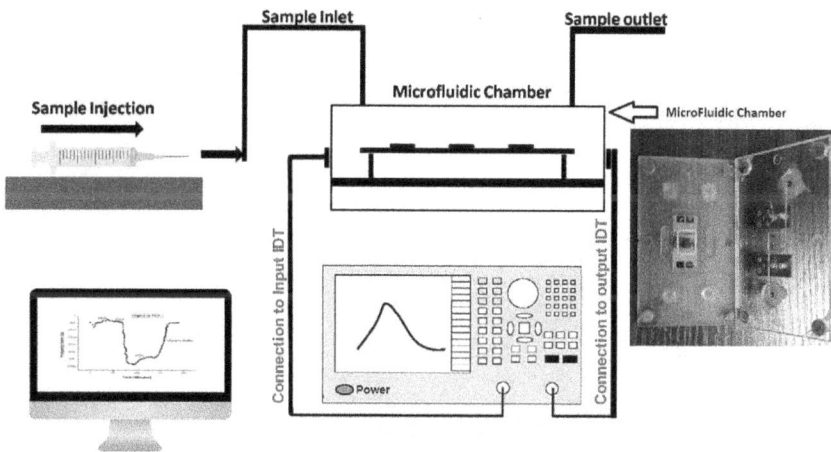

FIGURE 4.5 Biosensing set-up for CEA detection. Reprinted with permission from Reference (52).

concentration. The decline of frequency is likely due to the predominance of mass loading effects. Furthermore, the time-dependent responses of LW-SAW biosensors to the same amount (5 ppm) of ammonia and ethylene depicted that the response and the time required to reach 90% of the recovery. Thus, the gold-modified and non-modified LW-SAW-based biosensors had similar recovery times and responses for the same analyte and also concentration. It has also been reported that response and recovery times to ethylene are longer than that of ammonia (Figure 4.6).

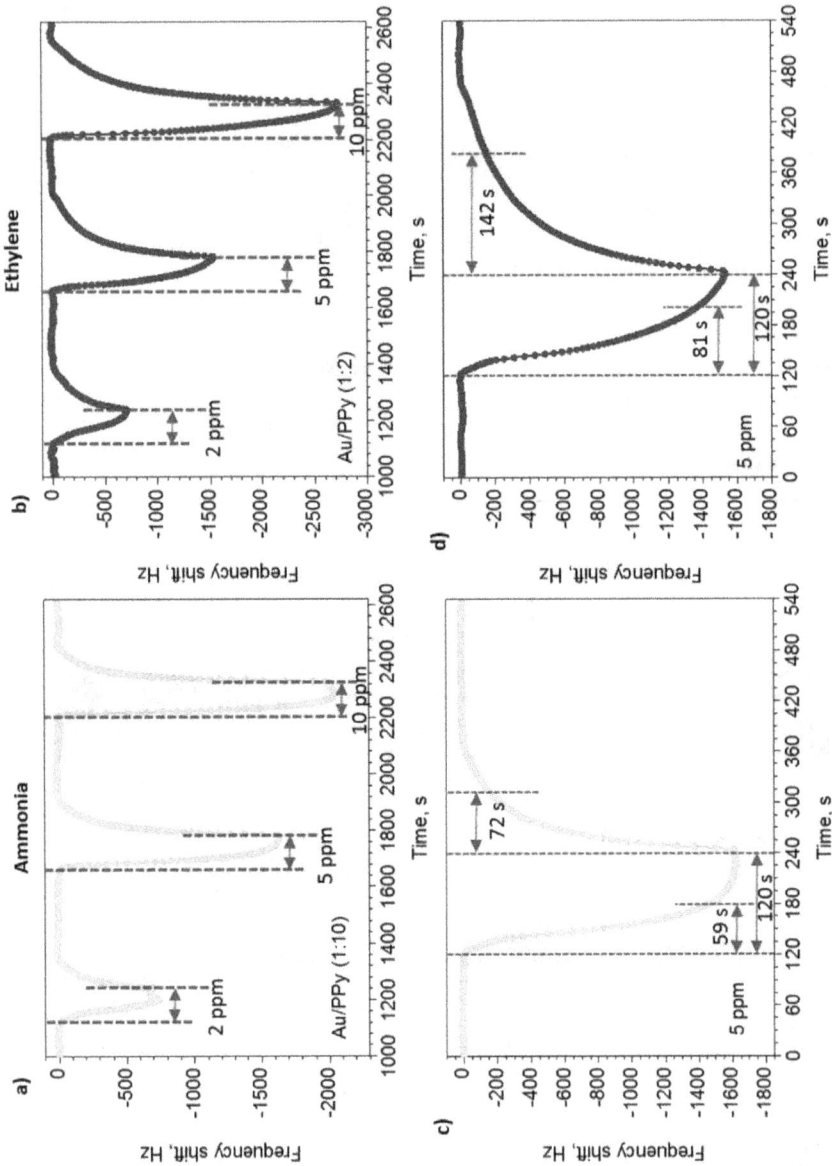

FIGURE 4.6 (a, b) Responses of LW-SAW biosensors to different amounts of ammonia and ethylene and (c, d) time-dependent responses of LW-SAW biosensors to the same amount of ammonia and ethylene. Reprinted with permission from Reference (120).

4.3 CONCLUSION AND FUTURE REMARKS

Mass-based biosensors are phenomenon platforms for monitoring of interactions and the analysis can be designed as label-free which means that interaction between biosensor chip metal surface and the target analyte is responded directly without application of any specific reagent or sample processing (121). Quartz is the specific material having piezoelectric features that means talent to create potency when mechanically squeezed or mechanically deformed when oriented dipole is given on crystal surface. These biosensors are deeply mass-sensitive and can monitor changes of mass at the nanogram level. QCM has appeared as the most common sensor platform in the quartz form of biosensors tools. The mass-sensitive based biosensors are highly sensitive to the mass change on the crystal biosensor surface; their detection for a target has pretty perfect excellence with designed special nanomaterials (122).

The spread of the pandemic has had a huge impact on economies and societies and has changed the lives of people around the world (123). The development of smart biosensors that can further increase detection sensitivity in monitoring currently known and future threats is an extremely important step (124). Mass sensitive-base biosensors are carried out for the quick detection and real-time monitoring of diverse molecules that are significant for the medical, security, food, and environment fields. As a result, the development of portable and reliable biosensors capable of simultaneously and continuously monitoring against analytes has become an important area of research. For this reason, mass sensitive-based biosensors will become tools on which more research will be done in this field.

REFERENCES

1. Zhang J, Zhang X, Wei X, Xue Y, Wan H, Wang P. 2021. Recent advances in acoustic wave biosensors for the detection of disease-related biomarkers: A review. *Analytica Chimica Acta*. 338321
2. Zhou Q, Tang D. 2020. Recent advances in photoelectrochemical biosensors for analysis of mycotoxins in food. *TrAC Trends in Analytical Chemistry*. 124: 115814
3. Lange K. 2019. Bulk and surface acoustic wave sensor arrays for multi-analyte detection: A review. *Sensors*. 19(24): 5382
4. Turner AP. 2013. Biosensors: Sense and sensibility. *Chemical Society Reviews*. 42(8): 3184–3196
5. Andryukov BG, Besednova NN, Romashko RV, Zaporozhets TS, Efimov TA. 2020. Label-free biosensors for laboratory-based diagnostics of infections: Current achievements and new trends. *Biosensors*. 10(2): 11
6. Yoo H, Jo H, Oh, SS. 2020. Detection and beyond: Challenges and advances in aptamer-based biosensors. *Materials Advances*. 1(8): 2663–2687
7. Saylan Y, Erdem Ö, Cihangir N, Denizli A. 2019. Detecting fingerprints of waterborne bacteria on a sensor. *Chemosensors*. 7(3): 33
8. Ren R, Cai G, Yu Z, Tang D. 2018. Glucose-loaded liposomes for amplified colorimetric immunoassay of streptomycin based on enzyme-induced iron (II) chelation reaction with phenanthroline. *Sensors and Actuators B: Chemical*. 265: 174–181
9. Akgönüllü S, Yavuz H, Denizli A. 2020. SPR nanosensor based on molecularly imprinted polymer film with gold nanoparticles for sensitive detection of aflatoxin B1. *Talanta*. 219: 121219

10. Yang G, Xiao Z, Tang C, Deng Y, Huang H, He Z. 2019. Recent advances in biosensor for detection of lung cancer biomarkers. *Biosensors and Bioelectronics*. 141: 111416
11. Dibekkaya H, Saylan Y, Yılmaz F, Derazshamshir A, Denizli A. 2016. Surface plasmon resonance sensors for real-time detection of cyclic citrullinated peptide antibodies. *Journal of Macromolecular Science, Part A*. 53(9): 585–594
12. Rocha GMI, March IC, Montoya BA, Arnau VA. 2009. Surface generated acoustic wave biosensors for the detection of pathogens: A review. *Sensors*. 9(7): 5740–5769
13. Lu X, Cui M, Yi Q. 2020. Detection of mutant genes with different types of biosensor methods. *TrAC Trends in Analytical Chemistry*. 126: 115860
14. Oprea A, Weimar U. 2019. Gas sensors based on mass-sensitive transducers part 1: Transducers and receptors—basic understanding. *Analytical and Bioanalytical Chemistry*. 411(9): 1761–1787
15. Nazemi H, Joseph A, Park J, Emadi A. 2019. Advanced micro-and nano-gas sensor technology: A review. *Sensors*. 19(6): 1285
16. Ferreira GN, Da SAC, Tome B. 2009. Acoustic wave biosensors: Physical models and biological applications of quartz crystal microbalance. *Trends in Biotechnology*. 27(12): 689–697
17. Janshoff A, Galla HJ, Steinem C. 2000. Piezoelectric mass-sensing devices as biosensors—an alternative to optical biosensors *Angewandte Chemie International Edition*. 39(22): 4004–4032
18. Tombelli S. 2012. Piezoelectric biosensors for medical applications. In *Biosensors for Medical Applications*, pp. 41–64. Woodhead Publishing
19. Curie J, Curie P. 1880. An oscillating quartz crystal mass detector. *Rendu*. 91(5): 294–297
20. Cavic BA, Thompson M, Hayward GL. 1999. Acoustic waves and the study of biochemical macromolecules and cells at the sensor—liquid interface. *Analyst*. 124(10): 1405–1420
21. Fu YQ, Luo JK, Nguyen NT, Walton AJ, Flewitt AJ, Zu XT, Milne WI. 2017. Advances in piezoelectric thin films for acoustic biosensors, acoustofluidics and lab-on-chip applications. *Progress in Materials Science*. 89: 31–91
22. Galipeau DW, Story PR, Vetelino KA, Mileham RD. 1997. Surface acoustic wave microsensors and applications. *Smart Materials and Structures*. 6(6): 658
23. Lakin KM. 2003. Thin film resonator technology. In *IEEE International Frequency Control Symposium and PDA Exhibition Jointly with the 17th European Frequency and Time Forum, 2003. Proceedings of the 2003*, pp. 765–778. IEEE
24. Campanella H, Esteve J, Montserrat J, Uranga A, Abadal G, Barniol N, Romano RA. 2006. Localized and distributed mass detectors with high sensitivity based on thin-film bulk acoustic resonators. *Applied Physics Letters*. 89(3): 033507
25. Du J, Harding GL, Collings AF, Dencher PR. 1997. An experimental study of love-wave acoustic sensors operating in liquids. *Sensors and Actuators A: Physical*. 60(1–3): 54–61
26. Ballantine JDS, White RM, Martin SJ, Ricco AJ, Zellers ET, Frye GC, Wohltjen H. 1996. *Acoustic Wave Sensors: Theory, Design and Physico-Chemical Applications*. Elsevier
27. Campbell C. 1998. *Surface Acoustic Wave Devices for Mobile and Wireless Communications, Four-Volume Set*. Academic Press
28. Gizeli E, Stevenson AC, Goddard NJ, Lowe CR. 1992. A novel Love-plate acoustic sensor utilizing polymer overlayers. *IEEE Transactions on Ultrasonics, Ferroelectrics, and Frequency Control*. 39(5): 657–659
29. Kovacs G, Venema A. 1992. Theoretical comparison of sensitivities of acoustic shear wave modes for (bio) chemical sensing in liquids. *Applied Physics Letters*. 61(6): 639–641

30. Martin SJ, Ricco AJ, Niemczyk TM, Frye GC. 1989. Characterization of SH acoustic plate mode liquid sensors. *Sensors and Actuators*. 20(3): 253–268

31. Rocha GMI, March IC, Montoya BA, Arnau VA. 2009. Surface generated acoustic wave biosensors for the detection of pathogens: A review. *Sensors*. 9(7): 5740–5769

32. Mujahid A, Afzal A, Dickert FL. 2019. An overview of high frequency acoustic sensors—QCMs, SAWs and FBARs—chemical and biochemical applications. *Sensors*. 19(20): 4395

33. Grammoustianou A, Gizeli E. 2018. Acoustic wave—based immunoassays. In *Handbook of Immunoassay Technologies*, pp. 203–239. Academic Press

34. Marx KA. 2003. Quartz crystal microbalance: A useful tool for studying thin polymer films and complex biomolecular systems at the solution– surface interface. *Biomacromolecules*. 4(5): 1099–1120

35. Sauerbrey G. 1959. Verwendung von Schwingquarzen zur Wägung dünner Schichten und zur Mikrowägung. *Zeitschrift für physik*. 155(2): 206–222

36. Sohna JES, Cooper MA. 2016. Does the Sauerbrey equation hold true for binding of peptides and globular proteins to a QCM: A systematic study of mass dependence of peptide and protein binding with a piezoelectric sensor. *Sensing and Bio-Sensing Research*. 11: 71–77

37. Ballantine DS, White RM, Martin SJ, Ricco AJ, Zellers ET, Frye GC, Wohltjen H. 1996. *Acoustic Wave Sensors: Theory, Design and Physico-Chemical Applications*. Elsevier

38. Tellechea E, Johannsmann D, Steinmetz NF, Richter RP, Reviakine I. 2009. Model-independent analysis of QCM data on colloidal particle adsorption. *Langmuir*. 25(9): 5177–5184

39. Olsson AL, Quevedo IR, He D, Basnet M, Tufenkji N. 2013. Using the quartz crystal microbalance with dissipation monitoring to evaluate the size of nanoparticles deposited on surfaces. *ACS Nano*. 7(9): 7833–7843

40. Reviakine I, Johannsmann D, Richter RP. 2011. Hearing what you cannot see and visualizing what you hear: Interpreting quartz crystal microbalance data from solvated interfaces. *Analytical Chemistry*. 8838–8848

41. Molino PJ, Higgins MJ, Innis PC, Kapsa RM, Wallace GG. 2012. Fibronectin and bovine serum albumin adsorption and conformational dynamics on inherently conducting polymers: A QCM-D study. *Langmuir*. 28(22): 8433–8445

42. Kushiro K, Lee CH, Takai M. 2016. Simultaneous characterization of protein—material and cell—protein interactions using dynamic QCM-D analysis on SAM surfaces. *Biomaterials Science*. 4(6): 989–997

43. Martin SP, Lamb DJ, Lynch JM, Reddy SM. 2003. Enzyme-based determination of cholesterol using the quartz crystal acoustic wave sensor. *Analytica Chimica Acta*. 487(1): 91–100

44. He F, Zhao J, Zhang L, Su X. 2003. A rapid method for determining Mycobacterium tuberculosis based on a bulk acoustic wave impedance biosensor. *Talanta*. 59(5): 935–941

45. Meszaros G, Akbarzadeh S, DeLa FB, Keresztes Z, Thompson M. 2021. Advances in electromagnetic piezoelectric acoustic sensor technology for biosensor-based detection. *Chemosensors*. 9(3): 58

46. Chang K, Pi Y, Lu W, Wang F, Pan F, Li F, Chen M. 2014. Label-free and high-sensitive detection of human breast cancer cells by aptamer-based leaky surface acoustic wave biosensor array. *Biosensors and Bioelectronics*. 60: 318–324

47. Deng Y, Zheng H, Yi X, Shao C, Xiang B, Wang S, Hui G. 2019. Paralytic shellfish poisoning toxin detection based on cell-based sensor and non-linear signal processing model. *International Journal of Food Properties*. 22(1): 890–897

48. Drouvalakis KA, Bangsaruntip S, Hueber W, Kozar LG, Utz PJ, Dai H. 2008. Peptide-coated nanotube-based biosensor for the detection of disease-specific autoantibodies in human serum. *Biosensors and Bioelectronics*. 23: 1413–1421

49. Sener G, Ozgur E, Yılmaz E, Uzun L, Say R, Denizli A. 2010. Quartz crystal micro-balance based nanosensor for lysozyme detection with lysozyme imprinted nanoparticles. *Biosensors and Bioelectronics*. 26(2): 815–821

50. Latif U, Mujahid A, Afzal A, Sikorski R, Lieberzeit PA, Dickert FL. 2011. Dual and tetraelectrode QCMs using imprinted polymers as receptors for ions and neutral analytes. *Analytical and Bioanalytical Chemistry*. 400(8): 2507–2515

51. Kidakova A, Boroznjak R, Reut J, Öpik A, Saarma M, Syritski V. 2020. Molecularly imprinted polymer-based SAW sensor for label-free detection of cerebral dopamine neurotrophic factor protein. *Sensors and Actuators B: Chemical*. 308: 127708

52. Jandas PJ, Luo J, Prabakaran K, Chen F, Fu YQ. 2020. Highly stable, love-mode surface acoustic wave biosensor using Au nanoparticle-MoS2-rGO nano-cluster doped polyimide nanocomposite for the selective detection of carcinoembryonic antigen. *Materials Chemistry and Physics*. 246: 122800

53. Zhang Y, Yu K, Xu R, Jiang D, Luo L, Zhu Z. 2005. Quartz crystal microbalance coated with carbon nanotube films used as humidity sensor. *Sensors and Actuators A: Physical*. 120(1): 142–146

54. Rianjanu A, Roto R, Julian T, Hidayat SN, Kusumaatmaja A, Suyono EA, Triyana K. 2018. Polyacrylonitrile nanofiber-based quartz crystal microbalance for sensitive detection of safrole. *Sensors*. 18(4): 1150

55. Li W, Guo YJ, Tang QB, Zu XT, Ma JY, Wang L, Fu YQ. 2019. Highly sensitive ultraviolet sensor based on ZnO nanorod film deposited on ST-cut quartz surface acoustic wave devices. *Surface and Coatings Technology*. 363: 419–425

56. Wu J, Yin C, Zhou J, Li H, Liu Y, Shen Y, Duan, H. 2020. Ultrathin glass-based flexible, transparent, and ultrasensitive surface acoustic wave humidity sensor with ZnO nanowires and graphene quantum dots. *ACS Applied Materials & Interfaces*. 12(35): 39817–39825

57. Montagut Y, García JV, Jimenez Y, March C, Montoya A, Arnau A. 2011. Validation of a phase-mass characterization concept and interface for acoustic biosensors. *Sensors*. 11(5): 4702–4720

58. Fernandez R, García P, García M, García JV, Jiménez Y, Arnau A. (2017). Design and validation of a 150 MHz HFFQCM sensor for bio-sensing applications. *Sensors*. 17(9): 2057

59. He H, Zhou L, Wang Y, Li C, Yao J, Zhang W, Dong WF.2015. Detection of trace micro-cystin-LR on a 20 MHz QCM sensor coated with in situ self-assembled MIPs. *Talanta*. 131: 8–13

60. Özgür E, Parlak O, Beni V, Turner AP, Uzun L. 2017. Bioinspired design of a poly-mer-based biohybrid sensor interface. *Sensors and Actuators B: Chemical*. 251: 674–682

61. Marx KA. 2003. Quartz crystal microbalance: A useful tool for studying thin polymer films and complex biomolecular systems at the solution– surface interface. *Biomacro-molecules*. 4(5): 1099–1120

62. Uludağ Y, Piletsky SA, Turner AP, Cooper MA. 2007. Piezoelectric sensors based on molecular imprinted polymers for detection of low molecular mass analytes. *The FEBS Journal*. 274(21): 5471–5480

63. Kanazawa KK, Gordon JG. 1985. The oscillation frequency of a quartz resonator in contact with liquid. *Analytica Chimica Acta*. 175: 99–105

64. Kelley SO. 2017. What are clinically relevant levels of cellular and biomolecular analytes. *ACS Sensors*. 2(2): 193–197

65. Rossetti C, Switnicka PMA, Halvorsen TG, Cormack PA, Sellergren B, Reubsaet L. 2017. Automated protein biomarker analysis: On-line extraction of clinical samples by molecularly imprinted polymers. *Scientific Reports*. 7(1): 1–11

66. Lim HJ, Saha T, Tey BT, Tan WS, Ooi CW. 2020. Quartz crystal microbalance-based biosensors as rapid diagnostic devices for infectious diseases. *Biosensors and Bioelectronics*. 112513

67. Atay S, Pişkin K, Yılmaz F, Çakır C, Yavuz H, Denizli A. 2016. Quartz crystal microbalance based biosensors for detecting highly metastatic breast cancer cells via their transferrin receptors. *Analytical Methods*. 8(1): 153–161

68. Zhang S, Bai H, Luo J, Yang P, Cai J. 2014. A recyclable chitosan-based QCM biosensor for sensitive and selective detection of breast cancer cells in real time. *Analyst*. 139(23): 6259–6265

69. Shan W, Pan Y, Fang H, Guo M, Nie Z, Huang Y, Yao S. 2014. An aptamer-based quartz crystal microbalance biosensor for sensitive and selective detection of leukemia cells using silver-enhanced gold nanoparticle label. *Talanta*. 126: 130–135

70. Chi L, Xu C, Li S, Wang X, Tang D, Xue F. 2020. In situ amplified QCM immunoassay for carcinoembryonic antigen with colorectal cancer using horseradish peroxidase nanospheres and enzymatic biocatalytic precipitation. *Analyst*. 145: 6111–6118

71. Hwang SS, Chan H, Sorci M, Van Deventer J, Wittrup D, Belfort G, Walt D. 2019. Detection of amyloid β oligomers toward early diagnosis of Alzheimer's disease. *Analytical Biochemistry*. 566: 40–45

72. Yao C, Qi Y, Zhao Y, Xiang Y, Chen Q, Fu W. 2009. Aptamer-based piezoelectric quartz crystal microbalance biosensor array for the quantification of IgE. *Biosensors and Bioelectronics*. 24: 2499–2503

73. Chou, SF, Hsu WL, Hwang JM, Chen CY. 2002. Determination of α-fetoprotein in human serum by a quartz crystal microbalance-based immunosensor. *Clinical Chemistry*. 48: 913–918

74. Wang X, Yu H, Lu D, Zhang J, Deng W. 2014. Label free detection of the breast cancer biomarker CA15.3 using ZnO nanorods coated quartz crystal microbalance. *Sensors and Actuators B: Chemical*. 195: 630–634

75. Crivianu GV, Aamer M, Posaratnanathan RT, Romaschin A, Thompson M. 2016. Acoustic wave biosensor for the detection of the breast and prostate cancer metastasis biomarker protein PTHrP. *Biosensors and Bioelectronics*. 78: 92–99

76. Drouvalakis KA, Bangsaruntip S, Hueber W, Kozar LG, Utz PJ, Dai H. 2008. Peptide-coated nanotube-based biosensor for the detection of disease-specific autoantibodies in human serum. *Biosensors and Bioelectronics*. 23: 1413–1421

77. Sakti SP, Hauptmann P, Zimmermann B, Bühling F, Ansorge S. 2001. Disposable HSA QCM-immunosensor for practical measurement in liquid. *Sensors and Actuators B: Chemical*. 78: 257–262

78. Saha S, Raje M, Suri C. R. 2002. Sandwich microgravimetric immunoassay: Sensitive and specific detection of low molecular weight analytes using piezoelectric quartz crystal. *Biotechnology Letters*. 24: 711–716

79. Ito T, Aoki N, Kaneko S, Suzuki K. 2014. Highly sensitive and rapid sequential cortisol detection using twin sensor QCM. *Analytical Methods*. 6: 7469–7474

80. Luo Y, Chen M, Wen Q, Zhao M, Zhang B, Li X, Fu W. 2006. Rapid and simultaneous quantification of 4 urinary proteins by piezoelectric quartz crystal microbalance immunosensor array. *Clinical Chemistry*. 52: 2273–2280

81. Fonseca RAS, Ramos JJ, Kubota LT, Dutra RF. 2011. A nanostructured piezoelectric immunosensor for detection of human cardiac troponin T. *Sensors*. 11: 10785–10797

82. Wangmaung N, Chomean S, Promptmas C, Masodi S, Tanyong D, Ittarat W. 2014. Silver quartz crystal microbalance for differential diagnosis of plasmodium falciparum and plasmodium vivax in single and mixed infection. *Biosensors and Bioelectronics*. 62: 295–301

83. Yao C, Zhu T, Tang J, Wu R, Chen Q, Chen M, Fu W. 2008. Hybridization assay of hepatitis B virus by QCM peptide nucleic acid biosensor. *Biosensors and Bioelectronics*. 23: 879–885

84. Li D, Wang J, Wang R, Li Y, Abi-Ghanem D, Berghman L, Lu H. 2011. A nanobeads amplified QCM immunosensor for the detection of avian influenza virus H5N1. *Biosensors and Bioelectronics*. 26: 4146–4154

85. Su CC, Wu TZ, Chen LK, Yang HH, Tai DF. 2003. Development of immunochips for the detection of dengue viral antigens. *Analytica Chimica Acta*. 479: 117–123

86. Wang R, Wang L, Callaway ZT, Lu H, Huang TJ, Li Y. 2017. A nanowell-based QCM aptasensor for rapid and sensitive detection of avian influenza virus. *Sensors and Actuators B: Chemical*. 240: 934–940

87. Wangchareansak T, Thitithanyanont A, Chuakheaw D, Gleeson MP, Lieberzeit PA, Sangma C. 2013. Influenza A virus molecularly imprinted polymers and their application in virus sub-type classification. *Journal of Materials Chemistry B*. 1: 2190–2197

88. Srinivasan B, Tung S. 2015. Development and applications of portable biosensors. *Journal of Laboratory Automation*. 20: 365–389

89. Afzal A, Mujahid A, Schirhagl R, Bajwa SZ, Latif U, Feroz S. 2017. Gravimetric viral diagnostics: QCM based biosensors for early detection of viruses. *Chemosensors*. 5: 7

90. Gao F, Al-Qahtani AM, Khelif A, Boussaid F, Benchabane S, Cheng Y, Bermak A. 2020. Towards acoustic radiation free lamb wave resonators for high-resolution gravimetric biosensing. *IEEE Sensors Journal*. 21: 2725–2733

91. Mirea T, Olivares J, Clement M, Iborra E. 2019. Impact of FBAR design on its sensitivity as in-liquid gravimetric sensor. *Sensors and Actuators A: Physical*. 289: 87–93

92. Flewitt AJ, Luo JK, Fu YQ, Garcia-Gancedo L, Du XY, Lu JR, Milne WI. 2015. ZnO based SAW and FBAR devices for bio-sensing applications. *Journal of Non-Newtonian Fluid Mechanics*. 222: 209–216

93. Wingqvist G, Bjurström J, Hellgren AC, Katardjiev I. 2007. Immunosensor utilizing a shear mode thin film bulk acoustic sensor. *Sensors and Actuators B: Chemical*. 127: 248–252

94. Zhao X, Pan F, Ashley GM, Garcia-Gancedo L, Luo J, Flewitt AJ, Lu JR. 2014. Label-free detection of human prostate-specific antigen (hPSA) using film bulk acoustic resonators (FBARs). *Sensors and Actuators B: Chemical*. 190: 946–953

95. Chen YC, Shih WC, Chang WT, Yang CH, Kao KS. Cheng CC. 2015. Biosensor for human IgE detection using shear-mode FBAR devices. *Nanoscale Research Letters*. 10: 1–8

96. Liu J, Chen D, Wang P, Song G, Zhang X, Li Z, Yang J. 2020. A microfabricated thickness shear mode electroacoustic resonator for the label-free detection of cardiac troponin in serum. *Talanta*. 215

97. Peng J, Song G, Niu H, Wang P, Zhang X, Zhang S, Chen D. 2020. Detection of cardiac biomarkers in serum using a micro-electromechanical film electroacoustic resonator. *Journal of Micromechanics and Microengineering*. 30

98. Zheng D, Guo P, Xiong J, Wang, S. 2016. Streptavidin modified ZnO film bulk acoustic resonator for detection of tumor marker mucin 1. *Nanoscale Research Letters*. 11: 1–8

99. Zheng D, Xiong J, Guo P, Li Y, Wang S, Gu, H. 2014. Detection of a carcinoembryonic antigen using aptamer-modified film bulk acoustic resonators. *Materials Research Bulletin*. 59: 411–415

100. Wajs E, Rughoobur G, Burling K, George A, Flewitt AJ, Gnanapragasam VJ. 2020. A novel split mode TFBAR device for quantitative measurements of prostate specific antigen in a small sample of whole blood. *Nanoscale*. 12: 9647–9652

101. Chen D, Wang JJ, Li DH, Li ZX. 2011. Film bulk acoustic resonator based biosensor for detection of cancer serological marker. *Electronics Letters.* 47: 1169–1170
102. White RM, Voltmer FW. 1965. Direct piezoelectric coupling to surface elastic waves. *Applied Physics Letters.* 7: 314–316
103. Ji J, Yang C, Zhang F, Shang Z, Xu Y, Chen Y, Mu X. 2019. A high sensitive SH-SAW biosensor based 36° YX black LiTaO3 for label-free detection of pseudomonas aeruginosa. *Sensors and Actuators B: Chemical.* 281: 757–764
104. Gronewold TM. 2007. Surface acoustic wave sensors in the bioanalytical field: Recent trends and challenges. *Analytica Chimica Acta.* 603: 119–128
105. Lu X, Mouthaan K, Soon YT. 2013. Wideband bandpass filters with SAW-filter-like selectivity using chip SAW resonators. *IEEE Transactions on Microwave Theory and Techniques.* 62: 28–36
106. Liu Y, Liu J, Wang Y, Lam CS. 2019. A novel structure to suppress transverse modes in radio frequency TC-SAW resonators and filters. *IEEE Microwave and Wireless Components Letters.* 29: 249–251
107. Lopez-Gomez A, Cerdan-Cartagena F, Suardiaz-Muro J, Boluda-Aguilar M, Hernandez-Hernandez ME, Lopez-Serrano MA, Lopez-Coronado J. 2015. Radiofrequency identification and surface acoustic wave technologies for developing the food intelligent packaging concept. *Food Engineering Reviews.* 7: 11–32
108. Di Pietrantonio F, Cannata D, Benetti M, Verona E, Varriale A, Staiano M, D'Auria S. 2013. Detection of odorant molecules via surface acoustic wave biosensor array based on odorant-binding proteins. *Biosensors and Bioelectronics.* 41: 328–334
109. Ji J, Pang Y, Li D, Huang Z, Zhang Z, Xue N, Mu X. 2020. An aptamer-based shear horizontal surface acoustic wave biosensor with a CVD-grown single-layered graphene film for high-sensitivity detection of a label-free endotoxin. *Microsystems and Nanoengineering.* 6: 1–11
110. Gray ER, Turbe V, Lawson VE, Page RH, Cook ZC, Ferns McKendry RA. 2018. Ultra-rapid, sensitive and specific digital diagnosis of HIV with a dual-channel SAW biosensor in a pilot clinical study. *NPJ Digital Medicine.* 1: 1–8
111. Wohltjen H, Dessy R. 1979. Surface acoustic wave probes for chemical analysis. II: Gas chromatography detector. *Analytical Chemistry.* 51: 1465–1470
112. Chang HW, Shih JS. 2007. Surface acoustic wave immunosensors based on immobilized C60-proteins. *Sensors and Actuators B: Chemical.* 121: 522–529
113. Krishnamoorthy S, Iliadis AA, Bei T, Chrousos GP. 2008. An interleukin-6 ZnO/SiO2/Si surface acoustic wave biosensor. *Biosensors and Bioelectronics.* 24: 313–318
114. Bisoffi M, Hjelle B, Brown DC, Branch DW, Edwards TL, Brozik SM, Larson RS. 2008. Detection of viral bioagents using a shear horizontal surface acoustic wave biosensor. *Biosensors and Bioelectronics.* 23: 1397–1403
115. Lee HJ, Namkoong K, Cho EC, Ko C, Park JC, Lee SS. 2009. Surface acoustic wave immunosensor for real-time detection of hepatitis B surface antibodies in whole blood samples. *Biosensors and Bioelectronics.* 24: 3120–3125
116. Bisoffi M, Severns V, Branch DW, Edwards TL, Larson RS. 2013. Rapid detection of human immunodeficiency virus types 1 and 2 by use of an improved piezoelectric biosensor. *Journal of Clinical Microbiology.* 51: 1685–1691
117. Baca JT, Severns V, Lovato D, Branch DW, Larson RS. 2015. Rapid detection of ebola virus with a reagent-free, point-of-care biosensor. *Sensors.* 15: 8605–8614
118. Howe E, Harding G. 2000. A comparison of protocols for the optimisation of detection of bacteria using a surface acoustic wave (SAW) biosensor. *Biosensors and Bioelectronics.* 15: 641–649

119. Rana L, Gupta R, Tomar M, Gupta V. 2018. Highly sensitive Love wave acoustic biosensor for uric acid. *Sensors and Actuators B: Chemical.* 261: 169–177
120. Setka M, Bahos FA, Matatagui D, Potocek M, Kral Z, Drbohlavova VS. 2020. Love wave sensors based on gold nanoparticle-modified polypyrrole and their properties to ammonia and ethylene. *Sensors and Actuators B: Chemical.* 304
121. Inci F, Celik U, Turken B, Özer HÖ, Kok FN. 2015. Construction of P-glycoprotein incorporated tethered lipid bilayer membranes. *Biochemistry and Biophysics Reports.* 2: 115–122
122. Battal D, Akgönüllü S, Yalcin MS, Yavuz H, Denizli A. 2018. Molecularly imprinted polymer based quartz crystal microbalance sensor system for sensitive and label-free detection of synthetic cannabinoids in urine. *Biosensors and Bioelectronics.* 111: 10–17
123. Saylan Y, Akgönüllü S, Yavuz H, Ünal S, Denizli A. 2019 Molecularly imprinted polymer based sensors for medical applications. *Sensors.* 19
124. Saylan Y, Yilmaz F, Özgür E, Derazshamshir A, Yavuz H, Denizli, A. 2017. Molecular imprinting of macromolecules for sensor applications. *Sensors.* 17: 898

Part III

**Biorecognition Elements
Used in Biosensors**

5 Biorecognition Elements in Biosensors

Michael López Mujica[1], Alejandro Tamborelli[1,2], Virginia Vaschetti[1,2], Pablo Gallay[1], Fabrizio Perrachione[1], Daiana Reartes[1], Rocío Delpino[1], Marcela Rodríguez[1], María D. Rubianes[1], Pablo Dalmasso[2], and Gustavo Rivas[1]

[1]INFIQC, CONICET
Universidad Nacional de Córdoba
Ciudad Universitaria
Departamento de Fisicoquímica
Facultad de Ciencias Químicas
Córdoba, Argentina

[2]CIQA, CONICET
Universidad Tecnológica Nacional
Departamento de Ingeniería Química
Facultad Regional Córdoba
Córdoba, Argentina

CONTENTS

DOI: 10.1201/9781003189435-8

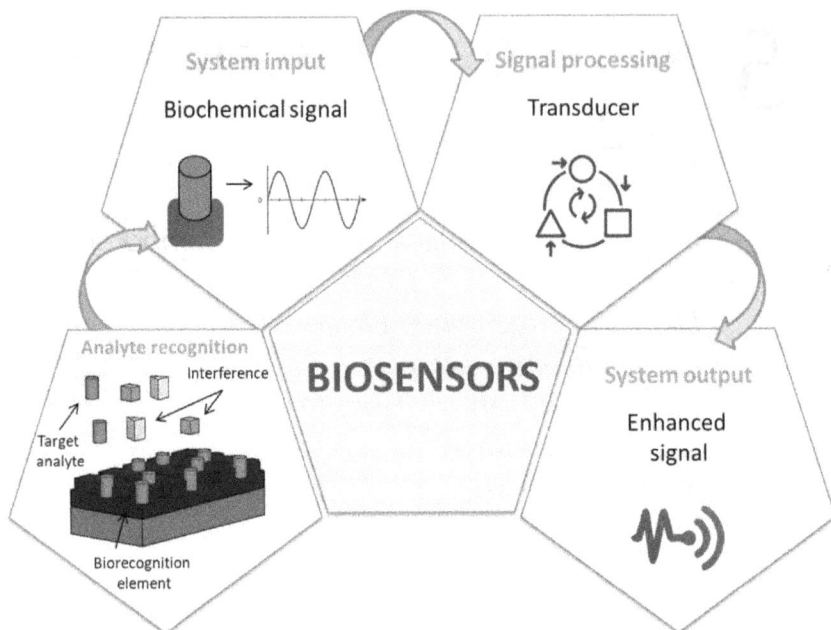

FIGURE 5.1 Schematic representation of a biosensor.

5.1 BIOSENSORS: COMPONENTS AND GENERAL ASPECTS

Biosensors are constituted by two basic components, the biorecognition element and the transducer, as shown in Figure 5.1. The biorecognition element is the "heart" of a biosensor since it is responsible for the specific recognition of the analyte in a given sample. The transducer has the key role of converting the physicochemical signal produced as a consequence of the biorecognition event into a measurable signal.

After developing of the first biosensor more than 60 years ago (1), the scientists noticed that some biorecognition events that occur in nature could be taken as models for the design of biosensing platforms and the synthesis of tailor-made molecules able to mimic the biorecognition capability of biomolecules. In the following sections, we discuss the most important elements/strategies for performing the biorecognition step.

5.2 BIORECOGNITION ELEMENTS

One of the most critical aspects when developing a biosensor is the immobilization of the biorecognition elements since they have to be in intimate contact with the transducer and exposed in an adequate way to facilitate the interaction with the analyte, at the same time that their recognition properties/biochemical activity are preserved as much as possible (2–3). The immobilization of biomolecules at the transducer by physical methods involve adsorption, entrapment in a polymer net or hydrogel, and inclusion within composite materials. Among them, the direct adsorption is the method that produces the smallest perturbation on the native structure of the biomolecule, although it offers the poorest stability. The covalent attachment is a highly used strategy and presents the great advantage of the stability; however, the denaturation/loss of activity

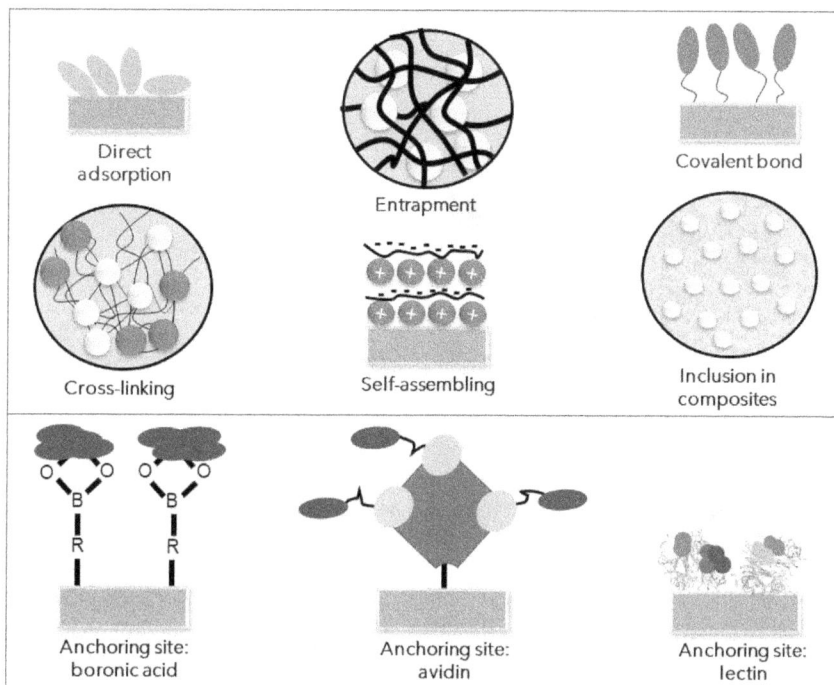

FIGURE 5.2 Schematic representation of different biorecognition elements immobilization strategies.

of the biomolecule is significant. The cross-linking using glutaraldehyde or similar is another interesting alternative. In the last decade of the 20th century appeared a very interesting and versatile alternative, the layer-by-layer self-assembling, that as shown in Figure 5.2, allows the organization of different layers in a supramolecular architecture for the development of a biosensing platform (4). Another interesting alternative that has received great attention in the last years, is the use of binding proteins as molecular glue, such as avidin (5) and lectins (or sugars) (6), to anchor the biorecognition element through the high affinity avidin/biotinylated biomolecules (6–7), lectins/glycoproteins (8), or sugars/lectins (9). In the same direction, derivatives of boronic acid have also been used successfully used for the immobilization of glycoproteins (6).

The biorecognition elements can be separated into three main groups, as illustrated in Figure 5.3: i) *biological molecules*, such as enzymes, antibodies, nucleic acids, lectins/sugars, antimicrobial peptides, and phages, ii) *synthetic molecules*, such as aptamers, nucleic acid analogues (peptide nucleic acids -PNAs- and locked nucleic acids -LNAs-), and molecularly imprinted polymers (MIPs), and iii) *tissues and bacterial cells*.

5.2.1 BIOLOGICAL MOLECULES

5.2.1.1 Enzymes

Since the use of glucose oxidase immobilized at an oxygen electrode for developing the first biosensor, several enzymes have been involved in the development of

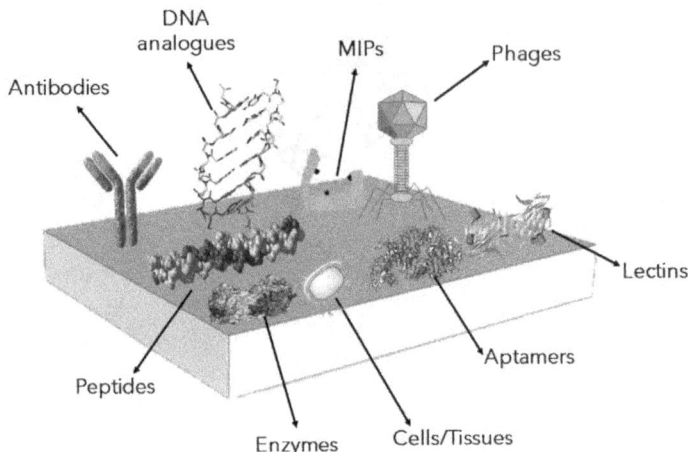

FIGURE 5.3   Biorecognition elements that can be immobilized at the transducer.

biosensing platforms like lyases, hydrolases, and oxidoreductases. Among them, the group of oxidoreductases has been the most widely used, being dehydrogenases, oxidases, reductases, peroxidases, oxygenases, catalases, and hydroxylases, the most representative (3, 10). All of them require cofactors, such as nicotinamide adenine dinucleotide (NAD), nicotinamide adenine dinucleotide phosphate (NADP), flavin mononucleotide (FMN), flavin adenine dinucleotide (FAD), pyrroloquinoline quinone (PQQ), coenzyme Q (CoQ), or hem centers (3).

5.2.1.2 Antibodies

Antibodies appeared in the scenario of biosensors in the 80s as a powerful tool for the development of immunosensors. Antibodies (Ab) or immunoglobulins (Ig) are produced in eukaryotes as a response to an antigen, which is any agent foreign to the organism (11). They are large glycoproteins of molecular weight around 150 kDa, that possess four chains, two small "light" polypeptide chains with an approximate molecular weight of 25 kDa each, and two large "heavy" polypeptide chains of 50 kDa each (12). These four chains are bound by disulfide bonds and form a Y-shape, as shown in Figure 5.4. They possess two distinct regions, the fragment crystallizable region (Fc fragment), which interacts with cell surface receptors and activates other immune system partners, and the antigen-binding region (Fab fragment) that recognizes and binds to antigens through a specific recognition domain called antigen determinant or epitope (12). The sequence of the amino-acids of these N-terminal ends determines the specific antigen-binding properties of the molecule. The complementary site on the antibody is called paratope. Binding interactions between antigen and antibody involve relatively weak noncovalent forces (electrostatic, hydrophobic, hydrogen bonding, and van der Waals interactions) (11–12).

Most antigens are highly complex and present several epitopes on their surface. Therefore, the immune response to an antigen generally involves the activation of multiple B-cells, all of which target a specific epitope. As a result, various antibodies with different epitope specificities are produced and they are known as polyclonal

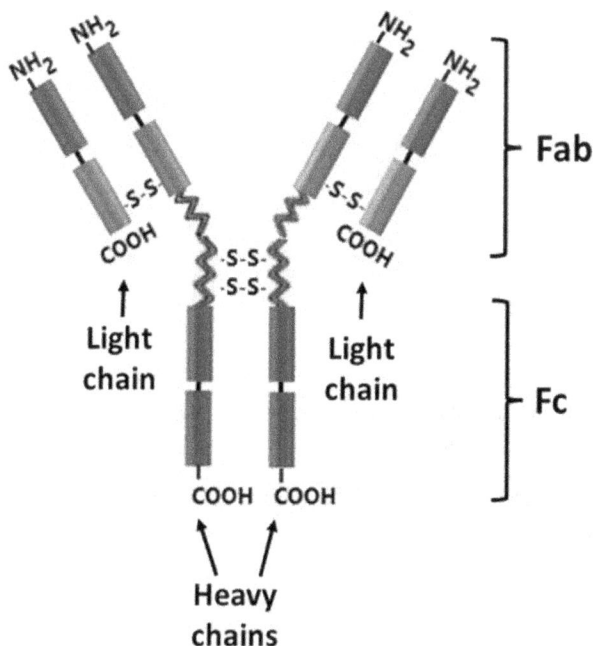

FIGURE 5.4 Scheme of an antibody, indicating the different chains and regions.

antibodies. In contrast, monoclonal antibodies are produced by a single B cell and contain a pool of the same antibody that binds to a unique specific epitope (13).

Due to their great affinity and specificity of interaction with a variety of molecular targets and universality, antibodies have become the most widely used biological recognition elements. Their broadest applications are in clinical analysis for the detection of a long list of analytes, mainly biomarkers or antigens associated with diverse pathologies (14–16).

Some special antibodies called single-domain antibodies, also known as nanobodies, are derived from heavy-chain antibodies (HCAbs), which are naturally present in all camelidae. These animals produce HCAbs in their serum, lacking Fc fragments and a canonical constant heavy chain 1 (CH1) domain in the heavy chain (17). These HCAbs are shown in Figure 5.5 and have an antigen recognition part composed of single variable domains, known as the variable domain of the heavy chain of heavy-chain antibody (VHH). Nanobodies are considered the smallest intact antigen-binding fragments (molecular weight of approximately 12–15 kDa). They are robust, resistant to denaturation/thermal degradation, easy to manipulate with aqueous solubility, and capable of reducing steric hindrance and recognizing inaccessible and cryptic epitopes (18–20).

5.2.1.3 Nucleic Acids

At the beginning of 90s, the analytical chemists realized that the unique recognition properties of nucleic acids could be exploited for the sequence-specific detection of DNA or RNA and the DNA-drugs/pollutants interaction or the DNA damage, representing the starting point of genosensors (21).

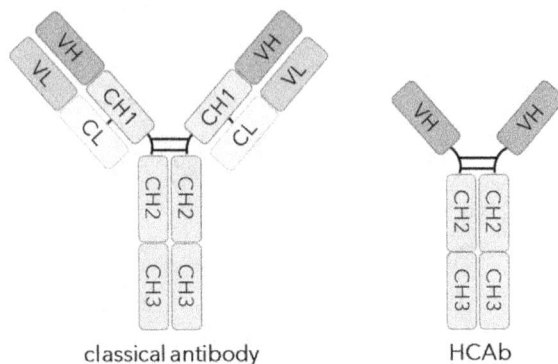

FIGURE 5.5 Comparison of classical antibodies and nanobodies.

Since there is well-documented information about nucleic acids in a long list of text-books, here, just few concepts should be mentioned. Nucleic acids are polymers constituted by simple units called nucleotides that are organized in a helicoidal strand. Double stranded nucleic acids are stabilized through hydrogen bonds between the complementary bases, hydrophobic interactions like van der Waals and π-stacking, that make the molecule highly stable despite the electrostatic repulsion of the phosphate groups (22).

Single-stranded DNAs are usually used as probes for developing hybridization biosensors (23), while double-stranded DNA is, in general, used for evaluating the interaction of DNA with drugs and pollutants or to evaluate the DNA damage (24–25).

5.2.1.4 Lectins

The increased knowledge about the importance of Glycobiology, the key role of different sugars and glycocompounds for the detection of microorganisms and cancer cells, in addition to their participation in several metabolic pathways, has also been taken by analytical chemists as model for the development of glycobiosensors based on the use of lectins as biorecognition elements (6, 26–27). Among lectin proteins, concanavalin A (Con A) is one of the most widely used biosensor design, which presents four sugar binding sites and specificity for the α-pyranose forms of D-glucose and D-mannose.

Lectin-modified surfaces have been mainly used as sensor platforms for the determination of sugars and glycoproteins. In addition, bacterial toxins can also be detected through lectin complexation because they often contain sugar moieties like the lipopolysaccharides located on the outer membrane of Gram-negative bacteria, and lipoteichoic acid found in Gram-positive bacteria (28). Lectin-modified sensors have also been used for detecting cells by exploiting the selective binding of lectins to hydrocarbon chains located on the surface of cells (29). Among cells, the detection of cancer cells is particularly important due to the presence of glycoproteins and glycolipids on their surface (27). Lectin-sensing bioplatforms are also important since lectins, in addition, play a prominent role in the immune system ranging from pathogen recognition and tuning of inflammation to cell adhesion or cellular signalling (30).

Lectin biosensors can also be prepared by using carbohydrates as recognition elements. In fact, a variety of devices have been developed by immobilizing carbohydrate chains on transducers surfaces. The spectrum of lectin-binding biomaterials covered ranges from glycosylated organic structures, calixarene, and fullerene cores over glycopeptides and glycoproteins, functionalized carbohydrate scaffolds of cyclodextrin or chitin to self-assembling glycopolymer clusters, gels, micelles, and liposomes (31).

5.2.1.5 Antimicrobial Peptides

Antimicrobial peptides (AMPs) were discovered in the 1920s and they are an integral part of the innate immune system in insects, amphibians, plants, microorganisms, and humans. They can be considered the first line of defence against a broad spectrum of pathogens including bacteria, viruses, fungi, and even cancer cells (30). Most AMPs interact with the cell membrane and produce disruption of cellular integrity, and, due to this interaction, AMPs are used for pathogen detection (32). Their high stability in extreme environment, their easy and low cost of synthesis, and their broad range of activities toward various microorganisms offer a promising approach for biodetection devices. AMPs are typically small (most of them contain less than 40 amino-acid residues), charged, amphipathic molecules. Although few AMPs are anionic, most of them (about 90%) present a cationic charge and an important proportion of hydrophobic residues (33). Three main families of cationic AMPs have been discovered to date: linear alpha-helical AMPs, cysteine-rich AMPs, and extended AMPs, which are enriched for specific amino-acids.

5.2.1.6 Phages

Bacteriophages are viruses that infect and replicate only within bacteria. As a consequence of their associated evolution with bacteria, bacteriophages possess highly specific mechanisms to recognize and subsequently infect their host for propagation. Two parts can usually be found, the head containing phage DNA and the tail responsible for bacterial recognition (30). Obtaining a correct orientation of the bacteriophages is one of the technical difficulties for biosensor applications, being the covalent attachment of the phage on the transducer is an efficient alternative.

Among various applications, bacteriophages have been used for bacterial detection. Since the bacteriophages can only recognize and infect alive bacteria, they offer the possibility to differentiate between viable and dead bacteria. In addition, they offer selective detection since some of them infect only one bacteria species (34–35).

5.2.2 SYNTHETIC BIOMOLECULES

5.2.2.1 Nucleic Acid Analogues

Many types of modifications have been introduced into native nucleic acids with high affinity toward DNA or RNA, such as peptide nucleic acid (PNA) and locked nucleic acid (LNA) (36). Both PNA and LNA have an outstanding affinity for natural nucleic acids, and the destabilizing effect of base mismatches in PNA- or LNA-containing heterodimers is much higher than in double-stranded DNA or RNA. Therefore,

PNA- and LNA-based biosensors have unprecedented sensitivity and specificity, with special applicability in DNA genotyping (36).

5.2.2.1.1 Peptide Nucleic Acids

At the end of the 20th century, the PNAs demonstrated to be extremely useful as biorecognition elements for hybridization biosensors due to the inherent stability of PNA/RNA or PNA/DNA duplexes (37). PNAs are a nucleic acid analogue of DNA, in which the phosphate backbone of DNA is replaced with a structurally homomorphous pseudopeptide backbone including repeating N-(2-aminoethyl)-glycine units as shown in Figure 5.6. PNAs hybridize with complementary strands of DNA or RNA, following Watson–Crick base pairing rules (38). The neutral backbone of PNA implies a lack of electrostatic repulsion between the PNA and DNA strands (compared to that existing between two negatively charged DNA oligomers), allowing a powerful binding ability even at low salt concentrations, and the consequent increase of the stability of PNA/DNA duplexes (38–39).

5.2.2.1.2 Locked Nucleic Acids

LNAs were prepared as an ideal oligomer for recognition of RNA (40). It is a negatively charged nucleic acid analogue of RNA, in which the furanose ring of the ribose

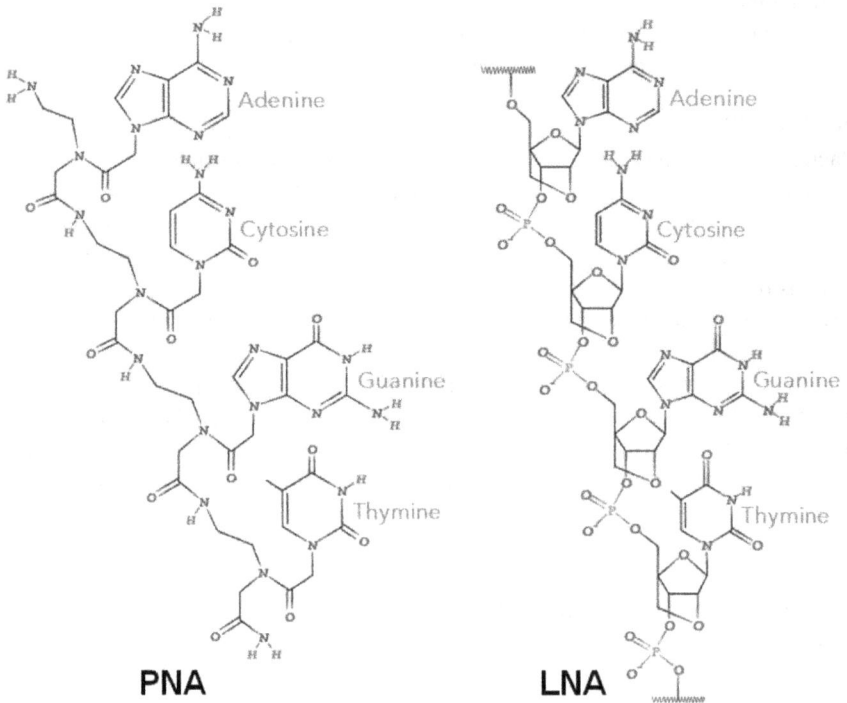

FIGURE 5.6 Scheme of the structure of peptide nucleic acids (PNAs) and locked nucleic acids (LNAs).

sugar is chemically locked by the introduction of a methylene linkage between O2' and C4' as shown in Figure 5.6. The methylene connection joins the 2' and the 4' position of the ribose sugar. The covalent bridge effectively "locks" the ribose in the N-type (3-endo) conformation that is dominant in A-form DNA and RNA (41). This conformation enhances the base stacking and the phosphate backbone pre-organization and results in improved affinity for complementary DNA or RNA sequences. Further, LNA residues confer a relative degree of resistance to exo- and endonucleases. They are characterized by reduced flexibility of the ribose residue and exist in a locked N-type conformation, which favors the formation of stable duplexes with DNA or RNA. Numerous structural and thermal stability studies on complexes formed by LNA oligomers and complementary DNA or RNA oligonucleotides have shown higher melting temperatures and specificity when compared to the unmodified isosequential compounds (42).

5.2.2.2 Aptamers

At the beginning of XXI century, the aptamers appeared on the horizon of biosensors as a very competitive alternative to antibodies, making them a very attractive option as biorecognition layer for the development of aptasensors (43). Due to their characteristics, they are called chemical antibodies. Even though they were simultaneously developed in the 1990s by two independent laboratories (43–44), the term "aptamer" was proposed by Tuerk in 1990 (43). Aptamers are short synthetic single-stranded nucleic acids, either ssDNA or RNA with typically less than 100 nucleotide residues, capable of binding to a specific target with high affinity and selectivity. The robustness of the phosphodiester backbone helped aptamers to exhibit improved stability over the antibodies. Aptamers can be combined with the target through non-covalent bonding to form a complex three-dimensional shape, such as stem, loop, hairpin, and G-quadruplex structure (44), as illustrated in Figure 5.7A. Aptamers are selected from highly complex libraries of nucleic acids manufactured by combinatorial synthesis, through an in vitro enrichment process called Systematic Evolution of Ligands by EXponential enrichment (SELEX) (45–47). A schematic diagram of the SELEX process steps is depicted in Fig. 5.7B. It is an iterative process that involves the following general steps: interaction between the starting library of randomized oligonucleotides and the selected target in a suitable milieu, separation of bound oligonucleotides, elution of the target-bound oligonucleotides to isolate the highest binders, and amplification of the obtained sequences by the polymerase chain reaction (PCR) to yield an enriched library for the next selection round, with lower variability than the starting mixture, which is enriched in molecules that recognize the target with high affinity and selectivity. The process is repeated for several rounds under increasingly stringent interaction conditions until the enrichment is considered sufficient. Usually, oligonucleotide sequences with good specificity and affinity can be enriched in the pool after 6–20 rounds and the whole progress can be monitored through some traceable labels.

The 3D folded structure permits the formation of stable complexes with various targets from ions and small molecules to proteins, nucleic acids, and viruses (47–49). Aptamers possess several advantages over natural antibodies. While the antibody production needs manipulation of whole host eukaryotes, the aptamers synthesis

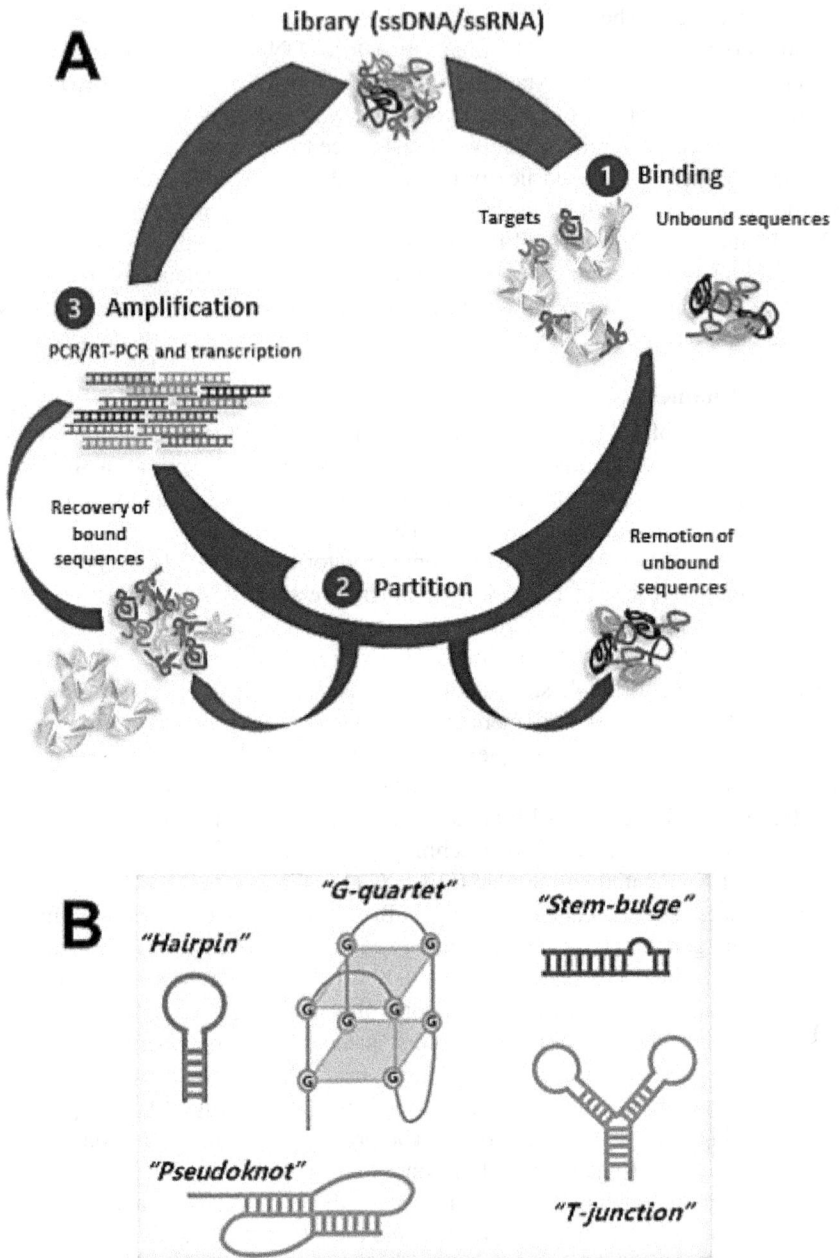

FIGURE 5.7   (A) Scheme of the conformation of the aptamer once established the interaction with the target. (B) Scheme of SELEX process.

is entirely chemical. Furthermore, aptamers are easily chemically modifiable, giving them additional properties, including nuclease resistance, an essential property for applications in complex media such as blood, a wide range of target molecules (inorganics, organics, cells, viruses, and bacteria, among others), facile preparation on a massive scale, low molecular weights, high pH stability, easier modification, and long-term storage stability. Moreover, in contrast to antibodies, aptamers can be exposed to elevated temperatures without being irreversibly denatured and simply refold when the temperature goes down. These properties, associated with affinities for their target close to those obtained for antibodies make aptamers an interesting alternative ligand for biosensing. Table 5.1 compares the characteristics of antibodies and aptamers.

5.2.2.3   Molecular Imprinted Polymers

Molecular imprinting is an interesting approach to mimic natural molecular recognition through the preparation of synthetic molecules called molecularly imprinted polymers (MIPs) with recognition sites selective for a given analyte (50). The target analyte is used as a template molecule that interacts with monomers by covalent or noncovalent bonding during the polymerization process. Therefore, the resulting macroporous polymers contain recognition sites with high affinity for the print molecule, long-term stability, and resistance to harsh environments. Due to the selectivity obtained in this process, the recognition of MIPs approaches the biological recognition of antibodies are also called synthetic antibodies (51). In another chapter there is an extensive discussion about MIPs.

5.2.3   Cells and Tissues

Sensors based on cells or intact tissue have received considerable attention as an alternative to molecular biosensors (52). Whole-cell biosensors (WCBs) are

TABLE 5.1
Comparison between Aptamers and Antibodies

Characteristic	Aptamers	Antibodies
Size	5–25 kDa	150 kDa
Production time	< 8 weeks	> 10 weeks
Control of the process	Highly controllable, in-vitro, free selection of the conditions	Dependent on the immune system of the animal
Possibility of chemical modifications	Yes synthetic production	No biological production
Targets	Very wide from ions, small molecules to proteins, cells	Limited only immunogenic compounds
Target binding site	Cavities, clefts	Epitopes
Thermal stability	Reversible	Irreversible
Target affinity	High	High

based on the use of living organisms, such as algae, bacteria, fungi, yeast, and even tissue slices as recognition elements by determining their general metabolic status (53). Even when, in principle, WCBs are less sensitive to environmental signals compared to molecular biosensors, modification by simple genetic engineering make them able to measure multiple responses within a live cell (54). The performance of the resulting WCB is highly dependent on the selected reporter gene, since the sensitivity and selectivity of the biosensor will be highly connected with the molecular recognition that will occur when the regulator proteins bind to their target analytes (55). Both reporter genes and the type of regulatory protein contribute significantly to the functional capability of WCBs. The genetically encoded microbial-derived biosensors are fabricated by integrating responsive genes to a promoter-less bio-recognition gene. The resulting recombinant genes can then be cloned into microbial host either by direct integration into the chromosome or by means of an appropriate plasmid. Figure 5.8 shows that in the presence of the target compound, the regulator stimulates the promoter which, in turn, produces the transcription of the reporter gene and, consequently, a quantifiable signal. Therefore, the sensitivity and specificity of microbial-derived biosensors are primarily determined by the responsiveness of the regulatory gene to its projected target (56). The high selectivity, sensitivity, and high-throughput in-situ detection capability of WCBs have made them very attractive for use in environmental, pharmaceutical, and food analysis and medical diagnostics (55, 57–59).

Another type of biosensors involves tissues as an enzymatic sources, like plan tissues, and in this way, they solve the problem of cost, stability or even availability of isolated enzymes. The emblematic bananatrode used banana as a rich source of polyphenol oxidase, for detecting dopamine (60–61). Other bioplatforms involved eggplant, potato, peach, pear, apple, and mushroom as polyphenol oxidase sources incorporated within carbon composites to obtain biosensors for phenols and catechols (62). They have shown a very interesting correlation between the plant-tissue and the selectivity and activity of the resulting biosensor toward a given phenol or catechol. These biocatalytic electrodes work in a similar way to enzymatic biosensors using the conventional pure enzyme.

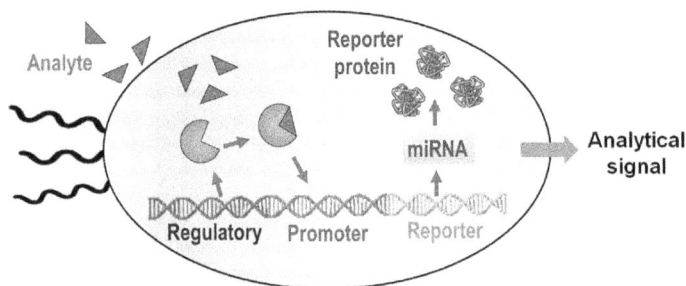

FIGURE 5.8   Scheme of the biorecognition in cells-based biosensors.

5.3   CONCLUSIONS

Biosensors have received great interest in the field of modern bioanalysis due to the high demand and opportunities that are emerging especially in diagnostics, food, drugs, environmental analysis, and quality and safety control. Even after six decades from the first enzymatic biosensor, enzymes still attract enormous attention. However, the successful use of antibodies for preparing immunosensors has made them the workhorse for the development of biosensing platforms able to quantify each new biomarker/bioanalyte of relevance. In spite of this preference for antibodies-based biosensors, new synthetic molecules that mimic natural biorecognition processes have been demonstrated to be extremely useful for the design of bioplatforms due to the multiple advantages that they present, mainly connected with the possibility of large scale production, robustness, and stability. In addition, their easy chemical modification gives them enormous possibilities to design new immobilization strategies and to build novel supramolecular architectures with biosensing applications. Surely, the future of the biorecognition event will be closely connected with the possible massive production of nanobodies, the use of new lectins and cell receptors, and the possibility of synthesizing new molecules analogues to biomolecules.

ACKNOWLEDGMENTS

Authors are grateful to CONICET, ANPCyT-FONCyT, SECyT-UNC, and SCTyP-UTN for the financial support. Also, doctoral and postdoctoral fellowships from CONICET and ANPCyT-FONCyT are gratefully acknowledged.

REFERENCES

1. Clark Jr LC, Lyons C. 1962. Electrode systems for continuous monitoring in cardiovascular surgery. *Annals of the New York Academy of Sciences*. 102(1): 29–45
2. Karadurmus L, Kaya S, Ozkan SA. 2021. Recent advances of enzyme biosensors for pesticide detection in foods. *Journal of Food Measurement and Characterization*. 15(5): 4582–4595
3. Nguyen HH, Lee SH, Lee UJ, Fermin CD, Kim M. 2019. Immobilized enzymes in biosensor applications. *Materials*. 12(1): 121
4. Mandler D, Kraus-Ophir S. 2011. Self-assembled monolayers (SAMs) for electrochemical sensing. *Journal of Solid State Electrochemistry*. 15(7–8): 1535
5. Livnah O, Bayer EA, Wilchek M, Sussman JL. 1993. Three-dimensional structures of avidin and the avidin-biotin complex. *Proceedings of the National Academy of Sciences*. 90(11): 5076–5080
6. Wang B, Anzai J. 2015. Recent progress in lectin-based biosensors. *Materials*. 8(12): 8590–8607
7. Gutierrez FA, Rubianes MD, Rivas GA. 2019. New bioanalytical platform based on the use of avidin for the successful exfoliation of multi-walled carbon nanotubes and the robust anchoring of biomolecules: Application for hydrogen peroxide biosensing. *Analytica Chimica Acta*. 1065: 12–20
8. Gallay P, Eguílaz M, Rivas G. 2020. Designing electrochemical interfaces based on nanohybrids of avidin functionalized-carbon nanotubes and ruthenium nanoparticles as peroxidase-like nanozyme with supramolecular recognition properties for site-specific anchoring of biotinylated residues. *Biosensors and Bioelectronics*. 148: 111764.

9. Ortiz E, Gallay P, Galicia L, Eguílaz M, Rivas G. 2019. Nanoarchitectures based on multi-walled carbon nanotubes non-covalently functionalized with concanavalin A: A new building-block with supramolecular recognition properties for the development of electrochemical biosensors. *Sensors and Actuators B: Chemical*. 292: 254–262

10. Xie Y, Liu T, Chu Z, Jin W. 2021. Recent advances in electrochemical enzymatic biosensors based on regular nanostructured materials. *Journal of Electroanalytical Chemistry*. 893: 115328

11. Harshavardhan S, Rajadas SE, Vijayakumar KK, Durai WA, Ramu A, Mariappan R. 2019. Electrochemical Immunosensors: Working principle, types, scope, applications, and future prospects. *Bioelectrochemical Interface Engineering*. 343–369

12. Gao S, Guisán JM, Rocha-Martin J. 2021. Oriented immobilization of antibodies onto sensing platforms-A critical review. *Analytica Chimica Acta*. 338907

13. Murphy, K. 2012. *Janeway's Immunobiology*, 8th edition. Garland Science.

14. Ahirwar R. 2021. Recent advances in nanomaterials-based electrochemical immunosensors and aptasensors for HER2 assessment in breast cancer. *Microchimica Acta*. 188(10): 1–18

15. Crivianu-Gaita V, Thompson M. 2016. Aptamers, antibody scFv, and antibody Fab' fragments: An overview and comparison of three of the most versatile biosensor biorecognition elements. *Biosensors and Bioelectronics*. 85: 32–45

16. Ranjan P, Singhal A, Yadav S, Kumar N, Murali S, Sanghi SK, Khan R. 2021. Rapid diagnosis of SARS-CoV-2 using potential point-of-care electrochemical immunosensor: Toward the future prospects. *International Reviews of Immunology*. 40(1–2): 126–142

17. Hamers-Casterman CTSG, Atarhouch T, Muyldermans SA, Robinson G, Hammers C, Songa EB, Hammers R. 1993. Naturally occurring antibodies devoid of light chains. *Nature*. 363(6428): 446–448

18. Bazin I, Tria SA, Hayat A, Marty JL. 2017. New biorecognition molecules in biosensors for the detection of toxins. *Biosensors and Bioelectronics*. 87: 285–298

19. Liu W, Song H, Chen Q, Yu J, Xian M, Nian R, Feng D. 2018. Recent advances in the selection and identification of antigen-specific nanobodies. *Molecular Immunology*. 96: 37–47

20. Siontorou CG. 2013. Nanobodies as novel agents for disease diagnosis and therapy. *International Journal of Nanomedicine*. 8: 4215

21. Paleček E, Fojta M, Tomschik M, Wang J. 1998. Electrochemical biosensors for DNA hybridization and DNA damage. *Biosensors and Bioelectronics*. 13(6): 621–628

22. Palecek E, Fojta M. 2005. Electrochemical DNA sensors. *Bioelectronics*. 127–192

23. Wang J. 1999. Towards genoelectronics: Electrochemical biosensing of DNA hybridization. *Chemistry—A European Journal*. 5(6): 1681–1685

24. Fojta M. 2002. Electrochemical sensors for DNA interactions and damage. *Electroanalysis*. 14(21): 1449–1463

25. Fojta M, Daňhel A, Havran L, Vyskočil V. 2016. Recent progress in electrochemical sensors and assays for DNA damage and repair. *TrAC Trends in Analytical Chemistry*. 79: 160–167

26. Bojarová P, Křen V. 2016. Sugared biomaterial binding lectins: Achievements and perspectives. *Biomaterials Science*. 4(8): 1142–1160

27. Sadighbayan D, Sadighbayan K, Tohid-Kia MR, Khosroushahi AY, Hasanzadeh M. 2019. Development of electrochemical biosensors for tumor marker determination towards cancer diagnosis: Recent progress. *TrAC Trends in Analytical Chemistry*. 118: 73–88

28. Mi F, Guan M, Hu C, Peng F, Sun S, Wang X. 2021. Application of lectin-based biosensor technology in the detection of foodborne pathogenic bacteria: A review. *Analyst*. 146(2): 429–443

29. Hendrickson OD, Zherdev AV. 2018. Analytical application of lectins. *Critical Reviews in Analytical Chemistry.* 48(4): 279–292

30. Templier V, Roux A, Roupioz Y, Livache T. 2016. Ligands for label-free detection of whole bacteria on biosensors: A review. *TrAC Trends in Analytical Chemistry.* 79: 71–79

31. Adak AK, Li BY, Lin CC. 2015. Advances in multifunctional glycosylated nanomaterials: Preparation and applications in glycoscience. *Carbohydrate Research.* 405: 2–12

32. Pardoux É, Boturyn D, Roupioz Y. 2020. Antimicrobial peptides as probes in biosensors detecting whole bacteria: A review. *Molecules.* 25(8): 1998

33. Wieland T, Assmann J, Bethe A, Fidelak C, Gmoser H, Janßen T, Urban GA. 2021. A Real-time thermal sensor system for quantifying the inhibitory effect of antimicrobial peptides on bacterial adhesion and biofilm formation. *Sensors.* 21(8): 2771

34. Li Y, Xie G, Qiu J, Zhou D, Gou D, Tao Y, Chen H. 2018. A new biosensor based on the recognition of phages and the signal amplification of organic-inorganic hybrid nanoflowers for discriminating and quantitating live pathogenic bacteria in urine. *Sensors and Actuators B: Chemical.* 258: 803–812

35. Tawil N, Sacher E, Mandeville R, Meunier M. 2014. Bacteriophages: Biosensing tools for multi-drug resistant pathogens. *Analyst.* 139(6): 1224–1236

36. Laschi S, Palchetti I, Marrazza G, Mascini M. 2009. Enzyme-amplified electrochemical hybridization assay based on PNA, LNA and DNA probe-modified micro-magnetic beads. *Bioelectrochemistry.* 76(1–2): 214–220

37. Wang J, Rivas G, Cai X, Chicharro M, Parrado C, Dontha N, Nielsen PE. 1997. Detection of point mutation in the p53 gene using a peptide nucleic acid biosensor. *Analytica Chimica Acta.* 344(1–2): 111–118

38. Wang J, Palecek E, Nielsen PE, Rivas G, Cai X, Shiraishi H, Farias PA. 1996. Peptide nucleic acid probes for sequence-specific DNA biosensors. *Journal of the American Chemical Society.* 118(33): 7667–7670

39. Bidar N, Amini M, Oroojalian F, Baradaran B, Hosseini SS, Shahbazi MA, de la Guardia M. 2021. Molecular beacon strategies for sensing purpose. *TrAC Trends in Analytical Chemistry.* 134: 116143

40. Briones C, Moreno M. 2012. Applications of peptide nucleic acids (PNAs) and locked nucleic acids (LNAs) in biosensor development. *Analytical and Bioanalytical Chemistry.* 402(10): 3071–3089

41. Soler-Bistué A, Zorreguieta A, Tolmasky ME. 2019. Bridged nucleic acids reloaded. *Molecules.* 24(12): 2297

42. Hagedorn PH, Persson R, Funder ED, Albæk N, Diemer SL, Hansen DJ, Koch T. 2018. Locked nucleic acid: Modality, diversity, and drug discovery. *Drug Discovery Today.* 23(1): 101–114

43. Tuerk C, Gold L. 1990. Systematic evolution of ligands by exponential enrichment: RNA ligands to bacteriophage T4 DNA polymerase. *Science.* 249(4968): 505–510

44. Ellington AD, Szostak JW. 1990. In vitro selection of RNA molecules that bind specific ligands. *Nature.* 346(6287): 818–822

45. Ellington AD, Szostak JW. 1992. Selection in vitro of single-stranded DNA molecules that fold into specific ligand-binding structures. *Nature.* 355(6363): 850–852

46. Stoltenburg R, Reinemann C, Strehlitz B. 2007. SELEX—a (r) evolutionary method to generate high-affinity nucleic acid ligands. *Biomolecular Engineering.* 24(4): 381–403

47. Zhuo Z, Yu Y, Wang M, Li J, Zhang Z, Liu J, Zhang, B. 2017. Recent advances in SELEX technology and aptamer applications in biomedicine. *International Journal of Molecular Sciences.* 18(10): 2142

48. Qi X, Yan X, Zhao Y, Li L, Wang S. 2020. Highly sensitive and specific detection of small molecules using advanced aptasensors based on split aptamers: A review. *TrAC Trends in Analytical Chemistry.* 116069

49. Shaban SM, Kim DH. 2021. Recent advances in aptamer sensors. *Sensors.* 21(3): 979

50. Ali GK, Omer KM. 2021. Molecular imprinted polymer combined with aptamer (MIP-aptamer) as a hybrid dual recognition element for bio (chemical) sensing applications: Review. *Talanta.* 122878

51. Lowdon JW, Diliën H, Singla P, Peeters M, Cleij TJ, van Grinsven B, Eersels K. 2020. MIPs for commercial application in low-cost sensors and assays—an overview of the current status quo. *Sensors and Actuators B: Chemical.* 128973

52. Liu Q, Wu C, Cai H, Hu N, Zhou J, Wang P. 2014. Cell-based biosensors and their application in biomedicine. *Chemical Reviews.* 114(12): 6423–6461

53. Gui Q, Lawson T, Shan S, Yan L, Liu Y. 2017. The application of whole cell-based biosensors for use in environmental analysis and in medical diagnostics. *Sensors.* 17(7): 1623

54. Xu X, Ying Y. 2011. Microbial biosensors for environmental monitoring and food analysis. *Food Reviews International.* 27(3): 300–329

55. Raut N, O'Connor G, Pasini P, Daunert S. 2012. Engineered cells as biosensing systems in biomedical analysis. *Analytical and Bioanalytical Chemistry.* 402(10): 3147–3159

56. Shin HJ. 2011. Genetically engineered microbial biosensors for in situ monitoring of environmental pollution. *Applied Microbiology and Biotechnology.* 89(4): 867–877

57. Bilal M, Iqbal HM. 2019. Microbial-derived biosensors for monitoring environmental contaminants: Recent advances and future outlook. *Process Safety and Environmental Protection.* 124: 8–17

58. Moraskie M, Roshid H, O'Connor G, Dikici E, Zingg JM, Deo S, Daunert S. 2021. Microbial whole-cell biosensors: Current applications, challenges, and future perspectives. *Biosensors and Bioelectronics.* 113359

59. Xie X, Stüben D, Berner Z, Albers J, Hintsche R, Jantzen E. 2004. Development of an ultramicroelectrode arrays (UMEAs) sensor for trace heavy metal measurement in water. *Sensors and actuators B: Chemical.* 97(2–3): 168–173

60. Wang J, Lin MS. 1988. Mixed plant tissue carbon paste bioelectrode. *Analytical Chemistry.* 60(15): 1545–1548

61. Wang J, Naser N, Ozsoz M. 1990. Plant tissue-based amperometric electrode for eliminating ascorbic acid interferences. *Analytica Chimica Acta.* 234: 315–320

62. Forzani ES, Rivas GA, Solis VM. 1997. Amperometric determination of dopamine on vegetal-tissue enzymatic electrodes. Analysis of interferents and enzymatic selectivity. *Journal of Electroanalytical Chemistry.* 435(1–2): 77–84

6 Imprinted Polymers in Biosensors

Yeşeren Saylan[1], Semra Akgönüllü[1],
Nilay Bereli[1], and Adil Denizli[1]
[1]Hacettepe University
Department of Chemistry
Ankara, Turkey

CONTENTS

6.1 INTRODUCTION

Molecularly imprinted polymers (MIPs) are synthetic antibodies and biological receptors, helpful to separate and analyze complex samples such as biological fluids and environmental samples (1). MIPs are becoming significant analytical materials (2) and they have seen a continuous improvement as a recognition element in sensing systems since the late 1990s. Molecular imprinting is favorably encouraged by the ideas of Günter Wulff (3) and Klaus Mosbach (4). In this way, molecularly imprinted and non-imprinted polymers (NIPs-prepared without template) that mimic nature by biomimetic strategies are synthesized and designed. These smart synthetic polymers have the advantages of simple synthesis based on a wide variety of inexpensive, easily accessible starting materials, and do not require time-consuming effort (5). MIPs are preferred due to their close recognition features to biological receptors and their usability for the wide variety of targets. Moreover, it is notable for its excellent physical and chemical stability compared to biological receptors (6–11). These perfect advantages have provided the application of MIPs in different areas such as purification (12), separation sciences (13), decontamination (14), chemical sensing (15), immunoassays (16), therapy (17), drug delivery (18), and cell imaging (19).

The main advantages of MIPs are their high selectivity and affinity for the template molecule chosen in the molecular imprinting technique procedure (20). MIPs compared to biological receptors have higher strength, resistance to elevated temperature and pressure, physical robustness, and inertness towards bases, acids, and organic solvents (21). Additionally, they are also cheap to synthesize and the storage life of the polymers can be very great, keeping their specific recognition capacity

DOI: 10.1201/9781003189435-9

also at room temperature for several years (1). This method is considered a versatile and hopeful technique that can recognize both biological and chemical molecules such as amino acids (22), enzymes (23), proteins (24), nucleotides (25), bacteria (26), viruses (27), pollutants (28), toxins (29), metal ions (30), and drugs (31–32). Various molecularly imprinted polymers are developed to use as biosensors, chromatographic stationary phases, or materials for solid-phase extraction.

In this chapter, we define the present-day state of the research field on MIPs-based optical, electrochemical, and piezoelectric biosensors. Herein, we also focused on recent advances and developments in the molecular imprinting technique field with special emphasis on the application of molecularly imprinted polymers in biosensors.

6.2 FUNDAMENTAL OF MOLECULARLY IMPRINTED POLYMERS (MIPs)

Molecular imprinting technology is at present a current synthetic method to synthesis strong molecular recognition polymeric materials able to mimic biological receptors, such as enzymes, and antibodies (1). The molecular imprinting technique is artificial polymers, which is a process where functional monomers and cross-linking monomers are co-polymerized in the presence of the template molecule. In this method, the imprinted molecule plays a role as a template. The functional monomers firstly form a pre-complex with the template analyte/molecule. Functional groups after polymerization are kept in place by the highly cross-linked polymeric matrix. As a final process, removal of the template creates binding sites that are complementary in shape and size to the template molecule. Finally, a polymer molecular memory is added that can rebind the template molecule with very high specificity (4). Different methods have been utilized in the synthesis of MIPs, but all follow the same steps; (i) Pre-complex is formed with functional groups bonded by covalent or non-covalent interactions with the template or target molecule (ii) the polymer is synthesized around the template molecule with suitable monomers and initiators (iii) the imprinted molecule is removed from the polymer and cavities are formed for back-bonding (iv) MIPs are interacted with the template molecule in the complex sample and the imprinted molecule is selectively placed in the cavities (Figure 6.1). In recent years, various molecular imprinted

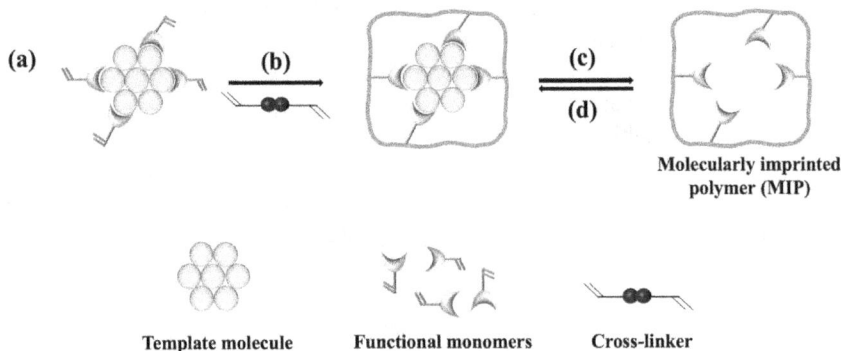

FIGURE 6.1   Steps of molecular imprinting technique: (a) Pre-complex combination, (b) polymerization, template molecule (c) removal and (d) rebinding.

polymers have been developed with application in the separation and/or detection of various analytes in different matrices with high selectivity (33–34).

MIPs are a thrilling choice for biosensor design owing to their high chemical and physical stability, low cost, robustness, and simplicity of preparation. Molecular imprinting provides the fabrication of custom-made polymeric materials containing specific binding sites with selective affinity for the analyte. For MIP synthesis, the target molecule chosen for imprinting interacts with covalent or non-covalent bonds with functional monomers that can polymerize in the presence of a crosslinker (35–37). Non-covalent imprinting has a great range of applications owing to the technical lack of limitations on the shape, size, or chemical character of the target molecule (2). When the target molecule is removed, the resulting three-dimensional supramolecular structure contains specific recognition regions that are complementary to the target molecule (2, 35–37).

In the most commonly selected non-covalent imprinting technique, the template molecule is complexed with one or several functional monomers that are polymerized in the presence of more crosslinking monomers. Finally, it can be removed from the imprinted polymer and recycled. The correct selection and design of functional monomers and templates are very important in the successful imprinting process. The functional monomer–template molecule complexes then need to be incorporated in a polymer scaffold by copolymerization with comonomers and cross-linking monomers (38). Some common functional monomers: acrylamide, methacrylic acid, methyl methacrylate, pyrrole, and aniline. Afterwards, the appropriate dimethacrylate and a polymerization initiator, typically azobis(isobutyronitrile), are added, polymerization is initiated by heating or ultraviolet radiation (34). From a common point of view, receptor design is probably the very important step in the process of biosensor design because it must compare the needs of high selectivity and sensitivity for a template molecule with the particular characteristics of a given readout. Thus, the straightforward engineering, fine-tuning, and simplicity of integration into the industrial process of MIPs make them ideal candidates for recognition elements (39). While natural systems can generate antibodies toward a series of strange bodies, the operation of such receptors in chemical processes faces many barriers such as cost and sensitivity to environmental conditions. The goal of modern biosensor research is to create synthetic receptors that mimic natural antibody-antigen behavior with similar specificity and sensitivity. This molecular recognition promises selective precision biosensors that, when combined with modern technologies to monitor changes in recognition elements, can track and identify targets without intervention (34). There are great hopes for improving a novel generation of biosensors using these synthetic smart materials as recognition elements (2). In recent years, MIPs-based biosensors have been applied to a wide variety of complex systems such as biomolecular interaction monitoring (40), medical diagnosis (24), food testing (29), homeland security (41), and environmental monitoring (42).

The biosensor consists of two main elements (Figure 6.2): The recognition element, called the receptor, and a physicochemical transducer. The receptor transforms the binding incident of the target molecule into a form of energy that can be measured by the transducer. For chemical biosensors, this binding event involves chemical species and the generation of a signal upon the change of a physicochemical

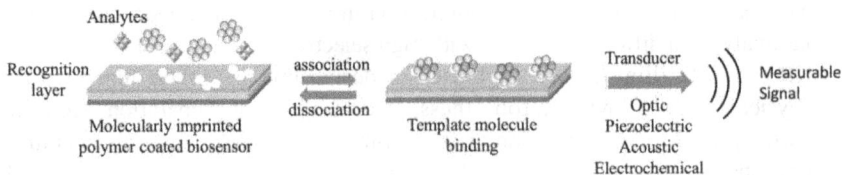

FIGURE 6.2 Scheme of the preparation of a MIPs-based biosensor.

parameter (e.g., the formation/breaking of a bond, exchange of electrons, and mass change of refractive index modification). Subsequently, the transducer transforms the chemical information received from the receptor into a useful analytical signal (39). Biosensors offer sensitivity, fast response time, robustness, and portability. It is a potential alternative to traditional grade evaluation devices with easier protocols. Biosensors must supply the needs of reproducibility, sensitivity, response correctness, nontoxicity, high specificity toward the wanted template molecule, and cost-effectiveness (43). Biosensors can be separated into diverse transducer groups for the generation of the output signal, such as electrochemical, optical, magnetic, gravimetric, and piezoelectric (43–46). The selection of a transduction tool is primarily based on natural and physicochemical features of the surface layer that change when interacting with the target molecule (47).

6.3 RECENT APPLICATIONS OF IMPRINTED POLYMERS IN BIOSENSORS

As one of the trendy paths for specific identification is molecular imprinting method has charmed rising notice to plan parts of different template molecules (26, 48–49). For instance, Xu et al. designed a molecularly imprinted photoelectrochemical biosensor for bisphenol A (BPA) detection using a nanosheet that contains In_2S_3/Cd^{2+} sensitized $AgBiS_2$ that provides the distinguished structure with a suitable surface accidence for another photoactive material. They mentioned that the In_2S_3 nanoparticles are employed for $AgBiS_2$ sensitizing, and a biosensor response is reached owing to electron delivery synergy and paired bandgap. They also dropped Cd^{2+} to achieve a steadier and higher photocurrent on the electrode to produce CdS and obtain sufficient biosensor response (Figure 6.3). Following the optimization of detection condition, the biosensor shows a broad range from 0.0005 to 50 µM for BPA with a 0.18 nM of detection limit (50).

Another example belongs to Wang et al. They developed a *Staphylococcus aureus* (*S. aureus*)-imprinted poly(3-thiophene acetic acid) film-depended impedimetric biosensor for *S. aureus* detection (26). They referred that the preparation of the imprinted film was eco-friendly and appropriate, which was in situ invested on the gold electrode out of the cross-linkers and organic solvents usage. After the characterization of the imprinted film, they observed that it had a new formation without *S. aureus*-shaped cavities obtained in the matrices (Figure 6.4). Then, they investigated the optimum monitoring performance with different factors and acquired a highly fast recognition (10 min) and a low detection limit value (2 CFU/mL) in a broad range ($10–10^8$ CFU/mL). Su et al. prepared an imprinted magnetic membrane-based

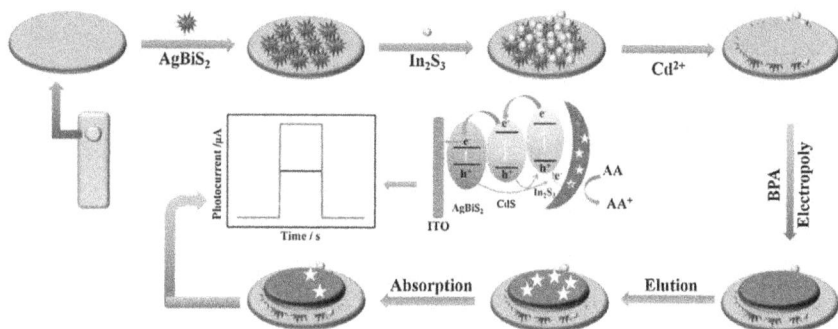

FIGURE 6.3 Fabrication steps of the MIPs-based biosensor for BPA detection. Reprinted from ref. (50) with permission.

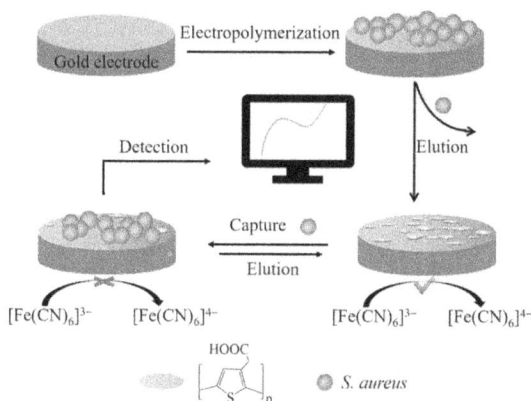

FIGURE 6.4 Preparation orders of imprinted film-based biosensor for *S. aureus* detection. Reprinted from ref. (26) with permission.

biosensor for acetaminophen determination (51). They first screened the available monomers and solvents to prepare polymer and calculated the electrostatic potentials, and then synthesized magnetic molecularly imprinted membrane with core-shell structure-based $Fe_3O_4@SiO_2$ nanoparticles. Following this process, they loaded with neodymium-iron-boron to obtain a magnet to the carbon paste electrode and got a magnetic biosensor via adding a polymer to the electrode surface (Figure 6.5). They reported that the response of the biosensor demonstrated a linear behavior on the various acetaminophen concentration (6×10^{-8} to 5×10^{-5} mol/L and 5×10^{-5} to 2×10^{-4} mol/L) with a low limit of detection (1.73×10^{-8} mol/L). Finally, they used actual samples with high recovery values (95.80–103.76%) and low relative standard deviation values (0.78%-3.05%).

Tran et al. reported a biosensor that is based on single-walled carbon nanotubes with picky antibodies towards a secreted protein biomarker (Sec-delivered effector-SDE 1) for citrus greening detection (52). The biosensor can determine SDE1

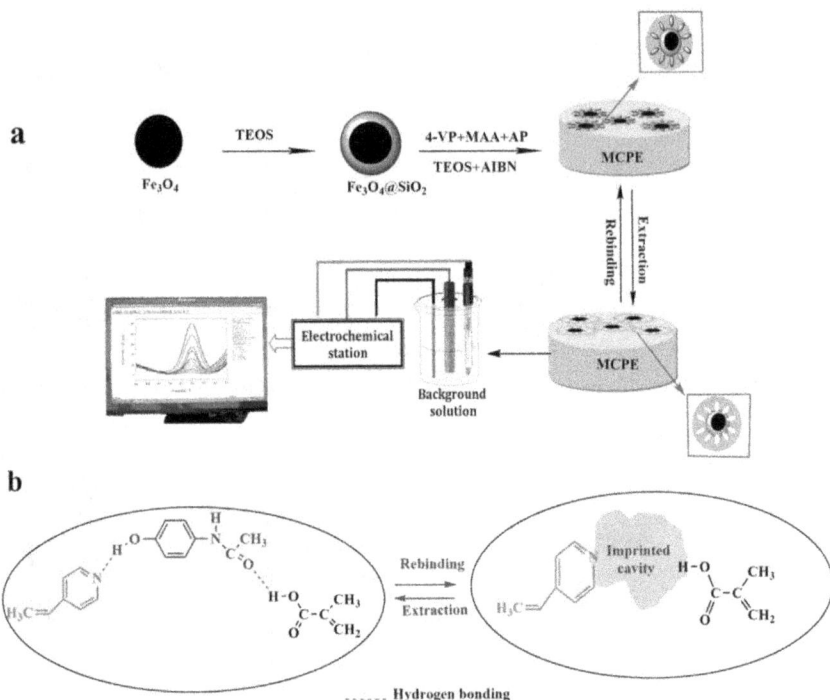

FIGURE 6.5 (a) Schematic diagram of MMIP/MCPE; (b) template extraction-rebinding mechanism.

biomarkers for citrus greening in plant tissue extracts with a wide concentration range and calculate the detection limit as 5 nM. They also showed the usage of the standard assay method with this biosensor to reach a 90% recovery of signals in concentrated extract, allowing quantitative detection from external calibration. Bakhshpour and Denizli proposed an optical biosensor platform that is able to detect Cd(II) ions in a short time (53). As depicted in Figure 6.6, they modified gold surfaces with ally mercaptan and then prepared different biosensors surfaces including nanofilm, polymeric nanoparticles, and gold nanoparticles, and utilized them for detection of Cd(II) ions, respectively. They reported that the combination of the signal improving properties of nanoparticles and imprinting method supplied a selective and sensitive detection platform with a comparatively low detection limit value (0.01 μg/L) which is less than the value declined by the World Health Organization. They also performed the detection of Cd(II) ions from wastewater solutions in real-time.

Dinu and Apetrei described the screen-printed carbon electrodes-based biosensor altered by different electroactive compounds (cobalt phthalocyanine, Meldola's blue, and Prussian blue) for phenylalanine detection, one by one (54). They depicted the Prussian blue-based biosensors' signals with the maximum performances regarding the sensitivity and electrochemical kinetics. They obtained a linear calibration in a range between 0.33 and 14.5 μM with a low limit of detection ($1.23 \ 10^{-8}$ M) and a low standard deviation (3%). They also used this biosensor for the determination

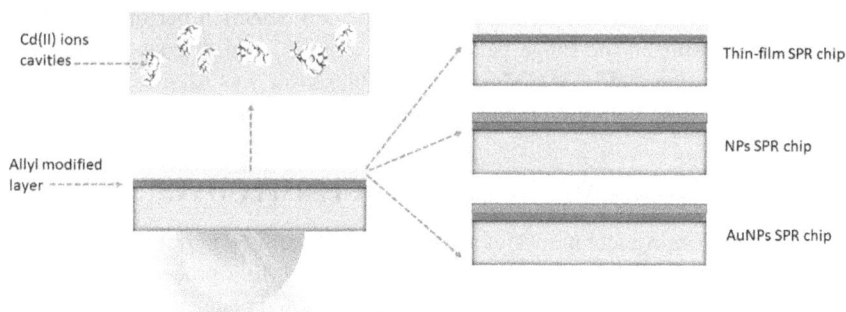

FIGURE 6.6 Representation of different biosensor surfaces for Cd(II) ion detection. Reprinted from ref. (53) with permission.

of phenylalanine through calibration and other additional methods on pure and/or multi-component pharmaceutical samples. Rebelo et al. developed an imprinted polymer-based electrochemical biosensor for the quantification of azithromycin. Before performing experiments, they selected the functional monomer by molecular modelling utilizing quantum mechanics calculations (55). They used electropolymerization of MIPs via cyclic voltammetry on a screen-printed carbon electrode using a 4-aminobenzoic acid solution. The azithromycin detection was carried out by differential pulse voltammetry in the range of 0.5–10.0 mM with a limit of detection (0.08 mM) and quantification (0.3 mM). The biosensor had high selectivity to successfully monitor azithromycin in the tap and river water. Chen et al. developed a fluorescence biosensor that depended on the protamine-induced carbon quantum dots aggregation for trypsin detection (56). They observed that the negatively charged carbon quantum dots fluorescence was quenched with protamine owing to the electrostatic interaction of dots and protamine. Then, the positively charged protamine was chosen to be hydrolyzed via trypsin. Trypsin can stimulate the dots and protamine deaggregation, so it enables carbon quantum dots release to restore the intensity. The biosensor can detect trypsin in the range from 25 to 500 ng/mL with a limit of detection value (8.08 ng/mL). Lv and Gao structured an optical biosensor depending on the electron transfer between quantum dots and phenol (57). This biosensor has a room-temperature phosphorescence feature of the quantum dots. They used double crosslinkers (divinylbenzene, and ethylene glycol dimethacrylate) for polymerization without stabilizer and additive for the preparation of the biosensor to obtain a high addition to the polymerization with hydrogen-bond interaction. After that, a carboxylic acid was transfused onto the surface of quantum dots and MIPs owing to the functional monomer (methacrylic acid). They obtained a high correlation coefficient (0.99) in the phenol detection in the range from 5.0 to 55 mmol/L with a high imprinting factor (3.43), as well. Zhao et al. constructed a fluorescent biosensor utilizing the surface molecular imprinting method for bovine hemoglobin detection (58). They employed a thermo-sensitive imprinted polymer that combined with the silanized carbon dots, and N-isopropyl acrylamide. Thanks to the composition of the strong fluorescence property of carbon dots and the high selectivity of the imprinted shell, the biosensor demonstrated high detection performance to the bovine hemoglobin in the range between 0.31 and 1.55 µM of concentration with a limit of detection (1.55 µM). Moreover, the biosensor was used to sense bovine hemoglobin in urine with high recovery value (98.6–100.5%).

6.4 CLOSING REMARKS AND FUTURE PERSPECTIVES

Molecular imprinting techniques can form highly specific and durable recognition sites (specific cavities) on the biosensor surface, owing to a procedure where chosen functional monomers are polymerized around a template molecule. Here, we reviewed the molecular imprinting technique and studies based on biosensor applications. Significant progress in the MIPs synthesis and its various applications have been published nonstop in the literature in the latest years. The perfect stability, low cost, and continuous development of the performance of MIPs make these smart polymers the most encouraging synthetic materials for specific recognition in different application fields. Molecular imprinting-based biosensor recognition surfaces can be designed as thin films, nanoparticles, or different combinations. Molecularly imprinted studies with biosensors that can control the thickness of the polymer film and ultimately provide sensitivity detection have been reported. Recently, important attention has been conducted to the growth of MIPs as highly selective materials for biosensors surface design. By combining molecular imprinting technology with sensing testing to achieve target detection, it offers further possibilities to improve the sensitivity of optical, electrochemical, and piezoelectric-based biosensor systems.

REFERENCES

1. Vasapollo G, Sole R Del, Mergola L, Lazzoi MR, Scardino A, et al. 2011. Molecularly imprinted polymers: Present and future prospective. *International Journal of Molecular Sciences*. 12: 5008–5945
2. Blanco-López MC, Lobo-Castañón MJ, Miranda-Ordieres AJ, Tuñón-Blanco P. 2004. Electrochemical sensors based on molecularly imprinted polymers. *TrAC Trends in Analytical Chemistry*. 23(1): 36–48
3. Wulff G. 1995. Molecular imprinting in cross-linked materials with the aid of molecular templates— a way towards artificial antibodies. *Angewandte Chemie International Edition English*. 34(17): 1812–1832
4. Haupt K, Mosbach K. 2000. Molecularly imprinted polymers and their use in biomimetic sensors. *Chemical Reviews*. 100(7): 2495–2504
5. Dickert FL. 2018. Molecular imprinting and functional polymers for all transducers and applications. *Sensors (Switzerland)*. 18: 327
6. Hayden O. 2016. One binder to bind them all. *Sensors (Switzerland)*. 16: 1665
7. Hayden O, Lieberzeit PA, Blaas D, Dickert FL. 2006. Artificial antibodies for bioanalyte detection—sensing viruses and proteins. *Advanced Functional Materials*. 16(10): 1269–1278
8. Uzun L, Turner APF. 2016. Molecularly-imprinted polymer sensors: Realising their potential. *Biosensors and Bioelectronics*. 76: 131–144
9. Lowdon JW, Diliën H, Singla P, Peeters M, Cleij TJ, et al. 2020. MIPs for commercial application in low-cost sensors and assays—an overview of the current status quo. *Sensors & Actuators, B: Chemical*. 325: 128973

10. Ye L, Haupt K. 2004. Molecularly imprinted polymers as antibody and receptor mimics for assays, sensors and drug discovery. *Analytical and Bioanalytical Chemistry*. 378: 1887–1897

11. Saylan Y, Akgönüllü S, Yavuz H, Ünal S, Denizli A. 2019. Molecularly imprinted polymer based sensors for medical applications. *Sensors (Switzerland)*. 19(6): 1279

12. Gómez-Arribas LN, Urraca JL, Benito-Penìa E, Moreno-Bondi MC. 2019. Tag-specific affinity purification of recombinant proteins by using molecularly imprinted polymers. *Analytical Chemistry*. 91(6): 4100–4106

13. Donato L, Drioli E. 2021. Imprinted membranes for sustainable separation processes. *Frontiers of Chemical Science and Engineering*. 15

14. Mohamed S, Balieu S, Petit E, Galas L, Schapman D, et al. 2019. A versatile and recyclable molecularly imprinted polymer as an oxidative catalyst of sulfur derivatives: A new possible method for mustard gas and v nerve agent decontamination. *Chemical Communications*. 55(88): 13243–1346

15. Wackerlig J, Lieberzeit PA. 2015. Molecularly imprinted polymer nanoparticles in chemical sensing—synthesis, characterisation and application. *Sensors & Actuators, B: Chemical*. 207(Part A): 144–157

16. Muhammad P, Tu X, Liu J, Wang Y, Liu Z. 2017. Molecularly imprinted plasmonic substrates for specific and ultrasensitive immunoassay of trace glycoproteins in biological samples. *ACS Applied Materials & Interfaces*. 9(13): 12082–12091

17. Parisi OI, Morelli C, Puoci F, Saturnino C, Caruso A, et al. 2014. Magnetic molecularly imprinted polymers (MMIPs) for carbazole derivative release in targeted cancer therapy. *Journal of Materials Chemistry B*. 2: 6619–6625

18. Tamahkar E, Bakhshpour M, Denizli A. 2019. Molecularly imprinted composite bacterial cellulose nanofibers for antibiotic release. *Journal of Biomaterials Science, Polymer Edition*. 30(6): 450–461

19. Piletsky S, Canfarotta F, Poma A, Bossi AM, Piletsky S. 2020. Molecularly Imprinted Polymers for Cell Recognition. *Trends in Biotechnology*. 38(4): 368–387

20. Saylan Y, Tamahkar E, Denizli A. 2017. Recognition of lysozyme using surface imprinted bacterial cellulose nanofibers. *Journal of Biomaterials Science, Polymer Edition*. 28(16): 1950–1965

21. Saylan Y, Erdem Ö, Cihangir N, Denizli A. 2019. Detecting fingerprints of waterborne bacteria on a sensor. *Chemosensors*. 7(3): 33

22. Zhang L, Liu Z, Xiong C, Zheng L, Ding Y, Lu H. 2018. Selective recognition of Histidine enantiomers using novel molecularly imprinted organic transistor sensor. *Organic Electronics*. 61: 254–260

23. Fang M, Zhuo K, Chen Y, Zhao Y, Bai G, Wang J. 2019. Fluorescent probe based on carbon dots/silica/molecularly imprinted polymer for lysozyme detection and cell imaging. *Analytical and Bioanalytical Chemistry*. 411: 5799–5807

24. Esentürk MK, Akgönüllü S, Yılmaz F, Denizli A. 2019. Molecularly imprinted based surface plasmon resonance nanosensors for microalbumin detection. *Journal of Biomaterials Science, Polymer Edition*. 30(8): 646–661

25. Bartold K, Pietrzyk-le A, Huynh T, Sosnowska M, Noworyta K, et al. 2017. Programmed transfer of sequence information into a molecularly imprinted polymer for Hexakis (2, 2′ -bithien-5-yl) DNA analogue formation toward single-nucleotide-polymorphism detection. *ACS Applied Materials & Interfaces*. 9: 3948–3958

26. Wang R, Wang L, Yan J, Luan D, Wu J, Bian X. 2021. Rapid, sensitive and label-free detection of pathogenic bacteria using a bacteria-imprinted conducting polymer film-based electrochemical sensor. *Talanta*. 226: 122135

27. Zhao X, He Y, Wang Y, Wang S, Wang J. 2020. Hollow molecularly imprinted polymer based quartz crystal microbalance sensor for rapid detection of methimazole in food samples. *Food Chemistry*. 309: 125787

28. Zarejousheghani M, Rahimi P, Borsdorf H, Zimmermann S, Joseph Y. 2021. Molecularly imprinted polymer-based sensors for priority pollutants. *Sensors.* 21: 2406

29. Akgönüllü S, Yavuz H, Denizli A. 2020. SPR nanosensor based on molecularly imprinted polymer film with gold nanoparticles for sensitive detection of aflatoxin B1. *Talanta.* 219(May): 121219

30. Safran V, Göktürk I, Derazshamshir A, Yılmaz F, Sağlam N. 2019. Rapid sensing of Cu + 2 in water and biological samples by sensitive molecularly imprinted based plasmonic biosensor. *Microchemical Journal.* 148: 141–150

31. Battal D, Akgönüllü S, Yalcin MS, Yavuz H, Denizli A. 2018. Molecularly imprinted polymer based quartz crystal microbalance sensor system for sensitive and label-free detection of synthetic cannabinoids in urine. *Biosensors and Bioelectronics.* 111: 10–17

32. Akgönüllü S, Battal D, Yalcin MS, Yavuz H, Denizli A. 2020. Rapid and sensitive detection of synthetic cannabinoids JWH-018, JWH-073 and their metabolites using molecularly imprinted polymer-coated QCM nanosensor in artificial saliva. *Microchemical Journal.* 153: 104454

33. Goyal G, Bhakta S, Mishra P. 2019. Surface Molecularly Imprinted Biomimetic Magnetic Nanoparticles for Enantioseparation. *ACS Applied Nano Materials.* 2(10): 6747–6756

34. Belbruno JJ. 2019. Molecularly Imprinted Polymers. *Chemical Reviews.* 119(1): 94–119

35. Wanekaya AK, Chen W, Mulchandani A. 2008. Recent biosensing developments in environmental security. *Journal of Environmental Monitoring.* 10: 703–712

36. Moreno-Bondi M, Navarro-Villoslada F, Benito-Pena E, Urraca J. 2008. Molecularly imprinted polymers as selective recognition elements in optical sensing. *Current Analytical Chemistry.* 4(4): 316–340

37. Peltomaa R, Glahn-Martínez B, Benito-Peña E, Moreno-Bondi MC. 2018. Optical biosensors for label-free detection of small molecules. *Sensors (Basel).* 18: 4126

38. Sellergren BJ, Hall A. 2012. Molecularly imprinted polymers. In *Supramolecular Chemistry: From Molecules to Nanomaterials*, ed. PA Gale, JW Steed. John Wiley & Sons, Ltd.

39. Leibl N, Haupt K, Gonzato C, Duma L. 2021. Molecularly imprinted polymers for chemical sensing: A tutorial review. *Chemosensors.* 9: 123.

40. Saylan Y, Denizli A. 2018. Molecular fingerprints of hemoglobin on a nanofilm chip. *Sensors (Switzerland).* 18: 3016

41. Prathish KP, Vishnuvardhan V, Rao TP. 2009. Rational ldesign of in situ monolithic imprinted polymer membranes for the potentiometric sensing of diethyl chlorophosphate—a chemical warfare agent simulant. *Electroanalysis.* 21(9): 1048–1056

42. Li H, Wang Z, Wu B, Liu X, Xue Z, Lu X. 2012. Rapid and sensitive detection of methyl-parathion pesticide with an electropolymerized, molecularly imprinted polymer capacitive sensor. *Electrochimica Acta.* 62: 319–326

43. Rezaei Z, Mahmoudifard M. 2019. Pivotal role of electrospun nanofibers in microfluidic diagnostic systems-a review. *Journal of Materials Chemistry B.* 7(30): 4602–4619

44. Qureshi A, Gurbuz Y, Niazi JH. 2012. Biosensors for cardiac biomarkers detection: A review. *Sensors & Actuators, B: Chemical.* 171–172: 62–76

45. Tibuzzi A, Rea G, Pezzotti G, Esposito D, Johanningmeier U, Giardi MT. 2007. A new miniaturized multiarray biosensor system for fluorescence detection. *Journal of Physics: Condensed Matter.* 19: 395006

46. Rahtuvanoğlu A, Akgönüllü S, Karacan S, Denizli A. 2020. Biomimetic nanoparticles based surface plasmon resonance biosensors for histamine detection in foods. *ChemistrySelect.* 5(19): 5683–5692

47. Heller DA, Baik S, Eurell TE, Strano MS. 2005. Single-walled carbon nanotube spectroscopy in live cells: Towards long-term labels and optical sensors. *Advanced Materials.* 17(23): 2793–2799

48. Zhang J, Wang Y, Lu X. 2021. Molecular imprinting technology for sensing food-borne pathogenic bacteria. *Analytical and Bioanalytical Chemistry*. doi:10.1007/s00216-020-03138-x

49. Siqueira M, Patricia A, Tavares M, Felipe L, Coelho L, et al. 2021. Biosensors and Bioelectronics rational selection of hidden epitopes for a molecularly imprinted electrochemical sensor in the recognition of heat-denatured dengue NS1 protein. *Biosensors and Bioelectronics*. 191: 113419

50. Xu R, Du Y, Liu L, Fan D, Yan L, et al. 2021. Molecular imprinted photoelectrochemical sensor for bisphenol A supported by flower-like AgBiS 2/In 2 S 3 matrix. *Sensors & Actuators, B: Chemical*. 330: 129387

51. Su C, Li Z, Zhang D, Wang Z, Zhou X, et al. 2020. A highly sensitive sensor based on a computer-designed magnetic molecularly imprinted membrane for the determination of acetaminophen. *Biosensors and Bioelectronics*. 148: 111819

52. Tran T, Clark K, Ma W, Mulchandani A. 2020. Biosensors and bioelectronics detection of a secreted protein biomarker for citrus Huanglongbing using a single-walled carbon nanotubes-based chemiresistive biosensor. *Biosensors and Bioelectronics*. 147: 111766

53. Bakhshpour M, Denizli A. 2020. Highly sensitive detection of Cd(II) ions using ion-imprinted surface plasmon resonance sensors. *Microchemical Journal*. 159: 105572

54. Dinu A, Apetrei C. 2020. Voltammetric determination of phenylalanine using chemically modified screen-printed based sensors. *Chemosensors*. 8(4): 113

55. Rebelo P, Pacheco J̃ o G, Natalia, M.Cordeiro DS, Melo Á, Delerue-Matos C. 2020. Analytical methods azithromycin electrochemical detection using a molecularly imprinted polymer prepared on a disposable screen-printed electrode. *Analytical Methods*. 12: 1486–1494

56. Chen Y, Lin Z, Miao C, Cai Q, Li F, et al. 2020. A simple fluorescence assay for trypsin through a protamine-induced carbon quantum dot-quenching aggregation platform. *RSC*. 10: 26765–26770

57. Lv X, Gao P. 2020. An optical sensor for selective detection of phenol via double cross-linker precipitation polymerization. *RSC Advances*. 10: 25402–25407

58. Zhuo YZ, Chen Y, Fang M, Tian Y, Bai G, Kelei Z. 2020. Silanized carbon dot-based thermo-sensitive molecularly imprinted fluorescent sensor for bovine hemoglobin detection. *Analytical and Bioanalytical Chemistry*. 412: 5811–5817

Part IV

Applications of Biosensors

7 Unprecedented Innovations in Electrochemical Biosensing Approaches for Medical Applications

Susana Campuzano[1], María Pedrero[1], Maria Gamella[1], Rebeca M. Torrente-Rodríguez[1], and José M. Pingarrón[1]
[1]Universidad Complutense de Madrid
Departamento de Química Analítica
Facultad de CC. Químicas
Madrid, Spain

CONTENTS

7.1 INTRODUCTION

Disease management is gradually shifting from a diagnosis and treatment strategy based on empirical clinicopathological observations to one that focuses more on molecular profiling by targeting specific candidate biomarkers to move toward a more efficient and personalized therapeutic intervention, applicable not only in

DOI: 10.1201/9781003189435-11

hospitals, but also in outpatient and even home settings. However, this transition and the clinical adoption of these candidate biomarkers requires, apart from their standardization, the development of new technologies for their determination that surpass the limitations of conventional ones for their application in different settings and that use simple and cost-effective methods within clinically actionable timeframes.

Recognizing that the evolution of human diseases involves a highly dynamic and interactive system of multiple layers of molecular markers, precision medicine aims to provide a detailed characterization of each disease to personalize healthcare. Moreover, it is now fully accepted that the simultaneous analysis of molecular markers from different omics layers (e.g. genetics, epigenetics, mRNA transcripts and proteins) leads to novel strategies for early detection of predisposition to prevalent and high-mortality diseases, thus improving their prevention and treatment. In this regard, features such as the versatility to profile multiple biomarkers at different omics levels even independently of healthcare institutions, simplicity, affordable cost, significantly shorter analysis time and the smaller amount of sample required for analysis compared to conventional or state-of-the-art omics methodologies, make electrochemical biosensing and bioplatforms particularly promising alternatives for this purpose.

With this in mind, this chapter provides a critical and up-to-date overview of the remarkable innovations in electrochemical biosensors in terms of the environment/ format in which they are used, mode of operation, other notable attributes already demonstrated, and success in particularly challenging and pioneering applications to decisively assist in personalized early diagnosis and/or staging of the diseases of our time.

The giant steps being taken at the forefront of electrochemical biosensing, echoed in this chapter, proving the development of electroanalytical bioplatforms potentially transferable to the clinic due to their simplicity, cost, assay time, versatility, multiplexing capacity and decentralized character, invite to expect the birth of new simple, versatile and affordable devices, applicable even in outpatient or home environments, that will play a leading role both in clinical routine and in our daily lives. These bio-devices are ready to validate candidate biomarkers, manage human diseases or address unexpected global health challenges in record time, such as the current coronavirus pandemic, in a personalized and early manner. This will bring unprecedented benefits in minimizing the spread of infections, improving patient statistics and quality of life, as well as alleviating the cost of treatment for healthcare systems and the emotional burden on families.

7.1.1 ELECTROCHEMICAL BIOSENSING IN THE CURRENT SOCIETY AND CLINICS

The development of electrochemical biosensors for clinical analysis is nowadays a booming discipline. Conventional methods used for the determination of clinical biomarkers or viruses are mostly lab-based and imply laborious sample preparation, high-cost non-portable instrumentation, skilled personnel and time-consuming procedures. Meanwhile, electrochemical platforms stand out for exploitable characteristics such as their simplicity, low cost easily miniaturized instrumentation and rapid response, together with the possibility to work with small sample amounts and their easy integration with enzymes, antibodies/antigens, other proteins and nucleic acids. All these advantages result in a great potential to develop outstanding devices for next generation medical diagnosis and prognosis.

Currently, electrochemical bioplatforms have demonstrated their usefulness for a wide range of clinical applications such as cancer detection (1), the determination of salivary biomarkers (2), tumor cells (3), cholesterol (4), viruses (5–6) etc. Such electrochemical bioplatforms also show diverse configurations involving screen-printed electrodes (7), wearable sensors (8–9), including micro-needle-based devices (10), field-effect transistors (11), microfluidic devices (12), different nanomaterials (1, 4, 13–14), graphene (12) and multienzyme systems (15). The main challenges these devices need to face is the way to make them easily available and affordable everywhere in the world, and user-friendly for unexperienced people. In this sense, the development of wearable systems in the form of tattoos and/or using Bluetooth to be connected to mobile phones apps have garnered great interest. This chapter describes and discusses novel approaches used in the development of electrochemical biosensing devices with application in the medical field.

7.2 UNPRECEDENTED INNOVATIONS IN ELECTROCHEMICAL BIOSENSORS

7.2.1 ENVIRONMENT/FORMAT

Wearable and implantable devices or electronics have rapidly entered the area of health and biomedical applications, including monitoring, tracking, and recording of vital signs with the aim of improving people's health. These devices can be worn on different parts of the body such as head, arm, and leg/foot skin or they can be implantable, such as smart pills, subcutaneous sensors, or artificial organs. They can be defined as on-body portable electronic devices (which can be flexible or not) that allow real-time monitoring of physiological signals. Many wearable devices are constructed for the measurement and sensing of biophysical parameters (16), including heart rate, body temperature or blood pressure (17). However, the track of wearable devices has branched out to the development of clinical tools able to recognize target analytes in complex biological samples, to determine the patterns of a particular disease, thus providing its improved understanding, and to monitor the health information of users (18). Therefore, these biodevices have the potential to revolutionize healthcare and disease management since they can assist in the measurement of circulating analyte concentrations in a dynamic and minimally invasive manner, enhancing the management of diseases, and alerting the user or medical professionals of abnormal situations to follow-up on the wearer´s health status (19–20).

Among biosensors, electrochemical biosensors are very promising to build wearable bioplatforms since they can be manufactured at production scale, are compatible with a great variety of materials and can operate with cheap battery powered electronic instrumentation (21). Over the past decades, electrochemical wearable sensors have been developed for detecting toxic chemicals (22), clinical biomarkers (23–24) and chemical warfare agents (25). In this context, the development of wearable and implantable devices has ushered the development of new devices mated with the human body for healthcare, entertainment, and sports applications, where electrochemical sensors play a key role (26).

Wearable and implantable devices can be classified into invasive and non-invasive devices (Figure 7.1). Invasive sensors involve the analysis of body fluids, such as blood or plasma, obtained by incising into the body. However, their use for

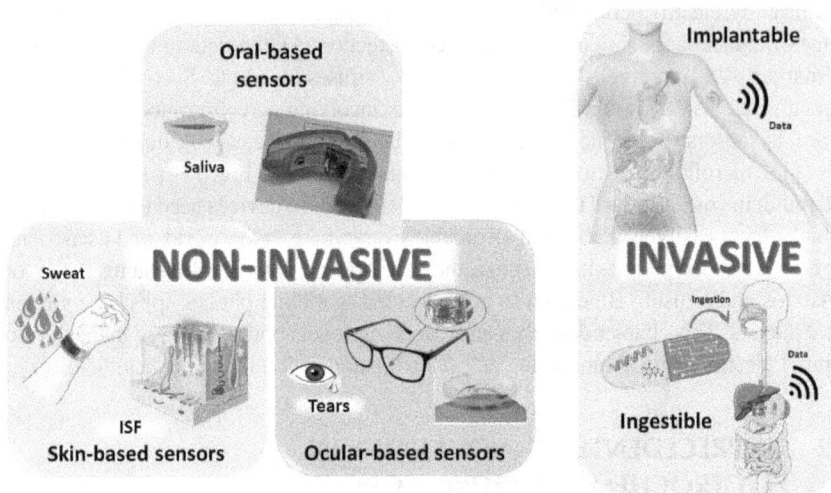

FIGURE 7.1 Overview of wearable and implantable devices. Adapted from (30–32) with permission from Elsevier.

continuous monitoring of parameters such as glucose, athletes' fitness monitoring, oxygen saturation, cholesterol tracking, and drug efficacy is not always suitable since blood contamination is probable, as needles can be misused (27). Nevertheless, the increasing demand and interest in developing such sensors have led to intravascular and subcutaneous applications, which, compared with the formers, allow continuous monitoring, simple operation, high reliability and less discomfort for the patient. Conversely, non-invasive sensors are more attractive to the user due to their ability to avoid painful blood sampling procedures since they rely on the analysis of body fluids such as saliva, sweat, tears, urine or skin interstitial fluid (28), whose extraction poses minimal risk of harm or infection, and they are generally more user friendly. Although the analysis of these biofluids is also carried out in vitro, the advent of flexible electronics has led to an emerging development of wearable biosensors capable of non-invasive continuous real-time monitoring of biomarkers that can be related to the wearer's health and performance (19, 27, 29).

7.2.1.1 Non-Invasive Wearable Electrochemical Biosensors

Non-invasive wearable biosensors include watches, clothing, bandages, glasses, contact lenses, tattoos and rings, which are conveniently attached to a person's body, and target analytes in tears, saliva, sweat and skin interstitial fluid. These kinds of devices can be classified according to the part of the body on which they are used, in skin, ocular and oral cavity devices.

Skin Wearable Biosensors

Skin-based electrochemical devices are the most interesting wearable sensors since the skin has the largest organ interface. In addition to the physical parameters measured from existing skin-worn biodevices, skin is a rich source of biomarkers valuable in diagnosis and monitoring of diseases. Moreover, it allows non-invasive access to sweat and interstitial fluid (ISF), which are very different in nature but provide a wealth of biochemical information (33).

Sweat contains a great variety of biomarkers such as electrolytes, metabolites, small molecules (glucose, lactate, hormones, urea), small proteins (cytokines) and peptides as well as environmental contaminants. Therefore, sweat is one of the most targeted biofluids for the development of non-invasive wearable biosensors (34). Sweat flow rate is very low for practical sensing applications, so most skin-worn electrochemical sensors sample the sweat using iontophoretic extraction or through exercise activity. Initially, sweat biosensors focused on the development of platforms able to capture sweat during exercise and detect a single bioanalyte (35). However, recently, sweat biosensing has relied on the use of various detection mechanisms, substrates, nanomaterials and the analysis of different target analytes (33), performed directly or using microfluidic systems, and even providing multiplexed information (36–37).

When less active physical states make sweat insufficient, iontophoresis is used. It consists of applying a mild electric current across the skin, facilitating the release of sweat-inducing small molecules into the dermis. Electrochemical sensing combined with iontophoresis has been applied for monitoring of cystic fibrosis (38), alcohol (39), lactate (36), glucose (36, 40–41) and urea (42).

On the other hand, ISF is an attractive alternative for biomedical applications. ISF is the fluid surrounding the cells and it is in constant equilibrium with blood capillaries through diffusion, therefore it has the closest composition to the blood (43). There are basically two approaches to access this biofluid: i) Reverse iontophoresis (RI) or ii) the use of epidermal microneedle devices. The current applied during RI induces ions to migrate from the inside of the skin to the surface carrying small non-charged molecules, such as glucose, within the electro-osmotic flow. RI extraction and amperometric detection were used for tracking ISF glucose levels (40) and, simultaneously, glucose and alcohol levels (44). Microneedles are used to puncture the outer skin layer to access ISF with no damage, providing minimally invasive wearable platforms for monitoring ISF biomarkers (45–46). Microneedles were used for the electrochemical detection of lactate (47–49), alcohol (50), L-dopa (51), tyrosinase (52) and glucose (53–55).

Skin flexible wearable devices have been implemented through direct transfer of sensors onto the skin, using electronic skin (e-skin) (56–57) or printed temporary tattoos (58–60), by sensor incorporation into wristbands (61) and patches (62) or embedding sensors directly into textiles (21, 29, 63).

Ocular Wearable Biosensors

Tears are a biological fluid that can be used for monitoring physiological status since biomarkers diffuse from the blood and their levels can be correlated with concentrations in blood (64–65). Human tears are a complex fluid containing a variety of compounds such as proteins, peptides, lipids, metabolites, enzymes, electrolytes, and water (66). However, it contains fewer proteins than other biofluids due to the blood-tear barrier filtering process, which makes tears an attractive candidate fluid for non-invasive monitoring since sensor surface biofouling may be lower (45).

Traditional tear sensors rely on the extraction of human tear samples followed for in vitro analysis (56). However, sampling of tears for in vitro diagnoses shows several limitations such as small sample volumes, easy evaporation during sample collection, and the need to develop collection methods (19, 28). In this context, smart contact lens-based systems are an attractive solution to tear collection as they can be worn without eye irritation (66–67). They do not require any surgery and can integrate biosensing, data processing and power sources within the contact lens (68–70). Biomedical applications of smart lenses include the detection of glucose (30, 68, 70–72), lactate (73), and monitoring the glaucoma progression (17, 74).

Eyeglasses-based sensors have also been developed for non-invasive detection of biomarkers in tears (75). Sempionato et al. developed a device on eyeglasses involving an integrated microfluidic system built on the nose bridge pad to collect and analyze glucose, alcohol and vitamins in stimulated tear samples (32).

Oral Cavity Wearable Biosensors

Saliva is a complex oral fluid composed of metabolites, enzymes, hormones, proteins, microorganisms and ions. Several of these components have been targeted in clinical settings because they provide meaningful diagnostic information (76–77). Thus, the interest in saliva as a diagnostic fluid has advanced rapidly in recent years. Since saliva can be easily collected, it has been mostly used in the development of in vitro diagnostic biosensors on strips or portable platforms (2, 78).

In addition, the capabilities exhibited by oral sensors and the good correlations between blood and salivary metabolite levels (79–80), have led to the development of wearable in-mouth devices. Their development is still challenging due to potential biofouling, contamination (from food or beverages), and safety issues, as well as the need for effective encapsulation and the use of biocompatible materials (19). Nevertheless, several oral-cavity biosensing platforms have been described using the mouthguards for the detection of lactic acid (81), uric acid (31), glucose (82), intraoral dental accessories (83–84) or even foodstuff such as lollipops (85).

Fabrication Methods and Materials

Wearable chemical sensors achieve the best performance when they are in tight contact with the underlying tissue, which makes it important to fabricate sensors on flexible or even stretchable substrates, including plastics, textiles, tattoos, paper and stretchable elastomers (17).

Several fabrication methods have been developed for flexible and stretchable wearable electrochemical biosensors including lithographic and printing tools. Lithographic methods such as thin-film deposition and etching, photolithography and ion-beam lithography can be used to reproducibly fabricate high-performance devices (36, 86–87). Despite their attractive attributes, they come at a high cost due to the cleanroom setup, multiple equipment acquisitions, complex processes, and the unique materials required (26).

Printing technologies for electrochemical devices can be classified into template and non-template methodologies. Non-template methods are primarily inkjet and 3D printing. Inkjet printing is an easy-to-use and low-cost deposition technique to simultaneously deposit and pattern thin films from liquid materials. It does not need complicated equipment and is compatible with processing flexible and non-planar substrates (88). 3D printing techniques fabricate three-dimensional objects using computer-aided design models by depositing layer by layer the feedstock material until the part is complete (89). The 3D printing technologies for sensors fabrication are usually categorized into photopolymerization, sheet lamination, materials extrusion, binder jetting, power bed fusion, direct energy deposition and electrostatic force has driven (90).

Template-based printing processes can be further classified into screen-printing, gravure printing, flexography, and imprinting. Screen-printing is well known as a stencil printing process in which an emulsion screen is used as a template to produce a design pattern. It has been widely used owing to an easy fabrication at an industrial scale, low cost, robustness, and attractive electroanalytical performance of the resulting substrates (23, 91). However, some important factors such as properties of the substrate, ink composition and manufacturing conditions must be considered when the devices are used for electrochemical measurements because the performance can be compromised.

Flexography and gravure printing are also used. While the ink is on the ridges of the pattern on the printing cylinder in flexography (92–93), gravure printing relies on impressing the film into the cavities of the roll where the ink resides (94–95). These technologies are intrinsically robust and large-area compatible, which make them suitable for industrial-scale production devices (96). Additionally, while screen-printing technology has only been applied to flat surfaces, the use of stamp transfer technologies allows printing electrochemical sensors on non-planar substrates, which sometimes is necessary due to the surface irregularities characteristic of the required materials and substrates such as the skin (23).

The materials used for fabrication depend on the sensor application, the materials availability and the total cost of manufacturing. Common materials for flexible substrates include polymers, silicones and rubbers due to their inherent high stretchability, low toxicity, hydrophobicity, and good workability. Among them, polydimethylsiloxane (PDMS), polyethylene terephthalate (PET), polyimide (PI), parylene and polypyrrole have been used to develop flexible sensors (97). Electrode inks can be made of different conducting materials like carbon nanomaterials including graphene, carbon nanotubes, carbon fibers and graphite (98–99). These materials provide extraordinary conductivity and mechanical properties. Metals have also been used in wearable biosensors. Nanowires (NWs) or nanoparticles (NPs) are often used as fillers to prepare piezoresistive composites and conductive inks (100). Among metallic nanoparticles, silver, gold, and nickel are some of the most used in the preparation of flexible wearable sensors (99, 101). In addition, conductive polymers such as poly 3,4-ethylene dioxythiophene (PEDOT) and polyvinylidene difluoride (PVDF) have also been utilized due to their thermal stability, high transparency, and tunable conductivity (102).

Fabrics are attractive substrates since, in contrast to plastics, their mechanical properties are well understood, and they are relatively similar to the skin. However, textiles can keep continuous and intimate contact with skin only at locations where textiles can be tightly worn. Additionally, not all fabrics have functional chemical groups that facilitate efficient adherence of the overlaying printed electrodes. Moreover, washing can easily degrade textiles under laundry conditions (29).

Traditional fabrication methods of conventional textiles can be used in e-textiles production since flexible conductive yarns are like conventional textile yarns, enabling traditional fabrication methods to merge conductive threads with non-conductive threads (21). Conductive yarns are incorporated into the textile structure by different technologies such as weaving embroidering, knitting, printing, and chemical treatments (63). Coating non-conductive yarns with metals using methods such as electroless plating, chemical vapor deposition, sputtering, and conductive polymer coating also enables an e-textile production (29). Stamping conductive inks, usually made of metals such as silver, copper and gold, is also an alternative to embed conductive lines into textiles with high conductive metals. Conductive threads with the immobilized enzyme were embroidered into textiles to develop a textile electrochemical biosensor for glucose (103), and yarn coating with conductive inks was handloom-woven as electrodes into patches of fabric to construct arrays of sensors for the determination of hemoglobin and glucose (104). Moreover, carbonization of silk coated with multi-walled carbon nanotubes was used to prepare electrochemical biosensors for glucose (105).

Tattoo-like electrochemical biosensors are attractive because of their intimate contact with the human skin without causing much discomfort on the body (96). The tattoo sensors can be fabricated using inexpensive screen-printing processes and the incorporation of elastomeric binders and self-healing agent loaded microcapsules allow self-heal after scratching (59). The resulting sensors exhibit robust performance against various deformation modes, with promising applications as potentiometric (106–107) and amperometric sensors for lactate (108), and glucose (40). However, there are still significant challenges for the preparation of these devices such as the stability of biomolecular receptors and surface fouling.

7.2.1.2 Invasive Wearable Electrochemical Biosensors

Ingestible Electrochemical Biosensors
Ingestible devices are a semi-invasive approach for direct access to biofluids through the gastrointestinal (GI) tract, since the mucous membrane of the gut provides facile and rapid exchange, and hence, a wide range of biomarkers and therapeutic targets are available to measure (109). However, the variability of pH and composition (digestive enzymes, bile acids, mucus and other compounds involved in the degradation and digestion of food) along different sections of the GI tract is a major challenge, compromising the stability of the device and accelerating the biofouling, thus requiring separation of the sensing element from the environment (110).

There are many parameters to be considered in ingestible electronic devices such as size, power consumption, data storage/transmission and material compatibility. Among them, the size plays an important role in the design. The larger size, the more components and complex functionality; however, it must be small enough to be swallowed and pass through the GI without causing obstruction (111).

The swallowability also depends on other factors such as texture or shape. Therefore, the materials used to play important roles for functionality and biocompatibility. Several review articles on the materials used in ingestible electromechanical systems have been reported (112–114). Material selection should consider important features such as chemical stability in low-pH environments, mechanical properties, process-ability, and intrinsic toxicity. Regarding biocompatibility, there are guidelines to classify certain ingestible materials as "generally recognized as safe (GRAS)" (115). Synthetic polymers used as food additives and emulsifiers such as polyethylene glycol (PEG), poly(l-lactide-coglycolide) (PLGA), gelatine and cellulose can be used as protective coatings (113). Biopolymers isolated from foodstuffs can also serve as materials for dielectric layers and substrate materials. Moreover, 3D printing can be applied to integrate electrical circuits into materials using a tattoo-paper-like transfer approach (116) or pulsed laser to generate graphene on carbon-based materials, including coconuts and bread (117). Many synthetic polymers including poly(vinyl alcohol) (PVA), PEG, polyesters, polyanhydrides and their derivatives and natural biopolymers (e.g., silk fibroin, albumin), polysaccharides and DNA can be used due to their biodegradability. In addition, there are numerous biodegradable metals including Zn, Mg, Fe, Cu, and Zn–Ca alloys and bioinert noble metals such as Au, Pt, Pd, and Ag that can be used for crucial device components since they are nontoxic, benign, and are commonly ingested in food products (113).

In vitro incorporation of electrochemical sensors in ingestible capsule devices consisting of a multielectrode sensor with potentiostatic circuits that can operate voltammetry was reported (118). The capsule was tested in vitro on stool liquid, showing consistency in measurement. Kim et al. developed an edible electrochemical sensor made of carbon paste and olive oil that can be placed on a variety of foods (112). Moreover, Ruiz-Valdepeñas et al. reported an ingestible electrochemical glucose sensor to be used in regions of interest within the GI tract via pH responsive polymer coatings (119).

Implantable Electrochemical Biosensors

Over the past few decades, the tremendous advance in electronic, biocompatible materials and nanomaterials has led to the progress of on-body accessories to small body insertions and implants that enable diagnosis and prognosis and improve the quality and efficacy of health care.

Implantable subcutaneous e-skins have received wide attention for the development of medical devices. These kinds of biosensors are partially or fully introduced into the human body aiming to remain there for long periods of time in a minimally invasive way for continuous monitoring, thus minimizing the pain and discomfort of the patient (120).

The development of implantable biosensors requires the integration of heterogeneous elements, including electrodes, circuits for performing measurements and transmitting the data, and a power source. Since these devices operate directly in the body, the final shape and dimensions of the implantable biosensor must be biocompatible and well tolerated by the host, avoiding toxicity and chronic inflammation, and preventing adverse effects such as rejection of transplantation (121). In addition, the wireless remote capability of these devices is essential, not only for transferring the data of the monitored patient, but also for maintaining the device's battery and status and function upgrades (122).

The materials used to prepare this type of biosensors should be not only biocompatible but also should show stability under physiological conditions. The use of hybrid materials can lead to several strategic improvements such as better surface chemistry for the immobilization of biomolecules, and better electrical conduction capability (123). Enzyme electrodes coated with Nafion consisting of MWCNTs cross-linked with redox polymers have demonstrated an enhanced stability (124). The ability of poly(3,4-ethylenedioxythiophene):polystyrene sulfonate (PEDOT:PSS) to improve an assortment of bioelectronics interfaces coordinated for electro-conductive tissues was reported (125). The aniline pentamer integrated with poly (L-lactic acid) (PLA) yielded a triblock copolymer with a defined sequence that supports the adhesion and proliferation of cardiomyocytes cells (126). PEG coating was employed to reduce tissue inflammation and extend the sensor lifetime (127). Zwitterionic systems have also been popular due to their ability to bind water molecules, which resulted in antifouling properties (128). Alginate hydrogel encapsulation has been broadly employed to improve implant biocompatibility and mitigate the host response. However, although the strategies to improve biocompatibility together with the release of agents able to mitigate inflammation promise to overcome long-term implantation problems, no strategy has yet achieved suppression of the foreign body response (33).

There are several commercial electrochemical implantable glucose biosensors for managing diabetes using minimally invasive subcutaneous sensors able to measure the interstitial glucose: Medtronic MiniMed's (CA, USA), Dexcom's SEVEN® PLUS (CA, USA) and Abbott's FreeStyle Navigator® (CA, USA), Eversense (Senseonics Inc.), allow glucose continuous monitoring for periods of time ranging from 3 to 7 days (129–130). There are also long-term implantable glucose electrochemical devices that were implanted in the abdomen (size and shape of an AA battery) (131) and in subcutaneous tissues of pigs for more than 1 year (132).

7.2.1.3 Organ-on-a-Chip Devices

The organ on a chip (OOAC) is a biomimetic system that mimics the physiological organs environment, with the ability to regulate key parameters including concentration gradients, shear force, cell patterning, tissue-boundaries, and tissue–organ interactions (133). Therefore, OOAC is a promising tool for tissue engineering and drug screening applications. Although most OOAC relies off-chip analysis and imaging techniques, there is an urgent demand for continuous, noninvasive, and real-time monitoring of tissue constructs which requires the direct integration of biosensors (134).

The key aspect in the formation of functional tissues in *in vitro* OOAC systems is the cell culture environment. In this context, the integration of biosensors to OOAC platforms enable to achieve of controllable and reproducible cell culture environments. Specifically, different microsensors have been integrated into OOAC for monitoring the metabolic activity. Although optical sensors are preferred for monitoring O_2 levels in OOAC, glucose and lactose levels are mainly monitored using electrochemical biosensors due to their ease of integration in microfluidic devices (135–136). Electrochemical immunosensors have also been used for the determination of cytokines in OOAC using several strategies to ensure efficient antibody immobilization (137–139).

7.2.2 OPERATING MODE

Electrochemical biosensors, due to their miniaturization capabilities, constitute appropriate systems for their integration into small devices able to undertake reliable in vivo measurements of a wide variety of clinical biomarkers, even in a continuous mode, to allow not only precision diagnostics but also the study of different diseases progression and treatments efficacy. As discussed in the previous section, multidisciplinary advancements in flexible/wearable sensors result in the ability to achieve real-time personalized healthcare, although important drawbacks such as long-term resistance, reusability or charging still need to be addressed (9, 140). These platforms may be coupled with microneedle electrochemical sensors for quantifying or continuous monitoring of various metabolites, electrolytes and biomarkers in dermal interstitial fluid (10). As reviewed by Madden et al. and discussed in Section 7.2.1.1., some devices have been tested in preliminary clinical evaluation studies with human subjects and a few of them integrated microneedle sensors with a miniaturized potentiostat, a communication module and a power source, thus showing the possibility of wearable molecular monitoring for both personal and on-patient health care. Non-invasive devices able to be attached almost anywhere on the body for continuous multiplexed motorization of biomarkers in body fluids, and able to resist fouling and changes in parameters such as pH and/or temperature (Figure 7.2) (141), envisage the future of high precision in situ medicine.

To choose between in vitro or in vivo measurements, the specific application should be considered. In vitro systems are suitable for clinical applications while in vivo strategies are more adequate for continuous monitoring to follow the disease's state in real time. Banerjee et al. (142) have reported the importance of measuring neurotransmitters (NTs) levels in a continuous mode, with high temporal resolution and in real time to understand the brain functioning and disorders. The NTs liberated into the extracellular space by exocytosis can be determined in vivo by implanting or semi-implanting electrodes in targeted brain regions. To do this, carbon, metallic and polymeric nanocomposites were used to fabricate microelectrodes looking for systems with short response times and high electrocatalytic activity to allow sensitive and selective NTs determination. Fast scan cyclic voltammetry (FSCV) is recommended for the detection of NTs because it allows the rapid detection of NTs with high temporal and spatial resolution. In fact, FSCV has been used at the Mayo Clinic to detect NTs in humans during Deep Brain Stimulation (DBS) surgery (143).

An electrochemical microsystem involving chitosan–carbon nanotube-modified microelectrodes have been reported for the real-time monitoring of antipsychotic clozapine treatment in schizophrenia patients (144). Only a few microliters of finger-pricked whole blood samples were used with no pretreatment. However, the system showed person-to-person variability of the electrochemical signature, electrode fouling and the results were not comparable with those obtained by LC-MS/MS, which was attributed to the effect of compounds that inhibit or induce liver metabolism. An amperometric sensor using an AgAu nanoparticles decorated MWCNTs nanohybrid deposited on a glassy carbon electrode has been reported for real time in situ monitoring of the neurochemical dopamine and the reactive oxygen species H_2O_2 released from rat pheochromocytoma PC-12 cells upon external stimulation.

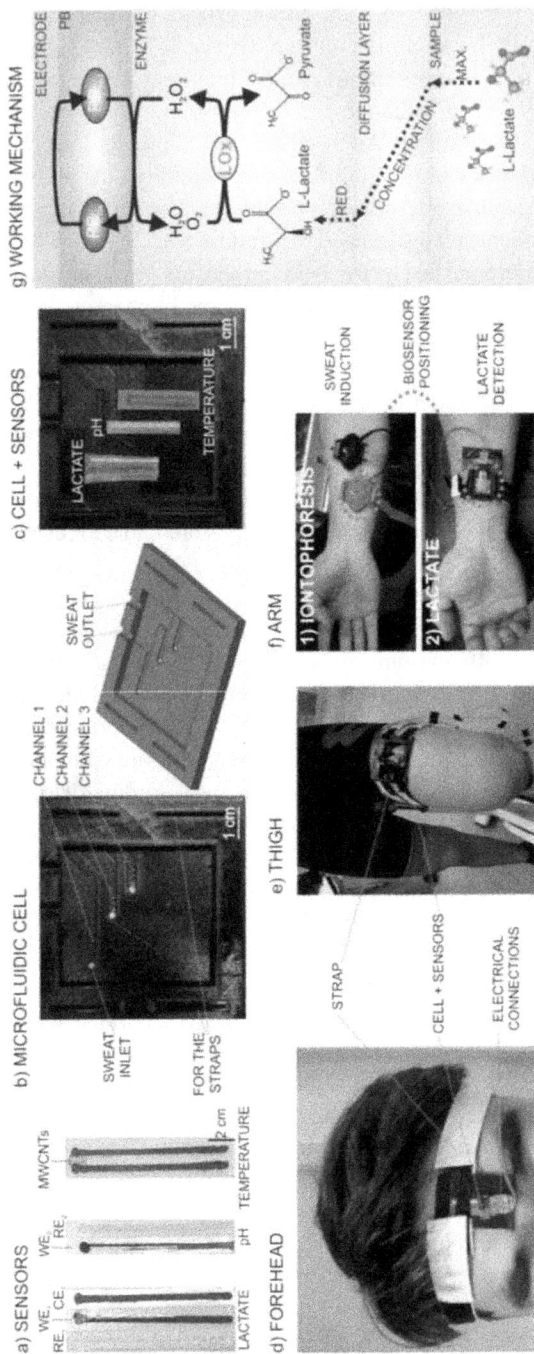

FIGURE 7.2 Pictures of (a) lactate, pH, and T sensors. (b) Epidermal patch with three microfluidic channels. (c) Epidermal patch with the sensors. Attachment of the epidermal patch in (d) the forehead and (e) thigh during cycling and (f) in the arm after iontophoresis. (g) Illustration of the sensing mechanism underlying lactate detection in sweat. Reprinted from (141) with permission (further permissions related to the material excerpted should be directed to the ACS).

The sensor showed low detection limits and good selectivity when a Nafion layer was added to the modified electrode (145). Point-of-care (POC) determination of lithium levels in serum would help to control lithium-based therapies. Constant exposure to lithium associated with therapeutic dose levels decreases the brain protein kinases functioning, affects kidneys and gene expression. Adverse effects appear for lithium concentrations in serum higher than 1.5 mM, and 2 mM concentrations are associated with high toxicity (146). A lithium sensor using a 11-mercaptoundecanoic acid self-assembled monolayer onto a disposable screen-printed gold electrode and physically adsorbed 6,6'-dibenzyl-14-crown-4 ether as lithium specific ionophore has been reported by Singh and Khumbhat (146). Li^+ ion was detected by differential pulse voltammetry using a hexaammineruthenium (III) chloride solution as redox probe. The sensor worked over the 0.1 to 2.5 mM lithium concentration range, and was selective to other ions such as sodium, potassium, or calcium. However, the sensor was only tested in serum samples spiked with lithium standards. Monitoring local cerebral tissue oxygen levels ($PbtO_2$) is used in neurological intensive care units to guide therapeutic strategies aimed at maintaining O_2 levels above threshold in patients suffering from severe acute brain conditions, and during certain neurosurgery procedures. An FDA-approved, commercially available, clinical grade depth recording electrode comprising 12 Pt contacts were characterized and tested by Ledo et al. (147) for in vivo $PbtO_2$ measurements in multiple brain sites to provide a more integrated vision of brain oxygenation. Again, only preliminary studies are reported and its translation to the real world is still a challenge to be confronted.

Real time measurement of targeted molecules in the body can allow therapeutic drug monitoring, i.e., the clinical practice in which dosing is adjusted in response to plasma drug detected levels to improve the safety and efficacy of medical treatments. An aptamer against the antibiotic vancomycin was selected by Dauphin-Ducharme et al. (148) and adapted to an electrochemical sensor able to measure vancomycin in a few seconds, using a calibration-free methodology, in finger-prick-scale samples of rata whole blood. Enough precision to identify statistically significant pharmacokinetic differences between individual animals was achieved. However, more work is needed to translate these sensors to longer term in vivo monitoring.

An electrochemical leucine-benzyl ferrocene carbamate (Leu-FC) based sensing platform has been reported for (i) in situ monitoring of Leucine aminopeptidase (LAP) activity, a potential biomarker for liver malignancy, in live cells, (ii) following the effects of chemotherapy toward liver carcinoma via monitoring LAP activity changes during clinical treatment, and (iii) studying drug resistance in HepG2 cancer cells against cisplatin from the real time assay of cellular LAP activity. All assays were done in vitro (149). Lee et al. (150) reported an electrochemical deposition and co-reduction methodology to prepare Pt nanoparticles (PtNP)-decorated, porous reduced graphene oxide (rGO)–CNT nanocomposites on a PtNP-deposited screen-printed carbon electrode for in vitro detection of H_2O_2 released from prostate cancer cells after phorbol 12-myristate 13-acetate (PMA) stimulation. Peritoneal carcinomatosis, induced by the intraperitoneal dissemination of several digestive and gynecological malignancies, constitutes a severe stage of cancers that need to be frequently monitored. Xu et al. (151) have recently reported the possibility to monitor it through in vivo and real-time peritoneal glucose detection using a Pt nanotree microelectrode with wireless, battery-free and flexible electrochemical patch. The system was tested in vivo by implanting the microelectrode in a minimally invasive way in the rat peritoneal cavity.

7.2.3 REMARKABLE ATTRIBUTES BEYOND SENSITIVITY AND SELECTIVITY

Despite the well-known intrinsic sensitivity and selectivity of electrochemical biosensors allowing sensing of a wide range of markers (152–153), their definitive integration into 'real healthcare scenarios' requires these biosensors to exhibit other attributes beyond sensitivity and selectivity. Ideal 'patient-friendly' sensing tools were defined by WHO based on the ASSURED criteria, comprising affordable, sensitive, specific, user-friendly, rapid and robust, equipment-free and delivered to the end-users' devices (154–155). In addition, the continuous tracking of molecular targets in real-time and in different and complex body fluids is highly desirable to redefine today's paradigm of patient care and general healthcare monitoring through advances in minimally or invasiveness, biocompatible and bioresorbable sensor technology (156).

The gradual passivation of the electrode interfaces as result of the non-specific adhesion and adsorption of fouling compounds and biomolecules circulating in biological matrices, known as biofouling, disrupts the reliable and continuous interrogation of high relevance target analytes (153, 157–158). Once the electrode surface is exposed to the biofluid sample, biofouling phenomena can be triggered upon the formation of an electrical double layer with the subsequent adsorption of biomolecules that react with the electrode surface by mass transfer (157). Non-specific adsorption processes are governed by thermodynamics, where the entropy increases while enthalpy decreases, thus resulting in energetically preferred adsorption events (159) yielding biofilms where foulants can adhere through hydrophobicity, charge attraction and surface nucleation (160). In general, biofouling strongly affects selectivity and binding affinity of biorecognition elements, as well as the electron transfer reactions involved in electrochemical readouts, resulting in a poor signal-to-noise ratio and high background signal (161).

To overcome this important limitation, biosensing compatible antifouling coatings have been proposed. Physical and chemical strategies have been designed to modify electrodes showing lower adsorption of fouling molecules. Mechanical coatings, physical adsorption, porous, nanoporous and superhydrophobic structures, SAMs, polymer brushes and novel peptides (161–163) are just a few examples of the physical and chemical-based strategies that can be implemented onto sensing surfaces to reach testing performance with negligible or minimal surface fouling adsorption.

Wei et al. reported a strategy to minimize electrode biofouling by masking carbon fiber electrodes (CFEs) with leukocyte membranes (LMs) taking advantage of the intrinsic properties of cell membranes (Figure 7.3). Cellular membrane fractions from leukocytes were isolated and layered onto previously APTES-coated-CFEs by immersion for 1 hour. The phospholipids carrying hydrophilic moieties that compose LMs exhibited high resistance to nonspecific protein adsorption while maintaining the electrochemical reactivity of the coated CFEs for monitoring neurochemical

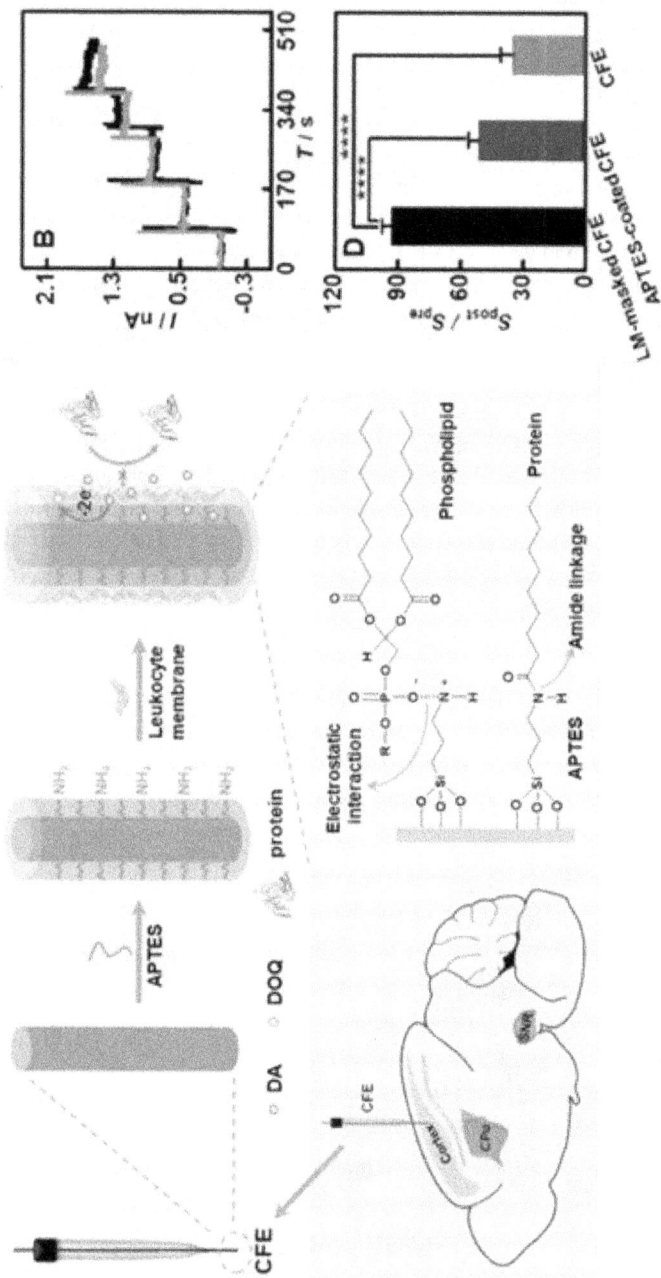

FIGURE 7.3 Scheme process for the CFEs modification with LMs and ratios of sensitivity obtained with the LM-masked CFE in artificial cerebrospinal fluid after implantation. Reprinted and adapted with permission from (164), copyright 2020 American Chemical Society.

dynamics directly in artificial cerebrospinal fluid after *in vivo* implantation in the brain cortex for 8 hours (164).

pH-responsive protective coatings have been amazingly exploited by Wangs's group for the direct monitoring of glucose in raw blood and saliva over prolonged periods of time without compromising the sensitivity of the biosensors. A pH-responsive electrochemical sensor array coated with a protective film of commercial pH-sensitive methacrylate polymers (Eudragit® L100 and Eudragit® EPO, which are dissolved under certain pH value conditions, pH>6 and pH<5, respectively) was developed for carrying out delayed detection approaches (Figure 7.4). The negligible fouling interference for targeting of salivary and blood glucose was demonstrated after coating a glucose oxidase (GOx)-Prussian Blue graphite electrode able to maintain extraordinary sensitivity (100%) over the uncoated electrode (29.6 and 34.6%, respectively) after 2 h of incubation in these biological media (165).

From a rigorous medical point of view, bioanalytical sensing tools enable the continuous, reliable, and precise monitoring of clinically important targets for prolonged periods of time (166). In this sense, and due to their reversible mechanism, electrochemical aptasensors (E-AB) readily support continuous target interrogation in complex and untreated biological scenarios (166). Successful applications of E-AB biosensing in both diluted and raw whole blood for the continuous, reversible, and multi-hour therapeutic drug monitoring in situ in live animals has been realized in the last years (167–169). In addition, E-Ab shows excellent attributes regarding the compatibility with feedback-controlled drug dosing in vivo (170–172), as well as for detecting small diagnostically useful molecules in less-explored biological scenarios, such as excreted perspiration.

Figure 7.5 displays a wash-free and reagentless E-AB sensor using square wave voltammetry (SWV) for the high frequency, real-time measurement of Neutrophil Gelatinase-Associated Lipocalin (NGAL) in urine, inside a urinary catheter with sub-minute time resolution and at concentration levels corresponding to the threshold

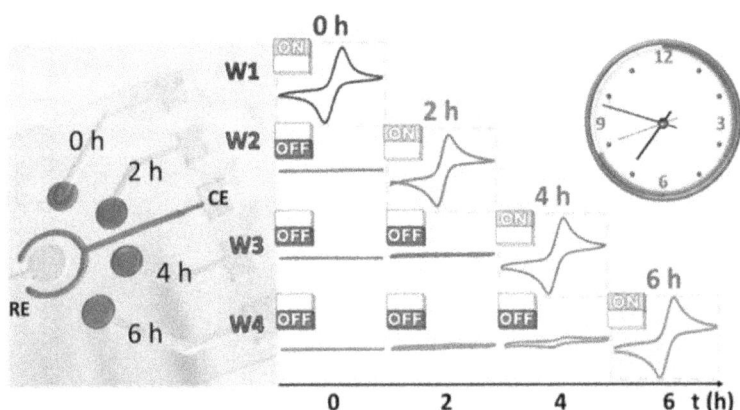

FIGURE 7.4 Four-electrode array design and general rationale of the delayed detection sensor based on transient coatings. Reprinted and adapted with permission from (165), copyright 2018 American Chemical Society.

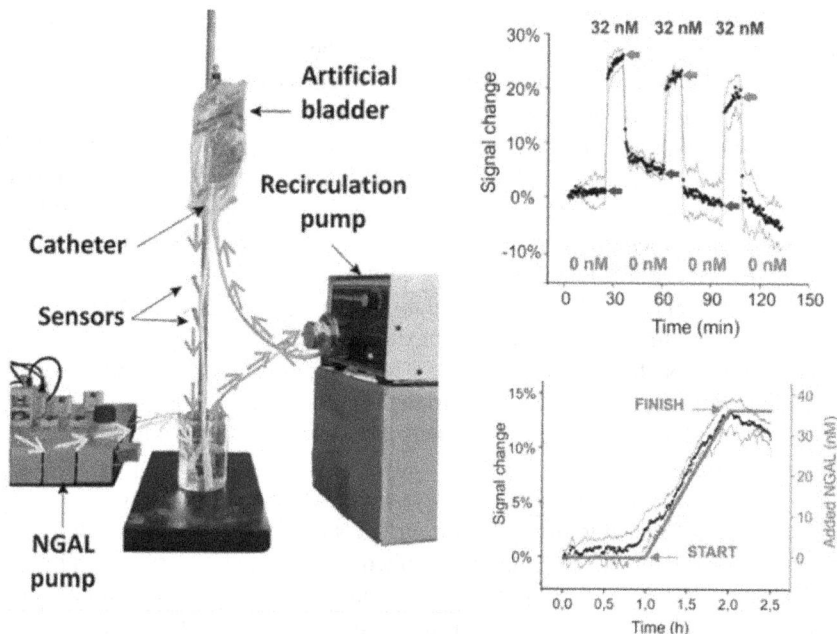

FIGURE 7.5 Prototype system able to pass artificial urine through a catheter containing two E-AB sensors (left); sensor response and continuous monitoring of spiked NGAL mimicking acute renal injury (right). Reprinted and adapted with permission from (173), copyright 2020 American Chemical Society.

for acute renal injury (173). Conversely to previous E-AB sensors, this work exemplifies that continuous monitoring of protein biomarkers (and not only small molecules) is a reality that may enable many early warning systems in healthcare.

In addition to more explored body fluids, sweat has shown to be highly useful for reliable deciphering of individuals' health states. In this context, cortisol has been determined due to its significance in the functioning of endocrine, nervous, cardiovascular, respiratory, etc. systems (174), as well as being considered the gold standard stress-related biomarker.

An autonomous, label-free impedimetric aptasensor for real-time monitoring of cortisol levels in sweat with sub-microliter sample volumes and free from active stimulation of sweat glands has been recently reported by Ganguly et al. (175). The aptasensor can operate for 8 h without interruption, in an 'accumulative' mode, i.e., the sensor can perform continuous sensing with no need for binding sites regeneration. The sensor substrate consists of a hydrophilic polyamide microporous membrane coated with nanoporous zinc oxide onto which silver screen-printed microelectrodes were deposited and further functionalized with SH-aptamers to measure the capacitive changes at the electrical double layer interfaces when cortisol is in the surrounding environments. The constructed device can determine the hormone in 5 min using 0.5–1.0 µL of sweat sample. The developed sensing system was tested as sweat wearable by dynamically tracking on-body cortisol during the stressful conditions of

a regular workday, from 8 am to 5 pm, in three healthy volunteers. Once the impedance data were collected, they were averaged and the trends showed that the sensing
system was able to monitor the diurnal trend in cortisol variation in a prolonged,
continuous and autonomous mode, with no required labelling.

It is widely assumed that the dynamic and continuous monitoring of protein biomarkers in vivo by self-generating sensing devices is a powerful tool for disease
monitoring and treatment response. In this context, O'Kelley's group reported an
elegant reagentless sensing concept with potential application in the analysis of physiological markers related to stress, allergy, cardiovascular health, inflammation, and
cancer. The concept suitable for biological fluids and in living animals is based on the
molecular pendulum (MP) approach that exhibits field-induced transport modulated
by the presence of the bound analyte. MP relies on checking the potential-depen-
dent modulation of DNA orientation onto the electrode. Using a redox active moiety
(ferrocene, Fc) that shows electrical activity at a potential compatible with the generation of a positive electrical field, a negatively charged linker composed of DNA
and analyte-specific antibody as recognition element in a bound or unbound state are
simultaneously attracted to the electrode surface and the electron-transfer dynamics
measured. When a positive potential is applied, the induced electric field provokes
that the negatively charged MP pivots down experiencing several forces with the surrounding MPs from the MP monolayer which modulates the time required for the MP
to reach the electrode surface. This time can be correlated to electrochemical currents measured through chronoamperometry. The approach was demonstrated using
Troponin I as model marker, which was dynamically and continuously monitored
into the blood stream and saliva from living animals. In addition, the authors tested
a panel of 10 different proteins of different sizes, charges and molecular weights just
by combining the corresponding specific antibody to each target analyte with the
MPs. This novel protein sensing concept provides a general approach for the development of implantable sensory units with broad applications to the direct monitoring
of physiologically relevant proteins and other target molecules in different types of
body fluids and biological matrices of high analytical complexity (Figure 7.6) (176).

Despite their great attributes, however, adaptation of electrochemical biosensors
to clinical practice can be limited by the requirement of calibration, which becomes
more complicated for continuous and in vivo target monitoring.

In this regard, Li et al. developed a dual-frequency calibration free operational
electrochemical strategy for the continuous measurement of two drugs in flowing, undiluted whole blood, during several hours. As sensing platform, the authors
employed an electrochemical aptamer-based design (E-AB) comprising of a
redox-reporter-modified DNA or RNA probe attached to gold disk electrodes previously modified with a 6-mercapto-hexanol layer and monitored the binding-induced
changes in electron transfer kinetics by SWV using methylene blue (MB) as redox
reporter (177). The method was successfully evaluated for cocaine and the cancer
chemotherapeutic drug doxorubicin in undiluted blood serum. In both cases, the
measured concentrations of each target molecule were as accurate as those obtained
with calibration-mode devices.

A self-calibration dual aptasensing electrode has been reported for the detection of the nucleoprotein (NP) from the avian influenza virus (AIVs), built on

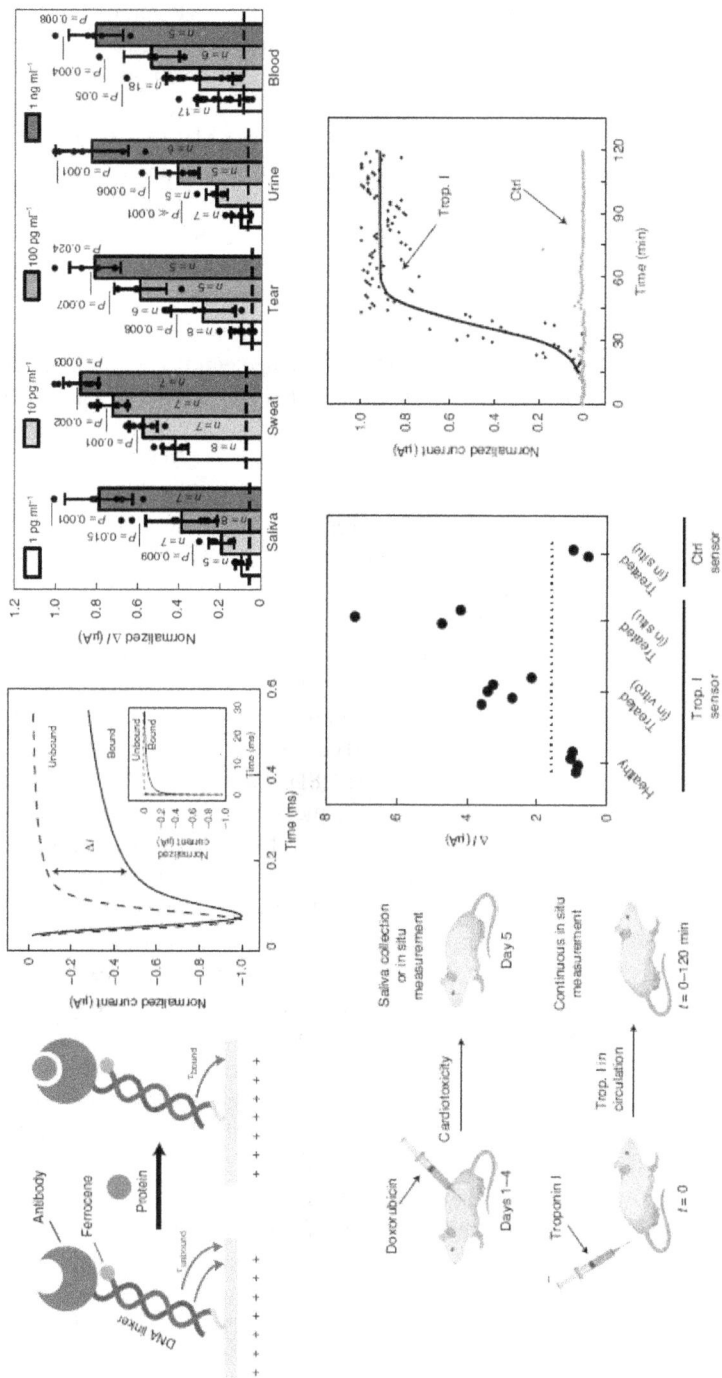

FIGURE 7.6 Protein-binding MP fundamental based on double stranded DNA and specific antibody against Troponin I (top left); detection of target cardiac marker Troponin I in biofluids from different nature, composition, and complexity (top right); cardiac dysfunction induction and Troponin I injection in mice (bottom left) for the ex situ and in situ salivary Troponin I measurement (bottom center), and in situ continuous monitoring of Troponin I in murine models (bottom right). Reprinted and adapted with permission from Springer Nature: (176), copyright 2021.

3D nanostructured porous silica films on the surface of tungsten rod electrodes (3DNRE). The platform is composed of a MB-loaded anti-AIV-NP-aptamer-capped electrode (Apt_{AIV}-MB@3DNRE) and a control-aptamer-capped electrode (Apt_{con}-MB@3DNRE) able to self-correcting drifted baseline signal. In the presence of target AIV NP, the interaction with the specific aptamer triggers its detachment from the outer electrode surface with the subsequent release of the MB thus resulting in a decreased MB redox current. On the other hand, signal from the control electrode eliminates false responses caused by unstable interactions between the aptamer or MB probes and the electrode surface. Normalized signals calculated as the difference of cathodic peak currents between the testing and the control electrodes provided a linear range from 2.0 to 12 nM and a LOD of 1.13 nM for the determination of AIV NP. The developed platform was applied in AIV lysates free from sample purification and washings processes, as well as in clinical specimens including oral and cloacal swab samples collected from chicken infected with H9N2 viruses (178). Although the application of this platform is not directly related to the medical field, its future exploitation in other fields including clinical diagnoses can be undoubtedly envisioned.

The emergence of OOAC as viable platforms for drug and biomarkers screening (179) has led to its selection as one of the Top Ten Emerging Technologies. These biomimetic systems built on a microfluidic chip that recapitulate the environment of a physiological human organs exhibit broad applications in precision medicine. Unlike in vivo experiments, OOAC can reveal in an acutely realistic manner the underlying mechanisms of physiological phenomena considering the interaction and adaptation of thousands of components and compounds, including tissues, cells, proteins and genes (133, 180), as well as the crucial cell-cell and cell-matrix interactions for studying the molecular basis of pathological responses (181). As the evolution of OOAC continues increasing, the need for integrating assessment methods into such devices also increases. In this sense, more sophisticated sensors able to continuously respond and provide information in real time about the viability of tissues constructs/organoids are highly demanded. Flexible or fluidic-based-sensors and biosensors using electrochemical transduction and in combination with a wide variety of nanomaterials, including metal nanoparticles and carbon-based nanomaterials, have emerged in this field due to their simple required instrumentation and the ability to generate sensitive and selective responses in a short assay time. An illustrative example is a work by Xiao et al., who constructed a flexible electrochemical sensor using Pt NPs on freestanding reduced graphene oxide paper carrying MnO_2 nanowires for the detection of H_2O_2 with a limit of detection of 1.0 μM, which was employed for monitoring oxidative damage to human liver cancer cells (182).

Much more recently, and with the aim of overcoming major hurdles related to automation capacity for chronological monitoring of relevant clinical biomarkers, electrochemical affinity biosensors integrated into microfluidic chips have been reported by Aleman et al. (181). The developed platform (Figure 7.7) achieved in-line microelectrode functionalization, biomarker detection and sensor regeneration, and it can quantify soluble biomarkers on OOAC platforms in a continuous, multiplexed (if desired), in situ and non-invasive mode. This versatile functionalization strategy relies on the use of three-electrode-configuration gold microelectrodes modified

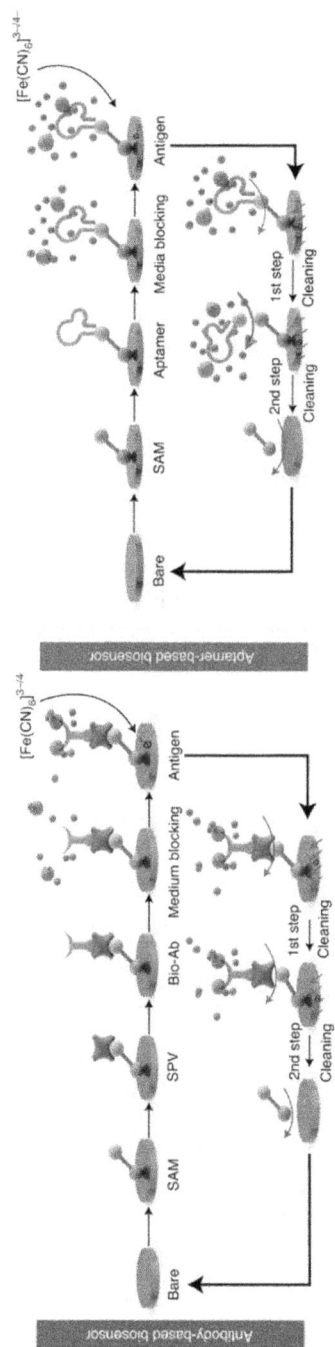

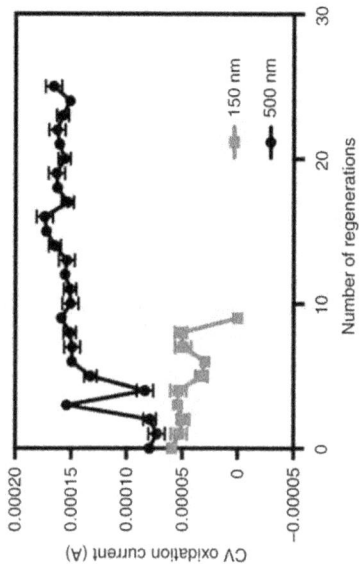

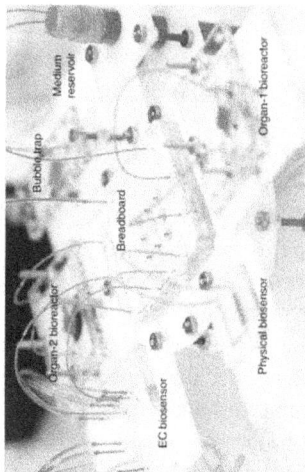

FIGURE 7.7 Schematic display of the two-stages experimental protocol for electrode fabrication and functionalization (top); multisensory-integrated multi-organ-on-chip platform (bottom left), and comparison of regeneration cycles in two differentially thicknesses Au microelectrodes by means of the changes in oxidation current values (bottom right). Reprinted with permission from Springer Nature: (181), copyright 2021.

with self-assembled monolayers that are subsequently treated with EDC and NHS crosslinkers to covalently incorporate streptavidin as uniform patterns to attach biotinylated capture probes, or amine-terminated aptamers as biorecognition elements for the selective capture of protein biomarkers secreted to the cell culture medium. Electrochemical impedance spectroscopy was employed to monitor the changes of interfacial electron-transfer kinetics using $(Fe(CN)_6)^{4-/3-}$ as redox probe upon the biorecognition event took place. The authors improved the methodology starting from an automatized antibody/aptamer-based single electrochemical biosensor with non-regeneration capabilities for cardiac biomarkers detection (139), to an affinity biosensor with regeneration capabilities for the continuous detection of liver-associated secretomes (183), as well as to a multisensory EC system integrated into a microfluidic breadboard controller with ability for the continuous and multiplexed detection of hepatic and cardiac biomarkers from microbioreactor modules (184).

Big efforts are also being made to construct electrochemical biosensing systems to be applied in personalized medicine concerning cancer diagnosis. In this context, both microRNAs (miRNAs) and messenger RNAs (mRNAs) are gaining enormous attention as biomarkers for the early diagnosis of high mortality-rate diseases such cancer (185–189). The low abundance of these biomarkers, varying from a single copy to more than 50,000 in the sample of interest, make the methods for their determination exhibit exceptional sensitivity, especially for early stage diagnosis. In this sense, electrochemical biosensors have shown ability for the sensitive detection of these biomolecules in different biological matrices (190–193).

The intrinsic versatility exhibited by electrochemical biosensors allows integration or coupling with innovative and powerful molecular systems, such as the CRISPR/Cas (clustered regularly interspaced short palindromic repeats and CRISPR-associated protein) for genome editing, transcription regulation and molecular diagnostics (194–196).

Through the combination of CRISPR/Cas 13a with catalytic hairpin DNA circuit (CHDC), Sheng et al., successfully developed the COMET, Cas-CHDC-powered electrochemical RNA sensing technology, for sequential non-small-cell lung carcinoma (NSCLC)-related miRNA and mRNA detection at ultralow levels. Assuming that single biomarker detection is insufficient for accurate and early stage cancer diagnosis (197–198), the authors selected a panel of six NSCLC-associated RNAs, miR-17, miR-155, TTF-1 mRNA, miR-19b, miR-210 and EGFR mRNA, which showed up-regulation in malignant nodules to be detected with the COMET chip (Figure 7.8) (199). The COMET chip involves a two-stage signal amplification, where the key component, a molecule with two functional domains (a uridine domain and an intermediate domain), upon release, initiates the CHDC mechanism. In the presence of the target RNA, hybridization occurs with the guide crRNA, Cas13a cleaves multiple triggers, numerous intermediary strands are released and, consequently, strand displacement processes and catalytic cycles are triggered. The sensing methodology involved one pot off-chip amplification, on-chip RNA detection, and SWV onto chips constructed with screen-printed Au working, Pt counter and silver/silver chloride reference electrodes and a pre-model polydimethylsiloxane (PDMS) plate. To avoid baseline drift, the authors employed a ratiometric baseline drift correction by using two independent redox reporters, MB as sensing reporter and ferrocene as

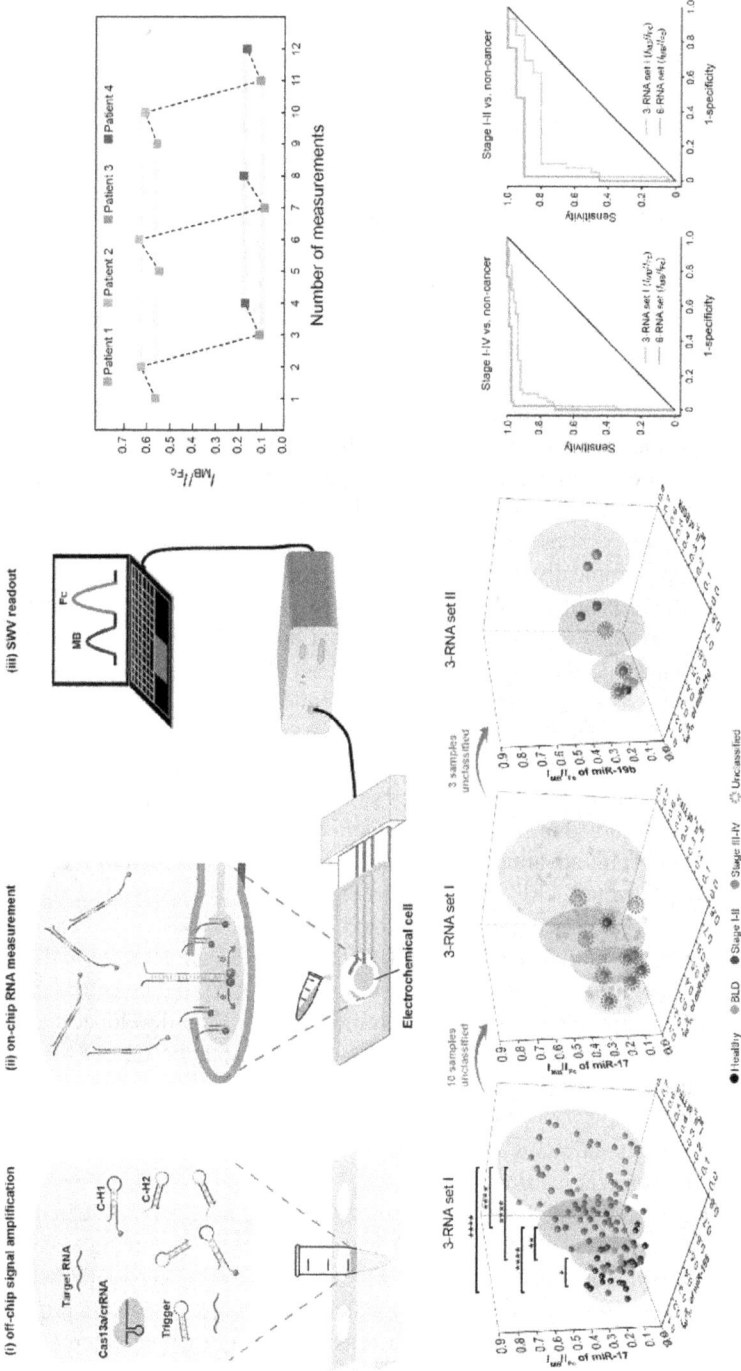

FIGURE 7.8 General operation of the COMET assay (up left); miR-17 analysis in four randomly serum samples of NSCLC patients (up right), dimensional dot chart for different RNA sets (bottom left) and receive operating characteristic (ROC) curves analysis for discriminating NSCLC at different disease stages (bottom right). Reprinted and adapted from (199) with permission from Elsevier, copyright 2021.

reference reporter to correct background effects. The correction method provided a dramatic reduction of the variability percentage from 34.8% to 2.2%. Multiple and consecutive RNA measurements with COMET required enzymatic regeneration of the sensing surface with a mixture of uracil-DNA glycosylase and endonuclease IV 45. Repeated measurements of 50 pM miR-17 carried out with a single chip proved the successful one-step reusability strategy. The COMET chip was able to detect attomole RNAs in a concentration range of 8 orders of magnitude, required less than 10 µL sample/reagent per reaction and the readout time was 6 min. The clinical usefulness of the COMET device was tested by measuring the abundance of miR-17, miR-155 and TTF-1 mRNA in total RNA samples extracted from sera of healthy subjects, 12 patients diagnosed with benign lung disease (BLD), 20 early-stage (I-II) and 55 large-stage (III-IV) lung cancer. The method allowed discrimination of early-stage NSCLC and non-cancer cohorts, with great correlation with the results obtained with qRT-PCR.

Although the detection of protein markers produced at gene level is extremely useful to get diagnostic information, other naturally produced proteins can manifest themselves even before the biological changes at gene level indicate some type of pathological event. This is the case of autoantibodies (AAs) synthesized from a triggered humoral immune response, whose presence in disease-related conditions can be detected in long prodrome diseases while no evident symptoms appear (200–202).

Electrochemical biosensing allowed the multiplexed determination of a panel of several tumor-associated antigens (TAAs) autoantibodies for the early diagnosis of colorectal cancer (CRC) (203). In this work, an electrochemical disposable multiplexed immunosensing platform for the POCT and sensitive determination of eight CRC-specific AAs against CRC TAAs was reported. The HaloTag technology was employed for the in vitro transcribing/translating of expressed fusion proteins coupled to commercial magnetic microparticles (MBs) onto the expressed HaloTag fused antigens self-assembled in situ with no protein degradation. The HaloTag technology provided important advantages in terms of high yield, purity, and recovery protein production. Selected CRC TAAs including difficult-to-express and purify antigens were in vitro expressed and immobilized onto MBs to selectively capture the corresponding AAs and the subsequent labelling with HRP-conjugated anti-human IgG. The biorecognition events, where specific TAAs, AAs and detector labels were involved, were amperometrically monitored on SPCEs. The developed methodology was applied to the determination of AAs against specific CRC-TAAs in human plasma, allowing discrimination of CRC and premalignant subjects from control individuals.

On the other hand, continuous progress is made to avoid biological fouling in multiplexed electrochemical reading.

Using a graphene oxide nanoflakes antifouling conductive nanocomposite, a multiplexed electrochemical sepsis POC method was implemented involving specific antibodies, phosphorylcholine (PC), and Fc-Mannose Binding Lectin (Fc-MBL) as capture bioreceptors for the detection of procalcitonin (PCT), C-reactive protein (CRP), and pathogen-associated molecular patterns (PAMPs) in whole blood (204). As it is shown in Figure 7.9, the conductive nanocomposite consisted of a mixture containing amine-functional reduced graphene oxide, denatured bovine serum

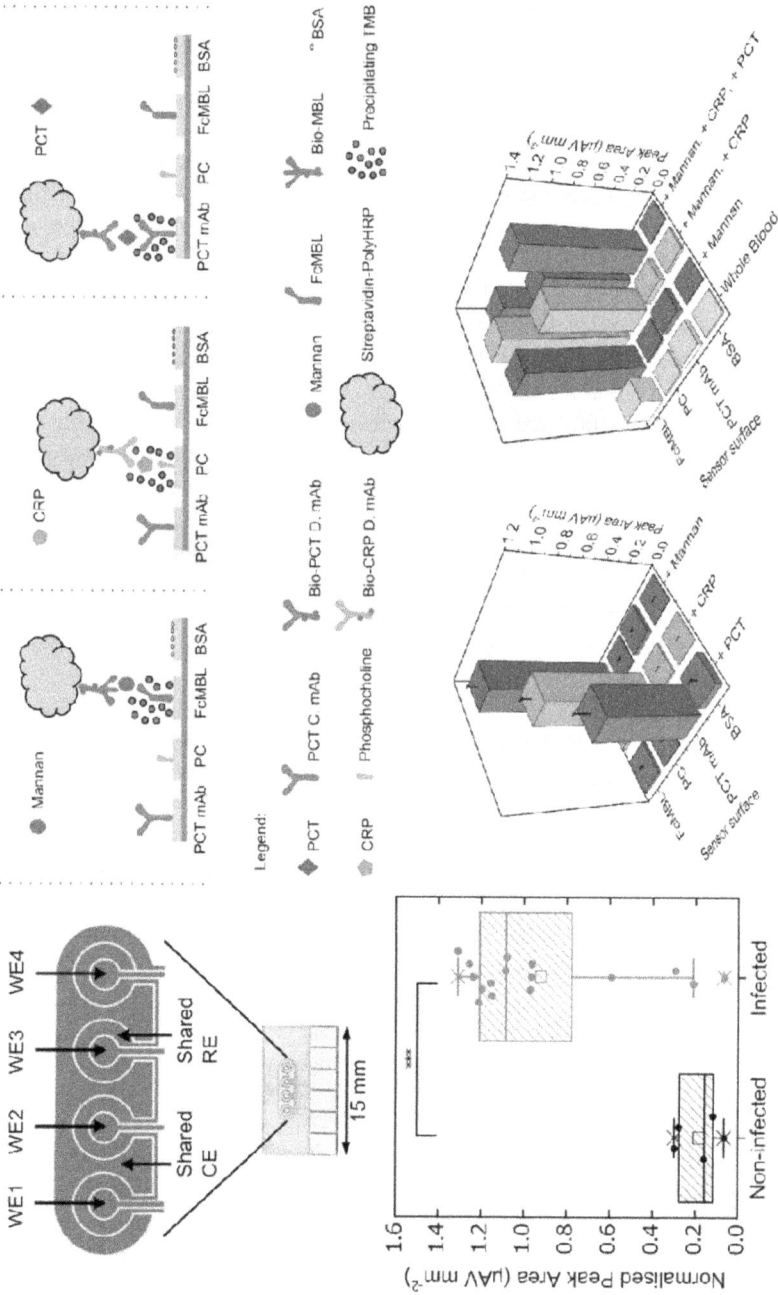

FIGURE 7.9 Gold electrode chip comprising four individual working electrodes, shared counter, and shared quasi reference electrode; schematics for monitoring target analytes (up left and right, respectively); clinical evaluation of PCT levels in blood stream non-infected and infected patients, cross-reactivity evaluation, and multiplexed detection of target biomarkers (bottom left, center, right, respectively). Reprinted with permission from (204).

albumin (BSA) and glutaraldehyde (GA) was drop-casted onto photolithographically prepared gold electrodes and subsequently functionalized with the corresponding capture receptor. Cyclic voltammetry was employed to determine each target sepsis biomarker using a sandwich configuration with biotinylated-detector receptors labeled with a streptavidin-PolyHRP conjugate. The multiplexed POC diagnostic capabilities of the prototype platform were demonstrated, allowing: i) combination of assays for the simultaneous interrogation of indirect infection biomarkers (PCT and CRP) with direct detection of pathogens in blood; ii) direct analysis in human raw blood; iii) parallel detection of CRP and PCT, taking into account that sepsis blood CRP concentrations are three to five order of magnitude larger than that of PCT, a previous depletion of CRP concentration in whole blood was carried out as well as using antibodies with higher dissociation constants; iv) reduction of the analysis time through coupling one of the developed EC chips to a microfluidic channel.

It is worth highlighting that, before this work, multiplexed analysis of raw blood was only doable by sequential blood prefilter, target analyte capture, and its further release to be detected. This fact denotes the giant strides made by electrochemical biosensing technology with versatility and adaptability to novel achievements from any scientific discipline to reach the development of effective precision and personalized medical devices.

The dynamic interplay that occurs between different layers of molecular markers throughout the development of human diseases must be taken into consideration in the development of effective devices contributing to precision medicine to insightfully typify and characterize a specific disease at a fully global scale (205). Therefore, the whole uncover of disease etiology is attainable only by interrogating multiple omics layers, including genomics, transcriptomics, proteomics, epigenomics or microbiomics with the goal to decipher their complementary effects but also their synchronous interactions (206–208).

Despite the numerous obstacles to achieving the previously mentioned goal, nascent incursion of electrochemical biosensing technology into omics field is already happening. Gao et al. have reported a flexible multiplexed immunosensing array system, nicknamed VeCare, for the simultaneous quantification of clinically relevant biomarkers at different omics layers, ranging from inflammatory mediators (TNF-alpha, IL-6, IL-8 and TGF-beta1), and microbial burden (*S. Aureus*), to physicochemical parameters (pH and temperature), coupled to a microfluidic wound exudate collector and involving flexible electronics for wireless, smartphone-based data readout (Figure 7.10) (209). The VeCare immunosensing platform used thiol- at one end and MB at the other end- functionalized aptamer-analyte affinity probes immobilized onto electrochemically exfoliated graphene-gold nanoparticles (AuNPs-GP) nanocomposite gold microelectrodes through covalent binding between thiol groups and AuNPs. The proximity of the redox probe to the sensing surface allowed electron transfer, which associated current measured by SWV. The system also included pH and temperature sensors based on polyaniline (PANI) polymer and a thermally responsive resistor. Initial studies in mouse models were undertaken for the in-situ interrogation of inflammation, colonization tendencies and physicochemical parameters, with the aim of proving both the real multibiomarker assessment and biocompatibility of the developed platform. Once these two were proven, VeCare

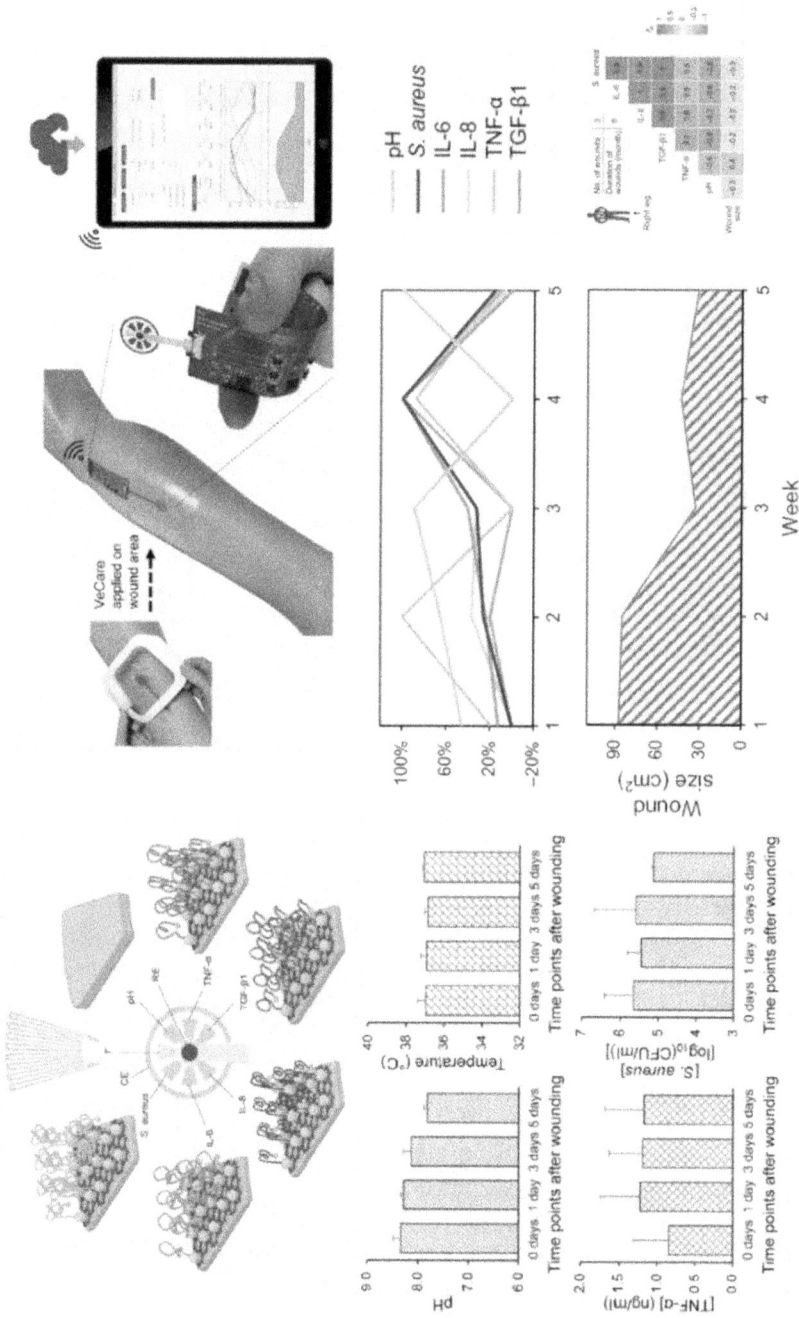

FIGURE 7.10 Schematic illustration of the POC immunosensing system for the determination of TNF-α, IL-6, IL-8, TGF-β1, *S. aureus*, pH, and temperature and VeCare prototype for chronic wound monitoring (up left and right, respectively); in situ assessment of pH, temperature, mouse TNF-α, and *S. aureus* in mouse wound model through the proposed multiplexed immunosensing device (down left), and sensor-derived data analysis of wound exudate samples from venous ulcer affected patient and patient's specific correlation matrices of parameters sensed by the VeCare platform. Reprinted with permission from (209).

immunosensor was used to analyze consecutively collected, once a week for 5 weeks, wound exudates from venous ulcers of 5 patients clinically diagnosed with non-healing venous ulcers. Data was obtained within minutes and with remarkable selectivity, sensitivity, reproducibility, linearity and scalability.

Importantly, the POCT VeCare platform can be considered as the first electrochemical POC device able to manage chronic non-healing ulcers in an accurate way in a robust, adaptable and personalized manner, which evidences the relevant role that electrochemical biosensors can play in the field of precision and personalized medicine.

7.3 PIONEERING CLINICAL APPLICATIONS USING ELECTROCHEMICAL BIOSENSORS

Due to the great progress that electrochemical biosensors have experienced over the past few years, they continue to unstoppably gain ground as suitable high-throughput and affordable tools easy to operate with little training and with the ability to perform multiplexed and/or multi-omics determinations in the field (non-laboratory or resource-limited environments). Moreover, they can provide accurate and reliable quantitative results in minimal actuation times and using small amounts of biological samples after minimal pretreatments (205, 210–211).

Focusing on the pioneering applications reported in the last two years, they have been primarily aimed to show the usefulness of electrochemical biosensors in the diagnosis, staging, prognosis, and monitoring of diseases such as cancer, cardiovascular, autoinmune and neurodegenerative diseases, unexpected global health challenges such as the ongoing SARS-CoV-2 coronavirus pandemic, and to determine analytes of relevance in fertility processes and chronic diseases such as diabetes.

This section provides only a brief overview of the huge variety of reported applications since 2019, so that the reader can have an updated information on the state of the art without going into too much detail on their rationale.

Cancer biomarkers are by far the analytes with more paid attention in the last years. Indeed, electrochemical affinity biosensors exhibit unique features to determine cancer-related transcriptomic (microRNAs, miRNAs, and long non-coding RNAs, lncRNAs), epigenetic (5-methylcytosine, 5-mC, 5-hydroxymethylcytosine, 5-hmC and N6-methyladenosine, m6A in nucleic acids) and proteomic biomarkers (proteins, glycosylated proteins, receptors, proteases and autoantibodies related to the presence of cancer and the tumoral hypoxia and metastasis events involved in its aggressiveness and evolution) (212–213). As a result of this wide research, different relevant applications can be mentioned:

i) Determination of the endogenous content of single (miRNA-21) (214–217) or multiple (miR-21, let-7a and miR-31) (218) miRNAs or lncRNAs (PCA3 (219)) in a small amount of total RNA extracted from cancer cells or tissues, or directly in serum with no need for prior extraction or amplification of the genetic material.

ii) Simultaneous determination of two RNA biomarkers (PCA3 lncRNA and PSA mRNA) directly in urine of prostate cancer patients (220).

iii) Discrimination of the metastatic potential of intact cancer cells through the determination of specific extracellular proteins (IL-13Rα2 and CDH-17) (221).

iv) Improvement of the diagnosing and staging metastatic cancer reliability by determining soluble and/or extracellular fractions of specific proteins (IL-13Rα2, E-CDH, CDH-17, RANKL, TNF-α) (222–225), proteases (trypsin and MMP-9) (226–227) or hypoxia-related biomarkers (HIF-1α and PD-L1) (228–229).

v) Use of aptamer biosensors (230) to target specific proteins (CEA (231–233) and PSA (234)) or their glycosylation patterns (PSA) (235–236).

vi) Use of a wide variety of electrode modifiers, such as polymers to impart antifouling properties, and silane, and ftalate derivatives to develop impedimetric immunosensors with enhanced simplicity, sensitivity, and assay time for a wide range of protein biomarkers: PAK 2 (237), SOX2 (238–239), CDH-22 (240–241), VEGF (242), RACK 1 (243), NSE (244), KLK 4 (245–246), and CCR4 (247–248).

vii) Assessing of the key role that pericytes play in triggering the migration and invasion of CRC cells through the immunosensing of TGF-β (249).

viii) Determination of methylated bases (5-mC, 5-hmC and m6A) in nucleic acids both globally and regionally with single-base sensitivity (250), as well as simultaneous determination of methylations in nucleic acids of different nature (DNA and RNA), using simple and rapid protocols (Figure 7.11) (251).

ix) Unmasking aggressiveness of cancer cells by targeting m6A at global level in a small amount of extracted raw unfragmented total RNA (252).

x) Discrimination between healthy and tumor tissues or healthy and CRC-affected individuals by targeting miRNAs and nucleic methylations in total RNA or genomic DNA without prior amplification, or in serum without genetic material extraction (214, 253).

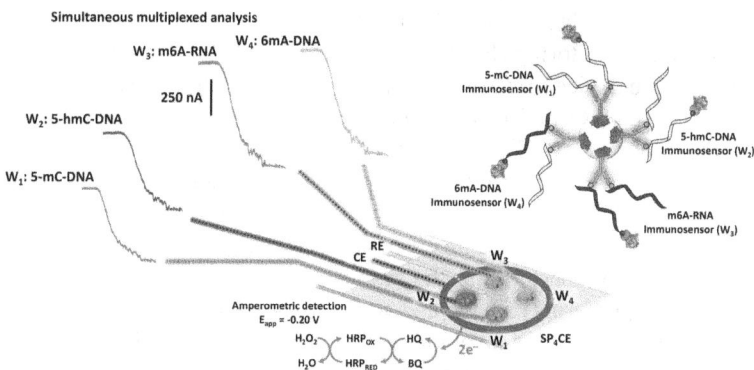

FIGURE 7.11 Multiplexed electrochemical bioplatforms for global detection of the most frequent and clinically relevant methylations (5-mC, 5-hmC and m6A) in nucleic acids. Reprinted and adapted from (251) with permission.

xi) Assessment of the clinical potential of specific AAbs signatures against TAAs in circulation (203) or from exosomes released by CRC cell lines and tissues from CRC patients (254), for the early and differential diagnosis of patients with premalignant lesions and CRC versus other prevalent neoplasias such as lung and breast.

In **cardiovascular diseases**, electrochemical affinity biosensors (mainly immunosensors (229, 255–258) and also aptasensors (259)) have been used to assess the diagnostic value of clinically accepted (cTnT (255, 259)) or emerging candidate protein biomarkers such as galectin-3 (256), Growth arrest-specific 6 (GAS6) (229) and ST2 (257, 258) in serum.

Regarding **autoimmune diseases**, electrochemical biosensors have been applied to the determination in serum of proteins (CXCL7 (260)), autoantibodies (RF, CCP-AAbs (261) and dsDNA-AAbs (262)).

Electrochemical biosensing has also been employed in the diagnosis of **Alzheimer's disease and other neurodegenerative disorders** (263). Electrochemical biodevices, unlike the long-adopted ELISA methodologies, provide the required sensitivity (due to the blood-brain barrier dilution) for the determination of biomarkers such as tau (264–266), TDP-43 (266)), NfL (267), SYN (268) and DJ-1/Park7 (269) proteins, in liquid biopsies other than cerebrospinal fluid such as serum and saliva (269). In addition, electrochemical biosensors are compatible with multiplexed determinations (266). It is important to highlight the pioneering application of the use of Phage-derived and frameshift aberrant HaloTag peptides as receptors for targeting specific serum AAbs (270) showing full diagnostic capability for Alzheimer's disease (Figure 7.12).

Electrochemical immunosensing has been also applied to monitor **fertility processes** through the single or multiplexed determination of salivary hormones (P4, LH, PRL, E2 and FSH) using custom-designed portable potentiostats fabricated with low-cost materials (271–275).

Regarding the detection of SARS-CoV-2 coronavirus, a representative appealing approach, among the many others that have arisen due to the seriousness of the situation we are suffering at global level, is the multiplexed telemedicine wireless platform reported by Gao's group (276) for the rapid, low-cost, and home COVID-19 diagnosis and monitoring (Figure 7.13). The multiplexed immunoplatform evaluates on the same device viral infection (nucleocapsid protein, NP, as viral carrier), immune response (IgM and IgG antibodies against viral S1), and disease severity (inflammatory biomarker CRP), and was successfully applied to the blood and saliva analysis of samples from infected and non-infected individuals.

Regarding the determination of clinically relevant metabolites (glucose, insulin and cortisol) for monitoring chronic diseases such as diabetes, electrochemical biosensing devices allow performing the on-the-spot simultaneous detection by integrating on single platforms enzyme and antibody-based assays (277), or competitive and sandwich immunoassays involving a single-step reaction with different enzyme tags (ALP and HRP) and detection potentials (278). These integrated electrochemical biosensing microchips, pioneered by Dr. Wang's group at UCSD, were able to operate directly in non-treated biofluids (whole blood, saliva, and serum), thus opening a new realm in POC multiplexed biomarker detection regardless of their clinical

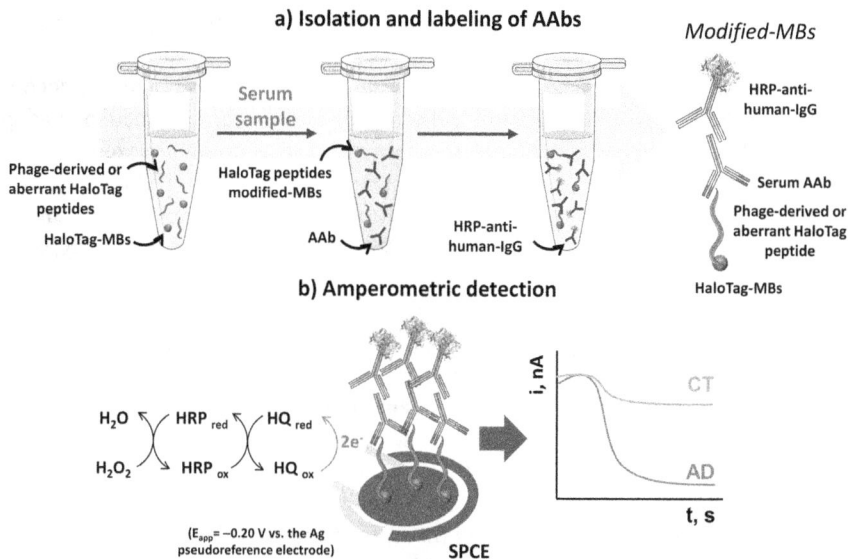

FIGURE 7.12 Schematic diagram of a phage-derived and aberrant peptides-based amperometric biosensing platform for serum autoantibodies with full Alzheimer's disease diagnostic capability. Reprinted and adapted from (270) with permission.

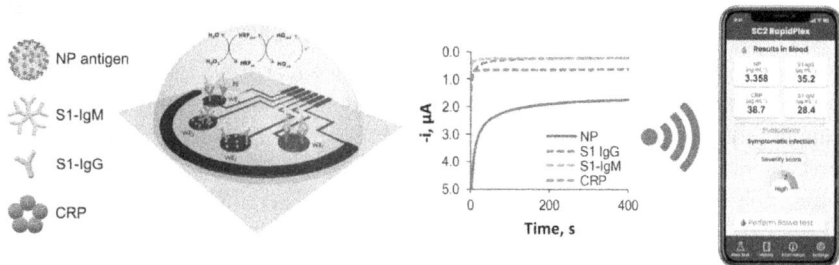

FIGURE 7.13 Multiplexed platform reported for the electrochemical detection of SARS-CoV-2 by targeting circulating levels of NP, IgG and IgM antibodies against S1, and CRP in blood and saliva. Reprinted and adapted from (276) with permission.

concentration range and nature to assist in the personalized management of patients. In addition, cost-effective disposable immunosensors have been recently reported for the impedimetric detection of adiponectin, involved in insulin, glucose and adipocyte metabolisms, in human serum (279).

7.4 EFFORTS TO BE MADE AND FUTURE DIRECTIONS WORTH PURSUING

The multiple innovations that electrochemical biosensors have undergone in terms of the different environments/formats in which they can be used, the way they operate, the remarkable attributes they have and the success demonstrated in particularly difficult applications encourage one to continue working in multiple directions.

We dare to highlight as particularly relevant, due to the boom and the benefits expected from the personalized and ideally decentralized precision medicine adoption, the enhancement of the application of these biodevices in multiplexed and multiomics determinations. This will make it possible, in addition to verifying the clinical potential of candidate biomarkers, to establish molecular signatures, of the same or different omics levels, characteristic of the type and state of a given pathology, at the same time providing society with competitive tools for their determination.

It is clear that these achievements require collaborative efforts to standardize protocols by establishing certified calibration standards, and to confront the resulting electrochemical devices with a sufficient number of samples in the hands of different users and in the framework of different scientific, health and private institutions to ensure, in addition to the robustness of the devices, that they combine the characteristics desired by the different user profiles and environments.

Scientists must continue to develop increasingly simple, competitive, ideally multifunctional, robust, cost-effective, high throughput and versatile methodologies able to be adapted to the resolution of new problems and to act in different environments with minimal modifications. Few would have bet a priori on the possibility of successfully employing electrochemical biodevices to: i) unmask the metastatic potential of whole cells; ii) discriminate between healthy and tumor tissues by targeting miRNAs and nucleic methylations directly in serum without genetic material extraction or in total RNA or genomic DNA without prior amplification, or iii) fully diagnose CRC or AD diseases by targeting specific serum autoantibodies. It is precisely the tremendous efforts made in recent years that have made it possible to develop in record time, in addition to vaccines, competitive methodologies for detecting infection and immunity to SARS-CoV-2, both of which will save millions of lives. The use of new bioreceptors (aptamers, peptides) and technologies that allow them to be expressed more simply, cheaply, in a tailored way, with improved stabilities and/or specificities and with less batch-to-batch variability (HaloTag and phage-display technologies) should not stop exploiting. It is essential to continue to do basic research on nanomaterials with improved properties, anti-fouling and biocompatible chemistries, assay formats and amplification strategies to be easily implemented in POC devices and less susceptible to environmental variations (such as nanozymes).

Healthcare institutions personnel must be receptive to both conveying their knowledge and needs to scientists and actively collaborating with them to demonstrate and validate the potential of the tools they develop by selecting and releasing appropriate patient cohort samples. They should take advantage of the potential provided by multiomics techniques to identify new molecular candidate biomarkers (such as those related to the extracellular matrix in cancer diseases) of interest for use in clinical practice.

And for their part, private companies should facilitate the work of both parties by providing them with the appropriate personal and material resources to facilitate and accelerate their efficient transfer to society. It is obvious that all this will involve titanic efforts that are sometimes mistakenly considered to be outside our daily duties. However, we must not forget that as scientists, healthcare professionals or entrepreneurs we also have a commitment to ensure that society obtains the best return on our knowledge and experience.

ACKNOWLEDGMENTS

The financial support of PID2019–103899RB-I00 (Spanish Ministerio de Ciencia e Innovación) and the TRANSNANOAVANSENS-CM Program from the Comunidad de Madrid (Grant S2018/NMT-4349) are gratefully acknowledged. Talento-Contract from Comunidad de Madrid (2019-T2/IND-15965, R.M. Torrente-Rodríguez) is also gratefully acknowledged.

REFERENCES

1. Gajdosova V, Lorencova L, Kasak P, Tkac J. 2020. Electrochemical nanobiosensors for detection of breast cancer biomarkers. *Sensors*. 20: 4022
2. Mani V, Beduk T, Khushaim W, Ceylan AE, Timur S, Wolfbeis OX, Salama KN. 2021. Electrochemical sensors targeting salivary biomarkers: A comprehensive review. *TrAC Trends in Analytical Chemistry*. 135: 116164
3. Zhang Z, Li Q, Du X, Liu M. 2020. Application of electrochemical biosensors in tumor cell detection. *Thoracic Cancer*. 11: 840–850
4. Yadav HM, Park JD, Kang HC, Lee JJ. 2021. Recent development in nanomaterial-based electrochemical sensors for cholesterol detection. *Chemosensors*. 9: 98
5. Ruiz de Eguilaz M, Cumba LR, Forster RJ. 2020. Electrochemical detection of viruses and antibodies: A mini review. *Electrochemistry Communications*. 116: 106762
6. Zhao Z, Huang C, Huang Z, Lin F, He Q, Tao D, Jaffrezic-Renault N, Guo Z. 2021. Advancements in electrochemical biosensing for respiratory virus detection: A review. *TrAC Trends in Analytical Chemistry*. 139: 116253
7. Yáñez-Sedeño P, Campuzano S, Pingarrón JM. 2020. Screen-printed electrodes: Promising paper and wearable transducers for (bio)sensing. *Biosensors*. 10: 76
8. Mohan AMV, Rajendran V, Mishra RK, Jayaraman M. 2020. Recent advances and perspectives in sweat based wearable electrochemical sensors. *TrAC Trends in Analytical Chemistry*. 131: 116024
9. Mathew M, Radhakrishnan S, Vaidyanathan A, Chakraborty B, Rout CS. 2021. Flexible and wearable electrochemical biosensors based on two-dimensional materials: Recent developments. *Analytical and Bioanalytical Chemistry*. 413: 727–762
10. Madden J, O'Mahony C, Thompson M, O'Riordan A, Galvin P. 2020. Biosensing in dermal interstitial fluid using microneedle based electrochemical devices. *Sensing and Bio-Sensing Research*. 29: 100348
11. Ahmadi A, Kabiri S, Omidfar K. 2021. Advances in HbA1c biosensor development based on field effect transistors: A review. *IEEE Sensors Journal*. 20(16): 8912–8921
12. Wu S, Wang X, Li Z, Zhang S, Xing F. 2020. Recent advances in the fabrication and application of graphene microfluidic sensors. *Micromachines*. 11: 1059
13. Jeerapan I, Sonsa-ard T, Nacapricha D. 2020. Applying nanomaterials to modern biomedical electrochemical detection of metabolites, electrolytes, and pathogens. *Chemosensors*. 8: 71
14. Madhurantakam S, Karnam JB, Brabazon D, Takai M, Ahad IU, Rayappan JBB, Krishnan UM. 2020. "Nano": An emerging avenue in electrochemical detection of neurotransmitters. *ACS Chemical Neuroscience*. 11: 4024–4047
15. Kucherenko IS, Soldatkin OO, Dzyadevych SV, Soldatkin AP. 2020. Electrochemical biosensors based on multienzyme systems: Main groups, advantages and limitations - a review. *Analytica Chimica Acta*. 1111: 114–131

16. Heikenfeld J, Jajack A, Rogers J, Gutruf P, Tian L, Pan T, Li R, Khine M, Kim J, Wang J, Kim J. 2018. Wearable sensors: Modalities, challenges, and prospects. *Lab on a Chip.* 18: 217–248

17. Koydemir HC, Ozcan A. 2018. Wearable and implantable sensors for biomedical applications. *Annual Review of Analytical Chemistry.* 11: 17

18. Tricoli A, Nasiri N, De S. 2017. Wearable and miniaturized sensor technologies for personalized and preventive medicine. *Advanced Functional Materials.* 27: 1605271

19. Kim J, Campbell AS, Esteban-Fernández de Ávila B, Wang J. 2019. Wearable biosensors for healthcare monitoring. *Nature Biotechnology.* 37: 389–406

20. Sharma A, Badea M, Tiwaris S, Marty JL. 2021. Wearable biosensors: An alternative and practical approach in healthcare and disease monitoring. *Molecules.* 26: 748

21. Hatamie A, Angizi S, Kumar S, Pandey SM, Simchi A, Willander M, Malhotra BD. 2020. Textile based chemical and physical sensors for healthcare monitoring. *Journal of the Electrochemical Society.* 167: 037546

22. Teymourian H, Parrilla M, Sempionatto JR, Felipe Montiel N, Barfidokht A, Van Echelpoel R, De Wael K, Wang J. 2020. Wearable electrochemical sensors for the monitoring and screening of drugs. *ACS Sensors.* 5: 2679–2700

23. Windmiller JR, Wang J. 2013. Wearable electrochemical sensors and biosensors: A review. *Electroanalysis.* 25: 29–46

24. Bandodkar AJ, Wang J. 2014. Non-invasive wearable electrochemical sensors: A review. *Trends in Biotechnology.* 32: 363–371

25. Arduini F, Scognamiglio V, Moscone D, Palleschi G. 2016. Electrochemical biosensors for chemical warfare agents. In *Biosensors for Security and Bioterrorism Applications,* pp. 115–139. Springer

26. Kim J, Kumar R, Bandodkar AJ, Wang J. 2017. Advanced materials for printed wearable electrochemical devices: A review. *Advanced Electronic Materials.* 3: 1600260

27. Islam T, Mukhopadhayay SC. 2017. Wearable sensors for physiological parameters measurement: Physics, characteristics, design and applications. In *Wearable Sensors: Applications, Design and Implementation,* pp. 1–31. IOP Publishing

28. Koralli P, Mouzakis DE. 2021. Advances in wearable chemosensors. *Chemosensors.* 9: 99

29. Gonçalves C, da Silva AF, Gomes J, Simoes R. 2018. Wearable e-textile technologies: A review on sensors, actuators and control elements. *Inventions.* 3: 14

30. Yao H, Shum AJ, Cowan M, Lähdesmäki I, Parviz BA. 2011. A contact lens with embedded sensor for monitoring tear glucose level. *Biosensors and Bioelectronics.* 26: 3290–3296

31. Kim J, Imani S, de Araujo WR, Warchall J, Valdés-Ramírez G, Paixão TRLC, Mercier PP, Wang J. 2015. Wearable salivary uric acid mouthguard biosensor with integrated wireless electronics. *Biosensors and Bioelectronics.* 74: 1061–1068

32. Sempionatto JR, Brazaca LC, García-Carmona L, Bolat G. Campbell AS, Martin A, Tang G, Shah R, Mishra RK, Kim J, Zucolotto V, Escarpa A, Wang J. 2019. Eyeglasses-based tear biosensing system: Non-invasive detection of alcohol, vitamins and glucose. *Biosensors and Bioelectronics.* 137: 161–170

33. Dervisevic M, Alba M, Prieto-Simon B, Voelcker NH. 2020. Skin in the diagnostics game: Wearable biosensor nano- and microsystems for medical diagnostics. *Nano Today.* 30: 100828

34. Sonner Z, Wilder E, Heikenfeld J, Kasting G, Beyette F, Swaile D, Sherman F, Joyce J, Hagen J, Kelley-Loughnane N, Naik R. 2015. The microfluidics of the eccrine sweat gland, including biomarker partitioning, transport, and biosensing implications. *Biomicrofluidics.* 9: 031301

35. Zhang L, Liu J, Fu Z, Qi L. 2020. A wearable biosensor based on bienzyme gel-membrane for sweat lactate monitoring by mounting on eyeglasses. *Journal of Nanoscience and Nanotechnology*. 20: 1495–1503

36. Gao W, Emaminejad S, Nyein HYY, Challa S, Chen K, Peck A, Fahad HM, Ota H, Shiraki H, Kiriya D, Lien DH, Brooks GA, Davis RW, Javey A. 2016. Fully integrated wearable sensor arrays for multiplexed in situ perspiration analysis. *Nature*. 529: 509–514

37. Gao W, Nyein HYY, Shahpar Z, Fahad HM, Chen K, Emaminejad S, Gao Y, Tai LC, Ota H, Wu E, Bullock E, Zeng Y, Lien DH, Javey A. 2016. Wearable microsensor array for multiplexed heavy metal monitoring of body fluids. *ACS Sensors*. 1: 866–874

38. Emaminejad S, Gao W, Wu E, Davies ZA, Nyein HYY, Challa S, Ryan SP, Fahad HM, Chen K, Shahpar Z, Talebi S, Milla C, Javey A, Davis RW. 2017. Autonomous sweat extraction and analysis applied to cystic fibrosis and glucose monitoring using a fully integrated wearable platform. *Proceedings of the National Academy of Sciences of the United States of America*. 114: 4625–4630

39. Kim J, Jeerapan I, Imani S, Cho TN, Bandodkar A, Cinti S, Mercier PP, Wang J. 2016. Noninvasive alcohol monitoring using a wearable tattoo-based iontophoretic-biosensing system. *ACS Sensors*. 1: 1011–1019

40. Bandodkar AJ, Jia W, Yardimci C, Wang X, Ramirez J, Wang J. 2015. Tattoo-based noninvasive glucose monitoring: A proof-of-concept study. *Analytical Chemistry*. 87: 394–398

41. Chen Y, Lu S, Zhang S, Li Y, Qu Z, Chen Y, Lu B, Wang X, Feng X. 2017. Skin-like biosensor system via electrochemical channels for noninvasive blood glucose monitoring. *Science Advances*. 3

42. Varadharaj EK, Jampana N. 2016. Non-invasive potentiometric sensor for measurement of blood urea in human subjects using reverse iontophoresis. *Journal of the Electrochemical Society*. 163: B340–B347

43. Heikenfeld J, Jajack A, Feldman B, Granger SW, Gaitonde S, Begtrup G, Katchman BA. 2019. Accessing analytes in biofluids for peripheral biochemical monitoring. *Nature Biotechnology*. 37: 407–419

44. Kim J, Sempionatto JR, Imani S, Hartel MC, Barfidokht A, Tang G, Campbell AS, Mercier PP, Wang J. 2018. Simultaneous monitoring of sweat and interstitial fluid using a single wearable biosensor platform. *Advanced Science*. 5: 1800880

45. Min J, Sempionatto JR, Teymourian H, Wang J, Gao W. 2021. Wearable electrochemical biosensors in North America. *Biosensors and Bioelectronics*. 172: 112750

46. Teymourian H, Tehrani F, Mahato K, Wang J. 2021. Lab under the skin: Microneedle based wearable devices. *Advanced Healthcare Materials*. 2002255

47. Windmiller JR, Zhou N, Chuang MC, Valdés-Ramírez G, Santhosh P, Miller PR, Narayan R, Wang J. 2011. Microneedle array-based carbon paste amperometric sensors and biosensors. *Analyst*. 136: 1846–1851

48. Caliò A, Dardano P, Di Palma V, Bevilacqua MF, Di Matteo A, Iuele H, De Stefano L. 2016. Polymeric microneedles based enzymatic electrodes for electrochemical biosensing of glucose and lactic acid. *Sensors & Actuators, B: Chemical*. 236: 343–349

49. Bollella P, Sharma S, Cass AEG, Antiochia R. 2019. Microneedle-based biosensor for minimally-invasive lactate detection. *Biosensors and Bioelectronics*. 123: 152–159

50. Mohan AMV, Windmiller JR, Mishra RK, Wang J. 2017. Continuous minimally-invasive alcohol monitoring using microneedle sensor arrays. *Biosensors and Bioelectronics*. 91: 574–579

51. Goud KY, Moonla C, Mishra RK, Yu C, Narayan R, Litvan I, Wang J. 2019. Wearable electrochemical microneedle sensor for continuous monitoring of levodopa: Toward Parkinson management. *ACS Sensors*. 23: 2196–2204

52. Ciui B, Martin A, Mishra RK, Brunetti B, Nakagawa T, Dawkins TJ, Lyu M, Cristea C, Sandulescu R, Wang J. 2018. Wearable wireless tyrosinase bandage and microneedle sensors: Toward melanoma screening. *Advanced Healthcare Materials*. 7

53. Invernale MA, Tang BC, York RL, Le L, Hou DY, Anderson DG. 2014. Microneedle electrodes toward an amperometric glucose-sensing smart patch. *Advanced Healthcare Materials*. 3: 38–342

54. Ribet F, Stemme G, Roxhed N. 2018. Real-time intradermal continuous glucose monitoring using a minimally invasive microneedle-based system. *Biomedical Microdevices*. 20: 101

55. Sharma S, El-Laboudi A, Reddy M, Jugnee N, Sivasubramaniyam S, El Sharkawy M, Georgiou P, Johnston D, Oliver N, Cass AEG. 2018. A pilot study in humans of microneedle sensor arrays for continuous glucose monitoring. *Analytical Methods*. 10: 2088–2095

56. Yan Q, Peng B, Su G, Cohan BE, Major TC, Meyerhoff ME. 2011. Measurement of tear glucose levels with amperometric glucose biosensor/capillary tube configuration. *Analytical Chemistry*. 83: 8341–8346

57. Wang L, Jiang K, Shen G. 2021. Wearable, implantable, and interventional medical devices based on smart electronic skins. *Advanced Materials Technologies*. 6: 2100107

58. Heo YJ, Takeuchi S. 2013. Towards smart tattoos: Implantable biosensors for continuous glucose monitoring. *Advanced Healthcare Materials*. 2: 43–56

59. Bandodkar AJ, Nuñez-Flores R, Jia W, Wang J. 2015. All-printed stretchable electrochemical devices. *Advanced Materials*. 27: 3060–3065

60. Bandodkar AJ, Jia WZ, Wang J. 2015. Tattoo-based wearable electrochemical devices: A review. *Electroanalysis*. 27: 562–572

61. Kamišalić A, Fister I, Turkanović M, Karakatič S. 2018. Wrist-wearable devices: A review. *Sensors*. 18: 1714

62. Meng L, Turner APF, Mak WC. 2020. Soft and flexible material-based affinity sensors. *Biotechnology Advances*. 39: 107398

63. Stoppa M, Chiolerio A. 2014. Wearable electronics and smart textiles: A critical review. *Sensors*. 14: 11957–11992

64. Farandos NM, Yetisen AK, Monteiro MJ, Lowe CR, Yun SH. 2015. Contact lens sensors in ocular diagnostics. *Advanced Healthcare Materials*. 4: 792–810

65. Baca JT, Finegold DN, Asher SA. 2016. Tear glucose analysis for the noninvasive detection and monitoring of diabetes mellitus. *Ocular Surface*. 5: 280–293

66. Pankratov D, González-Arribas E, Blum Z, Shleev S. 2016. Tear based bioelectronics. *Electroanalysis*. 28: 1250–1266

67. Mitsubayashi K, Arakawa T. 2016. Cavitas sensors: Contact lens type sensors & mouthguard sensors. *Electroanalysis*. 28: 1170–1187

68. Liao YT, Yao HF, Lingley A, Parviz B, Otis BP. 2012. A 3-mu W CMOS glucose sensor for wireless contact-lens tear glucose monitoring. *IEEE Journal of Solid-State Circuits*. 47: 335–344

69. Kim J, Kim M, Lee MS, Kim K, Ji S, Kim YT, Park J, Na K, Bae KH, Kim HK, Bien F, Lee CY, Park JU. 2017. Wearable smart sensor systems integrated on soft contact lenses for wireless ocular diagnostics. *Nature Communications*. 8: 14997

70. Park J, Kim SY, Cheong WH, Jang J, Park YG, Lee CY. 2018. Soft, smart contact lenses with integrations of wireless circuits, glucose sensors, and displays. *Science Advances*. 4

71. Shum AJ, Cowan M, Lahdesmaki I, Lingley A, Otis B, Parviz BA. 2009. Functional modular contact lens. *Proceedings of SPIE*. 7397: 73970

72. Kownacka AE, Vegelyte D, Joosse M, Anton N, Toebes BJ, Lauko J, Buzzacchera I, Lipinska K, Wilson DA, Geelhoed-Duijvestijn N, Wilson CJ. 2019. Clinical evidence for use of a noninvasive biosensor for tear glucose as an alternative to painful finger-prick for diabetes management utilizing biopolymer coating. *Biomacromolecules*. 19: 4504–4511

73. Thomas N, Lahdesmaki I, Parviz BA. 2012. A contact lens with an integrated lactate sensor. *Sensors & Actuators, B: Chemical*. 162: 128–134

74. Zubareva TV, Kiseleva ZM. 1977. Catecholamine content of the lacrimal fluid of healthy people and glaucoma patients. *Ophthalmologica*. 175: 339–344

75. Yu L, Yang Z, An M. 2019. Lab on the eye: A review of tear-based wearable devices for medical use and health management. *BioScience Trends*. 13: 308–313

76. Pfaffe T, Cooper-White J, Beyerlein P, Kostner K, Punyadeera C. 2011. Diagnostic potential of saliva: Current state and future applications. *Clinical Chemistry*. 57: 675–687

77. Goswami Y, Mishra R, Agrawal AP, Agrawal LA. 2015. Salivary biomarkers: A review of powerful diagnostic tool. *IOSR Journal of Dental and Medical Sciences*. 14: 80–87

78. Campuzano S, Yañez-Sedeño P, Pingarron JM. 2017. Electrochemical bioaffinity sensors for salivary biomarkers detection. *Trends in Analytical Chemistry*. 86: 14–24

79. Zagatto AM, Papoti M, Caputo F, Mendes ODC, Denadai BS, Baldissera V, Gobatto C.A. 2004. Comparison between the use of saliva and blood for the minimum lactate determination in arm ergometer and cycle ergometer in table tennis players. *Revista Brasileira de Medicina do Esporte*. 10: 475–480

80. Falk M, Psotta C, Cirovic S, Shleev S. 2020. Non-invasive electrochemical biosensors operating in human physiological fluids. *Sensors*. 20: 6352

81. Kim J, Valdes-Ramírez G, Bandodkar AJ, Jia W, Martinez AG, Ramirez J, Mercier PP, Wang J. 2014. Non-invasive mouthguard biosensor for continuous salivary monitoring of metabolites. *Analyst*. 139: 1632–1636

82. Arakawa T, Kuroki Y, Nitta H, Chouhan P, Toma K, Sawada S, Takeuchi S, Sekita T, Akiyoshi K, Minakuchi S, Mitsubayashi K. 2016. Mouthguard biosensor with telemetry system for monitoring of saliva glucose: A novel cavitas sensor. *Biosensors and Bioelectronics*. 84: 106–111

83. Mannoor MS, Tao H, Clayton JD, Sengupta A, Kaplan DL, Naik RR, Verma N, Omenetto FG, McAlpine MC. 2012. Graphene-based wireless bacteria detection on tooth enamel. *Nature Communications*. 3: 763

84. García-Carmona L, Martín A, Sempionatto JR, Moreto JR, Gonzalez MC, Wang J, Escarpa A. 2019. Pacifier biosensor: Toward noninvasive saliva biomarker monitoring. *Analytical Chemistry*. 91: 13883–13891

85. Li GP, Bachman M, Lee A. Google Patents. US20040220498A1 2004.

86. Nyein HYY, Gao W, Shahpar Z, Emaminejad S, Challa S, Chen K, Fahad HM, Tai L, Ota H, Davis RW, Javey A. 2016. A wearable electrochemical platform for noninvasive simultaneous monitoring of Ca^{2+} and pH. *ACS Nano*. 10: 7216–7224

87. Hondred JA, Johnson Z, Claussen JC. 2020. Nanoporous gold peel-and-stick biosensors created with etching inkjet maskless lithography for electrochemical pesticide monitoring with microfluidics. *Journal of Materials Chemistry C*. 8: 11376–11388

88. Singh M, Haverinen HM, Dhagat P, Jabbour G.E. 2010. Inkjet printing—process and its applications. *Advanced Materials*. 22: 673–685

89. Liu C, Huang N, Xu F, Tong J, Chen Z, Gui X, Fu Y, Lao C. 2018. 3D Printing technologies for flexible tactile sensors toward wearable electronics and electronic skin. *Polymers*. 10: 629

90. Xu Y, Wu X, Guo X, Kong B, Zhang M, Qian X, Mi S, Sun W. 2017. The boom in 3D-printed sensor technology. *Sensors*. 17: 1166

91. Ferreira PC, Ataíde VN, Chagas CLS, Angnes L, Coltro T, Paixao TRCL, Reis de Araujo W. 2013. Wearable electrochemical sensors for forensic and clinical applications. *Trends in Analytical Chemistry*. 119: 115

92. Weng B, Shepherd RL, Crowley K, Killard AJ, Wallace GG. 2010. Printing conducting polymers. *Analyst*. 135: 2779–2789

93. Kim S, Sojoudi H, Zhao H, Mariappan D, McKinley GH, Gleason KK, Hart AJ. 2016. Ultrathin high-resolution flexographic printing using nanoporous stamps. *Science Advances*. 2

94. Lee W, Koo H, Sun J, Noh J, Kwon KS, Yeom C, Choi Y, Chen K, Javey A, Cho G. 2015. A fully roll-to-roll gravure-printed carbon nanotube-based active matrix for multi-touch sensors. *Scientific Reports*. 5: 17707

95. Bariya M, Shahpar Z, Park H, Sun J, Jung Y, Gao W, Nyein HYY, Liaw TS, Tai LC, Ngo QP, Chao M, Zhao Y, Hettick M, Cho G, Javey A. 2019. Roll-to-roll gravure printed electrochemical sensors for wearable and medical devices. *ACS Nano*. 12: 6978–6987

96. Yang X, Cheng H. 2020. Recent developments of flexible and stretchable electrochemical biosensors. *Micromachines*. 11: 243

97. Yoon J, Cho HY, Shin M, Choi H.K, Lee T, Choi JW. 2020. Flexible electrochemical biosensors for healthcare monitoring. *Journal of Materials Chemistry*. 8: 7303–7318

98. Wang C, Xia K, Wang H, Liang X, Yin Z, Zhang Y. 2019. Advanced carbon for flexible and wearable electronics. *Advanced Materials*. 31: 1801072

99. Zhai Q, Cheng W. 2019. Soft and stretchable electrochemical biosensors. *Materials Today Nano*. 7: 100041

100. Amjadi M, Pichitpajongkit A, Lee S, Ryu S, Park I. 2014. Highly stretchable and sensitive strain sensor based on silver nanowire—elastomer nanocomposite. *ACS Nano*. 8: 5154–5163

101. Nag A, Mukhopadhyay SC, Kosel J. 2017. Wearable flexible sensors: A review. *IEEE Sensors Journal*. 17: 3949–3960

102. Eom J, Jaisutti R, Lee H, Lee W, Heo JS, Lee JY, Park SK, Kim YH. 2017. Highly sensitive textile strain sensors and wireless user-interface devices using all-polymeric conducting fibers. *ACS Applied Materials & Interfaces*. 9: 10190–10197

103. Liu X, Lillehoj PB. 2016. Embroidered electrochemical sensors for biomolecular detection. *Lab Chip*. 16: 2093–2098

104. Choudhary T, Rajamanickam GP, Dendukuri D. 2015. Woven electrochemical fabric-based test sensors (WEFTS): A new class of multiplexed electrochemical sensors. *Lab Chip*. 15: 2064–2072

105. Chen C, Ran R, Yang Z, Lv R, Shen W, Kang F, Huang ZH. 2018. An efficient flexible electrochemical glucose sensor based on carbon nanotubes/carbonized silk fabrics decorated with Pt microspheres. *Sensors & Actuators, B: Chemical*. 256: 63–70

106. Bandodkar AJ, Hung VWS, Jia W, Valdés-Ramírez G, Windmiller JR, Martinez AG, Ramírez J, Chan G, Kerman K, Wang J. 2013.Tattoo-based potentiometric ion-selective sensors for epidermal pH monitoring. *Analyst*. 138: 123–128

107. Guinovart T, Bandodkar AJ, Windmiller JR, Andrade FJ, Wang J. 2013. A potentiometric tattoo sensor for monitoring ammonium in sweat. *Analyst*. 138: 7031–7038

108. Jia W, Bandodkar AJ, Valdés-Ramírez G, Windmiller JR, Yang Z, Ramírez J, Chan G, Wang J. 2013. Electrochemical tattoo biosensors for real-time noninvasive lactate monitoring in human perspiration. *Analytical Chemistry*. 85: 6553–6560

109. Johnson L. 2012. *Physiology of the Gastrointestinal Tract*. Academic Press

110. Beardslee LA, Banis GE, Chu S, Liu S, Chapin AA, Stine JM, Pasricha PJ, Ghodssi R. 2020. Ingestible sensors and sensing systems for minimally invasive diagnosis and monitoring: The next frontier in minimally invasive screening. *ACS Sensors*. 5: 891–910

111. Li F, Gurudu SR, De Petri SG, Sharma V, Shiff AD, Heigh RI, Fleischer DE, Post J, Erickson P, Leighton JA. 2008. Retention of the capsule endoscope: A single-center experience of 1000 capsule endoscopy procedure. *Gastrointestinal Endoscopy*. 68: 174–180

112. Kim J, Jeerapan I, Ciui B, Hartel MzC, Martin A, Wang J. 2017. Edible electrochemistry: Food materials based electrochemical sensors. *Advanced Healthcare Materials*. 6: 1700770

113. Bettinger CJ. 2018. Advances in materials and structures for ingestible electromechanical medical devices. *Angewandte Chemie International Edition*. 57: 16946–16958

114. Steiger C, Abramson A, Nadeau P, Chandrakasan AP, Langer R, Traverso G. 2018. Ingestible electronics for diagnostics and therapy. *Nature Reviews Materials*. 4: 83–98

115. Burdock GA, Carabin IG. 2004. Generally recognized as safe (GRAS): History and description. *Toxicology Letters*. 5: 3–18

116. Bonacchini E, Bossio C, Greco F, Mattoli V, Kim YH, Lanzani G, Cairo M. 2018. Tattoo-paper transfer as a versatile platform for all-printed organic edible electronics. *Advanced Materials*. 30: 1706091

117. Chyan Y, Ye R, Li Y, Singh SP, Arnusch CJ, Tour JM. 2018. Laser- induced graphene by multiple lasing: Toward electronics on cloth, paper, and food. *ACS Nano*. 12: 2176–2183

118. McCaffrey C, Twomey K, Ogurtsov VI. 2015. Development of a wireless swallowable capsule with potentiostatic electrochemical sensor for gastrointestinal track investigation. *Sensors and Actuators B: Chemical*. 218: 8–15

119. Ruiz-Valdepeñas Montiel V, Sempionatto JR, Campuzano S, Pingarrón JM, Esteban Fernández de Ávila B, Wang J. 2019. Direct electrochemical biosensing in gastrointestinal fluids. *Analytical and Bioanalytical Chemistry*. 411: 4597–4604

120. Basaeri H, Christensen DB, Roundy S. 2016. A review of acoustic power transfer for biomedical implant. *Smart Materials and Structures*. 25: 123001

121. Vaddiraju S, Tomazos I, Burgess DJ, Jain FC, Papadimitrakopoulos F. 2010. Emerging synergy between nanotechnology and implantable biosensors: A review. *Biosensors and Bioelectronics*. 25: 1553–1565

122. Guk K, Han G, Lim J, Jeong K, Kang T, Lim EK, Jung J. 2019. Evolution of wearable devices with real-time disease monitoring for personalized healthcare. *Nanomaterials*. 9: 813

123. Pulugu P, Ghosh S, Rokade S, Choudhury K, Arya N, Kumar P. 2021. A perspective on implantable biomedical materials and devices for diagnostic applications. *Current Opinion in Biomedical Engineering*. 18: 100287

124. Bennett R, Leech D. 2020. Improved operational stability of mediated glucose enzyme electrodes for operation in human physiological solutions. *Bioelectrochemistry*. 133: 107460

125. Kim SMS, Kim N, Kim Y, Baik MS, Yoo M, Kim D, Lee WJ, Kang DH, Kim SMS, Lee K, Yoon MH. 2018. High-performance, polymer-based direct cellular interfaces for electrical stimulation and recording. *NPG Asia Materials*. 10: 255–265

126. Wang Y, Zhang W, Huang L, Ito Y, Wang Z, Shi X, Wei Y, Jing X, Zhang P. 2018. Intracellular calcium ions and morphological changes of cardiac myoblasts response to an intelligent biodegradable conducting copolymer. *Materials Science and Engineering C*. 90: 168–179

127. Hui N, Sun X, Niu S, Luo X. 2017. PEGylated polyaniline nanofibers: Antifouling and conducting biomaterial for electrochemical DNA sensing. *ACS Applied Materials & Interfaces*. 9: 2914–2923

128. Chou YN, Sun F, Hung HC, Jain P, Sinclair A, Zhang P, Bai T, Chang Y, Wen Q, Yu TC, Jiang S. 2016. Ultra-low fouling and high antibody loading zwitterionic hydrogel coatings for sensing and detection in complex media. *Acta Biomaterialia*. 40: 31–37

129. Zhou DD, Greenberg RJ. 2013. Implantable electrochemical biosensors for retinal prostheses. In *Biosensors and Their Applications in Healthcare*, ed. D Ozkan-Ariksoysal, pp. 24–36. Future Science

130. Scholten K, Meng E. 2018. A review of implantable biosensors for closed-loop glucose control and other drug delivery applications. *International Journal of Pharmaceutics*. 554: 319–334

131. Garg SK, Schwartz S, Edelman SV. 2004. Improved glucose excursions using an implantable real-time continuous glucose sensor in adults with type 1 diabetes. *Diabetes Care*. 27: 734–738

132. Gough DA, Kumosa LS, Routh TL, Lin JT, Lucisano JY. 2010. Function of an implanted tissue glucose sensor for more than 1 year in animals. *Science Translational Medicine*. 2: 1–8

133. Wu Q, Liu J, Wang X, Feng L, Wu J, Zhu X, Wen X, Gong X. 2020. Organ-on-a-chip: Recent breakthroughs and future prospects. *Biomedical Engineering*. 19: 9

134. Ferrari E, Palma C, Vesentini S, Occhetta P, Rasponi M. 2020. Integrating biosensors in organs-on-chip devices: A perspective on current strategies to monitor microphysiological systems. *Biosensors*. 10: 110

135. Bavli D, Prill S, Ezra E, Levy G, Cohen M, Vinken M, Vanfleteren J, Jaeger M, Nahmias Y. 2016. Real-time monitoring of metabolic function in liver-on-chip microdevices tracks the dynamics of mitochondrial dysfunction. *Proceedings of the National Academy of Sciences of the United States of America*. 113(16)

136. Dervisevic E, Tuck KL, Voelcker NH, Cadarso VJ. 2019. Recent progress in lab-on-a-chip systems for the monitoring of metabolites for mammalian and microbial cell research. *Sensors*. 19: 5027

137. Bange A, Halsall HB, Heineman WR. 2005. Microfluidic immunosensor systems. *Biosensors and Bioelectronics*. 20: 2488–2503

138. Riahi R, Shaegh SAM, Ghaderi M, Zhang YS, Shin SR, Aleman J, Massa S, Kim D, Dokmeci MR, Khademhosseini A. 2016. Automated microfluidic platform of bead-based electrochemical immunosensor integrated with bioreactor for continual monitoring of cell secreted biomarkers. *Scientific Reports*. 6: 24598

139. Shin SR, Zhang YS, Kim DJ, Manbohi A, Avci H, Silvestri A, Aleman J, Hu N, Kilic T, Keung W, Righi M, Assawes P, Alhadrami HA, Li RA, Dokmeci MR, Khademhosseini A. 2016. Aptamer-based microfluidic electrochemical biosensor for monitoring cell-secreted trace cardiac biomarkers. *Analytical Chemistry*. 88: 10019–10027

140. Fan R, Andrew TL. 2020. Perspective—challenges in developing wearable electrochemical sensors for longitudinal health monitoring. *Journal of the Electrochemical Society*. 167: 037542

141. Xuan X, Pérez-Ràfols C, Chen C, Cuartero M, Crespo GA. 2021. Lactate biosensing for reliable on-body sweat analysis. *ACS Sensors*. 6(7): 2763–2771

142. Banerjee S, McCracken S, Hossain F, Slaughter G. 2020. Electrochemical detection of neurotransmitters. *Biosensors*. 10: 101

143. Bah E, Hachmann J, Paek SB, Batton A, Min PK, Bennet K, Lee K. 2017. Wireless intraoperative real-time monitoring of neurotransmitters in humans. In *Proceedings of the 2017 IEEE International Symposium on Medical Measurements and Applications (MeMeA)*, pp. 123–128. Institute of Electrical and Electronics Engineers (IEEE)

144. Shukla RP, Rapier C, Glassmanb M, Liu F, Kelly DL, Ben-Yoav H. 2020. An integrated electrochemical microsystem for real-time treatment monitoring of clozapine in microliter volume samples from schizophrenia patients. *Electrochemistry Communications*. 120: 106850

145. Balasubramanian P, He SB, Jansirani A, Peng HP, Huang LL, Deng HH, Chen W. 2020. Bimetallic AgAu decorated MWCNTs enable robust nonenzyme electrochemical sensors for in-situ quantification of dopamine and H_2O_2 biomarkers expelled from PC-12 cells. *Journal of Electroanalytical Chemistry*. 878: 114554

146. Singh U, Kumbhat S. 2021. Ready to use electrochemical sensor strip for point-of-care monitoring of serum lithium. *Electroanalysis*. 33: 393–399

147. Ledo A, Fernandes E, Quintero JE, Gerhardt GA, Barbosa RM. 2020. Electrochemical evaluation of a multi-site clinical depth recording electrode for monitoring cerebral tissue oxygen. *Micromachines*. 11: 632

148. Dauphin-Ducharme P, Yang K, Arroyo-Currás N, Ploense KL, Zhang Y, Gerson J, Kurnik M, Kippin TE, Stojanovic MN, Plaxco K. W. 2019. Electrochemical aptamer-based

sensors for improved therapeutic drug monitoring and high-precision, feedback-controlled drug delivery. *ACS Sensors*. 4(10): 2832–2837

149. Balamurugan TST, Chen GZ, Kumaravel S, Lin CM, Huang ST, Lee YC, Chen CH, Luo GR. 2020. Electrochemical substrate for active profiling of cellular surface leucine aminopeptidase activity and drug resistance in cancer cells. *Biosensors and Bioelectronics*. 150: 111948

150. Lee S, Lee YJ, Kim JH, Lee GJ. 2020. Electrochemical detection of H_2O_2 released from prostate cancer cells using Pt nanoparticle-decorated rGO—CNT nanocomposite-modified screen-printed carbon electrodes. *Chemosensors*. 8: 63

151. Xu J, Cheng C, Li X, Lu Y, Hu Shen, Liu G, Zhu L, Wang N, Wang L, Cheng Pu, Su B, Liu Q. 2021. Implantable platinum nanotree microelectrode with a battery-free electrochemical patch for peritoneal carcinomatosis monitoring. *Biosensors and Bioelectronics*. 18: 113265

152. Bakker E. 2016. Can calibration-free sensors be realized? *ACS Sensors*. 1: 838–841

153. Liu N, Xu Z, Morrin A, Luo X. 2019. Low fouling strategies for electrochemical biosensors targeting disease biomarkers. *Analytical Methods*. 11: 702–711

154. Land KJ, Boeras DI, Chen XS, Ramsay AR, Peeling RW. 2019. Reassured diagnostics to inform disease control strategies, strengthen health systems and improve patient outcomes. *Nature Microbiology*. 4: 46–54

155. Li P, Lee GH, Kim SY, Kwon SY, Kim HR, Park R. 2021. From diagnosis to treatment: Recent advances in patient-friendly biosensors and implantable devices. *ACS Nano*. 15: 1960–2004

156. Li J, Liang JY, Laken SJ, Langer R, Traverso G. 2020. Clinical opportunities for continuous biosensing and closed-loop therapies. *Trends in Chemistry*. 2: 319–340

157. Campuzano S, Pedrero M, Yáñez-Sedeño P, Pingarrón JM. 2019. Antifouling (bio) materials for electrochemical (bio)sensing. *International Journal of Molecular Sciences*. 20: 423

158. Jiang C, Wang G, Hein R, Liu N, Luo X, Davis JJ. 2020. Antifouling strategies for selective in vitro and in vivo sensing. *Chemical Reviews*. 120: 3852–3889

159. Banerjee I, Pangule RC, Kane RS. 2011. Antifouling coatings: Recent developments in the design of surfaces that prevent fouling by proteins, bacteria, and marine organisms. *Advanced Materials*. 23: 690–718

160. Selim MS, Shenashen MA, El-Safty SA, Higazy SA, Selim MM, Isago H, Elmarakbi A. 2017. Recent progress in marine foul-release polymeric nanocomposite coatings. *Progress in Materials Science*. 87: 1–32

161. Lin PH, Li BR. 2020. Antifouling strategies in advanced electrochemical sensors and biosensors. *Analyst*. 145: 1110–1120

162. Zhao S, Lui N, Wang W, Xu Z, Wu Y, Luo X. 2021. An electrochemical biosensor for Alpha-fetoprotein detection in human serum based on peptides containing isomer D-amino acids with enhanced stability and antifouling property. *Biosensors and Bioelectronics*. 190: 113466

163. Sabaté del Río J, Henry OYF, Jolly P, Ingber DE. 2019. An antifouling coating that enables affinity-based electrochemical biosensing in complex biological fluids. *Nature Nanotechnology*. 14: 143–1149

164. Wei H, Wu F, Li L, Yang X, Xu C, Yu P, MF, Mao L. 2020. Natural leukocyte membrane-masked microelectrodes with an enhanced antifouling ability and biocompatibility for in vivo electrochemical sensing. *Analytical Chemistry*. 92: 11374–11379

165. Ruiz-Valdepeñas Montiel V, Sempionatto JR, Esteban-Fernández de Ávila B, Whitworth A, Campuzano S, Pingarrón JM, Wang J. 2018. Delayed sensor activation based on transient coatings: Biofouling protection in complex biofluids. *Journal of the American Chemical Society*. 140: 14050–14053

166. Arroyo-Currás N, Dauphin-Ducharme P, Scida K, Chávez JL. 2020. From the beaker to the body: Translational challenges for electrochemical, aptamer-based sensors. *Analytical Methods*. 12: 1288–1310

167. Ferguson BC, Hoggarth DA, Maliniak D, Ploense K, White RJ, Woodward N, Hsieh K, Bonham AJ, Eisenstein M, Kippin T, Plaxco KW, Soh HT. 2013. Real-time, aptamer-based tracking of circulating therapeutic agents in living animals. *Science Translational Medicine*. 5: 213ra165

168. Arroyo-Currás N, Somerson J, Vieira PA, Ploense KL, Kippin TE. 2017. Real-time measurement of small molecules directly in awake, ambulatory animals. *Proceedings of the National Academy of Sciences of the United States of America*. 14: 645–650

169. Vieira PA, Shin CB, Arroyo-currás N, Ortega G. 2019. Pharmacokinetic measurements highlight the need for and a route toward more highly personalized medicine. *Frontiers in Molecular Biosciences*. 6: 69

170. Mage PL, Ferguson BS, Maliniak D, Ploense KL, Kippin TE, Soh HT. 2017. Closed-loop control of circulating drug levels in live animals. *Nature Biomedical Engineering*. 1: 0070

171. Hespanha P, Plaxco KW, Plaxco ZA, Kippin TE. 2018. High-precision control of plasma drug levels using feedback- controlled dosing. *ACS Pharmacology & Translational Science*. 1: 110–118

172. Zhang Y, Gerson J, Kurnik M, Kippin TE, Stojanovic MN. 2019. Electrochemical aptamer-based sensors for improved therapeutic drug monitoring and high-precision, feedback-controlled drug delivery. *ACS Sensors*. 4: 2832–2833

173. Parolo C, Idili A, Ortega G, Csordas A, Hsu A, Arroyo-Currás N, Yang Q, Ferguson BS, Wang J, Plaxco KW. 2020. Real-time monitoring of a protein biomarker. *ACS Sensors*. 5: 1877–1881

174. Yaribeygi H, Panahi Y, Sahraei H, Johnston TP, Sahebkar A. 2017. The impact of stress on body function: A review. *EXCLI Journal*. 16: 1057–1072

175. Ganguly A, Lin KC, Muthukumar S, Prasad S. 2021. Autonomous, real-time monitoring electrochemical aptasensor for circadian tracking of cortisol hormone in sub-microliter volumes of passively eluted human sweat. *ACS Sensors*. 6: 63–72

176. Das J, Gomis S, Chen JB, Yousefi H, Ahmed S, Mahmud A, Zhou W, Sargent EH, Kelley SO. 2021. Reagentless biomolecular analysis using a molecular pendulum. *Nature Chemistry*. 13: 428–434

177. Li H, Dauphin-Ducharme P, Ortega G, Plaxco KW. 2017. Calibration-free electrochemical biosensors supporting accurate molecular measurements directly in undiluted whole blood. *Journal of the American Chemical Society*. 139: 11207–11213

178. Lee I, Kim SE, Lee J, Woo DH, Lee S, Song CS, Lee J. 2020. A self-calibrating electrochemical aptasensing platform: Correcting external interference errors for the reliable and stable detection of avian influenza viruses. *Biosensors and Bioelectronics*. 152: 112010

179. Van Den Berg A, Mummery CL, Passier R, Van der Meer AD. 2019. Personalised organs-on-chips: Functional testing for precision medicine. *Lab on a Chip*. 19(2): 198–205

180. Fuller HC, Wei T, Behrens MR, Ruder WC. 2020. The future application of organ-on-a-chip technologies as proving grounds for MicroBioRobots. *Micromachines (Basel)*. 11: 947

181. Aleman J, Kilic T, Mille LS, Shin SR, Zhang YS. 2021. Microfluidic integration of regeneratable electrochemical affinity-based biosensors for continual monitoring of organ-on-a-chip devices. *Nature Protocols*. 16: 2564–2593

182. Xiao F, Li Y, Zan X, Liao K, Xu R, Duan H. 2012. Growth of metal—metal oxide nanostructures on freestanding graphene paper for flexible biosensors. *Advanced Functional Materials*. 22: 2487–2494

183. Shin SR, Kilic T, Zhang YS, Avci H, Hu N, Kim D, Branco C, Aleman J, Massa S, Silvestri A, Kang J, Desalvo A, Hussaini MA, Chae SK, Polini A, Bhise N, Hussain MA, Lee H, Dokmeci MR, Khademhosseini A. 2017. Label-free and regenerative electrochemical microfluidic biosensors for continual monitoring of cell secretomes. *Advanced Science*. 4: 1–14

184. Zhang YS, Aleman J, Shin SR, Kilic T, Kim D, Shaegh SAM, Massa S, Riahi R, Chae S, Hu N, Avci H, Zhang W, Silvestri A, Nezhad AS, Manbohi A, De Ferrari F, Polini A, Calzone G, Shaikh N, Alerasool P, Budina E, Kang J, Bhise N, Ribas J, Pourmand A, Skardal A, Shupe T, Bishop CE, Dokmeci MR, Atala A, Khademhosseini A. 2017. Multisensor-integrated organs-on-chips platform for automated and continual in situ monitoring of organoid behaviors. *Proceedings of the National Academy of Sciences*. 114(12)

185. Ferreira de Souza M, Kuasne H, Barros-Filho MC, Cilião HL, Marchi FA, Fuganti PE, Paschoal AR, Rogatto SR, Cólus IMS. 2017. Circulating mRNAs and miRNAs as candidate markers for the diagnosis and prognosis of prostate cancer. *PLoS One*. 12(9): 1–16

186. Sakai Y, Honda M, Matsui S, Komori O, Murayama, T, Fujiwara T, Mizuno M, Imai Y, Yoshimura K, Nasti A, Wada T, Iida N, Kitahara Y, Horii R, Toshikatsu T, Nishikawa M, Okafuji H, Mizukoshi E, Yamashita T, Yamashita T, Arai K, Kitamura K, Kawaguchi K, Takatori H, Shimakami T, Terashima T, Hayashi T, Nio K, Kaneko, S. 2019. Development of novel diagnostic system for pancreatic cancer, including early stages, measuring mRNA of whole blood cells. *Cancer Science*. 110: 1364–1388

187. Costa C, Teodoro M, Rugolo CA, Alibrando C, Giambò Briguglio G, Fenga C. 2020. MicroRNAs alteration as early biomarkers for cancer and neurodegenerative diseases: New challenges in pesticides exposure *Toxicology Reports*. 7: 759–767

188. Oh SY, Kang SM, Kang SH, Lee HJ, Kwon TG, Kim JW, Lee ST, Choi SY, Hong SH. 2020. Potential salivary mRNA biomarkers for early detection of oral cancer. *Journal of Clinical Medicine*. 9: 243

189. Galvão-Lima LJ, Morais AHF, Valentim RAM, Barreto EJSS. 2021. MiRNAs as biomarkers for early cancer detection and their application in the development of new diagnostic tools. *BioMedical Engineering Online*. 20: 1–20

190. Cardoso AR, Moreira FT.C, Fernandes R, Sales MGF. 2016. Novel and simple electrochemical biosensor monitoring attomolar levels of miRNA-155 in breast cancer. *Biosensors and Bioelectronics*. 80: 621–630

191. Smith DA, Newbury LJ, Drago G, Bowen T, Redman JE. 2017. Electrochemical detection of urinary microRNAs via sulfonamide-bound antisense hybridisation. *Sensors & Actuators, B: Chemical*. 253: 335–341

192. Stobiecka M, Ratajczak K, Jakiela S. 2019. Toward early cancer detection: Focus on biosensing systems and biosensors for an anti-apoptotic protein survivin and survivin mRNA. *Biosensors and Bioelectronics*. 137: 58–71

193. Cheng YH, Liu SJ, Jiang JH. 2021. Enzyme-free electrochemical biosensor based on amplification of proximity-dependent surface hybridization chain reaction for ultrasensitive mRNA detection. *Talanta*. 222: 121536

194. Bruch R, Baaske J, Chatelle C, Meirich M, Madlener S, Weber W, Dincer C, Urban GA. 2019. CRISPR/Cas13a-powered electrochemical microfluidic biosensor for nucleic acid amplification-free miRNA diagnostics. *Advanced Materials*. 31: 1905311

195. Hajian R, Balderston S, Tran T, deBoer T, Etienne J, Sandhu M, Wauford NA, Chung JY, Nokes J, Athaiya M, Paredes J, Peytavi R, Goldsmith B, Murthy N, Conboy IM, Aran K. 2019. Detection of unamplified target genes via CRISPR—Cas9 immobilized on a graphene field-effect transistor. *Nature Biomedical Engineering*. 3: 427–437

196. Nouri R, Tang Z, Dong M, Liu T, Kshirsagar A, Guan W. 2021. CRISPR-based detection of SARS-CoV-2: A review from sample to result. *Biosensors and Bioelectronics*. 178: 113012

197. Liu C, Zhao, Tia, F, Cai L, Zhang W, Feng Q, Chang J, Wan F, Yang Y, Dai B, Cong Y, Ding B, Sun J, Tan W. 2019. Low-cost thermophoretic profiling of extracellular-vesicle surface proteins for the early detection and classification of cancers. *Nature Biomedical Engineering*. 3: 183–193

198. Whitwell H, Worthington J, Blyuss O, Gentry-Maharaj A, Ryan A, Gunu R, Kalsi J, Menon U, Jacobs I, Zaikin, A., and Timms, J.F. 2020. Improved early detection of ovarian cancer using longitudinal multimarker models. *British Journal of Cancer*. 122: 847–856

199. Sheng Y, Zhang T, Zhang S, Johnston M, Zheng X, Shan Y, Liu T, Huang Z, Qian F, Xie Z, Yiru A, Zhong H, Kuang T, Dincer C, Urban GA, Hu J. 2021. A CRISPR/Cas13a-powered catalytic electrochemical biosensor for successive and highly sensitive RNA diagnostics. *Biosensors and Bioelectronics*. 178: 113027

200. Leslie D, Lipsky P, Notkins AL. 2001. Autoantibodies as predictors of disease. *Journal of Clinical Investigation*. 108: 1417–1422

201. Barderas R, Villar-Vázquez R, Fernández-Aceñero MJ, Babel I, Peláez-García A, Torres S, Casal I. 2013. Sporadic colon cancer murine models demonstrate the value of autoantibody detection for preclinical cancer diagnosis. *Scientific Reports*. 3: 2938

202. Campuzano S, Pedrero M, González-Cortés A, Yáñez-Sedeño P, Pingarrón JM. 2019. Electrochemical biosensors for autoantibodies in autoimmune and cancer diseases. *Analytical Methods*. 11: 871–887

203. Garranzo-Asensio M, Guzmán-Aránguez A, Povedano E, Ruiz-Valdepeñas Montiel V, Poves C, Fernandez-Aceñero MJ, Montero-Calle A, Solís-Fernández G, Fernandez-Diez S, Camps J, Arenas M, Rodríguez-Tomàs E, Joven J, Sanchez-Martinez M, Rodriguez N, Dominguez, G, Yáñez-Sedeño P, Pingarrón JM, Campuzano S, Barderas R. 2020. Multiplexed monitoring of a novel autoantibody diagnostic signature of colorectal cancer using HaloTag technology-based electrochemical immunosensing platform. *Theranostics*. 10(7): 3022–3034

204. Zupančič U, Jolly P, Estrela P, Moschou D, Ingber DE. 2021. Graphene enabled low-noise surface chemistry for multiplexed sepsis biomarker detection in whole blood. *Advanced Functional Materials*. 31: 2010638

205. Campuzano S, Barderas R, Yáñez-Sedeño P, Pingarrón JM. 2021. Electrochemical biosensing to assist multiomics analysis in precision medicine. *Current Opinion in Electrochemistry*. 28: 100703

206. Sun YV, Hu YJ. 2016. Integrative analysis of multi-omics data for discovery and functional studies of complex human diseases. *Advanced Genetics*. 93: 147–190

207. Chakraborty S, Hosen MI, Ahmed M, Shekhar HU. 2018. Onco-multi-OMICS approach: A new frontier in cancer research. *BioMed Research International*. 9836256

208. Subramanian I, Verma S, Kumar S, Jere A, Anamika K. 2020. Multi-omics data integration, interpretation, and its application. *Bioinformatics and Biology Insights*. 14: 7–9

209. Gao Y, Nguyen D, Yeo T, Lim SB, Tan WX, Madden LE, Jin L, Long JYK, Aloweni FAB, Liew YJA, Tan MLL, Ang SY, Maniya S, Abdelwahab I, Loh KP, Chen CH, Becker DL Leavesley D, Ho JS, Lim CT. 2021. A flexible multiplexed immunosensor for point-of-care in situ wound monitoring. *Science Advances*. 7: 21

210. Vogiazi V, de la Cruz A, Mishra S, Shanov V, Heineman WR, Dionysiou DD. 2019. A comprehensive review: Development of electrochemical biosensors for detection of cyanotoxins in freshwater. *ACS Sensors*. 4: 1151–1173

211. Cui F, Zhou Z, Zhou HS. 2020. Review-measurement and analysis of cancer biomarkers based on electrochemical biosensors. *Electrochemical Society*. 167: 037525

212. Campuzano S, Barderas R, Pedrero M, Yánez-Sedeno P, Pingarron JM. 2020. Electrochemical biosensing to move forward in cancer epigenetics and metastasis: A review. *Analytica Chimica Acta*. 1109: 169–190

213. Díaz-Fernández A, Lorenzo-Gómez R, Miranda-Castro R, de-los-Santos-Álvarez N, Lobo-Castañón MJ. 2020. Electrochemical aptasensors for cancer diagnosis in biological fluids-A review. *Analytica Chimica Acta*. 1124: 1–19

214. Povedano E, Ruiz-Valdepeñas MV, Gamella M, Serafín V, Pedrero M, Moranova L, Bartosik M, Montoya JJ, Yáñez-Sedeño P, Campuzano SJM. 2020. A novel zinc finger protein—based amperometric biosensor for miRNA determination. *Analytical and Bioanalytical Chemistry*. 412: 5031–5041

215. Zouari M, Campuzano S, Pingarrón JM, Raouafi N. 2020. Femtomolar direct voltammetric determination of circulating miRNAs in sera of cancer patients using an enzymeless biosensor. *Analytica Chimica Acta*. 1104: 188–198

216. Zouari M, Campuzano S, Pingarrón JM, Raouafi N. 2020. Determination of miRNAs in serum of cancer patients with a label and enzyme-free voltammetric biosensor in a single 30-min step. *Microchimica Acta*. 187: 444

217. Zayani R, Rabti A, Aoun SB, Raouafi N. 2021. Fluorescent and electrochemical bimodal bioplatform for femtomolar detection of microRNAs in blood sera. *Sensors & Actuators, B: Chemical*. 237: 128950

218. Jirakova L, Hrstka R, Campuzano S, Pingarrón JM, Bartosik M. 2019. Multiplexed immunosensing platform coupled to hybridization chain reaction for electrochemical determination of microRNAs in clinical samples. *Electroanalysis*. 31: 293–302

219. Abardía-Serrano C, Miranda-Castro R, de-los-Santos-Álvarez N, Lobo-Castañón MJ. 2020. New uses for the personal glucose meter: Detection of nucleic acid biomarkers for prostate cancer screening. *Sensors*. 20: 5514

220. Moranova L, Stanik M, Hrstka R, Campuzano S, Bartosik M. 2021. Electrochemical LAMP-based assay for detection of RNA biomarkers in prostate cancer. *International Journal of Molecular Sciences*. in revision

221. Serafín V, Valverde A, Garranzo-Asensio M, Barderas R, Campuzano S, Yáñez-Sedeño Pingarrón JM. 2019. Simultaneous amperometric immunosensing of the metastasis-related biomarkers IL-13Rα2 and CDH-17 by using grafted screen-printed electrodes and a composite prepared from quantum dots and carbon nanotubes for signal amplification. *Microchimica Acta*. 186: 411

222. Muñoz-San Martín C, Pedrero M, Manuel de Villena FJ, Garranzo-Asensio M, Rodríguez N, Domínguez G, Barderas R, Campuzano S, Pingarrón JM. 2019. Disposable amperometric immunosensor for the determination of the E-cadherin tumor suppressor protein in cancer cells and human tissues. *Electroanalysis*. 31: 309–317

223. Valverde A, Serafín V, Montero-Calle A, González-Cortés A, Barderas R, Yáñez-Sedeño P, Campuzano S, Pingarrón JM. 2020. Carbon/inorganic hybrid nanoarchitectures as carriers for signaling elements in electrochemical immunosensors: First biosensor for the determination of the inflammatory and metastatic processes biomarker RANK-ligand. *ChemElectroChem*. 7: 810–820

224. Valverde A, Hassine A, Serafin V, Muñoz SMC, Pedrero M, Garranzo-Asensio M, Gamella M, Raouafi N, Barderas R, Yáñez-Sedeño P, Campuzano S, Pingarrón JM. 2020. Dual Amperometric immunosensor for improving cancer metastasis detection by the simultaneous determination of extracellular and soluble circulating fraction of emerging metastatic biomarkers. *Electroanalysis*. 32: 706–714

225. Valverde A, Serafín V, Garoz J, Montero-Calle A, Gonzalez-Cortes A, Arenas M, Camps J, Barderas R, Yáñez-Sedeño P, Campuzano S, Pingarrón JM. 2020. Electrochemical immuno platform to improve the reliability of breast cancer diagnosis through the simultaneous determination of RANKL and TNF in serum. *Sensors & Actuators, B: Chemical*. 314: 128096

226. Muñoz-San Martín C, Pedrero M, Gamella M, Montero-Calle A, Barderas R, Campuzano S, Pingarrón JM. 2020. A novel peptide-based electrochemical biosensor for the

determination of a metastasis-linked protease in pancreatic cancer cells. *Analytical and Bioanalytical Chemistry*. 412: 6177–6188

227. Arevalo B, Hassine A, Valverde A, Serafín MCA, Raouafi N, Camps J, Arenas M, Barderas R, Yánez-Sedeño P, Campuzano S, Pingarrón JM. 2021. Electrochemical immuno platform to assist in the diagnosis and classification of breast cancer through the determination of matrix-metalloproteinase-9. *Talanta*. 225: 122054

228. Muñoz-San MC, Gamella M, Pedrero M, Montero-Calle A, Barderas R, Campuzano S, Pingarrón JM. 2020. Magnetic beads-based electrochemical immunosensing of HIF-1α, a biomarker of tumoral hypoxia. *Sensors & Actuators, B: Chemical*. 307: 127623

229. Muñoz-San MC, Pérez-Ginés V, Torrente-Rodríguez RM, Gamella M, Solís-Fernández G, Montero-Calle A, Pedrero M, Serafín V, Martínez-Bosch N, Navarro P, García de Frutos P, Batlle M, Barderas R, Pingarrón JM, Campuzano S. 2021. Electrochemical immunosensing of Growth arrest-specific 6 in human plasma and tumor cell secretomes. *Electrochemical Science Advances*. 2021: e210096

230. Villalonga A, Pérez-Calabuig AM, Villalonga R. Electrochemical biosensors based on nucleic acid aptamers. 2020. *Analytical and Bioanalytical Chemistry*. 412: 55–72

231. Jiménez-Falcao S, Parra-Nieto J, Pérez-Cuadrado H, Martínez-Máñez R, Martínez-Ruiz P, Villalonga R. 2019. Avidin-gated mesoporous silica nanoparticles for signal amplification in electrochemical biosensor. *Electrochemistry Communications*. 108: 106556

232. Paniagua G, Villalonga A, Eguílaz M, Vegas B, Parrado C, Rivas G, Díez P, Villalonga R. 2019. Amperometric aptasensor for carcinoembryonic antigen based on the use of bifunctionalized Janus nanoparticles as biorecognition-signaling element. *Analytica Chimica Acta*. 1061: 84–91

233. Villalonga A, Vegas B, Paniagua G, Eguílaz M, Mayol B, Parrado C, Rivas G, Díez P, Villalonga R. 2020. Amperometric aptasensor for carcinoembryonic antigen based on a reduced graphene oxide/gold nanoparticles modified electrode. *Journal of Electroanalytical Chemistry*. 877: 114511

234. Raouafi A, Sánchez A, Raouafi N, Villalonga R. 2019. Electrochemical aptamer-based bioplatform for ultrasensitive detection of prostate specific antigen. *Sensors & Actuators, B: Chemical*. 297: 126762

235. Díaz-Fernández A, Miranda-Castro R, Díaz N, Suarez D, de-los-Santos-Álvarez N, Lobo-Castañón MJ. 2020. Aptamers targeting protein-specific glycosylation in tumor biomarkers: General selection, characterization and structural modeling. *Chemical Science*. 11: 9402

236. Díaz-Fernández A, Miranda-Castro R, de-los-Santos-Álvarez N, Lobo-Castañón MJ, Estrela P. 2021. Impedimetric aptamer-based glycan PSA score for discrimination of prostate cancer from other prostate diseases. *Biosensors and Bioelectronics*. 175: 112872

237. Demirbakan B, Sezgintürk MK. 2019. A novel electrochemical immunosensor based on disposable ITO-PET electrodes for sensitive detection of PAK 2 antigen. *Journal of Electroanalytical Chemistry*. 848: 113304

238. Özcan B, Sezgintürk MK. 2019. Highly sensitive and cost-effective ITO-based immunosensor system modified by 11-CUTMS: Analysis of SOX2 protein in real human serum. *International Journal of Biological Macromolecules*. 130: 245–252

239. Tarimeri N, Sezgintürk MK. 2020. A high sensitive, reproducible and disposable immunosensor for analysis of SOX2. *Electroanalysis*. 32: 1065–1074

240. Burcu AE, Aydın M, Sezgintürk MK. 2019. Electrochemical immunosensor for CDH22 biomarker based on benzaldehyde substituted poly(phosphazene) modified disposable ITO electrode: A new fabrication strategy for biosensors. *Biosensors and Bioelectronics*. 126: 230–239

241. Burcu AE, Aydın M, Sezgintürk MK. 2019. Ultrasensitive determination of cadherin-like protein 22 with a label-free electrochemical immunosensor using brush type

poly(thiophene-g-glycidylmethacrylate) modified disposable ITO electrode. *Talanta*. 200: 387–397

242. Akgün M, Sezgintürk MK. 2020. A novel biosensing system based on ITO-single use electrode for highly sensitive analysis of VEGF. *International Journal of Environmental Analytical Chemistry*. 100: 1–19

243. Burcu AE, Aydın M, Sezgintürk MK. 2020. A label-free immunosensor for sensitive detection of RACK 1 cancer biomarker based on conjugated polymer modified ITO electrode. *Journal of Pharmaceutical and Biomedical Analysis*. 190: 113517

244. Burcu AE, Aydın M, Sezgintürk MK. 2020. Selective and ultrasensitive electrochemical immunosensing of NSE cancer biomarker in human serum using epoxy-substituted poly (pyrrole) polymer modified disposable ITO electrode. *Sensors & Actuators, B: Chemical*. 306: 127613

245. Burcu Aydın E, Aydın M, Sezgintürk MK. 2020. Construction of succinimide group substituted polythiophene polymer functionalized sensing platform for ultrasensitive detection of KLK 4 cancer biomarker. *Sensors & Actuators, B: Chemical*. 325: 128788

246. Burcu AE, Aydın M, Yuzer A, Ince M, Ocakoğlu K, Sezgintürk MK. 2021. Detection of Kallikrein-related Peptidase 4 with a label-free electrochemical impedance biosensor based on a zinc (II) phthalocyanine tetracarboxylic acid-functionalized disposable indium tin oxide electrode. *ACS Biomaterials Science & Engineering*. 7: 1192–1201

247. Burcu AE, Aydın M, Sezgintürk MK. 2021. Electrochemical immunosensor for detection of CCR4 cancer biomarker in human serum: An alternative strategy for modification of disposable ITO electrode. *Macromolecular Bioscience*. 21: 2000267

248. Burcu AE, Aydın M, Sezgintürk MK. 2021. Fabrication of electrochemical immunosensor based on acid-substituted poly(pyrrole) polymer modified disposable ITO electrode for sensitive detection of CCR4 cancer biomarker in human serum. *Talanta*. 222: 121487

249. Navarro R, Tapia-Galisteo A, Martín-García L, Tarín C, Corbacho C, Gómez-López G, Sánchez-Tirado E, Campuzano S, González-Cortés A, Yáñez-Sedeño P, Compte M, Álvarez-Vallina L, Sanz L. 2020. TGF-β-induced IGFBP-3 is a key paracrine factor from activated pericytes that promotes colorectal cancer cell migration and invasion. *Molecular Oncology*. 14: 2609–2628

250. Povedano E, Ruiz-Valdepeñas as Montiel V, Gamella M, Pedrero M, Barderas R, Peláez-García A, Mendiola M, Hardisson D, Feliú J, Yáñez-Sedeño P, Campuzano S, Pingarrón JM. 2020. Amperometric bioplatforms to detect regional DNA methylation with single-base sensitivity. *Analytical Chemistry*. 92: 5604–5612

251. Povedano E, Gamella M, Torrente-Rodríguez RM, Ruiz-Valdepeñas Montiel V, Montero-Calle A, Solís-Fernández G, Navarro-Villoslada F, Pedrero M, Peláez-García, A, Mendiola M, Hardisson D, Feliú J, Barderas R, Pingarrón JM. Campuzano S. 2021. Multiplexed magnetic beads-assisted amperometric bioplatforms for global detection of methylations in nucleic acids. *Analytica Chimica Acta*. 1182: 338946

252. Povedano E, Gamella M, Torrente-Rodríguez RM, Montero-Calle A, Pedrero M, Solís-Fernández G, Navarro-Villoslada F, Barderas R, Campuzano S, Pingarrón JM. 2021. Magnetic microbeads-based amperometric immunoplatform for the rapid and sensitive detection of N6-methyladenosine to assist in metastatic cancer cells discrimination. *Biosensors and Bioelectronics*. 171: 112708

253. Povedano E, Ruiz-Valdepeñas MV, Valverde A, Navarro-Villoslada F, Yáñez-Sedeño P, Pedrero M, Montero-Calle A, Barderas R, Peláez-García A, Mendiola M, Hardisson D, Feliú J, Camps J, Rodríguez-Tomàs E, Joven J, Arenas M, Campuzano S, Pingarrón JM. 2019. Versatile electroanalytical bioplatforms for simultaneous determination of cancer-related DNA 5-methyl- and 5-hydroxymethyl-cytosines at global and gene-specific levels in human serum and tissues. *ACS Sensors*. 4: 227–234

254. Montero-Calle A, Aranguren-Abeigon I, Garranzo-Asensio M, Poves C, Fernández-Aceñero MJ, Martínez-Useros J, Sanz R, Dziaková J, Rodriguez-Cobos J, Solís-Fernández G, Povedano E, Gamella M, Torrente-Rodríguez RM, Alonso-Navarro M, Ríos V, Casal JI, Domínguez-Muñoz G, Guzmán-Aránguez A, Peláez-García A, Pingarrón JM, Campuzano S, Barderas R. 2021. Multiplexed biosensing diagnostic platforms detecting autoantibodies to tumor-associated antigens from exosomes released by CRC cells and tissue samples showed high diagnostic ability of colorectal cancer. *Engineering*. accepted

255. Demirbakan B, Sezgintürk MK. 2020. A novel ultrasensitive immunosensor based on disposable graphite paper electrodes for troponin T detection in cardiovascular disease. *Talanta*. 213: 120779

256. Piguillem SV, Gamella M, Garcia de Frutos P, Batlle M, Yañez-Sedeño P, Messina GA, Fernández-Baldo MA, Campuzano S, Pedrero M, Pingarrón JM. 2020. Easily Multiplexable immuno platform to assist heart failure diagnosis through amperometric determination of galectin-3. *Electroanalysis*. 32: 2775–2785

257. Demirbakan B, Sezgintürk MK. 2021. An impedimetric biosensor system based on disposable graphite paper electrodes: Detection of ST2 as a potential biomarker for cardiovascular disease in human serum. *Analytica Chimica Acta*. 1144: 43–52

258. Torrente-Rodríguez RM, Muñoz-San Martín C, Gamella M, Pedrero M, Martínez-Bosch N, Navarro P, García de Frutos P, Pingarrón JM, Campuzano S. 2021. Electrochemical immunosensing of ST2: A checkpoint target in cancer diseases. *Biosensors* 11: 202

259. Villalonga A, Estabiel I, Pérez-Calabuig AM, Mayol B, Parrado C, Villalonga R. 2021. Amperometric aptasensor with sandwich-type architecture for troponin I based on carboxyethylsilanetriol-modified graphene oxide coated electrodes. *Biosensors and Bioelectronics*. 183: 113203

260. Guerrero S, Cadano D, Agüí L, Barderas R, Campuzano S, Yañez-Sedeño P, Pingarrón JM. 2019. Click chemistry-assisted antibodies immobilization for immunosensing of CXCL7 chemokine in serum. *Journal of Electroanalytical Chemistry*. 837: 246–253

261. Guerrero S, Sánchez-Tirado E, Martínez-García G, González-Cortés A, Yáñez-Sedeño P, Pingarrón JM. 2020. Electrochemical biosensor for the simultaneous determination of rheumatoid factor and anti-cyclic citrullinated peptide antibodies in human serum. *Analyst*. 145: 4680–4687

262. Arévalo B, Serafín V, Sánchez-Paniagua M, Montero-Calle A, Barderas R, López-Ruíz B, Campuzano S, Yánez-Sedeño P, Pingarrón JM. 2020. Fast and sensitive diagnosis of autoimmune disorders through amperometric biosensing of serum anti-dsDNA autoantibodies. *Biosensors and Bioelectronics*. 160: 112233

263. Serafín V, Gamella M, Pedrero M, Montero-Calle A, Razzino CA, Yáñez-Sedeño P, Barderas R, Campuzano S, Pingarrón JM. 2020. Enlightening the advancements in electrochemical bioanalysis for the diagnosis of Alzheimer's disease and other neurodegenerative disorders. *Journal of Pharmaceutical and Biomedical Analysis*. 189: 113437

264. Razzino CA, Serafín V, Gamella M, Pedrero M, Montero-Calle A, Barderas R, Calero M, Lobo AO, Yánez-Sedeño P, Campuzano S, Pingarrón JM. 2020. An electrochemical immunosensor using gold nanoparticles-PAMAM-nanostructured screen-printed carbon electrodes for tau protein determination in plasma and brain tissues from Alzheimer patients. *Biosensors and Bioelectronics*. 163: 112238

265. Karaboga MNS, Sezgintürk MK. 2020. Analysis of Tau-441 protein in clinical samples using rGO/AuNP nanocomposite-supported disposable impedimetric neuro-biosensing platform: Towards Alzheimer's disease detection. *Talanta*. 219: 121257

266. Serafín V, Razzino CA, Gamella M, Pedrero M, Povedano E, Montero-Calle A, Barderas R, Calero M, Lobo AO, Yáñez-Sedeño P, Campuzano, S, Pingarrón JM. 2021. Disposable immunoplatforms for the simultaneous determination of biomarkers for

neurodegenerative disorders using poly(amidoamine) dendrimer/gold nanoparticle nanocomposite. *Analytical and Bioanalytical Chemistry*. 413: 799–811

267. Valverde A, Montero-Calle A, Barderas R, Calero M, Yáñez-Sedeño P, Campuzano S, Pingarrón JM. 2021. Electrochemical immunoplatform to unravel neurodegeneration and Alzheimer's disease through the determination of neurofilament light protein. *Electrochimica Acta*. 371: 137815

268. Karaboga MNS, Sezgintürk MK. 2019. Cerebrospinal fluid levels of alpha-synuclein measured using a poly-glutamic acid-modified gold nanoparticle-doped disposable neuro-biosensor system. *Analyst*. 144: 611–621

269. Karaboga MNS, Sezgintürk MK. 2021. A nano-composite based regenerative neuro biosensor sensitive to Parkinsonism-associated protein DJ-1/Park7 in cerebrospinal fluid and saliva. *Bioelectrochemistry*. 138: 107734

270. Valverde A, Montero-Calle A, Arevalo B, San Segundo-Acosta P, Serafín V, Alonso-Navarro M, Solís-Fernandez G, Pingarron JM, Campuzano S, Barderas R. 2021. Phage-derived and aberrant HaloTag peptides immobilized on magnetic microbeads for amperometric biosensing of serum autoantibodies and Alzheimer´s disease diagnosis. *Analysis & Sensing*. 1: 161–165

271. Serafín V, Martínez-García G, Aznar-Poveda J, Lopez-Pastor JA, Garcia-Sanchez AJ, Garcia-Haro J, Campuzano S, Yáñez-Sedeño, Pingarrón JM. 2019. Determination of progesterone in saliva using an electrochemical immunosensor and a COTS-based portable potentiostat. *Analytica Chimica Acta*. 1049: 65–73

272. Serafín V, Arevalo B, Martinez-Garcia G, Aznar-Poveda J, Lopez-Pastor JA, Beltran-Sanchez JF, Garcia-Sanchez AJ, Garcia-Haro J, Campuzano S, Yáñez-Sedeño, Pingarrón JM. 2019. Enhanced determination of fertility hormones in saliva at disposable immunosensing platforms using a custom designed field-portable dual potentiostat. *Sensors & Actuators, B: Chemical*. 299: 126934

273. Arevalo B, Serafín, V, Campuzano S, Yáñez-Sedeño P, Pingarrón JM. 2021. Multiplexed determination of fertility-related hormones in saliva using amperometric immunosensing. *Electroanalysis*. 33: 2096–2104.

274. Arevalo B, Serafín V, Campuzano S, Yánez-Sedeno P, Pingarron JM. 2021. Electrochemical immunosensor for the determination of salivary prolactin. *Microchimica Acta*. 169: 106589

275. Arevalo B, Serafín V, Francisco Beltran-Sanchez J, Aznar-Poveda J, López-Pastor JA, García-Sanchez AJ, García-Haro J, Campuzano S, Yáñez-Sedeño P, Pingarrón JM. 2021. Simultaneous determination of four fertility-related hormones in saliva using disposable multiplexed immunoplatforms coupled to a custom-designed and field-portable potentiostat. *Analytical Methods*. 13: 3471–3478

276. Torrente-Rodríguez RM, Lukas H, Tu J, Min J, Yang Y, Xu C, Rossiter HB, Gao W. 2020. SARS-CoV-2 RapidPlex: A graphene-based multiplexed telemedicine platform for rapid and low-cost COVID-19 diagnosis and monitoring. *Matter*. 3: 1981–1998

277. Vargas E, Teymourian H, Tehrani F, Eksin E, Sanchez-Tirado E, Warren P, Erdem A, Dassau E, Wang J. 2019. Enzymatic/immunoassay dual-biomarker sensing chip: Towards decentralized insulin/glucose detection. *Angewandte Chemie International Edition*. 58: 6376–6379

278. Vargas E, Povedano E, Krishnan S, Teymourian H, Tehrani F, Campuzano S, Dassau E, Wang J. 2020. Simultaneous cortisol/insulin microchip detection using dual enzyme tagging. *Biosensors and Bioelectronics*. 167: 112512

279. Ince B, Sezgintürk MK. 2021. A high sensitive and cost-effective disposable biosensor for adiponectin determination in real human serum samples. *Sensors & Actuators, B: Chemical*. 328: 129051

8 New Trends in Biosensors for Food and Water Safety Monitoring

Maroua Hamami[1], Sondes Ben Aissa[1], and Noureddine Raouafi[1]
[1]University of Tunis El Manar
Faculty of Sciences of Tunis
Laboratory of Analytical Chemistry and Electrochemistry
Tunis El Manar, Tunisia

CONTENTS

DOI: 10.1201/9781003189435-12

8.1 INTRODUCTION

Food and water contamination by pathogens and pollutants present a threat for food supply and animal production resulting in a critical economic impact (1). The ingestion of infected products contaminated by bacteria, viruses, toxins, or chemicals can lead to mild to severe intoxications and even death (2). It is considered as one of the major concerns of public health presenting a barrier to the socioeconomic growth across the world. According to WHO, more than 200 diseases have been identified, ranging from diarrhea to cancer, with 70% of emerging diseases coming from animal sources (3). Contaminated drinking water is estimated to cause 485,000 diarrheal deaths each year. In addition, 600 million people get sick after consuming contaminated food, thus resulting in serious health problems and even favorizing the spread of global pandemics (4). Therefore, the development of new tools that can be practically applied for the detection of contaminants in water and food samples and to trace their spread is highly required to prevent global threats in the environmental and healthcare fields. Conventional analytical methods used for the detection and identification of contaminants based on chromatographic separation techniques and molecular and culture methods are highly sensitive and specific, however they are complex, require expensive equipment and are only suitable for laboratory setting. Hence, there is still a need to search for rapid and affordable alternative methods to overcome these shortcomings.

In this context biosensors have been developed as rapid and sensitive tools for on-site monitoring of various targets such as toxins, heavy metals, bacteria etc. in water and food samples without using advanced instrumentation and trained personnel with ability to their exploitation in resource-limited settings (5). According to IUPAC, A biosensor can be defined as an analytical device based on the association of a bioreceptor, such as an antibody, a protein, an enzyme, a strand of nucleic acid or microorganisms, to a transducer that transforms molecular recognition into an analytically measurable signal (6). Different types of bioreceptors combined with electrochemical transducer have been used for the development of biosensors in view of their application in food and water safety monitoring (7).

To meet this big demand, many works focused on the design of suitable technological solutions that are small, economical, accurate and autonomous. Recent trends in electrochemical biosensors design will be addressed in the present chapter, with a special emphasis on the implementation of nanomaterials and their benefits in addition to the role of bioreceptors. New advances reported in the field during the last five years will be highlighted.

8.2 COMMON TARGETS

Food safety is of critical societal importance. The food production chain, "from farm to table", presents multiple opportunities for food contamination to occur. Therefore, there is growing demand to preserve food products against contaminants, particularly toxic substances harmful to human health (8). Among various types of contaminants, natural toxins, synthetic pesticides, and residues of antibiotics have been proven to present a severe threat.

8.2.1 SMALL ORGANIC MOLECULES

8.2.1.1 Toxins

Toxins are biologically produced poisonous molecules with diverse chemical structures. According to their origin, toxins are commonly classified into fungal toxins, bacterial toxins, and algal toxins. Contamination by toxins is unpredictable and unavoidable as it can occur during any phase of the food production chain, including harvesting, processing, shipping, and storage, mainly under favorable meteorologic conditions. The spread of toxins in foodstuffs can cause acute economic losses and critical public health issues. In fact, many toxins affect the nervous systems of mammals by interfering with the transmission of nerve signals. Others are responsible for blockage of central cellular metabolism, causing cellular death. Likewise, most toxins act immediately and are lethal in very low doses ($LD_{50} < 25$ mg/kg), making them sometimes more dangerous than chemical warfare agents (9). The following Table 8.1 summarizes the major mycotoxins identified for their approved toxicity.

8.2.1.2 Pesticides

Similarly, a wide variety of pesticides have been used in agriculture to increase the yield and quality and extend the storage life of crops. However, the abuse of pesticides due to the ever-increasing population and rapid urbanization leads to food contamination and presents a potential risk to human health (12). Among synthetic pesticides, the four major classes are organochlorines, organophosphates, carbamates and pyrethroids, which are frequently used thanks to their benefits in crop cultivation. In another classification, pesticides can be categorized based on their target organism. They are basically classed as (i) insecticide, (ii) herbicide, (iii) fungicide, (iv) rodenticide and (v) nematicide (12). Further, the risks of continuous exposure to pesticides include respiratory diseases, neurological deficits, and in extreme cases, fetal death, spontaneous abortion, genetic diseases, and cancer (13). Table 8.2 summarizes the classes, the chemical structure and the maximum residues limits (MRLs) and the use of the most pesticides.

8.2.1.3 Antibiotics

Another important class of residual contaminants in the food chain is antibiotics. Antibiotics are a group of pharmaceutical drugs widely used in human and veterinary medicine for preventing and/or treating many different infectious diseases (14). If overused or misused, such pharmacologically active substances can remain in animal-derived food (meat, milk, egg, fish, etc.). The most common adverse effect of antibiotic residues in foodstuffs is developing antimicrobial resistance. The resistant bacterial pathogens can be transferred to humans through the food chain and cause the inefficiency of antibiotic therapy (15). These residues may also cause various side effects such as immunopathological effects, allergy, mutagenicity, nephropathy, hepatotoxicity, reproductive disorders and even carcinogenicity (16). There are about twenty different classes of known antibiotics based on chemical structures, which are summarized in Table 8.3 with their mechanism of action.

TABLE 8.1

Chemical Structures, Toxicity and EU Limits of Major Mycotoxins (10–11)

Mycotoxin	Structure [a]	Toxicity Group	Food Matrix	Toxic Effects	EU MLs [c] (μg.kg^{-1})
Ochratoxins OTA[a], OTB, OTC		2B	Corn, barley, oats, rye, wheat, grapes, coffee beans, wine, beer, dried fruits	Nephrotoxic, teratogenic, immunosuppressive, potent teratogen	0.5–10
Aflatoxins AFB1[a], AFB2, AFG1, AFG2,		1	Nuts, cereals, maize, rice oilseeds	Highly toxic, carcinogenic, immunosuppressive, mutagenic, genotoxic, teratogenic	0.1–12 (AFB1) 4–15 (Total sum)
Aflatoxin M AFM1[a], AFM2		1	Milk and derivatives	Carcinogenic, stunting in children, mutagenic, genotoxic, teratogenic	0.025–0.05
Fumonisin FB1[a], FB2, FB3		2B	Soybeans, maize and sorghum	Hepatotoxic and carcinogenic, interferes with sphingolipid metabolism	200–4000
Patulin		3	Apple and derivatives, pears, cherries, and other fruits	Genotoxic, immunotoxic, neurotoxic, teratogenic	10–50

(Continued)

TABLE 8.1
(Continued)

Mycotoxin	Structure [a]	Toxicity Group	Food Matrix	Toxic Effects	EU MLs [c] (μg.kg^{-1})
Zearalenone		2A	Cereal crops such as wheat, maize, barley, and sorghum	Estrogenic	20–350
Citrinin		3	Wheat, oats, rye, corn, barley, rice	Cytotoxic and nephrotoxic	2000 (Rice)
T-2 toxin		3	Wheat, barley, oats, maize	Cytotoxic, immunosuppressive, potent inhibitors of eukaryotic protein synthesis	15–1000
Nivalenol		3	Oats, barley, maize, wheat, bread, pasta, cereals	Growth retardation, Leukopenia, Intrauterine growth delay	1.2
Deoxynivalenol		3	Grains such as wheat, barley, oats, rye, maize, rice, sorghum	Potent inhibitor of eukaryotic protein synthesis causes nausea, vomiting, and diarrhea	200–1750

a Only the main mycotoxin's structure within the corresponding groups is presented.
b A for Aspergillus, F for fusarium and P for Penicillium.
c Maximum limits (MLs) vary according to the food matrix.

TABLE 8.2

Classes, Chemical Structures, Lethal Doses, and Uses of Pesticides

Classes	Chemical Structure	MRLs (EU)[a]	Use
Organochlorines		0.02 to 0.05 mg/kg (for DDT)	Insecticide (banned)
Organophosphates		0.05 mg/kg (for Chlorpyrifos)	Acaricide Insecticide
Carbamates		0.1 to 0.2 mg/kg (for sum of methiocarb)	Insecticide Repellant
Pyrethroids		0.02 to 0.6 mg/kg (for deltamethrin)	Insecticide
Dinitroanilines		0.01 to 0.05 mg/kg (for trifluralin)	Herbicide
Urea pesticides		0.01 mg/kg (for fenuron)	Herbicide
Glyphosates		0.1 to 0.5 mg/kg (for glyphosate)	Herbicide
Triazines		0.05 to 0.1 mg/kg (for atrazine)	Herbicide
Chloracetanilides		0.02 mg/kg (for metazachlor)	Herbicide

(Continued)

TABLE 8.2
(Continued)

Classes	Chemical Structure	MRLs (EU)[a]	Use
Phenoxyalkanoic acids		0.05 to 1.0 mg/kg (for 2,4-D)	Herbicide Plant growth regulator
Neonicotinoids		0.01 to 2.0 mg/kg (for acetamiprid)	Insecticide
Carboxamides		0.01 mg/kg (for isoflucypram)	Fungicide
Fluorosulfone		0.01 to 4 mg/kg (for fluensulfone)[b]	Nematocide

a: https://ec.europa.eu/food/plant/pesticides/eu-pesticides-database/active-substances/?event=search.as
b: www.fao.org/fao-who-codexalimentarius/codex-texts/dbs/pestres/functional-classes/en/

8.2.2 METALLIC CATIONS

Pollution generated by heavy metals is mainly due to the release of untreated clean and poisonous industrial wastes, dumping of industrial wastewater, agriculture run-off from fields and even the use of cells and batteries. Almost all the heavy metals are toxic to human beings even at very low concentrations (17). For instance, cadmium, chromium, mercury and lead are the most concerning. Heavy metal ions are non-biodegradable, disrupt ecosystems, damage soils, can be harmful to aquatic organisms and accumulate in food chains (18). Furthermore, they cannot be eliminated by the organism and their accumulation can cause several diseases. The great diversity of chemicals discharged into the marine environment leads to serious consequences. Therefore, protection of the marine environment requires monitoring and regulation. Table 8.4 is shown some drinking water quality guidelines for heavy metals published by several organization, committees, and agencies throughout the world.

TABLE 8.3

Classes, Chemical Structures of Some Antibiotics, Their Generations and Mechanism of Action

Classes	Generic Structure	Generations[b]	Mechanism of Action[b]
Penicillins[a]		1 to 4	Inhibitors of bacterial cell wall synthesis by binding and inactivating proteins
Cephalosporins[a]		1 to 5	Inhibitors of bacterial cell wall synthesis by binding and inactivating proteins
Tetracyclines		1 to 3	Binding to 30S ribosomal subunit and blocking the attachment of aminoacyl t-RNA to A site
Chloramphenicols		NA	Vancomycin-resistant Enterococcus
Macrolides		1 and 2	Inhibitors of bacterial protein synthesis
Aminoglucosides		NA[c]	Inhibitors of bacterial cell wall synthesis by binding and inactivating proteins

(Continued)

TABLE 8.3
(Continued)

Classes	Generic Structure	Generations[b]	Mechanism of Action[b]
Lincosamides		NA	Inhibitors of bacterial protein synthesis
Sulphonamides		NA	Multiplication inhibitors by acting as competitive inhibitors of p-aminobenzoic acid in the folic acid metabolism cycle
Nitrofurans		NA	Inhibitors of the citric acid cycle and the synthesis of DNA, RNA, and protein
Nitroimidazoles		NA	Reducing the nitro function into nitro anion radical and hydroxylamine, which bind DNA to break it
Trimethoprim		NA	Inhibitors of dihydrofolate reductase, which reduces dihydrofolate into tetrahydrofolate (active form)
Polymyxins		NA	Disrupting the integrity of the outer membrane of gram-negative bacteria by binding with high affinity to the negatively charged Lipopoly-saccharides

TABLE 8.3
(Continued)

Classes	Generic Structure	Generations[b]	Mechanism of Action[b]
Quinolones		1 to 3	Interfering with the activity of DNA gyrase (bacterial topoisomerase II) and topoisomerase IV (the enzymes that cause DNA to rewind after copying)
Peptidomimetics		NA	Inhibitors of RNA synthesis by inhibiting DNA dependent RNA polymerase

a : β-lactams

b : Information from https://tmedweb.tulane.edu/pharmwiki/doku.php/start (Tulane University, School of Medicine)

c : NA: not applicable

TABLE 8.4
Current Drinking Water Quality Guidelines (µg.L⁻¹) (Data from (19–20))

Heavy Metal	WHO[a]	USEPA[b]	ECE[c]
Arsenic	10	10	10
Cadmium	3	5	5
Chromium	50	100	50
Lead	10	15	10
Mercury	6	2	1
Nickel	70	–	20

a: World Health Organization (WHO 2011)

b: United States Environmental Protection Agency (USEPA 2011)

c: European Commission Environment (ECE 1998)

8.2.3 Bacteria and Viruses

Several viruses and bacteria are known to contaminate animals and derived meat for human consumption. They can cause important economic losses to farmers when they infect the animals.

Pathogenic bacteria are the most common cause of foodborne diseases in the world; among them *Escherichia coli, Staphylococcus aureus, Salmonella, Listeria monocytogenes* and *mycobacterium bovis* are the most common (21). Since the infective dose of some of bacteria is low, pathogenic cells of some species must be totally absent from food, for example, the Salmonella case (22).

Viruses such as nodavirus (23), African swine fever virus (24), sheeppox virus (24), avian virus (24) or bluetongue virus (25) infect respectively fish, pigs, sheep, birds and cattle, they have high mortality rates, and it is important de detect them at early stages of the infection in order to mitigate the epidemics.

8.3 BIORECEPTORS

8.3.1 Antibodies

Antibodies or immunoglobulins is a family of glycoproteins produced as a response to the entry of foreign substances into the body. They are characterized by their Y-shape with two light and two heavy peptide chains connected by a single disulfide bond (26). The antibody arms contain constant and variable regions, with the variable region responsible for the antibody's selectivity toward different antigens or analytes (27). Antibodies that can recognize single or multiple binding sites (epitopes) on antigens, are called monoclonal or polyclonal, respectively. Polyclonal antibodies exhibit lower specificity than monoclonal antibodies, while monoclonal antibody production is more complex and expensive.

8.3.2 Nanobodies

Nanobodies, (Nbs), or stable single domain antibodies have emerged as useful molecular reagents to sense or track antigens. They are based on the variable domain of antibodies and could be found in camelids or in non-mammals such as sharks. These single-domain antibodies are the smallest domains of natural antibodies that are capable of antigen-binding, with a molecular weight ranging between 12 to 14 kDa. Nbs encoded by a single gene segment that can be obtained in bacteria and yeast with less cost than conventional monoclonal antibodies (28). In addition, recent studies have shown that nanobodies are well suited for the conception of immunosensors as they can be easily modified by site-specific functional groups, resulting in covalent binding with negligible loss of specificity and affinity. Recently, numerous nanobodies have been discovered and used for the detection of food contaminants (29). To the best of our knowledge, no electrochemical biosensor based on Nbs was reported for biosensing application in wastewater.

8.3.3 Aptamers

Aptamers are artificial single-stranded oligonucleotides (DNA or RNA), usually containing up to 90 nucleotides with high binding affinity. Such oligonucleotides are selected in vitro to specifically recognize a wide range of analytes, including organic molecules, proteins, viruses, bacteria or even living cells. This recognition is usually established by forming aptamer-target complexes with relatively stable sequence-dependent tertiary structures, thanks to the flexibility of the aptamer backbone. During recognition, the interaction between the host aptamer and its guest molecule is reversible without affecting the chemical nature of the analyte.

Aptamers are isolated from a large random library of synthetic nucleic acids (10^{14}-10^{15} variants) by an iterative process of binding, separation and amplification, commonly termed Systematic Evolution of Ligands by EXponential enrichment (SELEX). Through this iterative protocol (7 to 15 rounds), non-binding aptamers are discarded via elution and aptamers binding to the proposed target are expanded and amplified through the PCR technique. Initial positive selection rounds are sometimes followed by negative selection to improve the selectivity of the aptamer candidates. Subsequently, multiple rounds of SELEX are performed with increasing stringency to enrich further the oligonucleotide pool. Once the nucleic acid sequence is selected and validated, it can be well reproduced by a cost-effective routine synthesis compared to antibodies.

8.3.4 DNAzymes

DNAzymes are single-stranded (ss)DNA sequences that fold into complex tertiary structures and can catalyze several reactions, including degradation of target mRNA, ligation, or phosphorylation of DNA. They could be obtained in vitro by screening technology and show high structure recognition ability. Since their discovery in 1994, various DNAzymes have been proposed (30). Among these DNAzymes, the one with RNA-cleaving activity has attracted extensive research interests (31). Metal-dependent DNAzyme is composed of an enzyme strand and a substrate strand, which contains a single ribonucleobase (RNA) linkage (such as rA) that serves as the cleavage site (RNA phosphoester) followed by a G-T wobble pair (32).

8.3.5 Others

The development of cell-based biosensors also received considerable attention. Bacteria, algae and yeasts are the most used microorganisms for this purpose. Various types of strains have been discovered, ranging from genetically engineered organisms that are particularly designed to detect specific molecules or groups of molecules to environmental cells isolated from polluted sites offering higher robustness and more specific enzymatic properties. Cells are easier to produce in large quantities compared to other bioreceptors (enzymes, antibodies, or DNA), moreover they are more tolerant to temperature variations, pH, and ionic strength. A large variety of biosensors has been proposed for the detection of contaminants or for aquatic toxicity assessment. Several studies have been published on the development of electrochemical cell-based biosensors, some of them included the conception of yeast-based sensors for acute biotoxicity assessment of wastewater (33), also the determination of traces of arsenic and mercury in water, relying on the selective recognition from the bacterium (34), and finally the design of a lab-on-chip electrochemical biosensor based on microalgal photosynthesis for water quality (35).

8.4 DETECTION STRATEGIES

8.4.1 WATER SAFETY

While freshwater availability has been declining through past decades due to excessive anthropogenic activity, water demand has continued to increase. Entry points into surface waters include domestic discharge, agriculture runoff from fields, industrial discharge, etc. resulting in declining aquatic biodiversity and increasing water scarcity, which present a great risk to aquatic plants, animals, and humans. Monitoring of pollutants in wastewater is essential to identifying water pollution areas for treatment.

8.4.1.1 Amplification-Free Strategies

The fusion of microfluidic devices with biosensors enables the integration of biological and chemical components on a single platform and improves the portability, availability, and ability of biosensors to perform in-situ analyses. Due to the ease of operation and low cost, there is the possibility of application in less developed areas. For instance, Zhang et al. recently developed an electrochemical microfluidic paper-based analytical device used to monitor the inhibition of E. coli's respiratory activity after exposure to heavy metals, pesticides, and penicillin (36). The biosensor principle is shown in Figure 8.1. The half inhibitory concentration (IC_{50}) was determinate for Cu^{2+} solution: 13.5 µg/mL, Cu^{2+}-soil: 21.4 mg/kg and penicillin-soil: 85.1 mg/kg. The fabricated device was used to detect pesticide residues in vegetable juice.

Alawad et al. proposed a simple aptasensor based on intrinsic aptamer redox activity for the detection of tetracycline in water (37). Tetracyclines are a family of antibiotics frequently employed due to their broad spectrum of activity as well as their low cost. Figure 8.2 shows the fabrication process of the aptasensor and measurement procedure. The immobilized aptamer on SPCE revealed an unexpected electroactivity, therefore the binding of TET on the aptasensor induced a decrease

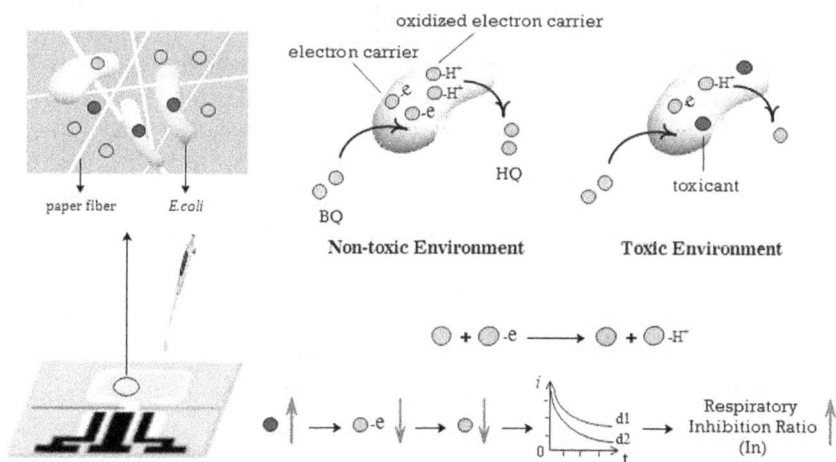

FIGURE 8.1 Schematic illustrations of the principle of BQ-mediated E. coli respiration method for electrochemical biotoxicity detection. (Reprinted with permission from Ref. (36)).

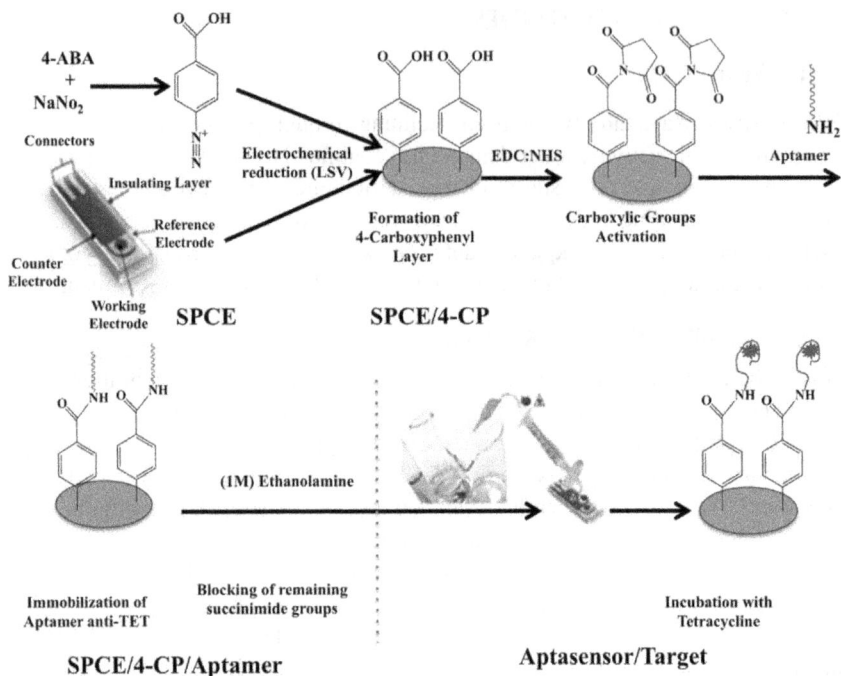

FIGURE 8.2 Schematic representation of aptasensor fabrication and measurement procedure. (Reprinted with permission from Ref. (37)).

of oxidation peak, allowing its detection in the dynamic range from 0.05 µg/L to 20 µg/L, with a detection limit of 0.035 µg/L.

An innovative amplification-free strategy was reported by Rengaraj et al. for bacteria detection in water (38). The functionalized paper-based electrodes were fabricated by screen-printing a conductive carbon ink onto a commercial hydrophobic paper. Concanavalin A, known for its ability to selectively interact with mono- and oligosaccharides on bacterial cells, was immobilized on the electrode as the bioreceptor. EIS measurements were performed for bacteria detection in a concentration range from 10^3 to 10^6 CFU/mL, with an estimated lower detection limit of 1.9×10^3 CFU/mL.

Very recently, Hamami et al. (39) reported an amplification-free aptameric biosensor for the detection of penicillin in river water using a screen-printed electrode modified with nanocomposite formed with PPyNPs and MoS_2 nanosheets, which are endowed with electrocatalytic properties and conductivity having a synergetic effect on the surface-tethered naphthoquinone redox reporter. The sensor allowed to detect down to 10 pg/L of ampicillin. Although the tedious preparation procedure, the device is a promising for the detecting antibiotics in the environment as a point-of-use system.

8.4.1.2 Amplification Based Strategies

Recent developments in nanomaterials and discovery of their interesting properties have attracted researchers toward their application in various sectors such as health, food, security, environment, technology etc. The integration of nanomaterials in biosensors design have shown great enhancement in analytical performances with high sensitivities and detection limits. For instance, Mehta et al. reported the development of an electrochemical immunosensor based on graphene quantum dots for the detection of the organophosphate pesticide parathion as represented in Figure 8.3 (40). The biosensor had a logarithmic linear range of 0.01 to 10^6 ng/L and a LOD of 46 pg/L and was highly specific toward parathion. The immunosensors showed high specificity and sensitivity toward parathion even in the presence of increased concentrations of other pesticides, namely, paraxon, malathion, and chlorpyrifos.

Liu et al. proposed a dual amplification signal strategy for Hg^{2+} detection based on peonylike Cu-MOFs in situ growth of PtPd NPs bimetallic label S1 and Mg^{2+} dependent DNAzyme-driven DNA Walker (41). The electrochemical biosensor was constructed following 2 steps as depicted in Figure 8.4. The first one is to self-assemble the hairpin DNA (HP) and signal label Cu-MOFs@PtPd NPs/S1 on the gold nanorods/poly(diallyldimethylammonium chloride) functionalized graphene modified electrode. Then S2, which contains two sequences, a substrate sequence of the Mg^{2+} dependent DNAzyme and another sequence complementary to HP, was immobilized on the electrode. In the presence of target Hg^{2+}, HP was opened to preferentially form (T-Hg^{2+}-T) with DNA Walker (S2). Then, DNA Walker exhibited catalytic activity of DNAzyme in the presence of Mg^{2+} to continuously cleave Cu-MOFs@PtPd NPs/S1, resulting in a decrease of electrochemical signal. Under optimal conditions, the change of current was linearly related to the negative logarithm of the Hg^{2+} concentration in the detection range of 0.001–100 nM with a low LOD of 0.52 pM.

Recently, N. Chérif et al. (42) reported the design of a simple and highly sensitive electrochemical biosensor for the direct detection of a conserved RNA of fish

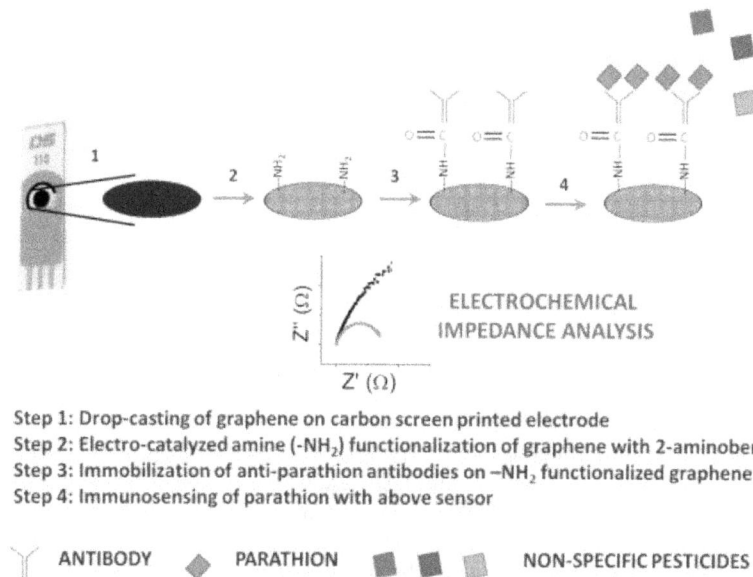

Step 1: Drop-casting of graphene on carbon screen printed electrode
Step 2: Electro-catalyzed amine (-NH₂) functionalization of graphene with 2-aminobenzylamine
Step 3: Immobilization of anti-parathion antibodies on –NH₂ functionalized graphene SPE
Step 4: Immunosensing of parathion with above sensor

ANTIBODY   PARATHION   NON-SPECIFIC PESTICIDES

FIGURE 8.3 Representation of the graphene-based screen-printed immunosensor for parathion detection. (Reprinted with permission from Ref. (40)).

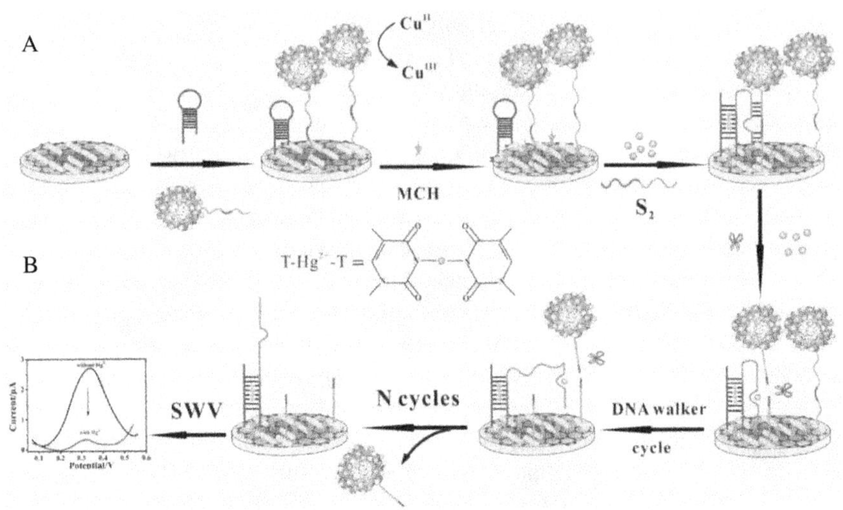

FIGURE 8.4 (A) Preparation process of Cu-MOFs@PtPd NPs/S1; (B) Schematic illustration of the fabrication of the electrochemical biosensor. (Reprinted with permission from Ref. (41)).

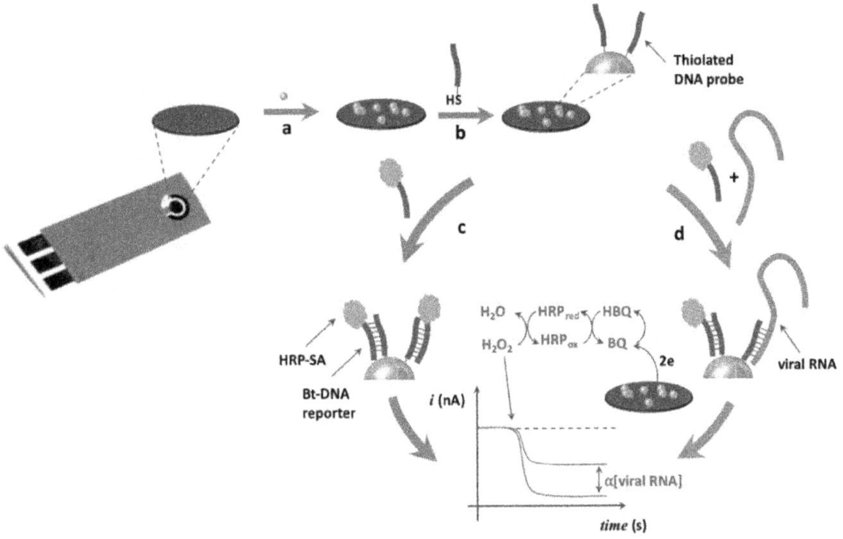

FIGURE 8.5 Schematic illustration of stepwise design of the electrochemical biosensor for the competition-based detection of nodavirus RNA. Steps: (a-b) gold electrochemical deposition and tethering of the thiolated DNA probe, (c) adding the biotinylated reporter probe and (d) adding the genomic RNA and the thiolated reporter probe. For case (c), a higher current is observed compared to case (d) where biotinylated probes are fewer due to the competition. Copyright (2021) 2021 by the authors. Licensee MDPI, Basel, Switzerland. This article is an open access article distributed under the terms and conditions of the Creative Commons Attribution (CC BY) license (https://creativecommons.org/licenses/by/4.0/).

nodavirus, making use of nanostructured disposable electrodes. As depicted in Figure 8.5, the device uses competitive hybridization between two oligonucleotides (biotinylated reporter DNA and a synthetic target RNA) complimentary to a thiolated DNA capture probe tethered to AuNPs-modified SPCE. The biosensor has a dynamic range varying from 0.1 to 25 pM with a detection limit of 20 fM. The method was further validated using extracted RNA samples from healthy carrier (Sparus aurata) and clinically infected (Dicentrarchus labrax). In parallel, the biosensor sensitivity was compared with a new devised real-time RT-PCR protocol. The results are in good agreement. This method provides a promising approach toward a more effective diagnosis and risk assessment of viral diseases in aquaculture.

8.4.2   FOOD SAFETY

Diverse electrochemical detection strategies were described for food safety monitoring depending on the target contaminant and using multiple bioreceptors. Based on the amplification of the electrochemical signal, these detection strategies can be classified into amplification-free and amplification-based biosensors, where a higher sensitivity can be achieved by means of the ingenious combination of nanomaterials and bio-elements. This part reviews concisely the figure-of-merits of such electrochemical biosensing approaches, underlining their corresponding virtues and limitations for on-site applications.

8.4.2.1   Amplification-Free Strategies

The most common and abundant toxins present in nature are mycotoxins. Therefore, the design of highly sensitive biosensors targeting mycotoxins and using electrochemical techniques is widely reported and reviewed (43–47). Particularly, amplification-free detection strategies of mycotoxins using antibodies and aptamers as affinity bioreceptors are well established (46). For instance, a competitive immunoassay for the voltammetric detection of two mycotoxins, OTA and AFM1, using modified gold SPCEs was reported (48). The immunosensor was constructed by the immobilization of the BSA-modified toxin onto the activated gold surface, followed by the injection of the mixture of primary antibody and sample containing free target (Figure 8.6). A secondary antibody was labeled with alkaline phosphatase enzyme to generate the detection signal associated to the enzymatically generated product. The biosensor was validated in red wine and milk samples with no need for pretreatment or preconcentration of the sample. The analytical response was linearly proportional to toxins concentrations ranging from 10^{-2} to 10^3 ng mL^{-1}, with a low LOD.

Among marine biotoxins, saxitoxin (STX) is one of the major toxins of paralytic shellfish poison and can cause shock, asphyxia and even death to fisheries and humans. Qi et al. developed a facile label-free electrochemical aptasensor assembled with nanotetrahedron and DNA triplex for the sensitive detection of STX (Figure 8.7) (49). The innovative combination between aptamer, DNA triplex and DNA nanotetrahedron scaffold was advantageous to assist the aptamer orientation and protect it from adsorption to the surface of screen-printed electrodes, allowing its full accessibility to STX. The aptasensor showed high sensitivity with a LOD of 0.92 nM and demonstrated good applicability to detect STX in seawater samples, with a recovery ranging from 94.4% to 111%.

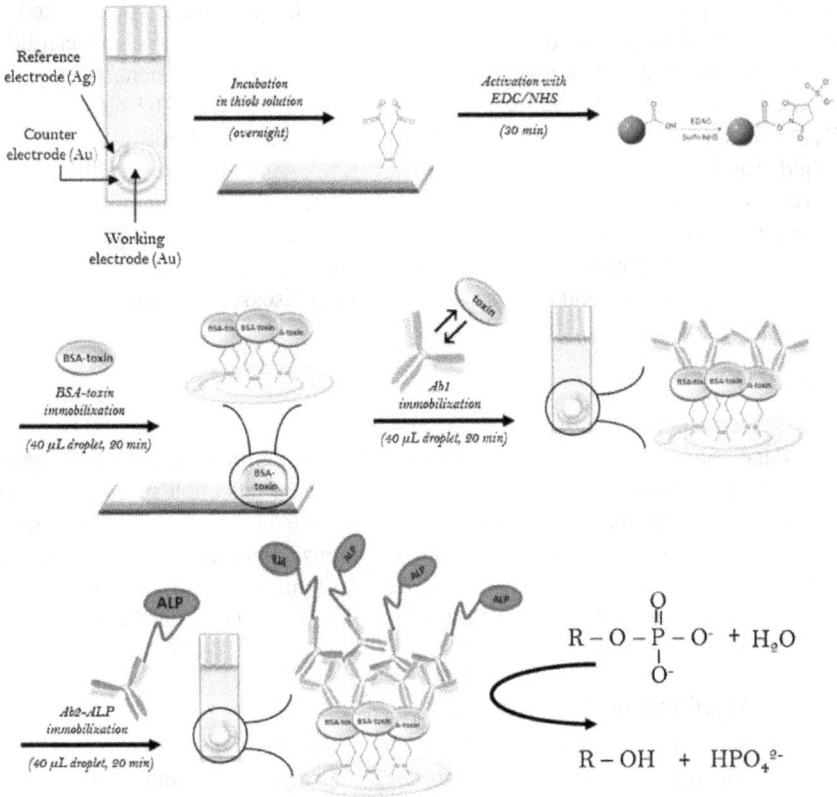

FIGURE 8.6 Schematic illustration of the preparation of the biosensor for toxin detection utilizing a competitive immunoassay format. (Reprinted with permission from Ref. (48)).

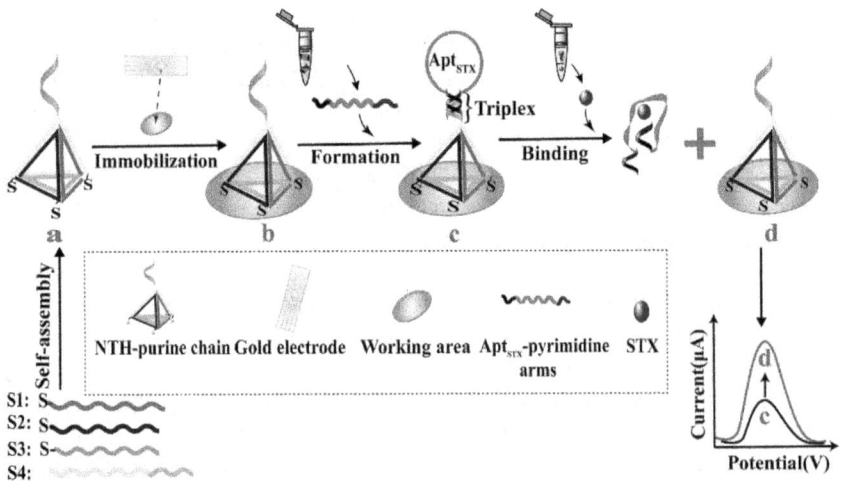

FIGURE 8.7 Schematic illustration of the electrochemical aptasensor constructed with the DNA nanotetrahedron and aptamer-triplex. (Reprinted with permission from Ref. (49)).

Furthermore, several electrochemical biosensors have been reported for the detection of foodborne pathogenic bacteria (50), such as *E. coli* (51), *Salmonella* (52) and *Listeria monocytogenes* (53). Among the most used electrochemical technique in the development of bacteria biosensors is the electrochemical impedance spectroscopy (EIS), as demonstrated in recent review by Leva-Bueno et al. (54). For instance, an impedimetric immunosensor based on magnetic capturing for detection of *E. coli* O157:H7 and *S. typhimurium* was recently described (55) using a screen-printed interdigitated microelectrode (Figure 8.8). Streptavidin-coated magnetic beads (S-MBs) were functionalized with corresponding biotinylated antibodies to capture the target bacteria, while the glucose oxidase enzyme was employed as label for secondary antibodies. Enzymatic oxidation glucose produced a redox signal that alters the system impedance after binding with complex of biotinylated rabbit anti-bacteria, S-MBs, and the target bacteria. The performance of immunosensor in culture medium reached a LOD at 3.9×10^2 CFU/mL and 1.66×10^3 CFU/mL for *E. coli* and S. *typhimurium*, respectively, while in food samples the LOD of 10^3 CFU/mL for both bacteria has been obtained without any enrichment process. The specificity of sensor was investigated with *E. coli* K12, *L. monocytogenes*, and *S. aureus*.

Very recently, Hamami et al. reported the detection of acetamiprid (ACT) using a SPCE modified with lipoic acid-functionalized MoS_2 nanosheets and aptamer (56). ACT is a systemic insecticide widely used to control sucking-type insects in vegetables (57). The sensor uses electrochemical capacitance spectroscopy (ECS) with the need of redox probe present in solution like in the case of EIS since the MoS_2 are endowed with pseudo-capacitive properties. The sensor shows a linear range from 50 to 450 fM and can detect down to 14 fM. The device was used to detect the herbicide in extract of tomatoes with high recovery values comprised between 95% and 104%. Similarly, Ben Aissa et al. designed a sensitive ECS-based biosensor using SiNPs

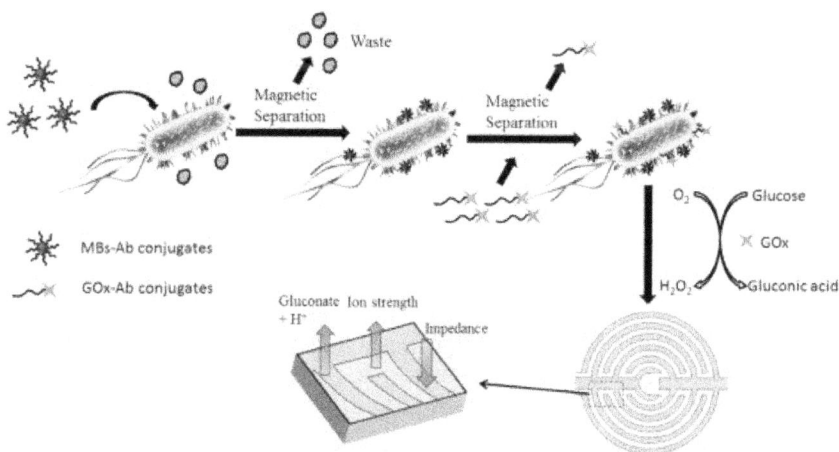

FIGURE 8.8 Rapid detection of E. coli O157:H7 and S. *typhimurium* in foods using an electrochemical immunosensor based on screen-printed interdigitated microelectrode and immunomagnetic separation. (Reprinted with permission from Ref. (55)).

and ferrocene tags for the sensitive detection of AFM1 in milk samples (58). The aptasensor is endowed with large dynamic range varying from 10 to 500 fM with a LOD of 4.53 fM. The results from the device were compared to those of ELISA measurements, showing good agreement.

8.4.2.2 Amplification Based Strategies

One of the relevant examples is a dual immunosensor developed to detect fumonisin B1 (FB1) and deoxynivalenol (DON) (59). A disposable SPCE was used as the sensing platform after its modification by AuNPs and a polypyrrole-electrochemically rGO nanocomposite film. This transducer was suitable for the effective immobilization of anti-toxins antibodies. Under optimized conditions, the LOD and linear range achieved for FB1 was 4.2 ppb and 0.2 to 4.5 ppm; and the corresponding values for DON were 8.6 ppb and 0.05 to 1 ppm. The immunosensor exhibited high sensitivity and low matrix interference when tested on spiked corn samples. Interestingly, it was possible to specifically detect the two target toxins even if present in the same sample, which makes this approach interesting for the rapid detection of different mycotoxins present in foodstuff.

More recently, a comparable approach was applied to detect ciprofloxacin (CFX), a commonly overused fluoroquinolone antibiotic. The aptasensing of ciprofloxacin residues in raw milk samples was described using rGO oxide and nanogold-functionalized poly(amidoamine) dendrimers (60). This nanopolymeric matrix, shown in Figure 8.9, not only provided an extensive active surface area, but also improved the loading capacity of aptamers in PAMAM-based nanocarriers, which enhanced the sensitivity for CFX monitoring. This aptasensor has a linear dynamic range from 1 mM to 1 nM in pasteurized milk samples by using SWV and DPV.

Parallelly to the boom of nanotechnology, the recent years have witnessed the rapid development of isothermal nucleic acid amplification technology with new principles and applications that are gradually emerging in biosensing, being fast, highly efficient, low cost, simple and highly sensitive. Among the isothermal amplification techniques applied in food biosensing, numerous recent papers describe the rolling circle amplification (RCA) strategy (61–62). For instance, Xie et al. (63) proposed a new impedimetric RCA-based biosensor to detect the lipopolysaccharide (LPS) in *E. coli*. Two aptamers (Apt I and Apt II) were used. Apt I was designed to be assembled on the surface of a gold-modified GCE via an Au-N bond to capture the target. Afterwards, AuNPs conjugated with primers and Apt II were added to the modified electrode. After LPS recognition, a sandwich structure was formed, and the exposed primers triggered the RCA reaction. Subsequently, the product was used to form copper nanoclusters and catalyze the generation of resistance signals. Following this strategy, the detection sensitivity reached femtomole level. The impedimetric aptasensor exhibited a wide dynamic working range of 0.01 pg/mL to 100 ng/mL with a detection limit of 4.8 fg/mL.

Hybridization chain reaction (HCR) is an attractive isothermal amplification technique to design E-biosensors given its enzyme-free exponential process. During this process, two metastable DNA hairpins coexist until the initiator is introduced triggering a cascade of hybridization events. Harnessing this strategy, Wang and coworkers (64) described a novel nanobody-based voltammetric immunosensor for

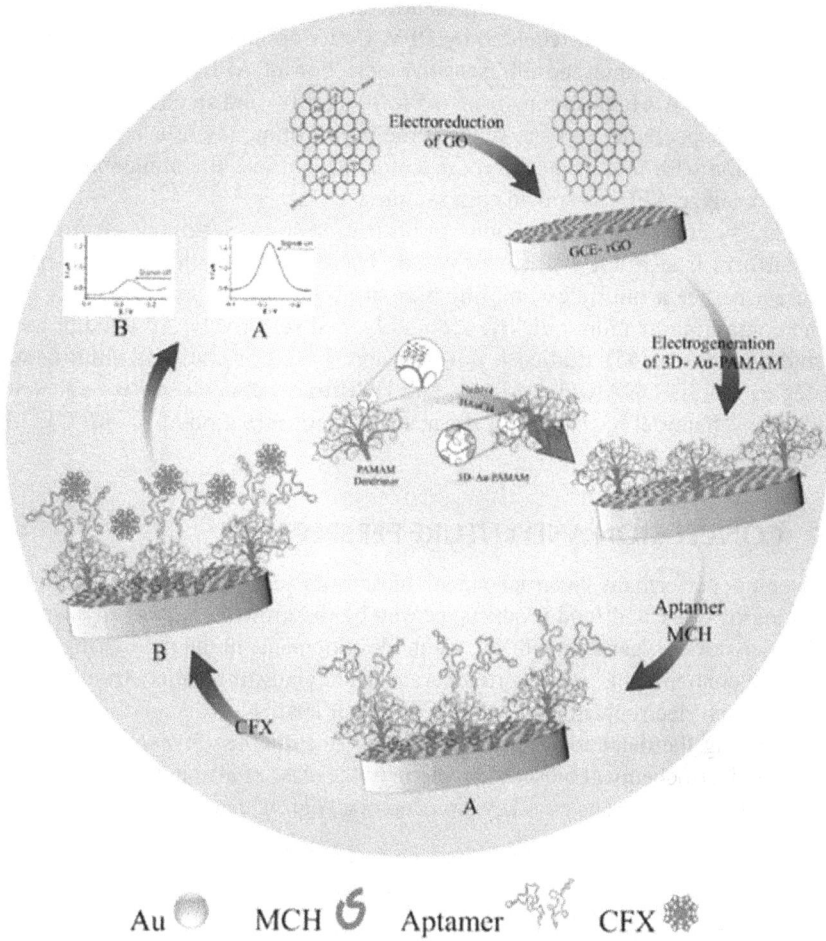

FIGURE 8.9 Aptasensing of ciprofloxacin residues in raw milk samples using rGO and nanogold-functionalized poly(amidoamine) dendrimer. (Reprinted with permission from Ref. (60)).

the detection of AFB1. The immunoassay relies on the competition between free AFB1 target and a previously prepared HCR-AFB1 system to be captured by the specific nanobody immobilized on nanostructured electrode surface. The HCR-AFB1 amplifying system acted as the competitor, amplifier, and signal reporter element. It consists of a conjugated AFB1 to a DNA primer via biotin-streptavidin (SA) linkage, that initiates the HCR of two biotinylated hairpins H1 and H2. Therefore, a long concatemer labeled with abundant biotins was obtained, offering a host to SA-modified polyHRP for signal conversion. Subsequently, in the presence of AFB1 analyte and prepared AFB1-HCR system, the latter was able to competitively combine with specific Nb. Eventually, owing to the strong binding between biotin of HCR-AFB1 concatemer and streptavidin of SA-polyHRP, numerous HRPs were assembled on the surface of the immunosensor, which could catalyze hydroquinone and H_2O_2 producing an amplified signal recorded by DPV. Under optimal conditions, the immunosensor displayed rapid and ultrasensitive detection of AFB1 with a good linear correlation toward AFB1 ranging from 0.5 to 10 ng mL^{-1} and an ultralow LOD of 68 fg mL^{-1}. The specificity test showed that the AFB1 immunosensor had no obvious cross-reaction with four co-occurrent mycotoxins. As well, the immunosensor was useful to analyze AFB1 in spiked corn samples.

So far, the combination of various amplification strategies in one electrochemical platform is still not well investigated. Therefore, it would be interesting to explore whether a multiplex amplification strategy can be used to detect food-borne contaminants more quickly, accurately, and sensitively. An attempt led by Li and coworkers (65) studied a novel system based on multiple amplification strategies via 3D DNA walker, RCA, and HCR to develop a sensitive and selective electrochemical biosensor for the accurate determination of *E. coli* O157:H7 (See Figure 8.10).

8.5 CONCLUSION AND FUTURE PERSPECTIVES

Developing performant electrochemical biosensors can significantly support the screening of water and food products, proven by the growing number of successful applications every year. Some of the recent developments in the monitoring of food and water contaminants are reported herein while analyzing the contributions produced by such electrochemical bioassays and their challenges.

Examining the detection approaches achieved in the last 5 years clearly shows that most electrochemical biosensors belong to the class of affinity biosensors, using either antibodies or aptamers as key bioreceptors. A growing effort was performed to

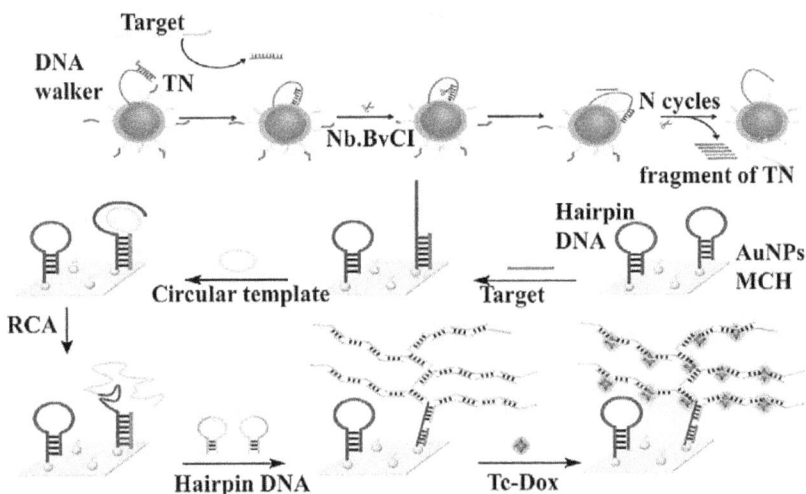

FIGURE 8.10 Schematic diagram of multiple amplification electrochemical biosensor for detecting E. coli O157:H7 (reproduced with permission (65)).

optimize the immobilization protocol of the biomolecule onto the electrode surface without affecting its biological activity. The improvements of this critical step can enhance the stability and overall life of biosensors.

In particular, recent trends have witnessed an exponential use of nanomaterials and signal amplification strategies to improve the analytical performances. However, several issues and challenges should be faced. Although there has been tremendous progress in real sample analysis, some uncertainties remain in the further application of portable E-biosensors outside chemical laboratories. Two major issues are associated with real matrices testing: the efficient extraction of target analytes from the heterogenous samples and the elimination of eventual electrochemical interferences naturally present.

REFERENCES

1. Gomes MP, Rocha DC, Moreira de Brito JC, Tavares DS, Marques RZ, et al. 2020. Emerging contaminants in water used for maize irrigation: Economic and food safety losses associated with ciprofloxacin and glyphosate. *Ecotoxicology and Environmental Safety*. 196: 110549
2. Kumar MD, Tortajada C. 2020. Health impacts of water pollution and contamination. In *Assessing Wastewater Management in India*, pp. 23–30. Springer
3. Wang L-F, Crameri G. 2014. Emerging zoonotic viral diseases. *Revue Scientifique et Technique (International Office of Epizootics)*. 33(2): 569–581
4. Haider N, Rothman-Ostrow P, Osman AY, Arruda LB, Macfarlane-Berry L, Elton L, Kock RA. 2020. COVID-19—zoonosis or emerging infectious disease? *Frontiers in Public Health*. 8: 763
5. Grieshaber D, MacKenzie R, Vörös J, Reimhult E. 2008. Electrochemical biosensors—sensor principles and architectures. *Sensors*. 8(3): 1400–1458
6. Thévenot DR, Toth K, Durst RA, Wilson GS. 2001. Electrochemical biosensors: Recommended definitions and classification. *Biosensors and Bioelectronics*. 16(1–2): 121–131
7. Reta N, Saint CP, Michelmore A, Prieto-Simon B, Voelcker NH. 2018. Nanostructured electrochemical biosensors for label-free detection of water- and food-borne pathogens. *ACS Applied Materials & Interfaces*. 10(7): 6055–6072
8. Rodriguez RS, O'Keefe TL, Froehlich C, Lewis RE, Sheldon TR, Haynes CL. 2021. Sensing food contaminants: Advances in analytical methods and techniques. *Analytical Chemistry*. 93(1): 23–40
9. Janik E, Ceremuga M, Saluk-Bijak J, Bijak M. 2019. Biological toxins as the potential tools for bioterrorism. *International Journal of Molecular Sciences*. 20(5): 1181
10. Ostry V, Malir F, Toman J, Grosse Y. 2017. Mycotoxins as human carcinogens—the IARC monographs classification. *Mycotoxin Research*. 33(1): 65–73
11. IARC. 2021. *Agents Classified by the IARC Monographs, Volumes 1–129—IARC Monographs on the Identification of Carcinogenic Hazards to Humans*. WHO
12. Narenderan ST, Meyyanathan SN, Babu B. 2020. Review of pesticide residue analysis in fruits and vegetables: Pre-treatment, extraction and detection techniques. *Food Research International*. 133: 109141
13. Ntzani EE, Ntritsos GCM, Evangelou E, Tzoulaki I. 2013. Literature review on epidemiological studies linking exposure to pesticides and health effects. *EFSA Supporting Publications*. 10(10): 497E
14. Ben Y, Fu C, Hu M, Liu L, Wong MH, Zheng C. 2019. Human health risk assessment of antibiotic resistance associated with antibiotic residues in the environment: A review. *Environmental Research*. 169: 483–493

15. MacGowan A, Macnaughton E. 2017. Antibiotic resistance. *Medicine (Baltimore).* 45(10): 622–628

16. Bacanlı M, Başaran N. 2019. Importance of antibiotic residues in animal food. *Food and Chemical Toxicology.* 125(January): 462–466

17. Sall ML, Diaw AKD, Gningue-Sall D, Efremova Aaron S, Aaron J-J. 2020. Toxic heavy metals: Impact on the environment and human health, and treatment with conducting organic polymers, a review. *Environmental Science and Pollution Research.* 27(24): 29927–29942

18. Ali H, Khan E, Ilahi I. 2019. Environmental chemistry and ecotoxicology of hazardous heavy metals: Environmental persistence, toxicity, and bioaccumulation. *Journal of Chemistry.* 6730305

19. Diarisso A, Fall M, Raouafi N. 2018. Elaboration of a chemical sensor based on polyaniline and sulfanilic acid diazonium salt for highly sensitive detection nitrite ions in acidified aqueous media. *Environmental Science: Water Research & Technology.* 4(7): 1024–1034

20. Sall ML, Fall B, Diédhiou I, Dièye EH, Lo M, et al. 2020. Toxicity and electrochemical detection of lead, cadmium and nitrite ions by organic conducting polymers: A review. *Chemistry Africa.* 3(3): 499–512

21. Curulli A. 2021. Electrochemical biosensors in food safety: Challenges and perspectives. *Molecules.* 26(10): 2940

22. Riu J, Giussani B. 2020. Electrochemical biosensors for the detection of pathogenic bacteria in food. *TrAC Trends in Analytical Chemistry.* 126: 115863

23. Chi S-C, Wu Y-C, Hong J. 2016. Nodaviruses of fish. *Aquaculture Virology.* 371–393

24. Murcia P, Donachie W, Palmarini M. 2009. Viral pathogens of domestic animals and their impact on biology, medicine and agriculture. In *Encyclopedia of Microbiology*, pp. 805–819. Elsevier

25. Schwartz-Cornil I, Mertens PCP, Contreras V, Hemati B, Pascale F, et al. 2008. Bluetongue virus: Virology, pathogenesis and immunity. *Veterinary Research.* 39(5)

26. Felix FS, Angnes L. 2018. Electrochemical immunosensors—a powerful tool for analytical applications. *Biosensors and Bioelectronics.* 102: 470–478

27. Chiu ML, Goulet DR, Teplyakov A, Gilliland GL. 2019. Antibody structure and function: The basis for engineering therapeutics. *Antibodies.* 8(4)

28. Li ZF, Dong JX, Vasylieva N, Cui YL, Wan D Bin, et al. 2021. Highly specific nanobody against herbicide 2,4-dichlorophenoxyacetic acid for monitoring of its contamination in environmental water. *Science of the Total Environment.* 753: 141950

29. Wang W, Yuan J, Jiang C. 2021. Applications of nanobodies in plant science and biotechnology. *Plant Molecular Biology.* 105(1): 43–53

30. Kumar S, Jain S, Dilbaghi N, Ahluwalia AS, Hassan AA, Kim KH. 2019. Advanced selection methodologies for DNAzymes in sensing and healthcare applications. *Trends in Biochemical Sciences.* 44(3): 190–213

31. Liu M, Chang D, Li Y. 2017. Discovery and biosensing applications of diverse RNA-cleaving DNAzymes. *Accounts of Chemical Research.* 50(9): 2273–2283

32. Lu Y. 2002. New transition-metal-dependent DNAzymes as efficient endonucleases and as selective metal biosensors. *Chemistry.* 8(20): 4588–4596

33. Gao G, Fang D, Yu Y, Wu L, Wang Y, Zhi J. 2017. A double-mediator based whole cell electrochemical biosensor for acute biotoxicity assessment of wastewater. *Talanta.* 167: 208–216

34. Sciuto EL, Petralia S, van der Meer JR, Conoci S. 2021. Miniaturized electrochemical biosensor based on whole-cell for heavy metal ions detection in water. *Biotechnology and Bioengineering.* 118(4): 1456–1465

35. Tsopela A, Laborde A, Salvagnac L, Ventalon V, Bedel-Pereira E, et al. 2016. Development of a lab-on-chip electrochemical biosensor for water quality analysis based on microalgal photosynthesis. *Biosensors and Bioelectronics.* 79: 568–573

36. Zhang J, Yang Z, Liu Q, Liang H. 2019. Electrochemical biotoxicity detection on a microfluidic paper-based analytical device via cellular respiratory inhibition. *Talanta.* 202: 384–391

37. Alawad A, Istamboulié G, Calas-Blanchard C, Noguer T. 2019. A reagentless aptasensor based on intrinsic aptamer redox activity for the detection of tetracycline in water. *Sensors & Actuators, B: Chemical.* 288: 141–146

38. Rengaraj S, Cruz-Izquierdo Á, Scott JL, Di Lorenzo M. 2018. Impedimetric paper-based biosensor for the detection of bacterial contamination in water. *Sensors & Actuators, B: Chemical.* 265: 50–58

39. Hamami M, Bouaziz M, Raouafi N, Bendounan A, Korri-Youssoufi H. 2021. MoS2/ PPy nanocomposite as a transducer for electrochemical aptasensor of ampicillin in river water. *Biosensors.* 11(9): 311

40. Mehta J, Vinayak P, Tuteja SK, Chhabra VA, Bhardwaj N, et al. 2016. Graphene modi-fied screen printed immunosensor for highly sensitive detection of parathion. *Biosensors and Bioelectronics.* 83: 339–346

41. Liu H, Wang J, Jin H, Wei M, Ren W, et al. 2021. Electrochemical biosensor for sensitive detection of Hg2+ baesd on clustered peonylike copper-based metal-organic frameworks and DNAzyme-driven DNA Walker dual amplification signal strategy. *Sensors & Actuators, B: Chemical.* 329: 129215

42. Chérif N, Zouari M, Amdouni F, Mefteh M, Ksouri A, et al. 2021. Direct amperometric sensing of fish nodavirus RNA using gold nanoparticle/DNA-based bioconjugates. *Pathogens.* 10(8): 932

43. Rhouati A, Bulbul G, Latif U, Hayat A, Li ZH, Marty JL. 2017. Nano-aptasensing in mycotoxin analysis: Recent updates and progress. *Toxins (Basel).* 9(11): 349

44. Chauhan R, Singh J, Sachdev T, Basu T, Malhotra BD. 2016. Recent advances in myco-toxins detection. *Biosensors and Bioelectronics.* 81: 532–545

45. Evtugyn G, Hianik T. 2019. Electrochemical immuno- and aptasensors for mycotoxin determination. *Chemosensors.* 7(1): 10

46. Jia M, Liao X, Fang L, Jia B, Liu M, et al. 2021. Recent advances on immunosensors for mycotoxins in foods and other commodities. *Trends in Analytical Chemistry.* 136: 116193

47. Evtugyn G, Hianik T. 2019. Aptamer-based biosensors for mycotoxin detection. In *Nanomycotoxicology: Treating Mycotoxins in the Nano Way*, pp. 35–70. Elsevier

48. Karczmarczyk A, Baeumner AJ, Feller KH. 2017. Rapid and sensitive inhibition-based assay for the electrochemical detection of Ochratoxin A and Aflatoxin M1 in red wine and milk. *Electrochimica Acta.* 243: 82–89

49. Qi X, Yan X, Zhao L, Huang Y, Wang S, Liang X. 2020. A facile label-free electrochem-ical aptasensor constructed with nanotetrahedron and aptamer-triplex for sensitive detec-tion of small molecule: Saxitoxin. *Journal of Electroanalytical Chemistry.* 858: 113805

50. Subjakova V, Oravczova V, Tatarko M, Hianik T. 2021. Advances in electrochemical aptasensors and immunosensors for detection of bacterial pathogens in food. *Electrochimica Acta.* 389: 138724

51. Razmi N, Hasanzadeh M, Willander M, Nur O. 2020. Recent progress on the electro-chemical biosensing of Escherichia coli O157:H7: Material and methods overview. *Biosensors.* 10(5): 1–18

52. Cinti S, Volpe G, Piermarini S, Delibato E, Palleschi G. 2017. Electrochemical biosen-sors for rapid detection of foodborne Salmonella: A critical overview. *Sensors (Switzerland).* 17(8): 1–22

53. Silva NFD, Neves MMPS, Magalhães JMCS, Freire C, Delerue-Matos C. 2020. Emerg-ing electrochemical biosensing approaches for detection of Listeria monocytogenes in food samples: An overview. *Trends in Food Science and Technology.* 99(March): 621–633

54. Leva-Bueno J, Peyman SA, Millner PA. 2020. A review on impedimetric immunosensors for pathogen and biomarker detection. *Medical Microbiology and Immunology.* 209(3): 343–362

55. Xu M, Wang R, Li Y. 2016. Rapid detection of Escherichia coli O157:H7 and Salmonella Typhimurium in foods using an electrochemical immunosensor based on screen-printed interdigitated microelectrode and immunomagnetic separation. *Talanta.* 148: 200–208

56. Hamami M, Raouafi N, Korri-youssoufi H. 2021. Self-assembled MoS2/ssDNA nanostructures for the capacitive aptasensing of acetamiprid insecticide. *Applied Sciences.* 11(4)

57. Verdian A. 2018. Apta-nanosensors for detection and quantitative determination of acetamiprid—a pesticide residue in food and environment. *Talanta.* 176(August 2017): 456–464

58. Aissa SB, Mars A, Catanante G, Marty JL, Raouafi N. 2019. Design of a redox-active surface for ultrasensitive redox capacitive aptasensing of aflatoxin M1 in milk. *Talanta.* 195: 525–532

59. Lu L, Seenivasan R, Wang YC, Yu JH, Gunasekaran S. 2016. An electrochemical immunosensor for rapid and sensitive detection of mycotoxins Fumonisin B1 and Deoxynivalenol. *Electrochimica Acta.* 213: 89–97

60. Mahmoudpour M, Kholafazad-kordasht H, Nazhad Dolatabadi JE, Hasanzadeh M, Rad AH, Torbati M. 2021. Sensitive aptasensing of ciprofloxacin residues in raw milk samples using reduced graphene oxide and nanogold-functionalized poly(amidoamine) dendrimer: An innovative apta-platform towards electroanalysis of antibiotics. *Analytica Chimica Acta.* 1174: 338736

61. Xu X, Su Y, Zhang Y, Wang X, Tian H, et al. 2021. Novel rolling circle amplification biosensors for food-borne microorganism detection. *Trends in Analytical Chemistry.* 141: 116293

62. Liu Y, Liu Y, Qiao L, Liu Y, Liu B. 2018. Advances in signal amplification strategies for electrochemical biosensing. *Current Opinion in Electrochemistry.* 12: 5–12

63. Xie S, Zhang J, Teng L, Yuan W, Tang Y, et al. 2019. Electrochemical detection of lipopolysaccharide based on rolling circle amplification assisted formation of copper nanoparticles for enhanced resistance generation. *Sensors & Actuators, B: Chemical.* 301(June): 127072

64. Liu X, Wen Y, Wang W, Zhao Z, Han Y, et al. 2020. Nanobody-based electrochemical competitive immunosensor for the detection of AFB1 through AFB1-HCR as signal amplifier. *Microchimica Acta.* 187(6): 352

65. Li Y, Liu H, Huang H, Deng J, Fang L, et al. 2020. A sensitive electrochemical strategy via multiple amplification reactions for the detection of E. coli O157: H7. *Biosensors and Bioelectronics.* 147(July 2019): 111752

9 Recent Progress on Biosensors Developed for Detecting Environmental Pollutants

Arzum Erdem[1], Huseyin Senturk[1], and Esma Yildiz[2]
[1]Ege University
Faculty of Pharmacy, Analytical Chemistry Department
Izmir, Turkey
[2]Ege University
The Institute of Natural and Applied Sciences,
Biotechnology Department
Izmir, Turkey
Corresponding author: (Prof. Dr. Arzum Erdem)

CONTENTS

9.1   INTRODUCTION

Despite efforts to ensure environmental cleanliness, environmental pollution continues to grow in many areas. In addition to the damage to the environment, the effects of environmental pollution, which are reflected in human and animal health, also pose a risk. Problems related to traditional environmental pollution, such as industrial emissions, inadequate waste management, polluted water resources and exposure to air pollution, still persist today.

DOI: 10.1201/9781003189435-13

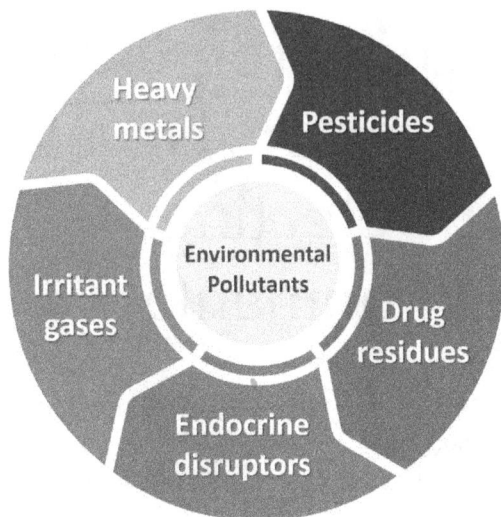

FIGURE 9.1   Schematic representation of various environmental pollutants.

Today, in addition to traditional pollutants, modern pollutants have started to take place in our lives. Various types of environmental pollutants (Figure 9.1), such as chemicals in food production, pesticide control chemicals and use of endocrine disrupting chemicals also have a negative impact on public health. There are different assays reported for fast screening of various environmental pollutants, pathogens and toxins, including biosensors (1–10).

Biosensors are complex systems that recognize the target analyte on a recognizing surface and convert the resulting biological response into an electrical signal (11). Biosensors basically consist of two main units: the recognition element on the recognition surface and the transducer. By immobilizing the recognition element at the surface, the recognizer and the target deposited on the surface form two affinity pairs. These affinity pairs can be antigen/antibody, enzyme/substrate, receptor/its specific ligand, DNA sequence/its specific aptamer or living cell/analyte. The transducer detects the biological response generated by these affinity pairs and converts it into a readable electrical signal (12). Biosensors can be broadly classified as follows: (i) electrochemical biosensors; (ii) mass-based biosensors; and (iii) optical-based biosensors.

Numerous biosensors have been developed from past to present for the determination of environmental pollutants. Electrochemical and optical biosensors have been frequently used for the analysis of environmental pollutants in the last decade. Herein, the determination of different environmental pollutants through by electrochemical and optical biosensors has been introduced and further discussed with the feature prospects.

9.2 DETECTION OF PESTICIDES BY ELECTROCHEMICAL AND OPTICAL BIOSENSORS

Pesticides are substances that are used to prevent or reduce harmful organisms. The term pesticide includes all herbicides, insecticides, fungicides, nematicides, rodenticides, bactericides. Pesticides are generally used for protection of plants. Although they have such beneficial effects, it is a disadvantage that they have potential toxic effects for humans and animals (13–14).

Therefore, the use of these substances is restricted by various authorized organizations such as Federal Insecticide, Fungicide, and Rodenticide Act (FIFRA), Federal Food, Drug, and Cosmetic Act (FFDCA), Food Quality Protection Act of 1996 (FQPA), Pesticide Registration Improvement Act (PRIA), Endangered Species Act (ESA) (15).

Detection of pesticides in water, soil and food is very crucial due to their potential toxic effects on human and animal health. In this section, electrochemical and optical biosensors developed for the detection of pesticides are summarized in Table 9.1 (16–24, 4, 25–29, 5, 30–33).

Organophosphorus (OP) compounds are known to be harmful to human health. Sensitive and reliable detection of these compounds is essential (35).

There are biosensors developed through the inhibition of acetylcholinesterase (AChE) activity of organophosphorus (OP) compounds. These biosensors provide various advantages with low cost, fast detection and reliability.

Sun and Wang (18) developed a Prussian blue modified glassy carbon electrode for electrochemical determination of organophosphorus (OP) compounds such as dichlorvos, omethoate, trichlorfon, phoxim through inhibition of acetylcholinesterase (AChE) activity. They performed pesticide determination with the AChE-PB/GCE electrode they developed after AChE immobilization on the Prussian blue modified glassy carbon electrode surface. After 10 minutes incubation time for pesticide on the electrode surface, acetylthiocholine iodide (ATChI) was dropped the surface and measurements were taken by using amperometry technique.

Atrazine is the one of the most widely used pesticides. Although it is known to adversely affect human health, it has been shown to damage systems such as the nervous system, immune system, endocrine system (36).

The immunosensor reported by Liu et al. (21) on atrazine determination was developed using an electrode in which gold nanoparticles were modified onto the surface of gold electrode. They reported that an atrazine specific procedure was developed by using an anti-atrazine monoclonal antibody immobilized on the surface. After the analyte was added onto the immunosensor surface, it was shown that the determination was made over the atrazine signal by differential pulse voltammetry.

Zhao et al. (19) reported a hydrolase biosensor for the determination of methyl parathion that is the one of the organophosphorus compounds. They dispersed the synthesized Fe_3O_4@Au magnetic nanocomposites in methyl parathion hydrolase (MPH). They performed the incubation of MPH- Fe_3O_4@Au on the surface of screen printed electrode (SPE). After analyte was added onto the electrode surface, it was shown that the determination of methyl parathion was performed by electrochemical impedance spectroscopy (EIS).

TABLE 9.1

Electrochemical and Optical Biosensors Developed for the Detection of Pesticides

Electrochemical detection of pesticides

Analyte	Method	Electrode	Detection Limit	Linear Concentration Range	Application to Real Sample	Reference
Atrazine	DPV	GNPs/gold electrode	0.016 ng/mL	0.05–0.5 ng/mL	Maize sample	(21)
Catechol Chlorpyrifos	Amperometry	SPCE/IrOx/Tyr	0.08 µM 0.003 µM	0.25–27.5 µM 0.01–0.1 µM	Spiked tap and river water	(22)
Chlortoluron Phenol	Amperometry	SPCE/ZnO/GA/Tyr	0.47 nM 20 nM	1–100 nM 0.1–14 µM	Well water Tap water River water	(20)
Dichlorvos	DPV	CS@TiO$_2$-CS/rGO/GCE	29 nM	0.036–22.6 µM	Cabbage juice	(16)
	EIS	Ionic liquids-gold nanoparticles porous carbon composite boron-doped diamond electrode	6.61×10^{-11} g/L	10^{-10}–10^{-6} g/L	Lettuce leaves	(17)
Dichlorvos, Omethoate, Trichlorfon, Phoxim	Amperometry	PB/GCE	2.5 ng/L 15 ng/L 5 ng/L 10 ng/L	0.01–10 g/L 0.05–10 g/L 0.03–5 g/L 0.05–10 g/L	-	(18)
MalathionAcephate	DPV	AChE—SiSG—CPE	0.058 µg/mL 0.044 µg/mL	0.07–1.3 µg/mL 0.1–0.85 µg/mL	Food sample	(23)
Methyl parathion	EIS	Fe$_3$O$_4$/AuNP/SPE	0.1 ng/mL	0–1500 ng/mL	-	(19)
Methyl ParathionParaoxon	Amperometry	NF/AChE/NF-NiCo$_2$S$_4$/CPE	4.2×10^{-13} g/mL 3.5×10^{-14} g/mL	1.0×10^{-12}–1.0×10^{-8} g/mL 1.0×10^{-13}–1.0×10^{-10} g/mL	-	(34)
Paraoxon	Amperometry	SPE/CBNPs-Glut-Nf-BSA-BChE	5 µg/L	5–30 µg/L	Waste water samples	(4)

(Continued)

TABLE 9.1
(Continued)

Optical Detection of Pesticides

Analyte	Method	Detection Limit	Linear Concentration Range	Application to Real Sample	Reference
Chlorpyrifos	SPR	0.056 ng/mL	0.25–50.0 ng/mL	Agricultural products	(33)
Fenitrothion	SPR	38 nM	0.25–4 μM	-	(32)
Irgarol 1051	Interferometry	3 ng/L	9–1190 ng/L	Seawater samples	(31)
Malathion	Colorimetric	0.06 pM	0.5–1000 pM	Lake water and apple samples	(25)
Methyl-parathion	Tapered-fiber optic	2.4×10^{-10} M	10^{-10}–5×10^{-10} M	-	(26)
	Fluorescent biosensor	4.8×10^{-11} M	1.0×10^{-10}– 1.0×10^{-4} M	Cabbage, milk and fruit juice samples	(27)
	Absorbance based biosensor	4 μM	0–10 μM	-	(28)
	Absorbance-based biosensor	0.1 μM	0.1–10 μM	Agricultural products and water samples	(29)
Paraoxon	Colorimetric	4.7 ppb (17 nmol/L)	10–80 ppb (36–290 nmol/L)	Environmental water samples	(5)
Paraoxon and parathion	Fluorescent biosensor	1.05×10^{-11} M for paraoxon, 4.47×10^{-12} M for parathion	10^{-12}–10^{-6} M for paraoxon	Vegetable and fruit samples	(30)

Abbreviations: AChE: Acetylcholinesterase, AuNP: Gold nanoparticles, BChE: butyrylcholinesterase, BSA: Bovine serum albumin, CBNPs: Carbon black nanoparticles, CPE: Carbon paste electrode, CS: Chitosan, DPV: Differential pulse voltammetry, EIS: Electrochemical impedance spectroscopy, GA: Glutaraldehyde, GCE: Glassy carbon electrode, Glut: glutaraldehyde, GNPs: Gold nanoparticles, IrOx: Iridium oxide, NF: Nafion, PB: Prussian blue, rGO: Reduced graphene oxide, SiSG: Silica sol-gel, SPCE: Screen printed carbon electrode, SPE: Screen printed electrode, SPR: Surface plasmon resonance, Tyr: Tyrosinase, ZnO: zinc oxide.

A colorimetric aptamer biosensor for the detection of organophosphorus pesticide malathion was developed by Bala and coworkers (25). According to Malathion detection mechanism, it is reported that polyelectrolyte Polydiallyldimethylammonium chloride (PDDA) binds to aptamer and gold nanoparticles cannot aggregate in the absence of malathion; thus, the red color can be seen. On the other hand, aptamer forms a complex with malathion when the malathion is added to the medium. Then PDDA is released and gold nanoparticles aggregated. As a result, the color changes from red to blue. Accordingly, they detected malathion in lake water and apple samples by using the aptamer based biosensor.

Paraoxon is also an organophosphorus pesticide and residues of paraoxon in food products threat public health. Therefore, Guo et al. (5) developed a colorimetric biosensor for the determination paraoxon. It was reported that the mechanism of this colorimetric biosensor based on iodine-starch color reaction and multi-enzyme cascade catalytic reactions. In the absence of the paraoxon, blue color could be seen in the solution, when the paraoxon is added, the solution becomes clear. Accordingly, the detection of paraoxon detection was performed in the spiked real samples by Guo et al.

Irgarol 1051 (also known as cybutryne) is an antifouling biocide and environmental contaminant. It inhibits the photosynthesis and prevents the growth of algae. The increased use of the Irgarol 1051 causes a decrease in biodiversity in water (37).

Chocarro-Ruiz et al. (31) fabricated a bimodal waveguide (BiMW) interferometric nanoimmunosensor for the detection of Irgarol 1051 in the seawater. In that study, they obtained the limit of detection as 3 ng/L that is found lower than the maximum allowable concentration for Irgarol 1051 in the seawater determined by the EU.

9.3 DETECTION OF HEAVY METALS BY ELECTROCHEMICAL AND OPTICAL BIOSENSORS

Heavy metals are generally defined as metals with high density, atomic weight and atomic number. Metals such as iron, cobalt, copper, manganese and zinc are essential for human health and insufficiency of these metals in the body causes various health problems. On the other hand, heavy metals such as cadmium, mercury and lead are very dangerous for human health. These metals accumulate in the human body through food and water and cause many health problems. Therefore, detection of these metals in water and food is essential for human and animal health (38).

In this section, electrochemical and optical biosensors developed for detection of heavy metals are summarized in Table 9.2 (39–44, 6, 45–61).

Mercury ions are known to be harmful to human health and the environment. It is identified as a toxic heavy metal even at low concentrations (63).

Shi et al. (39) developed an electrochemical biosensor for mercury determination. It was presented that the gold substrate is covered with SAM (self-assembled monolayer) and vertical single-walled carbon nanotubes and then the amino labeled DNA probe (A probe) was attached to the surface. After then, equivalent DNA probe (R probe) with methylene blue label was hybridized with the A probe. As a result of the partial base pairing, T-T base pairs were released at the terminal ends to recognize mercury ions. Thus, it is known that T-Hg^{2+}-T stable structure will be formed in the presence of mercury ion. After the addition of analyte, measurements were carried out by the square wave voltammetry (SWV) method.

Lead (Pb) is the one of the dangerous heavy metals in the environmental pollutant class. Taghdisi et al. (40) reported an aptasensor used for lead determination. They explained the experimental procedure as follows: The hairpin complementary conjugate (CS) of the aptamer is attached to the screen printed gold electrode (SPGE) surface. Then, CS, which is also the aptamer of lead, was incubated onto the surface. CS and aptamer formed a hybrid and the gold nanoparticles-thionine complex was attached to CS, whose hairpin structure was disrupted. After then the reduction-oxidation signals were investigated. The gold nanoparticles-thionine complex does not be able to bind to the complementary conjugate (CS) in the hairpin structure, as the lead and aptamer pair in the presence of lead.

Long et al. (53), fabricated an evanescent wave DNA biosensor for the detection of mercury ions (Hg^{2+}). Structure-competitive sensing mechanism was used the detection of Hg^{2+}. First, DNA probe that contains T-T mismatch pairs was immobilized the optical fiber sensor's surface. In the presence of mercury, a T-Hg^{2+}-T complex formed by folding of the probe segments into a hairpin structure. Then, different concentrations of Hg^{2+} solution and fluorescently labeled complementary DNA (cDNA) was mixed and added onto the sensor's surface. cDNA and Hg^{2+} were competed for binding to the DNA probe. It was reported that cDNA could not be attached to the DNA and fluorescence signal decreased but Hg^{2+} concentration increased. In addition, they reported that the sensor's surface can be regenerated over 100 times, however, there was no significant loss of signal.

Yin et al. (56) developed a fluorescent biosensor for the detection of copper ion. The mechanism of the biosensor based on pyoverdine copper binding. Copper ions were captured by the pyoverdine and the pyoverdine's fluorescence signal was quenched. In addition, they reported that the developed sensor successfully detected the copper ion in drinking water, seawater, artificial cerebrospinal fluid and shellfish samples.

Tian et al. (59) developed a DNAzyme-based SERS biosensor for lead ions detection including the recovery tests in the samples of buffer, tap water and human serum. In addition, the lead detection was performed in 2 hours in that study of Tian et al.

9.4 DETECTION OF ENDOCRINE DISRUPTORS BY ELECTROCHEMICAL AND OPTICAL BIOSENSORS

Endocrine disruptors are chemicals that change the functions of the endocrine system and thus they negatively affect human and animal health. For this reason, sensitive detection of endocrine disruptors in water and food is extremely important for human and animal health (64).

Table 9.3 presents the electrochemical and optical biosensors developed for detection of endocrine disruptors (65–67, 8, 68–77, 9, 78).

Matairesinol is a plant lignan found in flax seeds and rye. It has an estrogen-like structure. Therefore, in the gastrointestinal tract, it is converted into the metabolite enterodiol, which is known to have estrogenic properties. It is considered among the endocrine disruptors due to its endocrine disrupting effect on animal and human health (79–80).

Rather et al. (65) developed an electrochemical biosensor for the determination of matairesinol. The glassy carbon electrode surface has been modified with multi-walled carbon nanotubes and the thionine is attached to the surface. Tyrosinase was then immobilized onto the surface. Matairesinol is known to bind to tyrosinase to

TABLE 9.2

Electrochemical and Optical Biosensors Developed for the Detection of Heavy Metals

Electrochemical Detection of Heavy Metals

Analyte	Method	Electrode	Detection Limit	Linear Concentration Range	Application to Real Sample	Reference
Copper (Cu^{+2})	SWV	Gold	20 nM	20 nM–600 µM	real and spiked lake samples	(62)
	SWV	AuNPs/Gold	0.1 pM	0.1 nM–10 µM	lake water and tap water	(51)
Lead (Pb)	DPV, CV	SPGE	312 pM	0.6–50 nM	tap water serum	(40)
Mercury (Hg^{+2})	DPV	Gold	0.25 nM	0.5 nM–5 µM	lake water samples	(49)
	SWV	v-SWCNT/gold substrate	3 fM	10 fM–1 µM	lake water, human serum samples	(39)
	DPV	Gold electrode	60 pM	0.2–1 nM	river water	(41)
	CV	SPGE	0.6 nM	0.001–10 µM	-	(42)
	CV, EIS	Gold coated glass electrode	1 nM	1–300 nM	-	(43)
	CV, EIS, DPV	RGO/GCE	5 nM	8 nM–0.1 µM	river water	(44)
	DPV, CV, EIS	PGE	0.06 nM	0.1 nM–5 µM	-	(47)
	DPV	ITO coated glass chip	0.2 nM	0.5 nM–20 µM	-	(48)
Silver (Ag$^+$)	DPV	Gold	0.03 nM	0.1–120 nM	-	(6)
	DPV	MHA/Gold	1.3 nM	10–500 nM	lake water and river water	(45)
	SWV	NPG/GCE	0.048 nM	0.1 nM–1 µM	water samples	(46)

(Continued)

TABLE 9.2
(Continued)

Optical Detection of Heavy Metals

Analyte	Method	Detection Limit	Linear Concentration Range	Application to Real Sample	Reference
Arsenic (III)	Colorimetric	14.44 ppb at aqueous samples and 1.97 ppm at soil samples	-	Field soil samples	(61)
Chromium (III)	Colorimetric	300 nM	0.1–0.7 µM	Environmental water samples	(58)
Copper	Fluorescent biosensor	50 nM	0.2–10 µM	Drinking water, seawater, artificial cerebrospinal fluid, shellfish samples	(56)
	FRET	7 nM	0.04–11 µM	Purified water, tap water and waste water samples	(57)
Lead (II)	SERS	70 fM	1.0×10^{-13} M–1.0×10^{-7} M	Tap water and human serum	(59)
	DNAzyme based optical fiber sensor	1.03 nM	2–75 nM	Environmental water samples	(60)
Mercury (Hg^{2+})	Waveguide-based evanescent wave	0.2 µg/L in real water samples and 0.4 µg/L in ultrapure water	1.4–240.7 µg/L	Real water samples	(52)
	Fiber optic	2.1 nM at 10 nM cDNA; 5.0 nM at 20 nM cDNA	0–600 nM and 0–2 µM	Environmental water samples	(53)
	LSPR based optical fiber	0.1 ppb in tap water, 0.2 ppb sea fish and vegetable samples	0.1–540 ppb in tap water	Biological and environmental samples	(54)
	SERS	0.16 pM	1 pM–1 µM	Tap water and lake water	(55)

Abbreviations: CV: Cyclic voltammetry, DPV: Differential pulse voltammetry, EIS: Electrochemical impedance spectroscopy, FRET: Fluorescence resonance energy transfer, GCE: Glassy carbon electrode, ITO: Indium tin oxide, AuNPs: Gold nanoparticles, LSPR: Localized surface plasmon resonance, MHA: Mercaptohexadecanoic acid, NPG: Nanoporous gold, PGE: Pencil graphite electrode, RGO: Reduced graphene oxide, SERS: Surface enhanced Raman spectroscopy, SPGE: Screen printed gold electrodes, SWV: Square wave voltammetry, v-SWCNT: Vertical single-walled carbon nanotube.

TABLE 9.3

Electrochemical and Optical Biosensors Developed for the Detection of Endocrine Disruptors

Electrochemical Detection of Endocrine Disruptors

Analyte	Method	Electrode	Detection Limit	Linear Concentration Range	Application to Real Sample	Reference
17β-estradiol	DPV	cDNA/Aptamer/AuNPs/CoS/GCE	7.0×10^{-13} M	1.0×10^{-9}–1.0×10^{-12} M	-	(69)
	EIS	Aptamer/Au	2 pM	0.01–10 nM	Human urine	(70)
Bisphenol A	DPV	GCE/Aptamer-NPGF	5.6×10^{-11} mol/L	1.0×10^{-10}–1.0×10^{-7} M	Serum samples	(66)
	SWV	GCE/Tyr/APS/Au-GN	1.8 µg/L	0.01–10 mg/L	Olive oil extracts and chips samples	(67)
	Amperometry	SPE/Laccase—thionine—carbon black	0.2 µM	0.5–50 µM	Tomato juice from metallic cans	(8)
Estrone	EIS	PGE	1 µg/mL	1–4 µg/mL	-	(68)
matairesinol	SWV, CV	PTH//MWCNT/GCE	37 nM	180 nM–4.33 µM	-	(65)

Optical Detection of Endocrine Disruptors

Analyte	Method	Detection Limit	Linear Concentration Range	Application to Real Sample	Reference
17β-estradiol	Chemiluminescent fiber optic aptasensor	48 ng/L	0–1000 µg/L	Water samples	(9)
	Hydrogel optical waveguide spectroscopy	50 pg/mL	0–10000 ng/mL	-	(78)
Atrazine	Fiber-Optic	Less than 1 ng/mL	1–100 ng/mL	Soil column	(71)
Bisphenol A	SPR	5.2 pg/mL	0.01–100000 ng/mL	-	(74)
	Fiber optic aptasensor	1.86 nM	2–100 nM	Wastewater samples	(75)
	SPR	0.08 ng/mL in buffer medium and 0.14 ng/mL in wastewater	0.05–1000 ng/mL	Wastewater samples	(76)
	Colorimetric paper based biosensor	0.28 µg/g	0.05–3.87 µg/g	-	(77)

(Continued)

TABLE 9.3
(Continued)

Optical Detection of Endocrine Disruptors

Analyte	Method	Detection Limit	Linear Concentration Range	Application to Real Sample	Reference
Vitellogenin	SERS	5 pg/mL	5–25 ng/mL	-	(72)
	Fiber Optic SPR biosensor	1 ng/mL	0–25 ng/mL	-	(73)

Abbreviations: APS: Au-Pt@SiO$_2$, AuNPs: Gold nanoparticles cDNA: complementary DNA, CoS: Cobalt sulfide nanosheet, CV: Cyclic voltammetry, DPV: Differential pulse voltammetry, EIS: Electrochemical impedance spectroscopy, GCE: Glassy carbon electrode, GN: Graphene, MWCNT: Multi-walled carbon nanotube, NPGF: Nanoporous gold film, PGE: Pencil graphite electrode, PTH: Phenothiazine, SERS: Surface enhanced Raman spectroscopy, SPE: Screen printed electrodes, SPR: Surface plasmon resonance, SWV: Square wave voltammetry, Tyr: Tyrosinase.

TABLE 9.4
Electrochemical and Optical Biosensors Developed for the Other Pollutants

Electrochemical Detection of Other Pollutants

Analyte	Method	Electrode	Detection Limit	Linear Concentration Range	Application to Real Sample	Reference
2,2',4,4',6-pentabromodiphenyl ether (BDE-100)2-bromobiphenyl (PBB-1) 2-chlorobiphenyl (PCB-1)2,4,4'-trichlorobiphenyl (PCB-28)2,2',4,5,5'-pentachlorobiphenyl (PCB-101)	Amperometry	Pt/PANi/HRP	0.014 µg/L0.018 µg/L0.022 µg/L0.016 µg/L0.019 µg/L	0.424–25.8 µg/L0.862–13.4 µg/L0.930–18.1 µg/L0.730–15.7 µg/L0.930–27.1 µg/L	Waste water samples	(82)
Microcystis spp.	DPV	MWCNT-SPE	5.25 µg/mL	0–40 µg/mL	-	(1)

Optical Detection of Other Pollutants

Analyte	Method		Detection Limit	Linear Range	Application	Reference
Chloramphenicol	Aptamer based fluorescence biosensor		0.01 ng/mL	0.01–1 ng/mL	Contaminated milk samples	(83)
Kanamycin	Fluorescence quenching evanescent wave aptasensor (FQ-EWA)		26 nM	200 nM–200 µM	Milk	(7)
Triclosan	SPR		0.017 ng/mL	0.05–1.0 ng/mL	Wastewater samples	(84)

Abbreviations: DPV: Differential pulse voltammetry, HRP: Horse radish peroxidase, MWCNT: Multi-walled carbon nanotube, PANi: Polyaniline, PBB: Polybrominated biphenyls, PBDEs: polybrominated diphenyl ethers, PCB: Polychlorinated diphenyl ethers, Pt: Platinum, SPE: Screen printed electrode, SPR: Surface plasmon resonance.

form o-diphenolate. After the analyte was added to the electrode surface, the measurements were carried out by the square wave voltammetry (SWV) method.

Alkasir et al. (77) fabricated a portable colorimetric paper based biosensor for the detection of bisphenol A (BPA). The results obtained by using developed method based on biosensor was validated with gas chromatography (GC) and gas chromatography/mass spectroscopy (GC/MS) techniques.

Yang et al. (9) developed a fiber optic chemiluminescent aptamer based biosensor for 17β-estradiol detection. In the study performed with an indirect competitive assay, a less chemiluminescent signal was achieved with the increase of the 17β-estradiol. 17β-estradiol detection was performed in water samples such as bottled, tap and wastewater samples and recovery was calculated. It was also reported that the whole analysis process took less than 15 minutes.

SERS biosensor was developed by Srivastava and coworkers (72) for the detection of vitellogenin, which is a biomarker of endocrine disrupting compounds. A vitellogenin antibody (Anti-Vg) was immobilized on the 4-Aminothiophenol coated silver nano sculptured thin films on Si substrates and the SERS signals of the 4-Aminothiophenol were measured. In another study of same group (73), a fiber optic SPR biosensor was developed for the detection of vitellogenin.

9.5 DETECTION OF OTHER POLLUTANTS BY ELECTROCHEMICAL AND OPTICAL BIOSENSORS

Persistent organic pollutants (POP) in the water can pose a risk to human health and the environment. They can maintain their negative effect even in low concentration. Persistent organic pollutants are highly resistant to chemical and biological degradation (81). Table 9.4 presents the electrochemical and optical biosensors developed for detection of other pollutants (82, 1, 83, 7, 84).

It has a wide variety of structures such as polybrominated diphenyl ethers (PBDEs), polybrominated biphenyls (PBBs) and polychlorinated biphenyls (PCBs). Nomngongo et al. (82) developed an electrochemical biosensor for the determination of persistent organic pollutants. The surface of the platinum working electrode was modified with polyaniline and the horseradish peroxidase (HRP) enzyme was immobilized on the surface. Persistent organic pollutants determined in the study inhibited the enzyme activity. Thus, the inhibition in enzyme activity was used to determine these environmental contaminants. The electrochemical measurements were occurred by amperometry technique.

Triclosan is widely used in pharmaceuticals and cosmeceuticals as an antimicrobial agent. Due to the increased use of the triclosan, it accumulates into wastewater. The toxic nature of the triclosan is dangerous for aquatic organism and humans are also affected through the food chain (85).

Atar and coworkers (84) developed a molecular imprinted surface plasmon resonance nanosensor for the detection of triclosan. They used allymercaptane modified gold SPR chip and p(HEMEGA) nanofilm for the development of the nanosensor and characterization was performed via FTIR spectroscopy, AFM and contact angle measurements. Then, the detection of triclosan in wastewater samples was successfully performed by SPR technique.

9.6 CONCLUSION

It is known that environmental pollution, which is accepted as a global problem today, has a negative effect on all living creatures. Pesticides, heavy metals, endocrine disruptors and other environmental pollutants are considered as the main factors causing environmental pollution. The determination of these environmental pollutants has a key importance. Especially there have been many attempts for their on-site monitoring based on different methods and technologies, including biosensors. Electrochemical and optical biosensors are mostly the preferred ones developed for the determination of numerous environmental pollutants. This chapter summarized the recent studies presenting the electrochemical and optical biosensors that were developed in the last decade for monitoring various environmental pollutants while reporting that they are providing the cost-effective and on-site analysis in a good sensitivity, selectivity with reliability.

ACKNOWLEDGMENTS

A. E. would like to express her gratitude to the Turkish Academy of Sciences (TUBA) as a Principal member for its partial support.

REFERENCES

1. Erdem A, Karadeniz H, Canavar PE, Congur G. 2012. Single-use sensor platforms based on carbon nanotubes for electrochemical detection of DNA hybridization related to microcystis Spp. *Electroanalysis*. 24(3): 502–511. doi:10.1002/elan.201100369
2. Kerman K, Meric B, Ozkan D, Kara P, Erdem A, Ozsoz M. 2001. Electrochemical DNA biosensor for the determination of Benzo[a]Pyrene-DNA adducts. *Analytica Chimica Acta*. 450: 45–52
3. Kesici E, Erdem A. 2019. Impedimetric detection of fumonisin B1 and Its biointeraction with FsDNA." *International Journal of Biological Macromolecules*. doi:10.1016/j.ijbiomac.2019.08.024
4. Arduini F, Forchielli M, Amine A, Neagu D, Cacciotti I, Nanni F, Moscone D, Palleschi G. 2015. Screen-printed biosensor modified with carbon black nanoparticles for the determination of paraoxon based on the inhibition of butyrylcholinesterase. *Microchimica Acta*. 182(3–4): 643–651. doi:10.1007/s00604-014-1370-y
5. Guo L, Li Z, Chen H, Wu Y, Chen L, Song Z, Lin T. 2017. Colorimetric biosensor for the assay of paraoxon in environmental water samples based on the iodine-starch color reaction. *Analytica Chimica Acta*. 967: 59–63. doi:10.1016/j.aca.2017.02.028
6. Xu G, Wang G, He X, Zhu Y, Chen L, Zhang X. 2013. An ultrasensitive electrochemical method for detection of Ag+ based on cyclic amplification of exonuclease III activity on Cytosine-Ag +-Cytosine. *Analyst*. 138(22): 6900–6906. doi:10.1039/c3an01320k
7. Tang Y, Gu C, Wang C, Song B, Zhou X, Lou X, He M. 2018. Evanescent wave aptasensor for continuous and online aminoglycoside antibiotics detection based on target binding facilitated fluorescence quenching. *Biosensors and Bioelectronics*. 102(August 2017): 646–651. doi:10.1016/j.bios.2017.12.006
8. Portaccio M, Di Tuoro D, Cammarota M, Arduini F, Moscone D, Mita DG, Lepore M. 2013. Laccase biosensor based on screen-printed electrode modified with thionine-carbon black nanocomposite, for Bisphenol A detection. *Electrochimica Acta*. 109: 340–347. doi:10.1016/j.electacta.2013.07.129

9. Yang R, Liu J, Song D, Zhu A, Xu W, Wang H, Long F. 2019. Reusable chemilumi-nescent fiber optic aptasensor for the determination of 17β-estradiol in water samples. *Microchimica Acta.* 186(726): 1–9. doi:10.1007/s00604-019-3813-y

10. Ozsoz M, Erdem A, Ozkan D, Kerman K, Pinnavaia TJ. 2003. Clay/sol-gel-modified electrodes for the selective electrochemical monitoring of 2,4-dichlorophenol. *Lang-muir.* doi:10.1021/la034116v

11. Mehrotra P. 2016. Biosensors and their applications—a review. *Journal of Oral Biol-ogy and Craniofacial Research.* 6(2): 153–159. Craniofacial Research Foundation. doi:10.1016/j.jobcr.2015.12.002

12. Mungroo NA, Neethirajan S. 2014. Biosensors for the detection of antibiotics in poultry industry-a review. *Biosensors.* 4(4): 472–493. doi:10.3390/bios4040472

13. Syafrudin M, Kristanti RA, Yuniarto A, Hadibarata T, Rhee J, Al-Onazi WA, Algarni TS, Almarri AH, Al-Mohaimeed AM. 2021. Pesticides in drinking water-a review. *Inter-national Journal of Environmental Research and Public Health.* 18: 468. doi:10.3390/ijerph18020468

14. Rajmohan KS, Chandrasekaran R, Varjani S. 2020. A review on occurrence of pesticides in environment and current technologies for their remediation and management. *Indian Journal of Microbiology.* 60(2): 125–138. doi:10.1007/s12088-019-00841-x

15. Pesticide Laws and Regulations. 2021. Accessed July 28. http://npic.orst.edu/reg/laws.html

16. Cui HF, Wu W, Li M, Song X, Lv Y, Zhang T. 2018. A highly stable acetylcholinest-erase biosensor based on Chitosan-TiO2-graphene nanocomposites for detection of organophosphate pesticides. *Biosensors and Bioelectronics.* 99(May 2017): 223–229. doi:10.1016/j.bios.2017.07.068

17. Wei M, Wang J. 2015. A novel acetylcholinesterase biosensor based on ionic liq-uids-AuNPs-porous carbon composite matrix for detection of organophosphate pes-ticides. *Sensors and Actuators, B: Chemical.* 211: 290–296. doi:10.1016/j.snb.2015.01.112

18. Sun X, Wang X. 2010. Acetylcholinesterase biosensor based on prussian blue-modified electrode for detecting organophosphorous pesticides. *Biosensors and Bioelectronics.* 25(12): 2611–2614. doi:10.1016/j.bios.2010.04.028

19. Zhao Y, Zhang W, Lin Y, Du D. 2013. The vital function of Fe3O4@Au nanocompos-ites for hydrolase biosensor design and its application in detection of methyl parathion. *Nanoscale.* 5(3): 1121–1126. doi:10.1039/c2nr33107a

20. Haddaoui M, Raouafi N. 2015. Chlortoluron-induced enzymatic activity inhibition in tyrosinase/ZnO NPs/SPCE biosensor for the detection of Ppb levels of herbicide. *Sen-sors and Actuators, B: Chemical.* 219: 171–178. doi:10.1016/j.snb.2015.05.023

21. Liu X, Li WJ, Li L, Yang Y, Mao LG, Peng Z. 2014. A label-free electrochemical immu-nosensor based on gold nanoparticles for direct detection of atrazine. *Sensors and Actu-ators, B: Chemical.* 191: 408–414. doi:10.1016/j.snb.2013.10.033

22. Mayorga-Martinez CC, Pino F, Kurbanoglu S, Rivas L, Ozkan SA, Merkoçi A. 2014. Iridium oxide nanoparticle induced dual catalytic/inhibition based detection of phe-nol and pesticide compounds. *Journal of Materials Chemistry B.* 2(16): 2233–2239. doi:10.1039/c3tb21765e

23. Raghu P, Madhusudana Reddy T, Reddaiah K, Kumara Swamy BE, Sreedhar M. 2014. Acetylcholinesterase based biosensor for monitoring of malathion and acephate in food samples: A voltammetric study. *Food Chemistry.* 142: 188–196. doi:10.1016/j.foodchem.2013.07.047

24. Liu R, Bloom BP, Waldeck DH, Zhang P, Beratan DN. 2017. Controlling the elec-tron-transfer kinetics of quantum-dot assemblies. *Journal of Physical Chemistry C.* 121(27): 14401–14412. doi:10.1021/acs.jpcc.7b02261

25. Bala R, Kumar M, Bansal K, Sharma RK, Wangoo N. 2016. Ultrasensitive aptamer biosensor for malathion detection based on cationic polymer and gold nanoparticles. *Biosensors and Bioelectronics.* 85: 445–449. doi:10.1016/j.bios.2016.05.042

26. Arjmand M, Saghafifar H, Alijanianzadeh M, Soltanolkotabi M. 2017. A sensitive tapered-fiber optic biosensor for the label-free detection of organophosphate pesticides. *Sensors and Actuators, B: Chemical.* 249: 523–532. doi:10.1016/j.snb.2017.04.121

27. Hou J, Dong J, Zhu H, Teng X, Ai S, Mang M. 2015. A Simple and sensitive fluorescent sensor for methyl parathion based on L-tyrosine methyl ester functionalized carbon dots. *Biosensors and Bioelectronics.* 68: 20–26. doi:10.1016/j.bios.2014.12.037

28. Lan W, Chen G, Cui F, Tan F, Liu R, Yushupujiang M. 2012. Development of a novel optical biosensor for detection of organophoshorus pesticides based on methyl parathion hydrolase immobilized by metal-chelate affinity. *Sensors (Switzerland).* 12(7): 8477–8490. doi:10.3390/s120708477

29. Senbua W, Mearnchu J, Wichitwechkarn J. 2020. Easy-to-use and reliable absorbance-based MPH-GST biosensor for the detection of methyl parathion pesticide. *Biotechnology Reports.* 27: e00495. doi:10.1016/j.btre.2020.e00495

30. Zheng Z, Zhou Y, Li X, Liu S, Tang Z. 2011. Highly-sensitive organophosphorous pesticide biosensors based on nanostructured films of acetylcholinesterase and CdTe quantum dots. *Biosensors and Bioelectronics.* 26(6): 3081–3085. doi:10.1016/j.bios.2010.12.021

31. Chocarro-Ruiz B, Herranz S, Gavela AF, Sanchís J, Farré M, Pilar Marco M, Lechuga LM. 2018. Interferometric nanoimmunosensor for label-free and real-time monitoring of irgarol 1051 in seawater. *Biosensors and Bioelectronics.* 117(March): 47–52. doi:10.1016/j.bios.2018.05.044

32. Kant R. 2020. Surface plasmon resonance based fiber—optic nanosensor for the pesticide fenitrothion utilizing Ta2O5 nanostructures sequestered onto a reduced graphene oxide matrix. *Microchimica Acta.* 187(8): 1–11. doi:10.1007/s00604-019-4002-8

33. Li Q, Dou X, Zhang L, Zhao X, Luo J, Yang M. 2019. Oriented assembly of surface plasmon resonance biosensor through staphylococcal protein A for the chlorpyrifos detection. *Analytical and Bioanalytical Chemistry.* 411(23): 6057–6066. doi:10.1007/s00216-019-01990-0

34. Peng L, Dong S, Wei W, Yuan X, Huang T. 2017. Synthesis of reticulated hollow spheres structure NiCo2S4 and its application in organophosphate pesticides biosensor. *Biosensors and Bioelectronics.* 92(October 2016): 563–569. doi:10.1016/j.bios.2016.10.059

35. Tanimoto De Albuquerque YD, Ferreira LF. 2007. Amperometric biosensing of carbamate and organophosphate pesticides utilizing screen-printed tyrosinase-modified electrodes. *Analytica Chimica Acta.* 596: 210–221. doi:10.1016/j.aca.2007.06.013

36. Sagarkar S, Mukherjee S, Nousiainen A, Björklöf K, Purohit HJ, Jørgensen KS, Kapley A. 2013. Monitoring bioremediation of atrazine in soil microcosms using molecular tools. *Environmental Pollution.* 173: 108–115. doi:10.1016/j.envpol.2012.07.048

37. Thomas KV, Brooks S. 2009. The environmental fate and effects of antifouling paint biocides. *The Journal of Bioadhesion and Biofilm Research.* 26: 73–88. doi:10.1080/08927010903216564

38. Harsha Vardhan K, Kumar S, Panda RC. 2019. A review on heavy metal pollution, toxicity and remedial measures: Current trends and future perspectives. *Journal of Molecular Liquids.* 290: 111197. doi:10.1016/j.molliq.2019.111197

39. Shi L, Wang Y, Chu Z, Yin Y, Jiang D, Luo J, Ding S, Jin W. 2017. A highly sensitive and reusable electrochemical mercury biosensor based on tunable vertical single-walled carbon nanotubes and a target recycling strategy. *Journal of Materials Chemistry B.* 5(5): 1073–1080. Royal Society of Chemistry. doi:10.1039/C6TB02658C

40. Taghdisi SM, Danesh NM, Lavaee P, Ramezani M, Abnous K. 2016. An electrochemical aptasensor based on gold nanoparticles, thionine and hairpin structure of complementary

strand of aptamer for ultrasensitive detection of lead. *Sensors and Actuators, B: Chemical*. 234: 462–469. doi:10.1016/j.snb.2016.05.017

41. Wu D, Zhang Q, Chu X, Wang H, Shen G, Yu R. 2010. Ultrasensitive electrochemical sensor for mercury (II) based on target-induced structure-switching DNA. *Biosensors and Bioelectronics*. 25(5): 1025–1031. doi:10.1016/j.bios.2009.09.017

42. Niu X, Ding Y, Chen C, Zhao H, Lan M. 2011. A novel electrochemical biosensor for Hg2+ determination based on Hg2+-induced DNA hybridization. *Sensors and Actuators, B: Chemical*. 158(1): 383–387. doi:10.1016/j.snb.2011.06.040

43. Park H, Hwang SJ, Kim K. 2012. An electrochemical detection of Hg 2 + ion using graphene oxide as an electrochemically active indicator. *Electrochemistry Communications*. 24(1): 100–103. doi:10.1016/j.elecom.2012.08.027

44. Zhang Y, Zhao H, Wu Z, Xue Y, Zhang X, He Y, Li X, Yuan Z. 2013. A novel graphene-DNA biosensor for selective detection of mercury ions. *Biosensors and Bioelectronics*. 48: 180–187. doi:10.1016/j.bios.2013.04.013

45. Yan G, Wang Y, He X, Wang K, Su J, Chen Z, Qing Z. 2012. A highly sensitive electrochemical assay for silver ion detection based on un-labeled C-rich SsDNA probe and controlled assembly of MWCNTs. *Talanta*. 94: 178–183. doi:10.1016/j.talanta.2012.03.014

46. Zhou Y, Tang L, Zeng G, Zhu J, Dong H, Zhang Y, Xie X, Wang J, Deng Y. 2015. A novel biosensor for silver(i) ion detection based on nanoporous gold and duplex-like DNA scaffolds with anionic intercalator. *RSC Advances*. 5(85): 69738–69744. Royal Society of Chemistry. doi:10.1039/c5ra10686a

47. Wu J, Li L, Shen B, Cheng G, He P, Fang Y. 2010. Polythymine oligonucleotide-modified gold electrode for voltammetric determination of mercury(II) in aqueous solution. *Electroanalysis*. 22(4): 479–482. doi:10.1002/elan.200900441

48. Xuan F, Luo X, Ming Hsing I. 2013. Conformation-dependent exonuclease III activity mediated by metal ions reshuffling on thymine-rich DNA duplexes for an ultrasensitive electrochemical method for Hg2+ detection. *Analytical Chemistry*. 85(9): 4586–4593. doi:10.1021/ac400228q

49. Zhang Y, Xiao S, Li H, Liu H, Pang P, Wang H, Wu Z, Yang W. 2016. A Pb2+-ion electrochemical biosensor based on single-stranded DNAzyme catalytic beacon. *Sensors and Actuators, B: Chemical*. 222: 1083–1089. doi:10.1016/j.snb.2015.08.046

50. Lidong L, Chen Z-B, Zhao H-T, Guo L, Mu X. 2010. An aptamer-based biosensor for the detection of lysozyme with gold nanoparticles amplification. *Sensors & Actuators: B. Chemical*. 149(1): 110–115. doi:10.1016/j.snb.2010.06.015

51. Chen Z, Li L, Mu X, Zhao H, Guo L. 2011. Electrochemical aptasensor for detection of copper based on a reagentless signal-on architecture and amplification by gold nanoparticles. *Talanta*. 85(1): 730–735. doi:10.1016/j.talanta.2011.04.056

52. Chen Y, Zhu Q, Zhou X, Wang R, Yang Z. 2021. Reusable, facile, and rapid aptasensor capable of online determination of trace mercury. *Environment International*. 146: 106181. doi:10.1016/j.envint.2020.106181

53. Long F, Gao C, Shi HC, He M, Zhu AN, Klibanov AM, Gu AZ. 2011. Reusable evanescent wave DNA biosensor for rapid, highly sensitive, and selective detection of mercury ions. *Biosensors and Bioelectronics*. 26(10): 4018–4023. doi:10.1016/j.bios.2011.03.022

54. Sadani K, Nag P, Mukherji S. 2019. LSPR based optical fiber sensor with chitosan capped gold nanoparticles on BSA for trace detection of Hg (II) in water, soil and food samples. *Biosensors and Bioelectronics*. 134(March): 90–96. doi:10.1016/j.bios.2019.03.046

55. Song C, Yang B, Zhu Y, Yang Y, Wang L. 2017. Ultrasensitive sliver nanorods array SERS sensor for mercury ions. *Biosensors and Bioelectronics*. 87: 59–65. doi:10.1016/j.bios.2016.07.097

56. Yin K, Wu Y, Wang S, Chen L. 2016. A sensitive fluorescent biosensor for the detection of copper ion inspired by biological recognition element pyoverdine. *Sensors and Actuators, B: Chemical*. 232: 257–263. doi:10.1016/j.snb.2016.03.128

57. Zhao Y, Chen D, Yang J, Yang B. 2019. Visual and fast detection of trace copper ions using biosensor based on FRET. *Spectrochimica Acta—Part A: Molecular and Biomolecular Spectroscopy*. 217: 101–106. doi:10.1016/j.saa.2019.03.082

58. Ly NH, Oh CH, Joo SW. 2015. A submicromolar Cr(III) sensor with a complex of methionine using gold nanoparticles. *Sensors and Actuators, B: Chemical*. 219: 276–282. doi:10.1016/j.snb.2015.04.130

59. Tian A, Liu Y, Gao J. 2017. Sensitive SERS detection of lead ions via DNAzyme based quadratic signal amplification. *Talanta*. 171(February): 185–189. doi:10.1016/j.talanta.2017.04.049

60. Yildirim N, Long F, He M, Gao C, Shi HC, Gu AZ. 2014. A portable DNAzyme-based optical biosensor for highly sensitive and selective detection of lead (II) in water sample. *Talanta*. 129: 617–622. doi:10.1016/j.talanta.2014.03.062

61. Siddiqui MF, Khan ZA, Jeon H, Park S. 2020. SPE based soil processing and aptasensor integrated detection system for rapid on site screening of arsenic contamination in soil. *Ecotoxicology and Environmental Safety*. 196(March): 110559. doi:10.1016/j.ecoenv.2020.110559

62. Li-Dong Li, Luo L, Mu X, Sun T, Guo L. 2010. A reagentless signal-on architecture for electronic, real-time copper sensors based on self-cleavage of DNAzymes. *Analytical Methods*. 2(6): 627–630. doi:10.1039/c0ay00176g

63. Cui L, Wu J, Ju H. 2015. Electrochemical sensing of heavy metal ions with inorganic, organic and bio-materials. *Biosensors and Bioelectronics*. 63: 276–286. doi:10.1016/j.bios.2014.07.052

64. Rahman Kabir E, Sharfin Rahman M, Rahman I. 2015. A review on endocrine disruptors and their possible impacts on human health. *Environmental Toxicology and Pharmacology*. 40: 241–258. doi:10.1016/j.etap.2015.06.009

65. Rather JA, Pilehvar S, De Wael K. 2013. A biosensor fabricated by incorporation of a redox mediator into a carbon nanotube/nafion composite for tyrosinase immobilization: Detection of matairesinol, an endocrine disruptor. *Analyst*. 138(1): 204–210. doi:10.1039/c2an35959f

66. Zhu Y, Zhou C, Yan X, Yan Y, Wang Q. 2015. Aptamer-functionalized nanoporous gold film for high-performance direct electrochemical detection of Bisphenol A in human serum. *Analytica Chimica Acta*. 883: 81–89. doi:10.1016/j.aca.2015.05.002

67. Wu L, Yan H, Wang J, Liu G, Xie W. 2019. Tyrosinase incorporated with Au-Pt@SiO$_2$ nanospheres for electrochemical detection of bisphenol A. *Journal of the Electrochemical Society*. 166(8): B562–B568. doi:10.1149/2.0141908jes

68. Congur G, Senay H, Turkcan C, Canavar E, Erdem A, Akgol S. 2013. Estrone specific molecularly imprinted polymeric nanospheres: Synthesis, characterization and applications for electrochemical sensor development. *Combinatorial Chemistry & High Throughput Screening*. 16(7): 503–510. doi:10.2174/1386207311316070001

69. Huang KJ, Liu YJ, Zhang JZ, Cao JT, Liu YM. 2015. Aptamer/Au nanoparticles/cobalt sulfide nanosheets biosensor for 17β-estradiol detection using a guanine-rich complementary DNA sequence for signal amplification. *Biosensors and Bioelectronics*. 67: 184–191. doi:10.1016/j.bios.2014.08.010

70. Lin Z, Chen L, Zhang G, Liu Q, Qiu B, Cai Z, Chen G. 2012. Label-free aptamer-based electrochemical impedance biosensor for 17β-estradiol. *Analyst*. 137(4): 819–822. doi:10.1039/c1an15856b

71. Das N, Reardon KF. 2012. Fiber-optic biosensor for the detection of atrazine: Characterization and continuous measurements. *Analytical Letters*. 45(2–3): 251–261. doi:10.1080/00032719.2011.633192

72. Srivastava SK, Shalabney A, Khalaila I, Grüner C, Rauschenbach B, Abdulhalim I. 2014. SERS biosensor using metallic nano-sculptured thin films for the detection of endocrine disrupting compound biomarker vitellogenin. *Small*. 10(17): 3579–3587. doi:10.1002/smll.201303218

73. Srivastava SK, Verma R, Gupta Banshi D, Khalaila I, Abdulhalim I. 2015. SPR based fiber optic sensor for the detection of vitellogenin: An endocrine disruption biomarker in aquatic environments. *Biosensors Journal*. 4(1): 1–5. doi:10.4172/2090-4967.1000114

74. Xue CS, Erika G, Jiří H. 2019. Surface plasmon resonance biosensor for the ultrasensitive detection of bisphenol A. *Analytical and Bioanalytical Chemistry*. 411(22): 5655–5658. doi:10.1007/s00216-019-01996-8

75. Yildirim N, Long F, He M, Shi HC, Gu AZ. 2014. A portable optic fiber aptasensor for sensitive, specific and rapid detection of Bisphenol-A in water samples. *Environmental Sciences: Processes and Impacts*. 16(6): 1379–1386. doi:10.1039/c4em00046c

76. Hegnerová K, Piliarik M, Šteinbachová M, Flegelová Z, Černohorská H, Homola J. 2010. Detection of bisphenol A using a novel surface plasmon resonance biosensor. *Analytical and Bioanalytical Chemistry*. 398(5): 1963–1966. doi:10.1007/s00216-010-4067-z.

77. Alkasir RSJ, Rossner A, Andreescu S. 2015. Portable colorimetric paper-based biosensing device for the assessment of Bisphenol A in indoor dust. *Environmental Science and Technology*. 49(16): 9889–9897. doi:10.1021/acs.est.5b01588

78. Zhang Q, Wang Y, Mateescu A, Sergelen K, Kibrom A, Jonas U, Wei T, Dostalek J. 2013. Biosensor based on hydrogel optical waveguide spectroscopy for the detection of 17β-estradiol. *Talanta*. 104: 149–154. doi:10.1016/j.talanta.2012.11.017

79. Adlercreutz H, Mazur W. 1997. Phyto-oestrogens and western diseases. *Annals of Medicine*. 29(2): 95–120. doi:10.3109/07853899709113696

80. Heinonen S, Nurmi T, Liukkonen K, Poutanen K, Wähälä K, Deyama T, Nishibe S, Adlercreutz H. 2001. In vitro metabolism of plant lignans: New precursors of mammalian lignans enterolactone and enterodiol. *Journal of Agricultural and Food Chemistry*. 49: 3178–3186. doi:10.1021/jf010038a

81. Katsoyiannis A, Samara C. 2004. Persistent organic pollutants (POPs) in the sewage treatment plant of Thessaloniki, Northern Greece: Occurrence and removal. *Water Research*. 38: 2685–2698. doi:10.1016/j.watres.2004.03.027

82. Nomngongo PN, Catherine Ngila J, Msagati TAM, Gumbi BP, Iwuoha EI. 2012. Determination of selected persistent organic pollutants in wastewater from landfill leachates, using an amperometric biosensor. *Physics and Chemistry of the Earth*. 50–52: 252–261. doi:10.1016/j.pce.2012.08.001

83. Wu S, Zhang H, Shi Z, Duan N, Fang CC, Dai S, Wang Z. 2015. Aptamer-based fluorescence biosensor for chloramphenicol determination using upconversion nanoparticles. *Food Control*. 50: 597–604. doi:10.1016/j.foodcont.2014.10.003.

84. Atar N, Eren T, Yola ML, Wang S. 2015. A sensitive molecular imprinted surface plasmon resonance nanosensor for selective determination of trace triclosan in wastewater. *Sensors and Actuators, B: Chemical*. 216: 638–644. doi:10.1016/j.snb.2015.04.076

85. Ying G-G, Kookana RS. 2007. Triclosan in wastewaters and biosolids from Australian wastewater treatment plants. *Environment International*. 33: 199–205. doi:10.1016/j.envint.2006.09.008

10 Lateral Flow Assays

Kamil Żukowski[1], Marcin Drozd[1,2],
Robert Ziółkowski[2], Mariusz Pietrzak*[1,2],
Katarzyna Tokarska[1], Adam Nowiński[3],
and Elżbieta Malinowska*[1,2]

[1]Warsaw University of Technology
Centre for Advanced Materials and Technologies
Warsaw, Poland

[2]Warsaw University of Technology
Chair of Medical Biotechnology, Faculty of Chemistry
Warsaw, Poland

[3]Screenmed
Piaseczno, Poland

CONTENTS

10.1 INTRODUCTION

Immunochromatographic tests in the form of lateral flow assays (LFAs) are disposable bioanalytical tools that find application for analysis of various types of samples. Since the 1980s the market of LFAs has been growing continuously and it is envisioned that it will reach a value of more than $13 billion by 2027 (1). The initial phase of their development was driven by a need for cheap method of pregnancy (conception) confirmation, while the recent expansion is mostly due to a spread of COVID-19 (coronavirus disease 2019) and the need for cheap and rapid detection of SARS-CoV-2 (severe

DOI: 10.1201/9781003189435-14

acute respiratory syndrome coronavirus 2). However, due to a number of advantages including relative ease of production, versatility as well as user-friendliness, a list of analytes, which could be detected or determined using LFAs, is rapidly broadening. They enable qualitative and quantitative detection of proteins, hormones, drugs, toxins, nucleic acids, viruses and bacteria (2). Similarly to biosensors, they usually contain a receptor layer formed of antibodies or nucleic acids, however due to their construction and working mechanism they are always single-use devices.

LFA tests are used in many areas, but one of the most common is POCT (point of care testing) or, in a broader context, decentralized diagnostics. All of the previously mentioned features make LFAs requested in developing countries that often lack resources (expensive conventional clinical laboratories) and trained operators (3). It is noteworthy that LFAs are widely used not only in hospitals, laboratories or medical clinics but also many of them are commercially available for home use. Therefore, current trends in evolution of immunochromatographic assays cover both the advancements in analytical and biochemical principles as well as practical aspects, crucial from a point of view of an application by the end user. Both of these paths are discussed within this chapter. The goal behind the development of LFA technology is to make it a convenient and versatile tool for rapid diagnostics.

10.2 MATERIALS AND TECHNOLOGIES APPLIED FOR DEVELOPMENT OF LATERAL FLOW ASSAYS (LFAs)

Lateral flow assays belong to a group of most popular analytical tests, which can be purchased in pharmacies and drugstores worldwide, as they are cheap, portable, robust and most importantly user-friendly. While there is a number of variations of this technology, they all operate according to the same principle. LFAs are manufactured in the form of strips, usually 4–6 mm wide and 6–7 cm long. A typical lateral flow test strip (Figure 10.1) consists of overlapping membranes that are mounted on a backing card for better stability and handling (4–6).

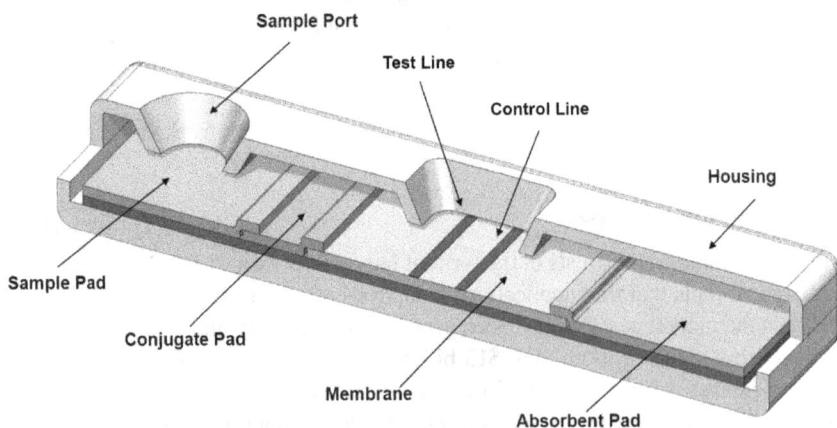

FIGURE 10.1   Scheme of a typical LFA strip.

A sample is applied at one end of the strip on the adsorbent sample pad. The major role of the sample pad is to allow a continuous supply and capillary flow of a liquid sample. If the concentration of an analyte is too high, the sample pad allows for both filtration and dilution of the sample (7). The sample pad is usually impregnated with buffer salts, proteins, surfactants and other agents to control the flow rate of the sample, make the conditions suitable for desired interactions/reactions and endow the compatibility with a detection system (8). The most commonly used materials for sample pads are cellulose, glass fiber, rayon or modified filter matrices (2). The sample pad should exhibit consistent absorbency, thickness and density so that uniform wicking rates ensure assay reproducibility.

When the sample application pad is saturated, the sample flows into a conjugate release pad where a detection reagent is then renatured. Subsequently, the detection reagent leaves the conjugate release pad and migrates with the sample to further areas of the test strip. The conjugate release pad plays a key role in the performance of the lateral flow assay. The material used for fabrication of conjugate pad should release a labeled receptor (reporter) immediately upon contact with the moving sample. The label-receptor conjugate should remain stable throughout the entire life span of the lateral flow strip. Any changes to the dosing, drying or release of the conjugate can significantly change the final result of the test. A substandard desorption and rehydration of the labeled conjugate, which implies the loss of its capability of target recognition, can adversely affect a sensitivity of the assay. It is important that the conjugate release pad has a consistent bed volume, as it ensures that the amount of detection reagent in each test strip remains constant. Materials commonly used to fabricate conjugate pads include glass fiber-, cellulose- and surface-treated (hydrophilic) polyester- or polypropylene filters.

In most cases, the analyte present in the sample after binding with a conjugated receptor migrates along the strip into the reaction membrane. This membrane is the most important element of each LFA. It must combine the appropriate bioreceptor

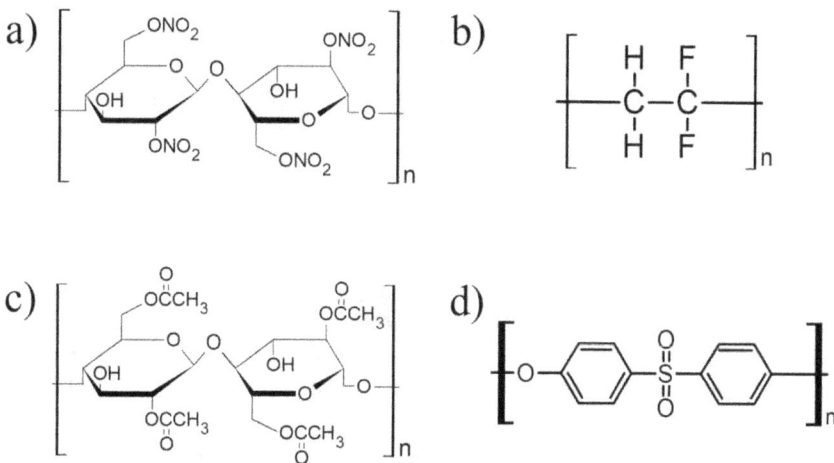

FIGURE 10.2 Molecular structures of polymers typically used for immobilization of receptors in LFAs: a) nitrocellulose, b) polyvinylidene fluoride, c) cellulose acetate, d) polyethersulfone.

binding capacity in the defined zones and the ability to transport liquid samples by means of capillary forces. The most commonly used material for its fabrication is nitrocellulose (NC) (9). Beyond NC membranes, also other materials such as cellulose acetate (10), hydrophilized polyvinylidene fluoride (PVDF) (11), charge modified nylon (12), and polyethersulfone (PES) can be also employed. Structures of typical membrane materials are shown in Figure 10.2.

One of the key parameters in selecting the membrane for a lateral flow assay is a capillary flow time. It is defined as the time required for the sample to pass into and completely fill the membrane. The capillary flow time depends on the physical properties of the membrane, such as: pore size, pore size distribution and porosity. Capillary flow time can affect a sensitivity and a specificity of the lateral flow assay as well as a consistency of a test line. In addition to physical characteristics of the membrane, the properties of a liquid, which flows in the membrane, also affect the capillary flow rate. Contact angle, viscosity and surface tension play important roles in regulating flow time, because the interaction between liquid and porous membrane is determined by the capillary forces.

The application of nitrocellulose brings a number of technological advantages compared to other types of reaction pads used in LFAs. Contrarily to hydrophobic PVDF membranes, their activation with organic solvents is not necessary, which facilitates automatic dispensing of biological components using printing techniques. Different pore sizes of NC pads (from 0.5 to 12 µM) allow finding a compromise between receptor binding capacity and interaction time (13–14). There are also pillar-based LFA capillary devices used to detect deoxyribonucleic acid (DNA) hybridization (15). The use of such microstructures provides more precise capillary flow control.

The last part of typical LFA is an absorbent pad, which is attached at the end of the strip. The role of this section is to absorb the liquid migrating from the reaction membrane. It allows the labeled analyte to reach the test lines and prevents backflow, which may cause false positive results. The absorbent pad supports the efficient and oriented transport of liquid sample along the test strip. Optimizing the overall volume absorbed by the test strip is best managed by changing the dimensions (usually the length) of the absorbent pad. Typically, absorbent pads are made from cellulose filters. The material should be selected on the basis of thickness, compressibility and uniformity of bed volume.

Most often, LFAs are placed in a disposable plastic cassettes (7). The cassette facilitates the use of the lateral flow assay. The cassette is profiled in such a way that facilitates sample application and supply and at the same time protects the other elements (conjugate release pad, membrane or absorbent pad) against undesirable, direct contact with the sample. The plastic housing also provides protection against mechanical damage and dirt. Additionally, on the plastic housing there are markers, which show the position of the test and control lines, which facilitates the readout of the result. When the result is evaluated using instrumental methods, the plastic housing is typically designed as an element of the cassette-reader interface, which allows for aligning a position of the cassette in the reading device holder.

There are two general approaches to the LFA manufacturing, namely batch processing and *in-line* or *roll-to-roll* assembly. The batch processing approach allows for the use of low-cost equipment and it is suitable for relatively low production volumes, however its disadvantages are that it involves a high degree of manual labor and can be prone to product variability. During *in-line* or *roll-to-roll* manufacturing, lateral flow assay strips are machine-processed in continuous rolls. This strategy results in higher throughput and considerably lower product variability.

The type of the fabrication approach affects the dimensions of the input materials. Batch processing is usually based on previously cut strips or sheets. Rolls can also be used but usually they have to be cut into strips at some point during manufacturing. Continuous processes usually require rolls of each material. The roll length, which typically ranges from 50 to 150 meters, depends on many factors, including lot sizes of other components, production rates, final product lot size and manufacturing capacity.

10.3   LATERAL FLOW IMMUNOASSAYS AND IMMOBILIZATION OF RECEPTORS

Interaction between antibody and antigen is one of the most widely exploited mechanisms of biomolecular recognition in the construction of biosensors and diagnostic assays. Such analytical devices are classified as heterophasic because they are based on the binding and separation of the analyte in the form of immunocomplex on a solid substrate (16). Despite the continuous development of new classes of bioreceptors, antibodies are still considered as the unmatched gold standard of the bioreceptor due to their versatility, high affinity and specificity of interaction with the antigen (17). Increased availability of industrial methodologies for the production of high-quality monoclonal antibodies for in vitro diagnostics (IVD) in recent years turned out to be a milestone on the way to standardization of the procedures of lateral-flow immunoassays fabrication (18). Also the recent advances in the field of methods of isolation and recombination of proteins strongly stimulated their implementation in LFAs.

10.3.1   Biorecognition Elements Immobilization for Applications in LFAs

Tests, in case of which immunological reaction takes place on a solid substrate, require immobilization of the selected biorecognition elements. In each immunochromatographic test key biochemical reactions are based on the antibody-antigen interaction. These reactions cover capturing of the analyte by functionalized label and the binding of labeled immunocomplex within the test and control zones of the strip. To meet the requirements of lateral flow immunoassays (LFIAs) as cheap and easy to use and handle POCT tools for rapid self-diagnostics, the assay design must provide the ability to store all biological components in a dry form. What is more, it is highly demanded to enable their easy hydration and thus renaturation by the sample. It is necessary for the transportation along the test strip by capillary flow (labeled receptors) or to regain the ability to capture the labeled analytes (immobilized receptors).

NC membranes offer high antibody/antigen binding capacity. Due to the introduction of nitrate residues, one-step, passive adsorption of antibodies directly from the solution is enhanced. The immobilization is driven by electrostatic and hydrophobic interactions as well as the formation of hydrogen bonds between peptide backbone or terminal groups and the porous substrate. Antibodies, as relatively large glycoproteins, have the ability to efficiently adsorb on polymer substrates without the need to employ labor-intensive methods of covalent immobilization. Also protein antigens can be directly adsorbed on polymeric supports. In turn, small molecule antigens such as peptides or low molecular weight organic compounds require prior conjugation with carrier proteins such as bovine serum albumin or ovalbumin (19). To minimize non-specific background adsorption, NC binding sites that are not involved in binding of antibodies should be additionally blocked with a cocktail of proteins or hydrophilic polymers such as poly(ethylene glycol) or poly(vinyl alcohol) (20). Despite significant differences in their chemical structure, also other membrane materials such as PVDF, cellulose acetate and PES show a high ability to bind proteins through non-covalent interactions. Thus, they found applications also in other immuno-techniques such as Western Blot.

To ensure a sufficiently long shelf life of all protein-based components within the test strip in a dry form, blocking solutions containing trehalose or sucrose are additionally applied to fix the proteins on the membrane. Working principle of such agents covers protection of antibodies against dehydration-induced denaturation. It is quite similar to solutions used for the preparation of dry-stored immunolayers in antibody-coated ELISA (enzyme-linked immunosorbent assay) plates. The additional advantage is the easy removal of readily soluble protectants during sample migration through the membrane.

10.3.2 FORMATS OF LATERAL FLOW IMMUNOASSAYS

The vast majority of lateral-flow immunoassays are based on the formats known from ELISA-type immunoassays, i.e. sandwich and competitive assays, as shown in Fig. 10.3. The role of bioreceptors in the test line is to generate a signal related to the presence of the analyte (appearance of a readable signal due to binding of the analyte labeled with reporter conjugate). Typical receptors responsible for the capturing of immunocomplexes within the test zone are analyte-specific IgG antibodies (sandwich format) or antibodies, antigens or their conjugates with the carrier protein (competitive format) (21). In turn, the receptor immobilized in the control zone (typically secondary antibodies) exhibits affinity to the antibodies/antigens of the conjugate. The appearance of control line indicates that the sample has successfully migrated along the reaction membrane, which confirms the validity of the assay.

Sandwich-type format is usually preferred for the detection/determination of large, multiepitope protein antigens. The sandwich format is more resistant to non-specific or cross interactions due to the double immuno-recognition mechanism, which takes place both at the step of labeling and immunocomplex binding to the membrane (see Fig. 10.3a) (22). The competitive format is beneficial in the case of small-molecule antigens detection—especially, when it is difficult to find two different epitopes for simultaneous binding with antibodies. The mechanism relies on the competition between unlabeled analyte and label-analyte complex for a limited number of binding sites of the immobilized receptor (see Fig. 10.3b). Competition for the binding site of the antibody-coated label between free analyte in the sample and analyte immobilized within test line is also possible (see Fig. 10.3c). It is worth emphasizing that the sandwich format offers a positive slope of dose-response curve (the more analyte, the more prominent signal), while the competitive format is characterized by gradual disappearance of the test line signal as the analyte concentration increases (16).

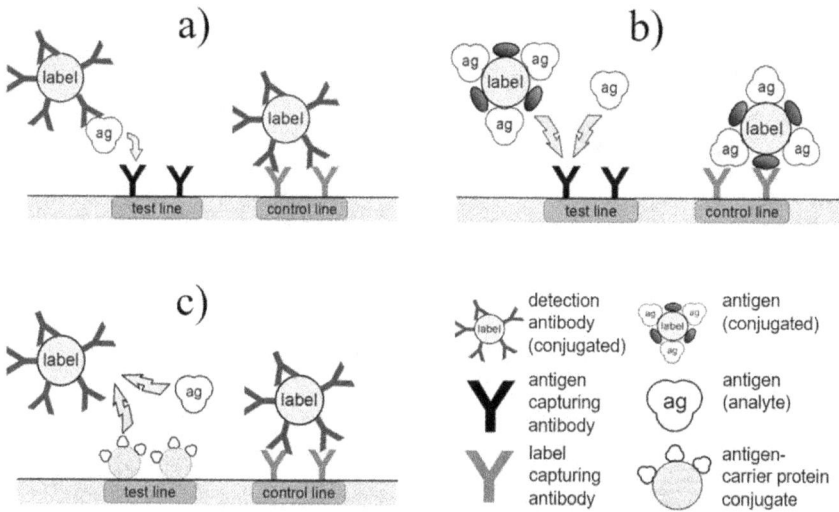

FIGURE 10.3 Schematic representation of the formation of immunocomplexes within test and control lines for various formats of lateral flow immunoassays: a) sandwich format, b) competitive format with immobilized antibody, c) competitive format with immobilized antigen-carrier conjugate.

Slightly modified formats of immunochromatographic tests can be also employed for the detection of antibodies. Several classes of antibodies (mainly IgG, IgM and IgA) are important group of analytes from the point of view of medical diagnostics e.g. as markers of viral or autoimmune diseases (23). Lateral flow-type serological tests based on specific interaction between immunoglobulins in the tested sample and membrane (24–25) or label (26–27) are getting more and more popular. The second reaction, leading to the formation of a labeled immunocomplex bound to the surface, can be accomplished by the use of secondary antibodies against the immunoglobulins of a chosen class. Thanks to this, the discrimination of different antibody classes within a single test can be carried out. A less common approach is the dual specificity serological assay, in which the bivalence of antibodies is exploited. In this assay format, both the labeling and membrane-binding step employ interaction of the antibody with the antigen. This format, thanks to double specific recognition, offers very good specificity. However, due to the exclusion of secondary antibodies, it is impossible to distinguish individual classes of antibodies. Detailed LFIAs formats for antibody detection are illustrated in Figure 10.4.

Current trends in the development of immunochromatographic tests in the field of bioreceptors and their implementation are focused mainly on the continuous increase of the range of their molecular targets (28). To meet the challenges of modern analytics, new types of monoclonal antibodies and recombinant antigens are constantly being developed. At the same time, solutions aimed at improvement of the immobilization efficiency are being introduced (29–30). Chemical immobilization of proteins and oriented immobilization of antibodies hold great promise for the further improvement of analytical parameters of immunochromatographic assays (31).

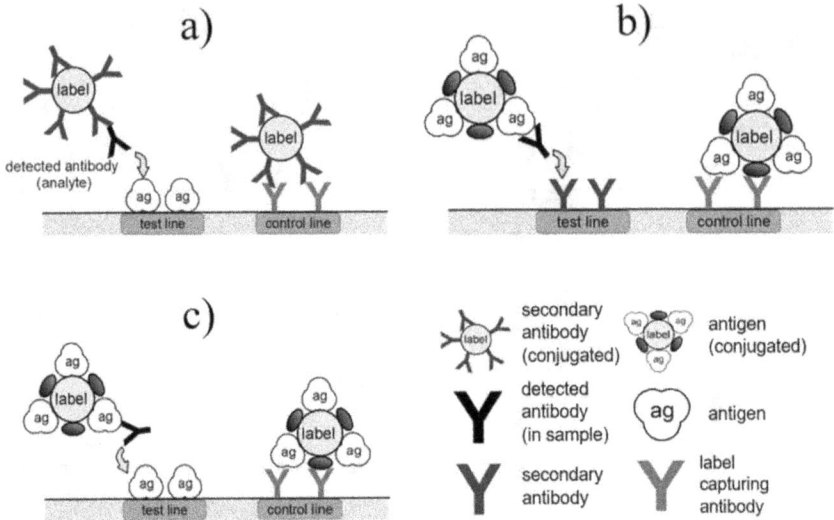

FIGURE 10.4 Various formats of serological immunoassays for the detection of specific antibodies: a) antigen bound with the membrane, secondary antibody conjugated with the label, b) secondary antibody bound with the membrane, antigen conjugated with the label, c) dual antibody-antigen recognition.

10.4 NUCLEIC ACID-BASED LATERAL FLOW ASSAYS AND IMMOBILIZATION OF RECEPTORS

To date numerous nucleic acid-based lateral flow assays (NALFA) have been developed for rapid and simple detection of various analytes, including DNA sequences (32–35). They are developed in two main forms, antibody-dependent (nucleic acid lateral flow immunoassay, NALFIA) and antibody-independent (nucleic acid lateral flow test, NALFT). In the case of the first mentioned, developed in 2000 and dedicated to detection of *Cryptosporidium parvum* (36), nucleic acid-antibody interaction and doubled stranded amplicon (ds-DNA) were used. In order to detect ds-DNA, it was previously amplified (in a reverse transcription-polymerase chain reaction-based assay) with two differently labeled primers i.e. with biotin and fluorescein isothiocyanate (FITC). The immobilized anti-fluorescein antibody on NC membrane bound to detect ds-DNA, whereas gold nanoparticles, modified with avidin, were used for the visualization (Fig. 10.4a). Soon after the first report on NALFIA, similar tests were developed, e.g. against *Staphylococcus aureus*, in case of which the immobilization was conducted with streptavidin/biotin interaction and visualization was carried out with anti-fluorescein antibodies labeled with gold nanoparticles (37). Since then, numerous reports describing similar detection mechanism where the immobilization and/or visualization was based on digoxigenin, carboxyfluorescein (FAM), FITC or biotin labels (38), were published. Le et al. presented a dual recognition type of LFA. The proposed method enabled particular strains detection due to the binding a nucleic acid aptamer with an antibody.

The constructed strips could be easily adapted to a wide range of different analytes, constituting a versatile diagnostic platform (39).

The antibody-independent format is based on two binding partners, biotinylated reporter probe or amplicon, and e.g. streptavidin-modified gold nanoparticles for a visualization. They show high affinity and irreversible linkage. The oligonucleotide capture probes can be immobilized on NC membrane via passive adsorption with (Fig. 10.4b) or without (Fig. 10.4c) conjugated BSA (bovine serum albumin) as a carrier. Because of much more complex hybridization kinetics in lateral flow test than the antibody/antigen or biotin/avidin interaction, the first antibody-independent format of LFA dedicated to DNA detection (NALFT) was developed quite recently, in 2003 (see Fig. 10.5) (40). A hybridization assay encompassed two sequence specific oligonucleotides (Fig. 10.6).

For the visualization the upconverting phosphor technology (UPT) was used, which was previously described for nucleic acids microarrays (41). Also in 2003 the NALFT with gold nanoparticles (NPs)-modified probes was developed (42). It should be underlined, that various formats of nucleic acids lateral flow assays were proposed to lower the limit of detection (LOD) (vast majority of them concerned the visualization) (43–44).

In real life applications the NALFA test are used for detection of DNA fragments amplified during PCR (45–47). It should be stressed, that in the case of classic, three temperature-based PCR reactions, the nucleic acids amplification step still is limited only to well-equipped laboratories with skilled staff. However, the reports dedicated to the use of LFA tests as detection element in the case isothermal nucleic acids

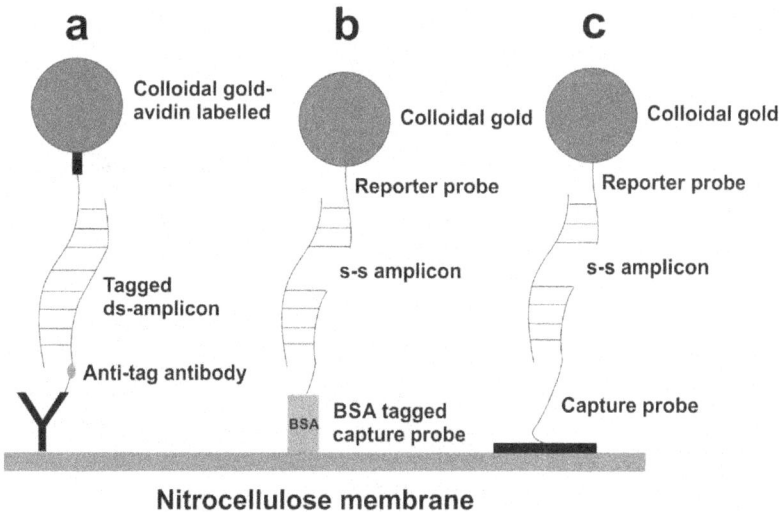

Nitrocellulose membrane

FIGURE 10.5   Scheme of basic detection principles in nucleic acid lateral flow test strips; a) NALFIA; b) NALFT using bovine serum albumin-oligonucleotide conjugate as capture probe; c) NALFT with oligonucleotide capture probe immobilized on the nitrocellulose membrane via passive adsorption.

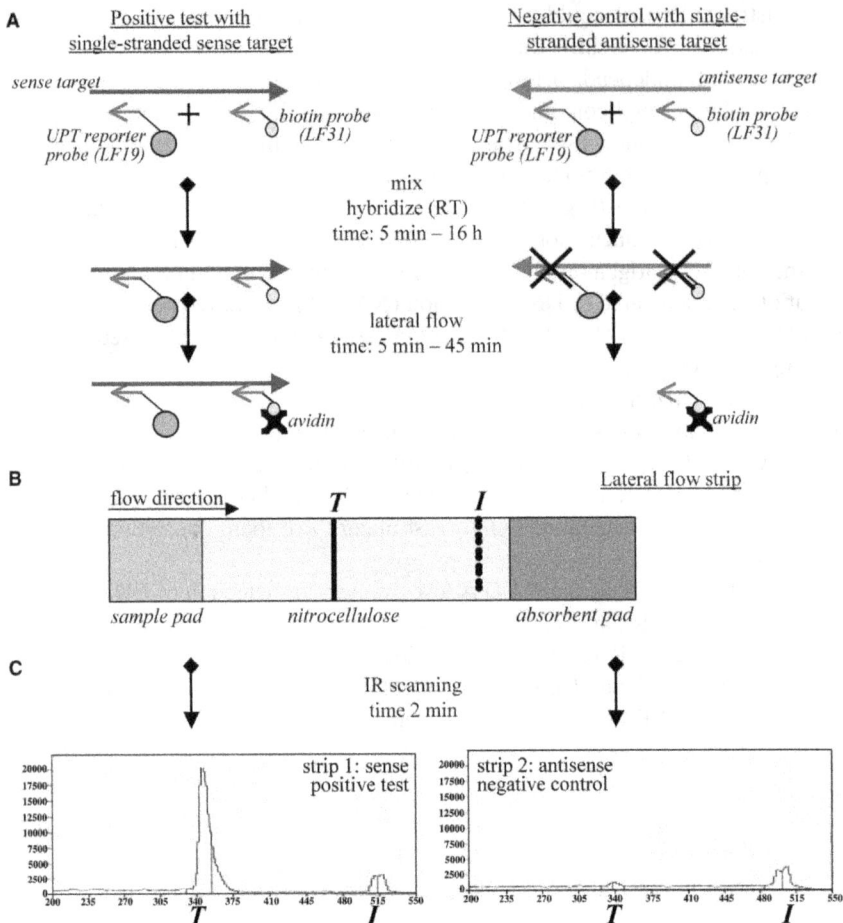

FIGURE 10.6 Schematic presentation of the sandwich-hybridization-based LF assay: A) left: positive tests; right: negative control; B) the LF strip consists of a sample pad, a nitrocellulose membrane, and an adsorbent pad with (T) target-capture line (avidin) and (I) a control line for scanning process validation; C) the result of a typical scan.

amplification methods (48–56), show the technological approach which truly could be appropriate for the point-of-need, thus fulfilling the ASSURED criteria (affordable, sensitive, specific, user-friendly, rapid and robust, equipment-free and deliverable to end-users). That is mainly because theoretically for the nucleic acid amplification or its detection no energy source is needed and the whole assay could be performed by unskilled persons. Importantly, this approach still offers high selectivity, low cost and satisfactory time of the analysis. Among different methods of isothermal DNA amplification the nucleic acid sequence based amplification (NASBA), transcription mediated amplification (TMA), self-sustained sequence replication (3SR), helicase dependent amplification (HAD), loop mediated isothermal amplification (LAMP) or recombinase polymerase amplification (RPA) should be mentioned. The latter seems to offer the most user friendly approach (57). This is because it requires conventionally designed primers and leads to exponential amplification with no need for pretreatment of DNA sample. Moreover, the reactions are sensitive, specific and rapid and the whole system works at constant low temperature (optimum in the range between 37 and 40° C) (57). It is a reason for multiple examples of LFAs dedicated to determination/detection of RPA products, in a form of both NALFIA and NALFT (35, 57–60) (see Fig. 10.7).

The most obvious analyte for single stranded DNA reporter probe is another nucleic acid strand. However, together with a development of aptamers (artificial short and single stranded oligonucleotides of RNA (ribonucleic acid) or DNA specific for a given analyte) and the methods of their selection the LFAs based on aptamers were constructed (61). The first test was dedicated to the protein detection (thrombin) and used aptamers, which could be easily synthesized and efficiently labeled both for thrombin capture (immobilized by biotin/streptavidin interactions) and for visualization (conjugated with AuNPs). To date, numerous examples of aptamer-based LFAs have been presented (62–64). Similarly, as for typical nucleic acids LFAs they are in two main formats, antibody-independent (Figs. 10.8a and 10.8b) and antibody-dependent (Fig. 10.8c) (65). However, what is worth mentioning, in contrary to lateral flow immunoassays or lateral flow assays dedicated to nucleic acids detection, the aptamer based strip tests are still in conceptual phase with no commercial examples.

10.5   GENERATION OF A SIGNAL AND ITS READING METHODS

A signal, in the case of lateral flow assays, is typically generated in a form of a colored line (rarely spot), which appears after carrying out all the steps suggested by the manufacturer. Next to the test line, usually the second line (control line) is formed, which informs about the correctness of assay execution. In both cases receptors carrying labels (reporters) are responsible for the line (spot) formation. Due to this phenomenon the final result of LFAs can be read by naked eye, however in such cases the assay allows only for qualitative analysis (positive/negative) or uncommonly for semiquantitative analysis (comparison with a scale). If the assay is to generate fully quantitative information a reading device (reader) has to be used. The number of colored lines, which are formed can be larger if more target species are to be analyzed (66–67). Such approach (multiplexed format) is usually applied when multiple analytes can be simultaneously detected/determined under the same conditions from the one sample (e.g. for clinical diagnostics).

In most of the LFAs the signal is generated directly, usually due to the presence of colorful labels (exhibiting strong absorption of radiation in a visible range), which are bound to antibodies (or other receptors). Most dominantly particles (nanoparticles above all) are used for this aim. Among them, AuNPs (colloidal gold) are the most popular as they can be easily synthesized in many shapes and sizes, and are chemically stable. What is more, spherical AuNPs of diameter between 40 and 100 nm are the most commonly used. Due to the presence of localized surface plasmon resonance AuNPs are characterized with strong absorption of light, which is manifested by ruby red or pink color of such NPs (and eventually formed lines or spots). To maximize the sensitivity of the analysis using LFA, labels should be able to produce the strongest signal possible (then even at a low concentration of an analyte its

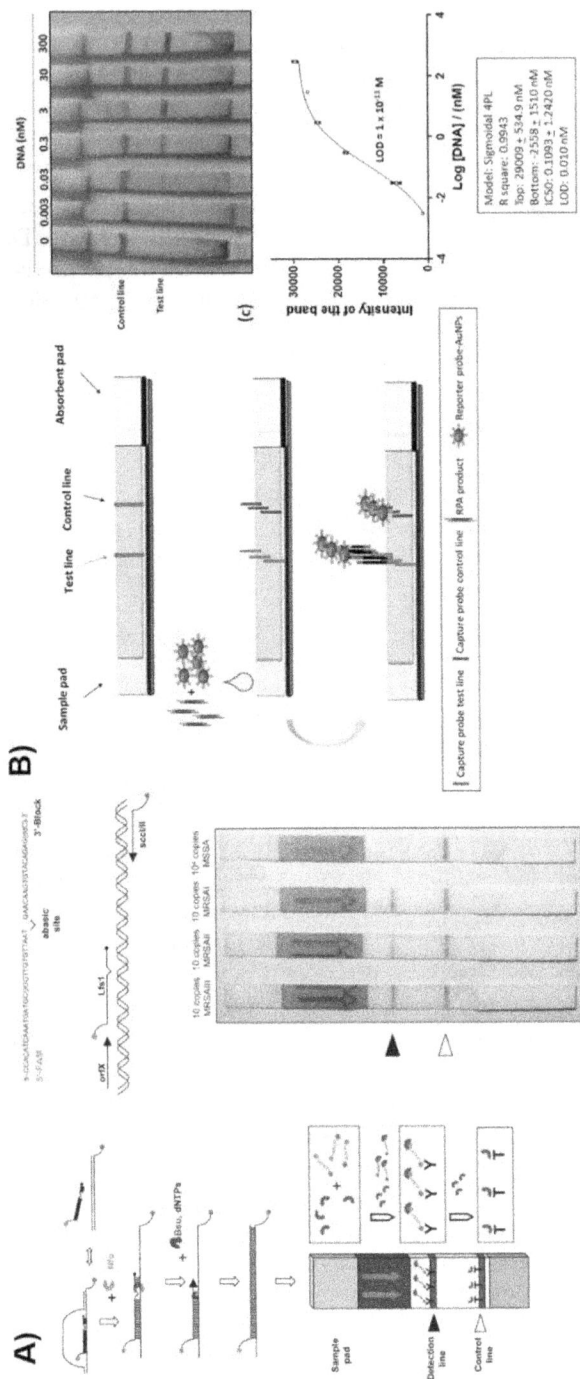

FIGURE 10.7 Examples of nucleic acids lateral flow tests based on A) immunoassay (57) and B) antibody independent approach (35) dedicated to recombinase polymerase amplification (RPA) product detection.

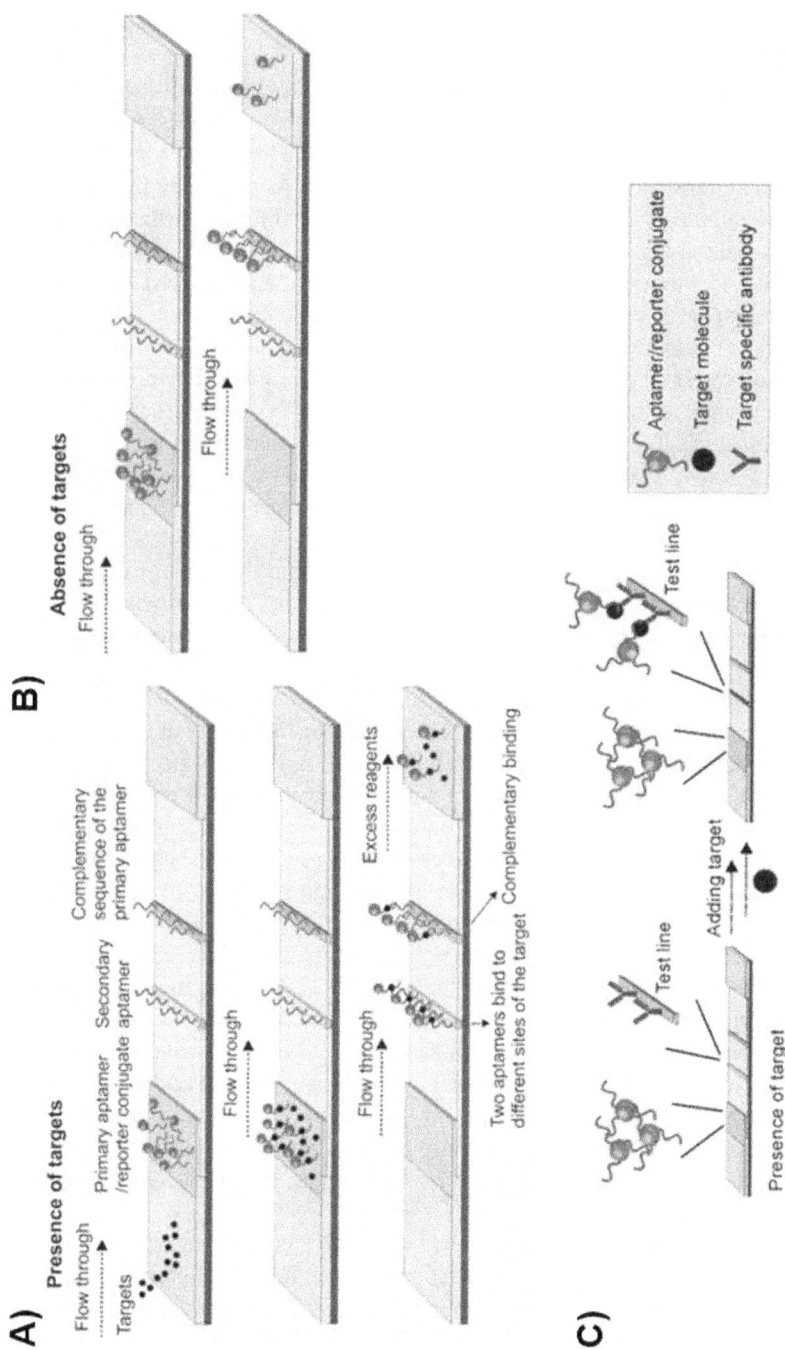

FIGURE 10.8 A) Scheme of antibody-independent, and B) antibody-dependent C) aptamer lateral flow tests. Reprinted from ref. (64) with permission.

presence can be visualized). It has to be underlined, that larger AuNPs are capable of generating stronger signals, but at the same time kinetic issues must be taken into account as large nanoparticles will not easily flow through the membrane and interaction of labeled receptor with an antigen or other receptor (depending on the format) will be slow and not efficient. The effect of spherical AuNPs size on lower detection limit of an assay has been tested and it has been found out that at least 6.7 10^7 particles·mm^{-2} must be bound in a case of 16 nm AuNPs while for the NPs of 115 nm 1.5 10^5 particles·mm^{-2} is enough to generate a visible signal. As suggested by Khlebtsov and co-workers, the intensity of the signal is proportional to the surface density of nanoparticles and their volume. Moreover, extinction has be pointed out as the main physical mechanism behind spot brightness rather than scattering (68). Other types of nanoparticles such as polystyrene or latex beads colored with organic dyes, or such organic or silica NPs decorated/covered with ultrasmall NPs or thin layer of a noble metal (preferably gold) are also used as reporters (69). They are advertised as exhibiting some advantages over solid metal-based NPs, as e.g. silica-based NPs covered with thin layer of gold (this way forming Au nanoshells) according to a manufacturer, provide a 3–20 fold increase in sensitivity of LFAs when compared to 40 nm gold particles and it is among others due to the fact that such nanoparticles are less dense than a solid gold particles. As a result they are able to flow unimpeded through the nitrocellulose membrane (70). One can also find examples of LFAs with carbon-based NPs, iron oxide (magnetic NPs) or quantum dots as labels. Other fluorescent labels also have been applied for this aim, however in contrast to colored ones the application of an additional device is not only optional but it is required to visualize the result of the assay in such a case.

As the success of LFA-based analysis is influenced by the quality of both receptors and labels, the efficient conjugation which does not significantly deteriorates their properties must be assured. Moreover, one must remember that if the homogenous line (spot) is to be generated (it assures both accuracy and precision of a quantitative analysis), NPs used as labels must be as monodispersive as possible. Regardless of the surface chemistry of label surface, the method of its conjugation with the immunoreceptor should meet a number of requirements, both in terms of conjugate durability and the ability to preserve the biological activity of the receptor. The simplest method—especially in the case of metal- and metal oxide nanoparticles—is a passive adsorption on the surface by means of hydrophobic interactions. Gold-based labels have the advantage, that there are many protocols available which lead to fully functional conjugates. To ensure spontaneous organization of antibodies on the surface of metal nanoparticles characterized by intrinsic negative charge (e.g. AuNPs), careful selection of appropriate pH (slightly higher than the pI of the adsorbed protein), low ionic strength of the immobilization medium as well as the appropriate purity of the protein is demanded.

However, a number of nanoparticle-based labels does not show intrinsic affinity to bioreceptors. In their case, the immobilization involves functional groups present on the surface, which are active in the coupling reactions. It can be carried out, among others, by amino coupling via carbodiimide chemistry or coupling via reduced sulfhydryl groups of proteins or DNA receptors with terminal amino- or thiol link (2, 71). An attractive alternative is the use of indirect methods, such as binding of antibodies

through surface-coated A/G protein or capturing of DNA-biotin conjugates by streptavidin. In the first approach, bacterial proteins are harnessed to bind Fc antibody fragments. Due to the oriented immobilization, Fab fragments are better exposed and they
can more easily interact with antigens. On the other hand, the interaction of streptavidin with biotinylated receptors combines a strong binding with a really small negative
effect of immobilization on the bioreceptor affinity. Selected methods of biorecognition elements conjugation with popular LFA labels are listed in Table 10.1.

TABLE 10.1
**The Overview of Antibody/Antigen Conjugation Methods with Common LFIA
Labels**

Receptor type	Label Type	Conjugation Chemistry	Analyte	Comments	Ref.
anti-HBsAg antibody	fluorescent quantum dots	hydrazide-aldehyde coupling	Hepatitis B surface antigen	oriented immobilization	(30)
anti-HBsAg antibody	fluorescent quantum dots	carbodiimide coupling (EDC)	Hepatitis B surface antigen	random immobilization	(30)
anti-hIgG antibody	lanthanide nanoparticles	carbodiimide coupling (EDC + sulfo-NHS)	anti-SARS-CoV-2 IgG	random immobilization	(24)
anti-hIgM antibody	Au nanoparticles	passive adsorption	anti-SARS-CoV-2 (NP) IgM	random immobilization	(25)
anti-CRP monoclonal antibody	Au nanoparticles	passive adsorption	C-reactive protein	random immobilization	(22)
anti-NSE antibody	silica-coated Au nanostars	3-(triethoxysilyl) propylsuccinic anhydride	neuron-specific enolase	amine coupling	(29)
anti-hAFP antibody	polystyrene latex particles@BSA	thiol coupling (via reduced disulfide bonds)	human α-fetoprotein	N-(6-maleimido-caproyloxy) succinimide linker	(71)
protein-hapten conjugate	Au nanoparticles	thiol coupling (via reduced disulfide bonds)	atrazine	direct chemisorption	(19)
nucleoprotein antigen	selenium nanoparticles	passive adsorption	*anti*-SARS CoV-2 IgG/IgM	random immobilization	(26)
nucleoprotein antigen	magnetic nanoparticles	carbodiimide coupling (EDC/NHS)	*anti*-SARS CoV-2 IgG/IgM	random immobilization	(27)
nucleic acid	up-converting phosphor particles	covalent coupling	specific RNA sequence	antibody-independent (NALFT)	(41)
RNA aptamer & *anti*-hemag-glutinin H3 antibody	Au nanoparticles	passive adsorption	influenza virus (specific sequence)	random immobilization dual recognition	(39)

It has to pointed out that signal in LFA can be generated indirectly by the catalytic reaction, which leads to occurrence of products which are characterized with interesting properties such as capability to absorb the visible light or emit radiation through fluorescence. Such an approach (as it is typical in the case of various immunoassays) usually leads to assays which are characterized with low detection limits and can be applied for fully quantitative analysis of antigens (analytes). One of the most recent achievements is the application of nanoparticles in a role of labels, which exhibit enzyme-like catalytic activity (nanozymes) or just exploiting of such properties of already applied NPs (e.g. AuNPs) (72). Such systems are called catalytic lateral flow immunoassays (cLFIA™) (73). Much lower detection limits can be achieved using such assays due to the fact of signal amplification (similar to ELISA as one labeling molecule/particle is capable of generation of many molecules which give rise to a signal). It has been reported that for such assays the LOD can be lower even 1,000 times in comparison to assays based on AuNPs with a direct generation of signal. It must be underlined, that nanozymes very often exhibit peroxidase-like activity and due to the popularity of peroxidases for various immunoassays, there is a wide range of substrates, which are compatible with nanozymes (74–75) and give products of different colors. Especially these substrates that are transformed into precipitated products are extremely useful for this type of LFAs (76).

When one wishes to obtain a fully quantitative information from LFA a suitable reader has to be applied. Nowadays there are several companies which propose various types of readers for LFAs. The dedicated devices such as DIGIVAL™ can capture, interpret, display and transmit the results of LFAs to a computer or a mobile phone (77–78). In some cases it is a camera based instrument that can detect the presence and determine the level of an analyte based on the coloration of a test line; it also analyzes the intensity of a control line and displays the results (positive, negative or invalid) on a touch screen. Some of the readers are just dedicated to LFAs manufactured by the same company, while there is also an option to purchase readers for general applications, which can work with any standard lateral-flow test format based on color change (e.g. the CUBE Lateral-Flow Reader) (79). The additional option offered by some readers is to scan a barcode, which identifies the kind of LFA and ascribes it to a certain batch. Development of such readers, which are usually small, inexpensive and portable widens up the opportunity for home use of point-of-care diagnostics instead of being based on stand-alone bench top readers available only at specialized laboratories. In a case of fluorescent labels, a fluorescence strip readers are used to record fluorescence intensity of test and control lines (2, 80). Magnetic strip readers (which work together with magnetic labels), electrochemical detectors and pressure meters (when PtNPs decompose hydrogen peroxide to oxygen according to a catalase like activity) were also reported as detection systems, but so far such systems were only developed strictly for scientific applications (81–83).

10.6   TYPES OF SAMPLES AND APPLICATIONS OF LATERAL FLOW ASSAYS

The most common commercially available tests utilize samples in a liquid state. LFAs have been implied for a wide range of applications including clinics, veterinary, environmental or agri-food. Their versatility allows to detect target analytes in a variety of sample matrices including: whole blood (84), nasal swab (85), sweat (86), saliva (87), milk (88), serum (89), urine (90), food (91), animal feed (92) or plant material (93).

Since latex agglutination assay was first reported in 1956 by Plotz and Singer (94), LFAs applications are boosting in various directions. However, the main development of solid phase rapid tests began with the human pregnancy test, which also sparked interest in urine testing for diagnostic purposes. This particular application, extensively developed by Vaitukaitis and co-workers (95), is based on a detection of human chorionic gonadotropin (hCG). Although hCG test origins date back to 1974, only multiple improvements in the field of antibody generation, nitrocellulose membrane manufacturing and flow technologies have finally led to the commercial application of the pregnancy test by Unipath, a Unilever spin out at Colworth (UK) in 1985. Nowadays, hCG immunoassays based on blood or urine testing are available in various formats including cassettes, pregnancy line strips and more complex analytic devices.

Considering the socio-economic aspect of the worldwide diagnostics market, LFAs play crucial role in detection of cardiac, infectious and chronic diseases among which geographically limited infections (e.g., malaria, dengue) or emerging diseases (e.g., H5N1 influenza) can be distinguished. Moreover, ageing population growth as well as high demand of portable POCT devices and home-diagnostics assays are the factors that drive the LFA market. One of the most relevant LFAs application is widely available serological (antibody-targeted) HIV (human immunodeficiency virus) screening. On the market there are several FDA (Food & Drug Administration)-approved self-test kits for detection of HIV-1 and HIV-2 from blood and oral swab (OraSure Technologies, BioSure, Alere Inc.) (85).

The previously mentioned emerging diseases with influenzas at the forefront over the years have resulted in high mortality among people and animals. Only a H1N1 strain, so-called Spanish flu, caused tens of millions deaths all over the world. Further outbreaks of 'swine flu' pandemic or H5N1 strains has resulted in mass deaths of animals. Facing the crisis, whereas PCR techniques fail to meet the requirements of field diagnostics, LFA-based multiple techniques that fulfill the WHO (World Health Organization) ASSURED criteria stand out. Recently, another urgent need in terms of pathogen detection was highlighted. The coronavirus disease 2019 pandemic has revealed a desire of a large-scale, reliable, rapid LFAs that could be easily implied in public places such as airports, workplaces or nursing homes. Although real time-PCR is considered as the gold standard for diagnosing SARS-CoV-2, it requires centralized diagnostic laboratories and highly skilled personnel. However, the FDA has authorized two antigen-based assays for SARS-CoV-2 rapid POCT detection under an Emergency Use Authorization for use by authorized laboratories (Sofia 2 SARS Antigen FIA test and BD Veritor System). Currently, there are many reports concerning the amplification-free SARS-CoV-2 RNA detection (96–98) as well as anti-viral antibodies detection (IgG, IgM) (24, 99–100) to observe seroprevalence in the population. Still, new solutions are constantly appearing for enhancing LFA sensitivity with application to COVID-19 (101–102).

Another globally prevalent viral infectious disease which can be detected using LFA is Hepatitis B. It is a dangerous infection which leads to liver failure and hepatocellular carcinoma. Hepatitis B surface antigen constitute a conventional serological marker for screening using radioimmunoassay, ELISA or chemiluminescence/electrochemiluminescence LFA tests. Especially the latter have been widely used as they facilitate self-examinations, health checks or blood donations in emergency situation. Nevertheless, the quantitative detection may be difficult due to the influence of sample matrix color and light reflection (103). In order to overcome these limitations, LFA is often combined with molecular methods (recombinase polymerase amplification) (104).

An important domain in LFA application that has attracted a lot of attention due to the vitally impact on public health is bacteria detection. Water contamination with infectious diseases cause around 2.2 million deaths per year worldwide, therefore the accurate monitoring of pathogenic bacteria in drinking water and food industry at low concentrations is of great of interest and could significantly reduce threat to public health. However, the major limitation in direct bacterial testing is insufficient detection limit. Therefore, a pre-concentration or pre-enrichment stage is often demanded, especially when large sample volume has to be rapidly tested. For this purpose Zhang et al. proposed a high performance automated bacterial concentration and recovery system that enabled *E. coli* detection with short processing time (less than 1 hour) using a ceramic membrane and special filtration system (105). Nevertheless, LFA tools for rapid (~ 20 min) and multiplexed detection of various foodborne pathogens, e.g. *E. coli, S. paratyphi, V. cholerae*, without the need to sample enrichment were also developed (106). This is a significant advantage over conventional methods that require sterile conditions, specialized personnel, costly equipment and above all are time-consuming (5–7 days of analysis) (107).

10.7 LATERAL FLOW ASSAYS—REQUIREMENTS, EXPECTATIONS, AND CHALLENGES

LFA diagnostic kits are used to detect many agents both in the healthcare systems and the industry. The decisions based on the results of LFA are often of key importance for a patient, e.g. initiation or withdrawal from a treatment, an admission to the hospital or a decision of patient isolation. This is the reason why during a research and development phase developers of LFA should be aware of the recommendations of the market regulators (108) and the industry organizations. The LFA test technology has been on the market for more than 30 years, therefore market regulators are gradually tightening the requirements for the analytical parameters of LFA tests, especially for the medical market. It is important to remember that every diagnostic kit targeted to the medical market should be a subject of the two type evaluations: the analytical and the clinical one. Analytical performance requirements for LFA tests have been described in detail (109), they concern accuracy, precision, selectivity, reproducibility, stability, the limit of detection, errors, turbulence, quality control and the hook effect.

Regarding analytical performance indices it is particularly important to consider several parameters. Since the most typical LFA (for example LFA based rapid antigen test) has no potentization of the analytical signal, the bioreceptor targeted the biological analyte does not amplify the signal generated by the measurement. So, for the typical end-user the key parameter of the LFA is LOD. Limit of detection determines the lowest detectable concentration of targeted analyte at which 100% of all (true positive) replicates give the positive result. The next analytical characteristic parameter directly influencing the suitability of the LFA in the real life is the cross-reactivity (often called analytical specificity). The analytical specificity describes interference of a given test against a panel of related chemicals, pathogens, high prevalence disease agents and normal or pathogenic flora including various microorganisms and viruses, and negative matrix that is reasonably likely to be encountered in the clinical sample and could potentially cross-react or interfere with the desired LFA test. An important effect in the sandwich immunological LFA tests is hook effect, where at very high concentrations of tested analyte, the biorecognition elements are saturated and the signal is off (110). In practice high dose hook effect determines the level at which false negative results can be seen when very high levels of the target are present in a tested sample.

In medical or diagnostic practice, the most important parameter assessing the value of the LFA test is the clinical characteristics, which better represent the real value of the test. The evaluation of clinical characteristics requires conducting a clinical trial in real conditions with natural samples containing the analyzed analyte. The trial should be performed by comparing the analyzed LFA test with the 'gold standard test' well established method in a laboratory practice. Main parameters of clinical characteristics needed for the developed LFA are: PPA—Positive Percent Agreement (Sensitivity), NPA—Negative Percent Agreement (Specificity), PPV—Positive Predictive Value, NPV- Negative Predictive Value, OPA—Overall Percent Agreement.

Classic LFIAs are usually characterized with very high specificity, which results from the antigen-antibody binding specificity and moderate sensitivity, due to the lack of the amplification of the measurement signal during analysis. Hence, numerous developments related to LFA often focus on improving the clinical sensitivity of the test. The sensitivity of the LFA test can be optimized in many ways. The first way is to optimize is the quality of the tested sample. The improvement of the sample quality can be achieved by better sample taking (e.g. nasopharyngeal or lower respiratory tract sample instead of a pharyngeal swab). This is usually the easiest way to improve the performance of the clinical LFA test. Another way to improve clinical performance of the LFA is the proper use of the test in the clinical context. This is widely accepted that LFA testing should be used in conjunction with specific disease symptom questionnaires (e.g. fever, cough) (111) or a broader context (such as the SARS-CoV-2 pandemic).

The landscape of immunological assays covers a wide range of diagnostic technologies, including methods based on recognized enzyme labelling technologies and their automated modifications, numerous experimental technologies that have not yet resulted in significant market implementations and many others. LFA technologies also underwent numerous modifications related to the way of labeling bio-sensing elements, reading methods and the degree of complexity of the test.

The ideal LFA used in POCT should follow the ASSURED principles of point of care diagnostics developed by the WHO for infectious diseases in 2017 year (112). The first category of this priciples is the availability. The availability term has different meanings depending on the context of the place of use. A test performed by a professional in a medical diagnostic laboratory must meet different requirements compared to a screening test performed at the workplace or at the airport. However, the availability of the test is always related to the quality/cost relation and should be compared to the same assessment of the other diagnostic methods. In the case of LFA tests, as we saw during the COVID-19 pandemic, the low price and ease of implementation of the tests under all conditions spoke for the usefulness of immunochromatographic methods.

An important factor that determines the availability of LFA tests is quite a long shelf life of LFA in relation to other immunoassays, in most cases it covers a period of 12–18 months. The use of dried bioreceptor conjugates (usually mono or polyclonal antibodies) with stable NPs significantly extends the life of the LFA kits, which are placed in individual foil packages with silicone desiccant. LFAs usually can be transported under non-refrigerated conditions, which also significantly improves the availability of the tests.

During the SARS-CoV-2 virus pandemic, LFA tests proved to be an indispensable diagnostic tool, being on the recommendation lists of both WHO and national regulators. The next ASSURED principle of LFA is the sensitivity and the clinical specificity, as discussed earlier. This is important to remember that as all other diagnostic tests, LFA tests should be interpreted in the light of clinical data. This was also proven during COVID-19 pandemic, SARS-CoV-2 antigen tests produced the most reliable results when respiratory samples were obtained between the 3rd and 8th day after infection, when the viral load in the upper respiratory tract is highest. From the point of view of public health, however, it turned out that the most expected feature of the availability of LFA tests for detecting COVID-19 antigens was the ability to perform tests in the community (airports, workplaces, borders, etc.). To note, LFA tests do not meet the next two parameters of the ideal test, i.e. sensitivity, because despite the high specificity, the sensitivity of COVID detection tests is variable and ranges from 40–95% but meet other ASSURED principles since they are rapid, robust and instrumentation free, etc.

One of the main challenges of the LFA tests is to enhance the sensitivity, especially in combination with RNA or DNA amplification procedures, often using quantitative assessment. A promising technique that allows for precise quantitative readings of LFA it the microarray lateral flow technique. Several companies started to produce microspot LFA tests that allow more precise quantification using dedicated readers. An element of this change is the potential possibility of using smartphones for such a quantitative assessment. On the other hand the rapid progress of the Internet of Things technology will lead to the fact that connected micro-automated LFA microarray test readers will read data better than the multipurpose smartphone. Such a near future for LFA research—the evolution toward sets read by low-cost readers that will connect directly to the Internet, may allow the established technology to persist for years.

ACKNOWLEDGEMENT

This work has been financially supported by the National Centre for Research and Development in Poland (grant no. POIR.04.01.04–00–0027/17)

REFERENCES

1. Lateral Flow Assays Market to Reach $13.85 Billion by 2027. Accessed August 2, 2021. www.globenewswire.com/en/news-release/2021/02/04/2169807/0/en/Lateral-Flow-Assays-Market-to-Reach-13-85-billion-by-2027-Exclusive-Report-Covering-Pre-and-Post-COVID-19-Market-Analysis-and-Forecasts-by-Meticulous-Research.html
2. Sajid M, Kawde AN, Daud M. 2015. Designs, formats and applications of lateral flow assay: A literature review. *Journal of Saudi Chemical Society.* 19(6): 689–705
3. Quesada-González D, Merkoçi A. 2015. Nanoparticle-based lateral flow biosensors. *Biosensors and Bioelectronics.* 73: 47–63
4. O'Farrell B. 2009. Evolution in lateral flow—based immunoassay systems. *Lateral Flow Immunoassy.* 1
5. Rosen S. 2009. Market trends in lateral flow immunoassays. *Lateral Flow Immunoassy.* 1–15
6. Ponti JS. 2009. Material platform for the assembly of lateral flow immunoassay test strips. *Lateral Flow Immunoassy.* 1–7
7. O'Farrell B. 2015. Lateral flow technology for field-based applications—basics and advanced developments. *Topics in Companion Animal Medicine.* 30(4): 139–147
8. Tsai T-T, Huang T-H, Chen C-A, Ho NY-J, Chou Y-J, Chen C-F. 2018. Development a stacking pad design for enhancing the sensitivity of lateral flow immunoassay. *Scientific Reports.* 8(1): 1–10
9. Rong-Hwa S, Shiao-Shek T, Der-Jiang C, Yao-Wen H. 2010. Gold nanoparticle-based lateral flow assay for detection of staphylococcal enterotoxin B. *Food Chemistry.* 118(2): 462–466
10. Nielsen K, Yu WL, Kelly L, Williams J, Dajer A, et al. 2009. Validation and field assessment of a rapid lateral flow assay for detection of bovine antibody to anaplasma marginale. *Journal of Immunoassay and Immunochemistry®.* 30(3): 313–321
11. He QH, Xu Y, Wang D, Kang M, Huang ZB, Li YP. 2012. Simultaneous multiresidue determination of mycotoxins in cereal samples by polyvinylidene fluoride membrane based dot immunoassay. *Food Chemistry.* 134(1): 507–512
12. Shaimi R, Ahmad AL, Low SC. 2012. Investigating membrane material and morphology for development of lateral flow biosensor. *Journal of Applied Membrane Science & Technology.* 15(1)
13. Koczula KM, Gallotta A. 2016. Lateral flow assays. *Essays in Biochemistry.* 60: 111–120
14. Kasetsirikul S, Shiddiky MJA, Nguyen NT. 2020. Challenges and perspectives in the development of paper-based lateral flow assays. *Microfluidics and Nanofluidics.* 24(2): 1–18
15. Huang C, Jones BJ, Bivragh M, Jans K, Lagae L, Peumans P. 2013. A capillary-driven microfluidic device for rapid DNA detection with extremely low sample consumption. *17th International Conference on Miniaturized Systems for Chemistry and Life Sciences (MicroTAS 2013).* 1: 191–193
16. Manz A, Dittrich PS, Pamme N, Iossifidis D. 2015. *Bioanalytical Chemistry,* 2nd edition, pp. 144–156. Imperial College Press.
17. Zherdev AV, Dzantiev BB. 2018. Ways to reach lower detection limits of lateral flow immunoassays. *Rapid Test—Advances in Design, Format and Diagnostic Applications; Laura Anfossi Editon.* 9–43

18. Qriouet Z, Cherrah Y, Sefrioui H, Qmichou Z. 2021. Monoclonal antibodies application in lateral flow immunochromatographic assays for drugs of abuse detection. *Molecules.* 26(4): 1058

19. Kaur J, Singh KV, Boro R, Thampi KR, Raje M, et al. 2007. Immunochromatographic dipstick assay format using gold nanoparticles labeled protein-hapten conjugate for the detection of atrazine. *Environmental Science & Technology.* 41(14): 5028–5036

20. Bahadır EB, Sezgintürk MK. 2016. Lateral flow assays: Principles, designs and labels. *TrAC—Trends in Analytical Chemistry.* 82: 286–306

21. Huang L, Tian S, Zhao W, Liu K, Ma X, Guo J. 2020. Multiplexed detection of biomarkers in lateral-flow immunoassays. *Analyst.* 145(8): 2828–2840

22. Panraksa Y, Apilux A, Jampasa S, Puthong S, Henry CS, et al. 2021. A facile one-step gold nanoparticles enhancement based on sequential patterned lateral flow immunoassay device for C-reactive protein detection. *Sensors & Actuators, B: Chemical.* 329: 129241

23. Mahmoudinobar F, Britton D, Montclare JK. 2021. Protein-based lateral flow assays for COVID-19 detection. *Protein Engineering Design & Selection.* 34: 1–10

24. Chen Z, Zhang Z, Zhai X, Li Y, Lin L, et al. 2020. Rapid and sensitive detection of anti-SARS-CoV-2 IgG, using lanthanide-doped nanoparticles-based lateral flow immunoassay. *Analytical Chemistry.* 92(10): 7226–7231

25. Huang C, Wen T, Shi FJ, Zeng XY, Jiao YJ. 2020. Rapid detection of IgM antibodies against the SARS-CoV-2 virus via colloidal gold nanoparticle-based lateral-flow assay. *ACS Omega.* 5(21): 12550–12556

26. Wang Z, Zheng Z, Hu H, Zhou Q, Liu W, et al. 2020. A point-of-care selenium nanoparticle-based test for the combined detection of anti-SARS-CoV-2 IgM and IgG in human serum and blood. *Lab Chip.* 20(22): 4255–4261

27. Bayin Q, Huang L, Ren C, Fu Y, Ma X, Guo J. 2021. Anti-SARS-CoV-2 IgG and IgM detection with a GMR based LFIA system. *Talanta.* 227: 122207

28. Pohanka M. 2021. Point-of-care diagnoses and assays based on lateral flow test. *International Journal of Analytical Chemistry.* 1–9

29. Gao X, Zheng P, Kasani S, Wu S, Yang F, et al. 2017. Paper-based surface-enhanced raman scattering lateral flow strip for detection of neuron-specific enolase in blood plasma. *Analytical Chemistry.* 89(18): 10104–10110

30. Hu J, Zhou S, Zeng L, Chen Q, Duan H, et al. 2021. Hydrazide mediated oriented coupling of antibodies on quantum dot beads for enhancing detection performance of immunochromatographic assay. *Talanta.* 223(58)

31. Bishop JD, Hsieh HV, Gasperino DJ, Weigl BH. 2019. Sensitivity enhancement in lateral flow assays: A systems perspective. *Lab Chip.* 19(15): 2486–2499

32. Doyle J, Uthicke S. 2021. Sensitive environmental DNA detection via lateral flow assay (dipstick)—a case study on corallivorous crown-of-thorns sea star (Acanthaster cf. solaris) detection. *Environmental DNA.* 3(2): 323–342

33. Choi JR, Hu J, Gong Y, Feng S, Bakar WA, et al. 2016. An integrated lateral flow assay for effective DNA amplification and detection at the point of care. *Analyst.* 141: 2930

34. Bai X, Ma X, Li M, Li X, Fan G, et al. 2020. Field applicable detection of hepatitis B virus using internal controlled duplex recombinase-aided amplification assay and lateral flow dipstick assay. *Journal of Medical Virology.* 92(12): 3344–3353

35. Jauset-Rubio M, Svobodová M, Mairal T, McNeil C, Keegan N, et al. 2016. Ultrasensitive, rapid and inexpensive detection of DNA using paper based lateral flow assay. *Scientific Reports.* 6(1): 1–10

36. Kozwich D, Johansen KA, Landau K, Roehl CA, Woronoff S, Roehl PA. 2000. Development of a novel, rapid integrated Cryptosporidium parvum detection assay. *Applied and Environmental Microbiology.* 66(7): 2711–2717

37. Fong WK, Modrusan Z, McNevin JP, Marostenmaki J, Zin B, Bekkaoui F. 2000. Rapid solid-phase immunoassay for detection of methicillin-resistant Staphylococcus aureus using cycling probe technology. *Journal of Clinical Microbiology*. 38(7): 2525–2529

38. Zheng C, Wang K, Zheng W, Cheng Y, Li T, et al. 2021. Rapid developments in lateral flow immunoassay for nucleic acid detection. *Analyst*. 146(5): 1514–1528

39. Le TT, Chang P, Benton DJ, McCauley JW, Iqbal M, Cass AEG. 2017. Dual recognition element lateral flow assay toward multiplex strain specific influenza virus detection. *Analytical Chemistry*. 89(12): 6781–6786

40. Corstjens PLAM, Zuiderwijk M, Nilsson M, Feindt H, Niedbala RS, Tanke HJ. 2003. Lateral-flow and up-converting phosphor reporters to detect single-stranded nucleic acids in a sandwich-hybridization assay. *Analytical Biochemistry*. 312(2): 191–200

41. Van De Rijke F, Zijlmans H, Li S, Vail T, Raap AK, et al. 2001. Up-converting phosphor reporters for nucleic acid microarrays. *Nature Biotechnology*. 19(3): 273–276

42. Glynou K, Ioannou PC, Christopoulos TK, Syriopoulou V. 2003. Oligonucleotide-functionalized gold nanoparticles as probes in a dry-reagent strip biosensor for DNA analysis by hybridization. *Analytical Chemistry*. 75(16): 4155–4160

43. Mao X, Ma Y, Zhang A, Zhang L, Zeng L, Liu G. 2009. Disposable nucleic acid biosensors based on gold nanoparticle probes and lateral flow strip. *Analytical Chemistry*. 81(4): 1660–1668

44. Qiu W, Xu H, Takalkar S, Gurung AS, Liu B, et al. 2015. Carbon nanotube-based lateral flow biosensor for sensitive and rapid detection of DNA sequence. *Biosensors and Bioelectronics*. 64: 367–372

45. Aveyard J, Mehrabi M, Cossins A, Braven H, Wilson R. 2007. One step visual detection of PCR products with gold nanoparticles and a nucleic acid lateral flow (NALF) device. *Chemical Communications*. 4251–4253

46. Nagatani N, Yamanaka K, Ushijima H, Koketsu R, Sasaki T, et al. 2012. Detection of influenza virus using a lateral flow immunoassay for amplified DNA by a microfluidic RT-PCR chip. *Analyst*. 137(15): 3422–3426

47. Nihonyanagi S, Kanoh Y, Okada K, Uozumi T, Kazuyama Y, et al. 2012. Clinical usefulness of multiplex PCR lateral flow in MRSA detection: A novel, rapid genetic testing method. *Inflammation*. 35(3): 927–934

48. Prompamorn P, Sithigorngul P, Rukpratanporn S, Longyant S, Sridulyakul P, Chaivisuthangkura P. 2011. The development of loop-mediated isothermal amplification combined with lateral flow dipstick for detection of Vibrio parahaemolyticus. *Letters in Applied Microbiology*. 52(4): 344–351

49. Roskos K, Hickerson AI, Lu HW, Ferguson TM, Shinde DN, et al. 2013. Simple system for isothermal DNA amplification coupled to lateral flow detection. *PLoS One*. 8(7)

50. Jaroenram W, Owens L. 2014. Recombinase polymerase amplification combined with a lateral flow dipstick for discriminating between infectious Penaeus stylirostris densovirus and virus-related sequences in shrimp genome. *Journal of Virological Methods*. 208: 144–151

51. Jaroenram W, Kiatpathomchai W, Flegel TW. 2009. Rapid and sensitive detection of white spot syndrome virus by loop-mediated isothermal amplification combined with a lateral flow dipstick. *Molecular and Cellular Probes*. 23(2): 65–70

52. Kersting S, Rausch V, Bier FF, von Nickisch-Rosenegk M. 2014. Rapid detection of plasmodium falciparum with isothermal recombinase polymerase amplification and lateral flow analysis. *Postprints der Universität Potsdam. Mathematisch-naturwissenschaftliche Reihe*. 948

53. Tomlinson JA, Dickinson MJ, Boonham N. 2010. Rapid detection of phytophthora ramorum and P. kernoviae by two-minute DNA extraction followed by isothermal amplification and amplicon detection by generic lateral flow device. *Phytopathology*. 100(2): 143–149

54. Puthawibool T, Senapin S, Flegel TW, Kiatpathomchai W. 2010. Rapid and sensitive detection of Macrobrachium rosenbergii nodavirus in giant freshwater prawns by reverse transcription loop-mediated isothermal amplification combined with a lateral flow dipstick. *Molecular and Cellular Probes.* 24(5): 244–249

55. Rosser A, Rollinson D, Forrest M, Webster BL. 2015. Isothermal recombinase polymerase amplification (RPA) of Schistosoma haematobium DNA and oligochromatographic lateral flow detection. *Parasites and Vectors.* 8(1): 1–5

56. Vincent M, Xu Y, Kong H. 2004. Helicase-dependent isothermal DNA amplification. *EMBO Reports.* 5(8): 795–800

57. Piepenburg O, Williams CH, Stemple DL, Armes NA. 2006. DNA detection using recombination proteins. *PLoS Biology.* 4(7): 1115–1121

58. Ghosh DK, Kokane SB, Kokane AD, Warghane AJ, Motghare MR, et al. 2018. Development of a recombinase polymerase based isothermal amplification combined with lateral flow assay (HLB-RPA-LFA) for rapid detection of "Candidatus Liberibacter asiaticus". *PLoS One.* 13(12)

59. Ivanov AV, Safenkova IV, Zherdev AV, Dzantiev BB. 2020. Nucleic acid lateral flow assay with recombinase polymerase amplification: Solutions for highly sensitive detection of RNA virus. *Talanta.* 210

60. Rani A, Ravindran VB, Surapaneni A, Shahsavari E, Haleyur N, et al. 2021. Evaluation and comparison of recombinase polymerase amplification coupled with lateral-flow bioassay for Escherichia coli O157:H7 detection using diifeerent genes. *Scientific Reports.* 11(1)

61. Xu H, Mao X, Zeng Q, Wang S, Kawde AN, Liu G. 2009. Aptamer-functionalized gold nanoparticles as probes in a dry-reagent strip biosensor for protein analysis. *Analytical Chemistry.* 81(2): 669–675

62. Carvalho D, Paulino M, Polticelli F, Arredondo F, Williams RJ, et al. 2019. Aptamer lateral flow assays for rapid and sensitive detection of cholera toxin. *Analyst.* 106(5): 393–403

63. Kaiser L, Weisser J, Kohl M, Deigner HP. 2018. Small molecule detection with aptamer based lateral flow assays: Applying aptamer-C-reactive protein cross-recognition for ampicillin detection. *Scientific Reports.* 8(1)

64. Huang L, Tian S, Zhao W, Liu K, Ma X, Guo J. 2021. Aptamer-based lateral flow assay on-site biosensors. *Biosensors and Bioelectronics.* 186: 113279

65. Wang T, Chen L, Chikkanna A, Chen S, Brusius I, et al. 2021. Development of nucleic acid aptamer-based lateral flow assays: A robust platform for cost-effective point-of-care diagnosis. *Theranostics.* 11(11): 5174–5196

66. Anfossi L, Di Nardo F, Cavalera S, Giovannoli C, Baggiani C. 2018. Multiplex lateral flow immunoassay: An overview of strategies towards high-throughput point-of-need testing. *Biosensors.* 9(1)

67. Taranova NA, Byzova NA, Zaiko VV, Starovoitova TA, Vengerov YY, et al. 2013. Integration of lateral flow and microarray technologies for multiplex immunoassay: Application to the determination of drugs of abuse. *Microchimica Acta.* 180(11–12): 1165–1172

68. Khlebtsov BN, Tumskiy RS, Burov AM, Pylaev TE, Khlebtsov NG. 2019. Quantifying the numbers of gold nanoparticles in the test zone of lateral flow immunoassay strips. *ACS Applied Nano Materials.* 2(8): 5020–5028

69. Matsumura Y, Enomoto Y, Takahashi M, Maenosono S. 2018. Metal (Au, Pt) Nanoparticle-latex nanocomposites as probes for immunochromatographic test strips with enhanced sensitivity. *ACS Applied Materials & Interfaces.* 10(38): 31977–31987

70. Reporter Nanoparticle Selection for Lateral Flow Immunoassays—nanoComposix. Accessed July 11, 2021. https://nanocomposix.com/pages/reporter-nanoparticle-selection-for-lateral-flow-immunoassays

71. Aikawa T, Mizuno A, Kohri M, Taniguchi T, Kishikawa K, Nakahira T. 2016. Polystyrene latex particles containing europium complexes prepared by miniemulsion polymerization using bovine serum albumin as a surfactant for biochemical diagnosis. *Colloids Surfaces B Biointerfaces*. 145: 152–159

72. Pietrzak M, Ivanova P. 2021. Bimetallic and multimetallic nanoparticles as nanozymes. *Sensors and Actuators B: Chemical*. 336: 129736

73. Mulvaney SP, Kidwell DA, Lanese JN, Lopez RP, Sumera ME, Wei E. 2020. Catalytic lateral flow immunoassays (cLFIA™): Amplified signal in a self-contained assay format. *Sensing and Bio-Sensing Research*. 30: 100390

74. Drozd M, Pietrzak M, Parzuchowski PG, Malinowska E. 2016. Pitfalls and capabilities of various hydrogen donors in evaluation of peroxidase-like activity of gold nanoparticles. *Analytical and Bioanalytical Chemistry*. 408(29): 8505

75. Drozd M, Pietrzak M, Pytlos J, Malinowska E. 2016. Revisiting catechol derivatives as robust chromogenic hydrogen donors working in alkaline media for peroxidase mimetics. *Analytica Chimica Acta*. 948: 80–89

76. Frey A, Meckelein B, Externest D, Schmidt MA. 2000. A stable and highly sensitive 3,3′,5,5′-tetramethylbenzidine-based substrate reagent for enzyme-linked immunosorbent assays. *Journal of Immunological Methods*. 233(1–2): 47–56

77. DIGIVAL | Abbott Point of Care testing. Accessed July 11, 2021. www.globalpointofcare.abbott/en/product-details/alere-reader-us.html

78. Sōna LFA Cube Reader Bench Top Analyser to Aid Interpretation of Results of the Sōna Aspergillus Galactomannan Lateral Flow Assay—Diagnostic Products. Accessed July 11, 2021. www.alphalabs.co.uk/lfardr

79. CUBE Lateral-Flow Reader—Bioassay Works LLC. Accessed July 11, 2021. https://bioassayworks.com/?product=cube-lateral-flow-reader

80. Joung HA, Oh YK, Kim MG. 2014. An automatic enzyme immunoassay based on a chemiluminescent lateral flow immunosensor. *Biosensors and Bioelectronics*. 53: 330–335

81. Wang Y, Xu H, Wei M, Gu H, Xu Q, Zhu W. 2009. Study of superparamagnetic nanoparticles as labels in the quantitative lateral flow immunoassay. *Materials Science and Engineering C*. 29(3): 714–718

82. Lin YY, Wang J, Liu G, Wu H, Wai CM, Lin Y. 2008. A nanoparticle label/immunochromatographic electrochemical biosensor for rapid and sensitive detection of prostate-specific antigen. *Biosensors and Bioelectronics*. 23(11): 1659–1665

83. Lin B, Guan Z, Song Y, Song E, Lu Z, et al. 2018. Lateral flow assay with pressure meter readout for rapid point-of-care detection of disease-associated protein. *Lab Chip*. 18(6): 965–970

84. Ang SH, Yu CY, Ang GY, Chan YY, Alias Y binti, Khor SM. 2015. A colloidal gold-based lateral flow immunoassay for direct determination of haemoglobin A1c in whole blood. *Analytical Methods*. 7(9): 3972–3980

85. Han M-Y, Xie T-A, Li J-X, Chen H-J, Yang X-H, Guo X-G. 2020. Evaluation of lateral-flow assay for rapid detection of influenza virus. *BioMed Research International*. 1

86. Dalirirad S, Steckl AJ. 2019. Aptamer-based lateral flow assay for point of care cortisol detection in sweat. *Sensors & Actuators, B: Chemical*. 283: 79–86

87. Carrio A, Sampedro C, Sanchez-Lopez JL, Pimienta M, Campoy P. 2015. Automated low-cost smartphone-based lateral flow saliva test reader for drugs-of-abuse detection. *Sensors*. 15(11): 29569–29593

88. Alhussien MN, Dang AK. 2020. Sensitive and rapid lateral-flow assay for early detection of subclinical mammary infection in dairy cows. *Scientific Reports*. 10(1): 1–12

89. Tsai TT, Huang TH, Chen CA, Ho NYJ, Chou YJ, Chen CF. 2018. Development a stacking pad design for enhancing the sensitivity of lateral flow immunoassay. *Scientific Reports*. 8(1): 1–10

90. Dalirirad S, Steckl AJ. 2020. Lateral flow assay using aptamer-based sensing for on-site detection of dopamine in urine. *Analytical Biochemistry*. 596: 113637

91. Hnasko RM, Jackson ES, Lin AV, Haff RP, McGarvey JA. 2021. A rapid and sensitive lateral flow immunoassay (LFIA) for the detection of gluten in foods. *Food Chemistry*. 355: 129514

92. Wang J, Zhou J, Chen Y, Zhang X, Jin Y, et al. 2019. Rapid one-step enzyme immunoassay and lateral flow immunochromatographic assay for colistin in animal feed and food. *Journal of Animal Science and Biotechnology*. 10(1): 1–10

93. López-Soriano P, Noguera P, Gorris MT, Puchades R, Maquieira Á, et al. 2017. Lateral flow immunoassay for on-site detection of Xanthomonas arboricola pv. pruni in symptomatic field samples. *PLoS One*. 12(4)

94. Singer JM, Plotz CM. 1956. The latex fixation test: I. Application to the serologic diagnosis of rheumatoid arthritis. *The American Journal of Medicine*. 21(6): 888–892

95. Vaitukaitis JL, Braunstein GD, Ross GT. 1972. A radioimmunoassay which specifically measures human chorionic gonadotropin in the presence of human luteinizing hormone. *American Journal of Obstetrics and Gynecology*. 113(6): 751–758

96. Wang D, He S, Wang X, Yan Y, Liu J, et al. 2020. Rapid lateral flow immunoassay for the fluorescence detection of SARS-CoV-2 RNA. *Nature Biomedical Engineering*. 4(12): 1150–1158

97. Grant BD, Anderson CE, Williford JR, Alonzo LF, Glukhova VA, et al. 2020. SARS-CoV-2 coronavirus nucleocapsid antigen-detecting half-strip lateral flow assay toward the development of point of care tests using commercially available reagents. *Analytical Chemistry*. 92(16): 11305–11309

98. COVID-19: Rapid antigen detection for SARS-CoV-2 by lateral flow assay: A national systematic evaluation of sensitivity and specificity for mass-testing | Elsevier Enhanced Reader. https://www.medrxiv.org/content/10.1101/2021.01.13.21249563v2

99. Ragnesola B, Jin D, Lamb CC, Shaz BH, Hillyer CD, Luchsinger LL. 2020. COVID19 antibody detection using lateral flow assay tests in a cohort of convalescent plasma donors. *BMC Research Notes*. 13(1): 1–7

100. Demey B, Daher N, François C, Lanoix J-P, Duverlie G, et al. 2020. Dynamic profile for the detection of anti-SARS-CoV-2 antibodies using four immunochromatographic assays. *Journal of Infection*. 81: 6–10

101. Kim K, Kashefi-Kheyrabadi L, Joung Y, Kim K, Dang H, et al. 2021. Recent advances in sensitive surface-enhanced Raman scattering-based lateral flow assay platforms for point-of-care diagnostics of infectious diseases. *Sensors & Actuators, B: Chemical*. 329: 129214

102. Peng T, Liu X, Adams LG, Agarwal G, Akey B, et al. 2020. Enhancing sensitivity of lateral flow assay with application to SARS-CoV-2. *Applied Physics Letters*. 117(12): 120601

103. Cai Y, Yan J, Zhu L, Wang H, Lu Y. A rapid immunochromatographic method based on a secondary antibody-labelled magnetic nanoprobe for the detection of hepatitis B preS2 surface antigen. *Biosensors*. 10(11)

104. Zhang B, Zhu Z, Li F, Xie X, Ding A. 2021. Rapid and sensitive detection of hepatitis B virus by lateral flow recombinase polymerase amplification assay. *Journal of Virological Methods*. 291: 114094

105. Magambo KA, Kalluvya SE, Kapoor SW, Seni J, Chofle AA, et al. 2014. Utility of urine and serum lateral flow assays to determine the prevalence and predictors of cryptococcal antigenemia in HIV-positive outpatients beginning antiretroviral therapy in Mwanza, Tanzania. *Journal of the International AIDS Society*. 17(1): 19040

106. Zhao Y, Wang H, Zhang P, Sun C, Wang X, Wang X, Zhou L. 2016. Rapid multiplex detection of 10 foodborne pathogens with an up-converting phosphor technology-based 10-channel lateral flow assay. *Scientific Reports*. 6(1): 1–8

107. Çam D. 2019. Lateral flow assay for salmonella detection and potential reagents. *New Insight into Brucella Infect Foodborne Diseases* (October). doi:10.5772/INTE CHOPEN.88827
108. FDA, CDER. 2018. *Bioanalytical Method Validation Guidance for Industry Biopharmaceutics Contains Nonbinding Recommendations.* FDA, CDER
109. Ellison SLR, Fearn T. 2005. Characterising the performance of qualitative analytical methods: Statistics and terminology. *TrAC Trends in Analytical Chemistry.* 24(6): 468–476
110. Amarasiri Fernando S, Wilson GS. 1992. Studies of the "hook" effect in the one-step sandwich immunoassay. *Journal of Immunological Methods.* 151(1–2): 47–66
111. Fine AM, Nizet V, Mandl KD. 2012. Large-scale validation of the centor and mcisaac scores to predict group A streptococcal pharyngitis. *Archives of Internal Medicine.* 172(11): 847–852
112. Kosack CS, Page AL, Klatser PR. 2017. A guide to aid the selection of diagnostic tests. *Bulletin of the World Health Organization.* 95(9): 639–645

11 Microfluidic Based Biosensors and Applications

Münteha Nur Sonuç Karaboğa[1],
and Mustafa Kemal Sezgintürk[2]
[1]Tekirdağ Namık Kemal University
School of Health, Nutrition and Dietetics Department
Tekirdağ, Turkey
[2]Çanakkale Onsekiz Mart University
Faculty of Engineering, Bioengineering Department
Çanakkale, Turkey

CONTENTS

11.1 INTRODUCTION

Improvements in environmental factors such as food, air, water and improvements in health play an important role in the welfare of societies. This is possible with the use of sensing technologies for the detection of chemical or biological molecules that directly or indirectly threaten health. Among the sensing technologies, biosensor applications shine with many remarkable features.

Biosensor technology, which has been working in many fields from the diagnosis of disease markers to agriculture for more than half a century, is still in the focus of researchers with its superior qualities. The combination of a biosensing unit and

DOI: 10.1201/9781003189435-15

a transducer, and many chemical modifications developed within the framework of this association, take sensor technologies further.

Microfluidics, which began to emerge in the early 1980s, first showed itself in inkjet printing, DNA chips and later LOC technologies (1). Microfluidics generally refers to the study of fluidic systems with critical operational lengths in the 1–100 µM range. Although the term "micro" is mentioned, microfluidics refers to small volumes such as nL, pL, fL and manipulations on a delicate scale in the micro field using channels which is the basis of LOC systems (2–4). The features that make microfluids systems more attractive and preferable over traditional methods, such as the ability to assay low volume sample(s) on a single platform, being portable and being complete in a short period of time, can be attributed to the size effect (5–6). The advantages of integrating microfluidic systems into biosensors include reduction in sample and reagent volumes resulting from miniaturization, reduction in assay times, and the most remarkable are ultra-low detectable concentration limits with high specificity (7). Changes in the behavior of liquids in microfluidics, such as energy loss, fluid resistance and surface tension—unlike those on the macro scale—and the orientations made to these changes play a leading role in biosensor integration (8–9).

Lab-on-a-chip (LOC) systems, which can be evaluated within the scope of microfluidics technologies, are based on the technique of manipulating and processing small volumes of liquids in at least one-dimensional microchannels. LOC is the miniaturization of macro-scale measurement methodologies. Therefore, these technologies are the successful integration of all processes applied in the laboratory (such as sample collection, pretreatment, chemical applications, discrimination, detection and data analysis) on a chip (10). Integrating microscale components such as valves, mixers and pumps into LOC platforms is accelerating the use of biosensors as point of care (POC) tools (9, 11).

In the following titles, the reasons and advantages of integrating microfluidics systems with biosensors and the studies that turned into lab-on-a-chip applications will be mentioned.

11.2   INTEGRATION OF MICROFLUIDICS INTO BIOSENSING PLATFORMS

A biosensor is expected to have some remarkable analytical characteristics when determining its target analyte. High sensitivity determination is the foremost and most important of these. The sensitivity of a biosensor is measured by its ability to capture the target analyte, characterize it, and convert it into output signals (12). On the other hand, the sensitivity of portable and open biosensor systems may be low due to external noise and interference. By integrating microfluidic systems, bioaccumulation processes become closed and relatively more stable. This provides high sensitivity and reliability in the determination of the target analyte (13–14).

In addition to the high sensitivity of the biosensor, the ability to achieve this sensitivity at lower analyte concentration and faster response times can be achieved significantly by reducing the sensing area with microfluidic channels (15). While there may be heterogeneous distributions on the surface of the biosensing element in general procedures, microfluidic channels can minimize this heterogeneity by eliminating physical parameters such as temperature and pressure (7). Especially in cases where the amount of target analyte is limited, very small amount of sample volume can be used by integrating microfluidic channels into biosensor platforms (13, 16).

Another reason why microfluidic structures save time is that all the equipment needed for analysis in the laboratory environment is gathered on a single platform. By integrating the designed microfluidic platform with the biosensor, sample injection, pretreatment and processing steps can be carried out practically. In addition, this single-device integration is the basis for accessing parameters such as binding affinity, rate and kinetics, which have an important place in bio-affinity processes, by providing volume and speed control (13, 17).

Microfluidics can perform multiple bonding reactions both separately and mutually in the presence of single or multiple samples. This indicates that there is more than one bioreceptor in a complex solution environment and the samples flowing from each of these bioreceptors may be different. This ordered passage of different samples is possible with different combinations of microfluidic channels, pneumatic valves and/or centrifugal forces (2, 18). Thanks to the microfluidic structures, control

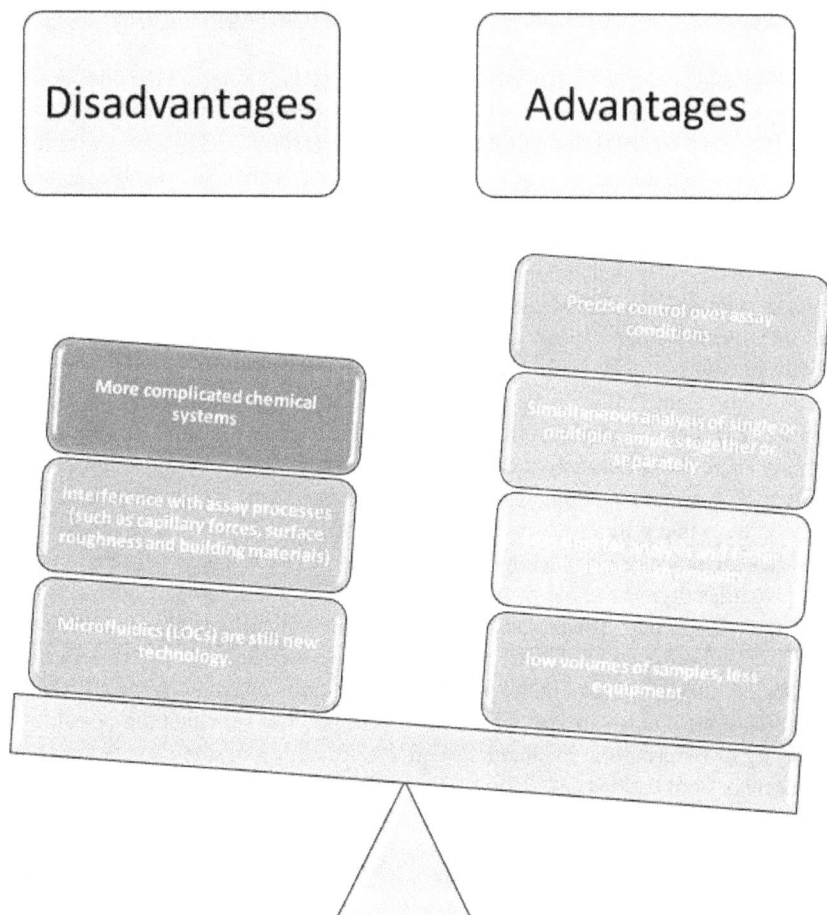

FIGURE 11.1 Some possible pros and cons of microfluidic based biosensors.

over the flow rate, sample volume, channel volume, channel height and reaction time can be achieved and thus a higher precision can be expressed about experimental conditions (13, 19). Figure 11.1 shows some of the possible advantages and disadvantages of microfluidics-based biosensors.

Microfluidic-based biosensors contain sensing surfaces integrated into the same platform so that fluid transport can be controlled and manipulated. Biomolecules in certain solutions can be immobilized by microarrays on the biosensor surface. The solvents are usually aqueous solutions to preserve the natural state, or after processing may be organic materials. In this combination, many biosensors have been developed. In the following titles, the materials and platforms used for the development of these biosensors will be discussed.

11.3 MATERIALS AND FABRICATION PROCESS OF MICROFLUIDICS BIOSENSORS

A large number of microfluidic devices are produced for many different biological and chemical applications. Materials used in the manufacture of microfluidic systems range from silicone, glass and ceramics to elastomers, thermosets, thermoplastics and paper-based polymers (20). Appropriate materials are selected depending on the application area, manipulations and integration. Properties such as flexibility, electrical conductivity, solvent compatibility, optical transparency and biocompatibility can be considered in material selection. In addition, cost effective and disposable materials are more preferred in biosensor applications. The most commonly used substrates among these materials are silicon and glass (21). Silicon is a common material that has been extensively studied in the semiconductor industry and therefore its properties are well known. However, the semiconductor nature of silicon can generate random noise in the sensor, especially in important electroconfigurations of complete insulation. Since silicon, which is transparent to infrared light, is not in the visible spectral region, difficulties may be encountered in fluorescence detection or liquid imaging (22). On the other hand, glass can be preferred especially in optical-based methods such as fluorescent or SPR, thanks to its transparency and high chemical stability. However, depending on the type of glass, there is a high probability of contamination that will affect the sensor operation. The glasses with the best optical properties are silica or quartz, but these materials are also quite expensive (8, 23).

Above all, polymers shine in microfluidic devices with their low cost, adaptability to many biosensor types and outstanding mechanical and physical properties. Poly(dimethyl siloxane) (PDMS) and poly(methyl methacrylate) (PMMA) are the most widely used polymers in the production processes of fluid channels for biosensor applications. Disadvantages of working with polymers include the possibility of collapsing or deformation in channels, and the possibility of non-specific adsorption on biosensor applications (24–25).

It is possible to collect fabrication of microfluidic systems in 3 methods in general. The classification is not limited to this. Rapid developments in microfluidic technologies allow diversification of materials and devices. Conventional microfluidic systems are based on continuous flow in micron-scale channels (1) (26–27). These microchannels are mainly fabricated using soft lithography methods (28). Droplet-based microfluidic systems (2) have been developed to reduce sample consumption and create more isolated reaction media (29). A new generation of droplet-based microfluidic systems called digital micro-fluidic (DMF) (30) (3) has been introduced to further reduce the sample consumption volume (31–32). Instead of a continuous stream of droplets, droplets are formed on a series of electrostatically actuated electrodes in DMF (32–33). Electro-wetting (EWOD) technique is used to move the droplets on the electrodes. EWOD-based DMF systems have many outstanding features such as lower power consumption and scalability. In addition, these systems have been trending in many biosensing applications in recent years, as they allow high-throughput parallel processing of large numbers of samples (34–36).

Continuous microfluidic systems have been successfully used in the study of many biological structures, from very small molecules to more complex structures, with increasing research in recent years. Among the reasons for this interest are the very high sensitivity detection of even very small analytes, the integration of all processes in a single miniature platform, reducing reagent and energy consumption, managing with less waste and low cost. Continuous microfluidic systems are widely designed in POC applications in environmental (37–40) and biomedical research (41–42), drug discovery (43–45) and clinical diagnostic (46–48).

Droplet-based microfluidics excel in integrating into biosensor technologies. It is frequently used as a new platform in agriculture (49–51), environment and biomedical (52–54) analyses due to its much better properties such as portable features, low energy and analyte volume (55–57).

In digital microfluidics, liquid samples are manipulated into microdroplets, which offers greater flexibility and more limited contamination during sample preparation and analysis. The absence of pumps, valves or channels used in continuous microfluidics in DMF has made it more flexible to integrate with electrochemical and optical based biosensors (58–60).

Liquid injection, mixing, separation, sensing, size and geometry perspective are important points in the design of microfluidic devices. The techniques used for production are UV-photolithography, X-ray photolithography, electron beam lithography and nanoimprinting. Soft lithography, which is frequently preferred due to its advantages such as high efficiency and cost sensitivity, is based on copying a topographically defined structure on a mold onto a soft elastomer (61–62).

The definition of biosensors includes having both a biorecognition element and a transducer. Biorecognition elements are structures designed to specifically capture the target bioanalyte, and then the transducers generate a measurable signal for analysis. Biorecognition unit and transduction, which are indispensable elements of a microfluidic based biosensor, can be shown as subsets of the microfluidic device (Figure 11.2). In the following section, examples of microfluidic based biosensors will be discussed from both the biorecognition element and transducer perspectives.

11.4 BIORECOGNITION ELEMENTS PERSPECTIVE ON MICROFLUIDIC BASED BIOSENSOR APPLICATIONS

Molecular diagnostics is at the heart of biosensor applications. The main purpose of the biorecognition element is to specifically capture the target analyte. This specificity can only be achieved if strong affinity between the biorecognition element and the target bioanalyte is achieved (63). There are different classes of biorecognition elements that greatly affect the design and performance of microfluidic-based biosensors.

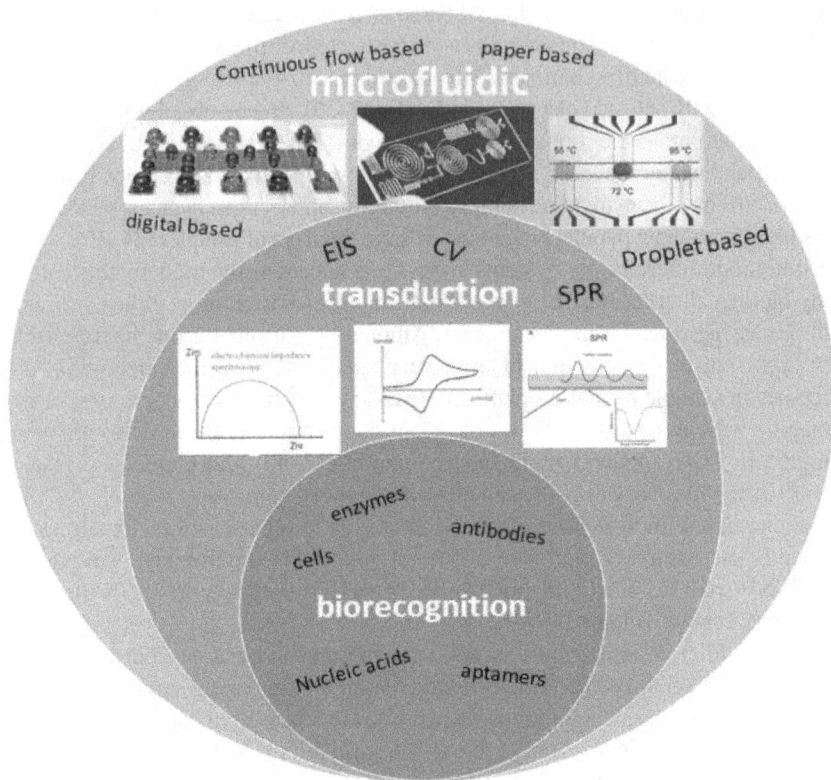

FIGURE 11.2   The building blocks of a microfluidic based biosensor.

11.4.1   ENZYME BASED

Enzyme-based biosensors are defined as systems in which specific enzyme molecules are immobilized on electrodes and catalyzed measurable changes in electroactive species in the sample in the presence of suitable substrate.

In the method developed by Yu et al. (64) for glucose determination, a biosensor system including the integration of glucose oxidase covalently immobilized on
single-walled carbon nanotube arrays into PDMS-based microfluidic channels was
used. This electrochemical based microfluidic device can detect up to 5×10^{-3} mol
L^{-1} glucose concentration. Glass wafers were used as the material and soft lithography approach was adopted. They argued that this method provides less consumption
in a shorter time, since very small amounts of liquid and enzymes are used in the
microchannels. In another study (65), the researchers produced a composite made of
nickel oxide nanoparticles (nNiO) and multi-walled carbon nanotubes (MWCNTs),
and integrated this nanocomposite into PDMS microchannels created using the pho-
tolithographic technique. Structural elucidation of this microfluidic nanochip was
performed in the presence of cholesterol oxidase (ChOx) and cholesterol esterase
bienzymes. The response studies on this nanobiochip reveal good reproducibility and

selectivity and a high sensitivity of 2.2 mA/mM/cm2. This integrated microfluidics biochip provides a promising low-cost platform for the rapid detection of biomolecules using minute samples.

While continuous microfluidic devices are less common in enzyme-based biosensor studies, droplet and digital microfluidics integrations are more common.

The freshness of fruits, vegetables, meat and fish and the determination of the degree of this freshness is among the subjects of biosensor studies. In their study for the determination of fish freshness, Itoh et al. developed a biosensor system based on the breakdown of ATP into uric acid as a result of a series of enzymatic reactions following the death of a fish. In the microfluidic device, in which ATP is used as a freshness indicator, extracts and reagents obtained from fish were processed as droplets. A large region is created in the flow channel to obtain a simpler procedure. A row of droplets was sent from each of the three branched flow channels, combined in the large region and transported to the detection region. ATP was determined using two enzymatic reactions involving glycerol kinase (GK) and glycerol-3-phosphate oxidase (G3PO). It has been determined that there is a correlation between the ATP concentrations obtained from the mackerel extracts and the resulting current (51).

The fact that paper, which is a preferred material especially in POC diagnostic studies, is included in the microfluidic integration, lies in its features such as being inexpensive, open to many chemical/biochemical applications and transporting liquids to capillaries without the need for external support forces (66).

Stating that paper is a powerful platform, Zhao et al. (67) in their study, proposed a microfluidic paper-based biosensor system for the multiple detections of some metabolic biomarkers. Unlike other paper-based devices, an array with eight sensors and a handheld potentiostat are available in this study. Enzymes corresponding to metabolic biomarkers (i.e. glucose oxidase, lactate oxidase and uricase) and an electron transfer mediator (potassium ferricyanide—$K_3(Fe(CN)_6)$) were located in the test regions of eight biosensing modules. This array of biosensors can detect several analytes in a sample and even perform multiple measurements for each analyte from a single run. It has been stated that simultaneous detection of glucose, lactate and uric acid in urine can be made using the device (Figure 11.3).

11.4.2 Antibody Based

Among the different types of bioreceptors, aptamers and antibodies are the most commonly used in the development of biosensors. Although aptamers have advantages such as being inexpensive, smaller in size and high affinity against antibodies, antibodies provide affinity only with targets for which they are highly immunogenic. The effectiveness of the antibody biorecognition element depends on the affinity of the immunocomplex formed between the antibody-antigen (63, 68).

In their study, Pinto et al. (69) developed a PDMS microfluidic immunosensor integrated into a metal-oxide-semiconductor (CMOS) optical detection system to detect cortisol in saliva rapidly and reliably. Immunosensors are structures in which antibodies are designed as bioreceptors for specific molecular recognition of antigens and a stable immunocomplex is formed as a result of interaction between them (70). Developed rapidly and non-invasively in saliva as an alternative to blood, this study

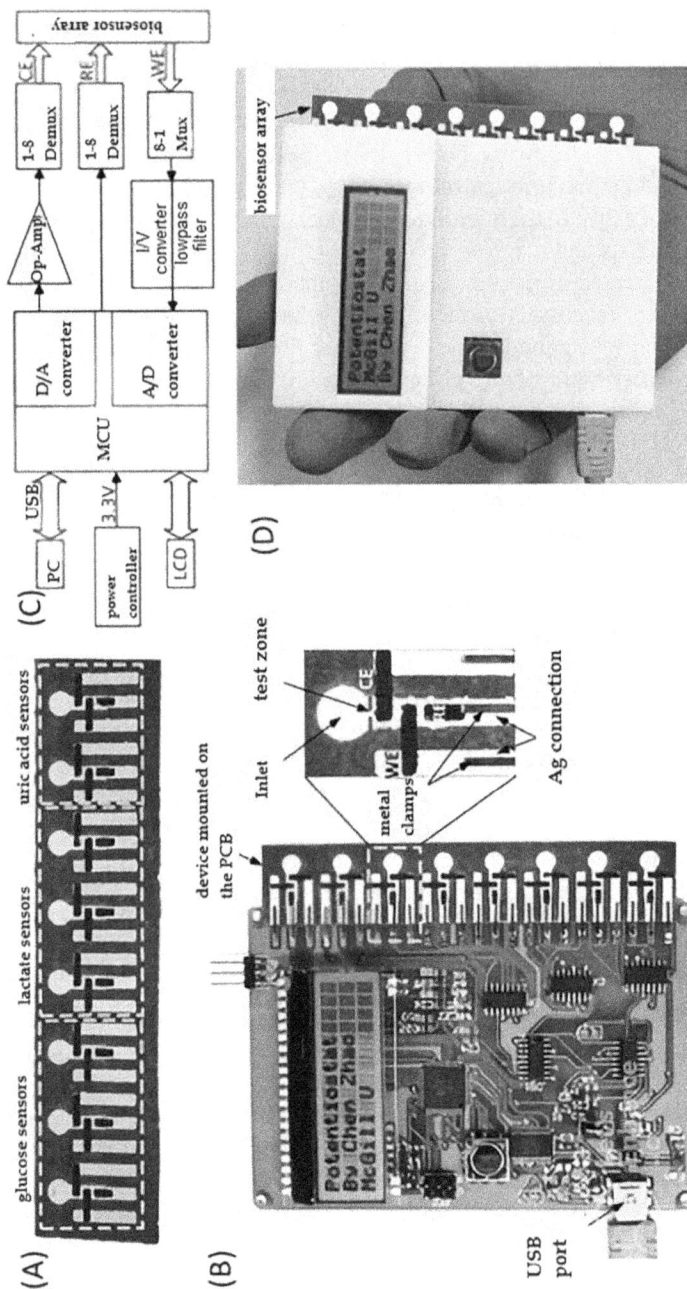

FIGURE 11.3 A) A paper-based electrochemical biosensor array. B) A microcontroller-based multiplexing potentiostat mounted with a paper-based biosensor array. C) Schematic diagram of the potentiostat architecture with eight measurement channels. D) A photograph of the paper-based biosensor array inserted into the potentiostat. Reprinted from ref. (67) with permission.

is based on the covalent immobilization of a polyclonal anti-IgG coating antibody (Ab) on a PDMS surface. Coating Ab binds capture Ab, a specific IgG for cortisol. Horseradish peroxidase (HRP)-labeled cortisol was added to the sample to compete with the cortisol in the sample. Cortisol measurement was performed by optical absorption at 450 nm. Researchers have optimized characteristics such as microfluidic geometry, immobilization parameters and immunoassay conditions. Under optimized conditions, cortisol could be analyzed in 35 minutes with a detection range of 0.01–20 ng/mL and a limit of detection (LOD) of 18 pg/mL.

Antibody-based microfluidic biosensors can also be designed to capture certain cancer cells in the circulation of patients. Circulating bladder tumor cells provide valuable information in the diagnosis and prognosis of the disease. Setting out for this purpose, Wang et al. (71) stated the difficulty of capturing bladder cancer cells and developed a microfluidic chip that captures this cell type with high specificity and efficiency. They showed that the microfluidic device they developed could be used to capture circulating bladder cancer cells based on antibody-BCMab1, a monoclonal antibody that binds to abnormally glycosylated integrin a3b1. The argument for this microfluidic platform is based on a polydimethylsiloxane (PDMS) chip functionalized with biotinylated BCMab1. With modifications, a herringbone or stripe channel pattern has also been added to the chip to increase cell-substrate contact and change the flow direction (Figure 11.4). Biotinylated antibodies and streptavidin were placed in the microchannel from the antibody inlets on either side of the sample inlets. The purpose of the outlets is to collect waste or samples. Herringbone patterns on the top layer increase the interaction between target cells and antibodies and increase cell capture efficiency. It has been stated that the microfluidic device designed in this way successfully captures 90% of cancer cells at flow rates of 10 μL/min and at various cell concentrations.

Among the various semiconductor electrodes, indium-tin oxide electrodes (ITO) are a promising material in devices developed for the characterization of biological systems. ITO surfaces maintain their stability under physiological conditions due to their polarizable properties (72). Due to its transparent structure, ITO enables multiple parameter measurements using optical and electrical techniques. ITO films have been successfully applied in many devices such as electrochromism, photovoltaics, sensors, plasmonic devices and light-emitting diodes (73). Seenivasan and colleagues (74) designed an optically clear indium tin oxide (ITO) three-electrode biosensor integrated with a microfluidic channel for immunosensing of prostate-specific membrane antigen (PSMA), a circulating prostate cancer biomarker expressed in prostate tissue. Cysteamine capped gold nanoparticles (N-AuNPs) covalently bonded with anti-PSMA antibody (Ab) were used to increase target analyte capture specificity. Polydimethylsiloxane (PDMS) microfluidic channel is preferred for high efficiency. Electrochemical changes in the redox probe as a result of binding of PSMA to its antibody on the N-AuNPs/ITO surface were followed by differential pulse voltammetry. The detection of PSMA-expressing cells and soluble PSMA was tested with the developed Microfluidic-integrated patterned ITO immunosensor. In conclusion, it was emphasized that the sensor has a suitable sensitivity and dynamic range for routine detection of circulating tumor cells of prostate cancer and can be adapted to detect other biomarkers/cancer cells.

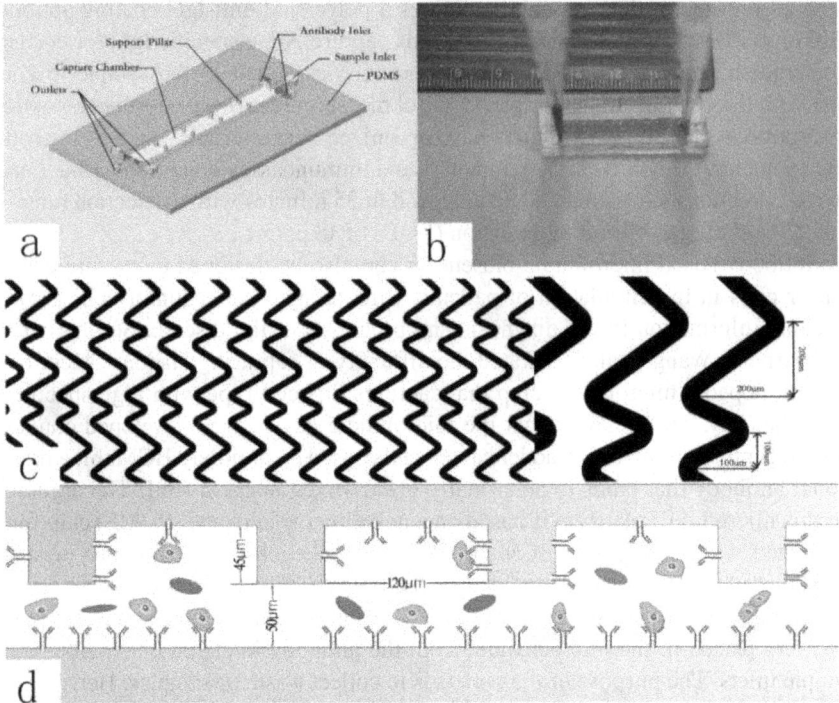

FIGURE 11.4 Microfluidic device diagram. a) Schematic presentation of the channel structure. b) The microfluidic device was used to capture target analyte. c) An array of the asymmetry and periodicity of the herringbone grooves. d) Illustration of chip capturing the target analyte. Reprinted from ref. (71) with permission.

11.4.3 APTAMER BASED

With their history of more than 30 years, aptamers are in increasing demand due to their advantages such as their stable three-dimensional structure and folding ability, which allows them to interact with target molecules through electrostatic interactions, hydrogen bonding, Van der Waals forces and electrostatic interactions. Aptamers, which are synthetic oligonucleotides (DNA or RNA) that bind to chemical and biological analyte targets through affinity interactions, can be developed for a variety of analytes such as small molecules, proteins and cells through an in vitro production process (75–77). Rapid advances in RNA and DNA synthesis have made possible the use of aptamers that offer high-affinity biomolecular recognition to a theoretically limitless variety of analytes. DNA and RNA aptamers compete with more established affinity ligands, including most immunoreceptors such as enzymes, lectins and antibodies, and are gaining more interest in biosensor studies. The high specificity of aptamers has also made it desirable for integration into microfluidic devices (77).

In their study, Nguyen et al. (78) developed a detection platform containing amine-terminated aptamer-modified gold electrodes for the specific, label-free detection of a lung cancer cell line (A549). This platform is a micro-device with coplanar electrode configuration and a simple microfluidic channel on a glass substrate and has been fabricated with the standard approach of photolithography and casting molding techniques. Gold substrates were modified with HS-PEG-COOH and a self-assembled monolayer was formed. After the carboxylic acid ends in the structure were activated with EDC/NHS, treatment with amine-labeled aptamers was carried out. (Figure 11.5).

Optical microscopic images and EIS data of the microfluidic biosensor device were obtained. Optical microscope observations and electrical impedance spectroscopy measurements have proven that the fabricated microchip can specifically and effectively identify A549 cells. The dominant capacitance element in the change of impedance is tuned to the appropriate frequency to evaluate the sensitivity of the biosensor. A linear relationship was found between capacitance change and cell concentration in the range of 1×10^5 to 5×10^5 cells/mL, with a correlation coefficient of up to 99%. It has been emphasized that this Aptmer-based Microfluidic biosensor has many superior features such as simplicity, rapidity, low cost, biocompatibility, selectivity and sensitivity for the diagnosis of lung cancer cells.

Paper is a frequently preferred material in the design of microfluidic platforms. Among the properties that make the paper an attractive material for microfluidic devices are the formation of microfluidic channels on the paper, trapping the fluid flow within the channels and therefore directing the fluid flow in a controlled manner (66). A paper-based aptamer-based microfluidic biosensor has been developed to detect different amounts of lead concentrations in water (79). The working principle of this aptamer-based biosensor is AuNP aggregation in the presence of lead ions, and the gold nanoparticles that form this aggregation show a color change from red to purple by interacting with NaCl. Two types of filter papers, Whatman No.1 and nylon filter papers, were used as the platform for this paper-based microfluidic test, and the properties of each were investigated separately. It was found that Pb2+ could be detected with linear trends at concentrations in the range of 10 nM to 1 mM on both Whatman No.1 and nylon filter papers. A detection limit of 1.2 nM and 0.7 nM was obtained for Whatman No. 1 and nylon filter papers, respectively. For practical purposes, Whatman No.1 filter paper was used for detection in real water samples and it was determined that it showed good potential for Pb2+ detection in environmental samples.

In another application-oriented study, which draws attention to the importance of monitoring biomarkers released in organ-on-a-chip models, a new aptamer-based electrochemical biosensing platform has been reported to monitor damage to cardiac organoids and to provide non-invasive and accurate information (80). The developed system is compatible with microfluidic platforms, which is low cost and easy to integrate with microfluidic bioreactors. Aptamers specific to CK-MB biomarkers secreted from heart tissue were used in this biosensor whose target analyte is creatine kinase (CK)-MB. Comparing the system they developed with antibody-based sensors, the researchers suggested that the proposed aptamer-based system is highly sensitive, selective and stable. It has been stated that it can analyze trace amounts of CK-MB secreted by cardiac organoids even in cases with after aptamer immobilization, surface characterization was done with AFM and electrochemical evaluation with EIS.

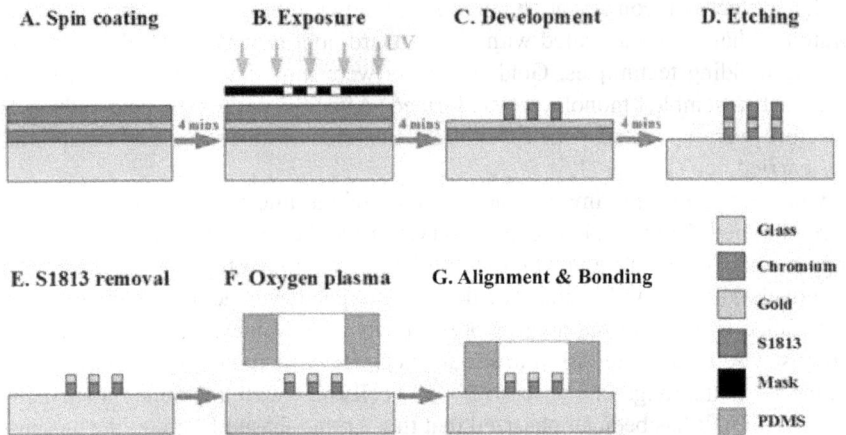

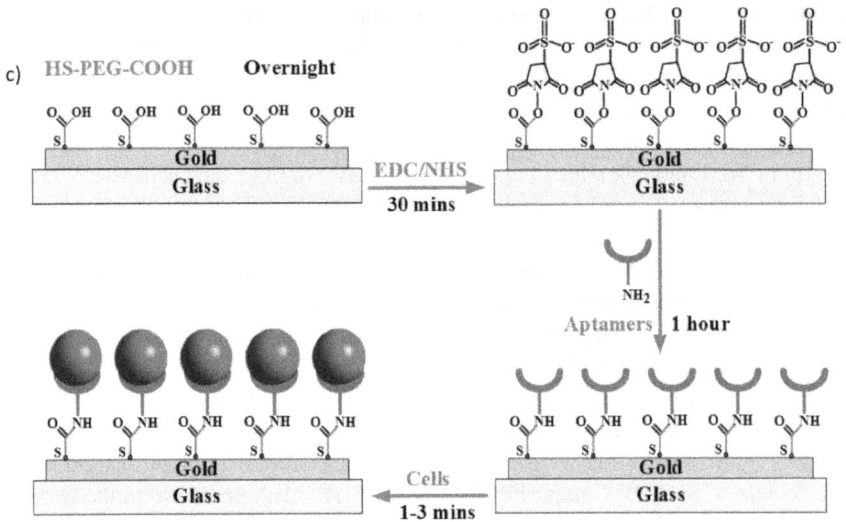

FIGURE 11.5 a) The fabrication process of the chip. b) Photograph of a fabricated microchannel device. c) Illustration of aptamer immobilization procedure onto gold substrate for binding of target cells. Reprinted from ref. (78) with permission.

What makes microfluidic-based biosensors relatively more attractive than their counterparts is that it aims to produce solutions for analyses that are closely related to public health. Noroviruses, which can be transmitted from person to person or through food/drink, are the main cause of viral gastroenteritis. There is an urgent need for new techniques due to the reasons such as norovirus, which is commonly detected by ELISA, the expensiveness of this method and the long time analysis. Therefore, a method developed for norovirus detection used 6-carboxyfluorescein-labeled norovirus-specific aptamer and multi-walled carbon nanotubes (MWCNT) and graphene oxide (GO) (81). In the presence of norovirus, fluorescence is recovered as a result of the release of the labeled aptamer from MWCNT or GO. An easy-to-make paper microfluidic platform based on nitrocellulose has been developed. Norovirus was successfully detected in the linear detection range from 13 ng·mL^{-1} to 13 µg·mL^{-1} with this aptamer-based, paper microfluidic biosensor. When using MWCNT or GO, the detection limits are 4.4 ng·mL^{-1} and 3.3 ng·mL^{-1}, respectively. This method, which is very practical and cost-effective, offers potential in the detection of noroviruses with high sensitivity and specificity. In the event of an epidemic that threatens public health, it will accelerate with its practicality of use.

11.5 TRANSDUCER PERSPECTIVE ON MICROFLUIDIC BASED BIOSENSORS APPLICATIONS

Classification of biosensors can also be made according to several different types of transducers. The transducer converts the biomolecule-analyte interaction into an optical (82) or electrical (83) signal. Transducer preference varies depending on the nature of the physicochemical change of the reaction occurring on the biosensing layer. Transducers are very important in increasing the sensitivity and detection limit of the biosensor. In the following sections, microfluidic-based biosensors will be discussed in terms of transducer selection.

11.5.1 ELECTROCHEMICAL BASED

Electrochemical transducers used in biosensors use either the redox activity of the analyte or the redox reflection of the interaction between the analyte and the biosensing unit. The electrons produced in the redox process are the current sense of the number of redox species involved in the process (84). Commonly electrochemical signal transducers are amperometry, voltammetry or EIS. Given an overview of the analytical techniques integrated with microfluidics, it is the electrochemical-based detections that are the most practical to implement and ideal for point-of-care studies. In addition to many advantages, microfluidic integrated electrochemical biosensors have limitations such as the limited number of biomarkers that can be used in a biosensor containing a disposable electrode for diagnosis, and the laborious and time-consuming production process. Lee et al. have eliminated many of these limitations in their study (85). Miniaturized microfluidic biosensors hold promise for POC analysis by integrating laboratory-on-a-chip technology and electrochemical analysis. Five chambers was used in the microfluidic electrochemical biosensing system developed to detect multiple biomarkers of pulmonary hypertension diseases in a single device. Each chamber is connected to an electrochemical sensor that will measure four different biomarkers and a reference. Manipulation of the flow path and sensitivity control are provided by micro valves. With this microfluidic device, pulmonary hypertension biomarkers called fibrinogen, adiponectin, low-density lipoprotein and 8-isoprostane can be detected simultaneously. Researchers emphasized that with the manufacturing approach of this device, a fast, miniature and sensitive diagnostic sensor is possible in a single device.

The electrochemical efficiency of the developed system was characterized by cyclic voltammetry (CV) and square wave voltammetry (SWV) after each channel was filled with $(K_3Fe(CN)_6)$. In cyclic voltammetry studies, the potential is observed against the electrode current density of the electrochemical cell system. When the analyte reacts with the biological component coated or immobilized on the electrode surface, a change in electric current occurs against the potential range. This current change indicates that there is an electron transfer in the electrochemical cell during the reaction of the biosensor electrode between the analyte and its biological component (86). SWV consists of a series of potential signals between the cathodic pulse and the anodic pulse of the same amplitude. The basic principle behind pulse techniques such as SWV is the difference in separation rates of charge and faradaic currents. The charge current decomposes much faster than faradaic current, decomposing as an exponential function, while the decomposition of faradaic current is inversely proportional to the square root of time. Therefore, as a result of each pulse, the capacitive current is negligible compared to that of faradaic current. Increasing the ratio of faradaic current to nonfaradaic current allows for a lower detection limit as well as a higher detection limit, making SWV very ideal for analytical purposes. SWV is a highly preferred technique because of its ability to operate at high frequency (87–88).

CV and SWV measurements were performed before and after the addition of the analyte. While no change was observed in chamber 1, which is the negative control, in other chambers, a decrease occurred in CV and SWV redox current signal, as evidence of immunocomplex, depending on the analyte concentrations in other chambers (85).

Droplet based electrochemical microfluidic device designs are also very common due to their small and portable size, low energy and reagent consumption, and rapid manipulation of liquids and suitable for various chemical and biological processes (51, 89–90).

In a study developed for the determination of 4-aminophenol (91), a droplet-based microfluidic device integrated with graphene-polyaniline (G-PANI/CPE) modified carbon paste electrode was used. The T-linked microfluidic platform was fabricated with microdroplets with an oil flow rate of 1.8 µL/min and an aqueous flow rate of 0.8 µL/min. The microchannel was expanded to cover the entire 3 electrodes, so that all electrochemical monitoring could be performed with a single droplet. Optimizing parameters such as flow rate, water ratio and applied sensing potential, the researchers emphasized that the use of G-PANI/CPE increased the 4-aminophenol response 2 times compared to the bare electrode. Under optimized conditions, the droplet-based system was able to detect 4-aminophenol in the detection range of 50–500 µM, with a limit of detection of 15.68 µM and a limit of quantification of

52.28 µM. The responses of the droplet based microfluidic biosensor were followed by both CV and chronoamperometry.

Paper, which we have repeated several times is a very popular material in micro-fluidic systems, plays the leading role in electrochemical integration studies, due to its features such as creating hydrophobic and hydrophilic models that allow the sample solution to move within microchannels. In addition to being an inexpensive material, paper is remarkable for its flexibility in electrochemical integration, rapid analysis and disposable nature. Paper based microfluidic biosensors can be produced with many different techniques such as photolithography, inkjet printing and wax printing (92–96). Fava et al. (97) developed a disposable microfluidic electrochemi-cal paper-based device consisting of 16 independent microfluidic channels for mul-tiplex analysis. The difference of this study is that the strategy of multiplexing both working and reference electrodes is used for the first time and the cumbersome and costly use of wax printer is avoided. 16 channels can be run at high efficiency and repeatability criteria by using more than one electrochemical technique. The elec-trochemical performance of the device was tested with cyclic voltammetry, square-wave voltammetry, differential-pulse voltammetry and chronoamperometry. This innovative detection platform offers wide application for clinical testing with safe and fast test time.

Another important point in analyzing biomarkers over long periods is the con-tinuity of the analysis. Any interruption can cause problems with the accuracy and applicability of the analysis. It is very important in situations where the functionality of organoids is evaluated as a result of their response to certain pharmaceutical com-pounds. In this study, Shin et al. (98) developed a regenerable, label-free microfluidic electrochemical biosensor for continuous measurement of cell-secreted biomarkers from an organoid culture (Figure 11.6). The developed electrochemical biosensor was integrated with a human liver tissue chip and the related metabolites (albumin and GST-α) could be analyzed for 7 days. Changes in biomarker level were success-fully measured with the developed device and the results are compatible with ELISA responses.

The researchers emphasize that this microfluidic EC biosensor, which can auto-matically and continuously detect soluble biomarkers, could be particularly useful for drug toxicity studies. In the study using a triple conventional electrode system, gold was preferred because of the high stability it offers in the working electrode. The electrical signal response of this biosensor based on antigen-antibody was fol-lowed by EIS. Table 11.1 presents information from the electrochemical biosensors integrated into the microfluidic device developed in the last 5 years.

11.5.2 Optical Based

Optical techniques detect the optical change caused by the interaction between the target analyte and the biological recognition element on the optical layer and convert the signal into measurable data that correlates with the analyte concentration in the sample (109). It has many remarkable features that make optical sensing one of the leading sensing methods in the biosensor field. These include low limit of detection, versatility, label-free, fast signal monitoring and detecting a wide variety of analytes

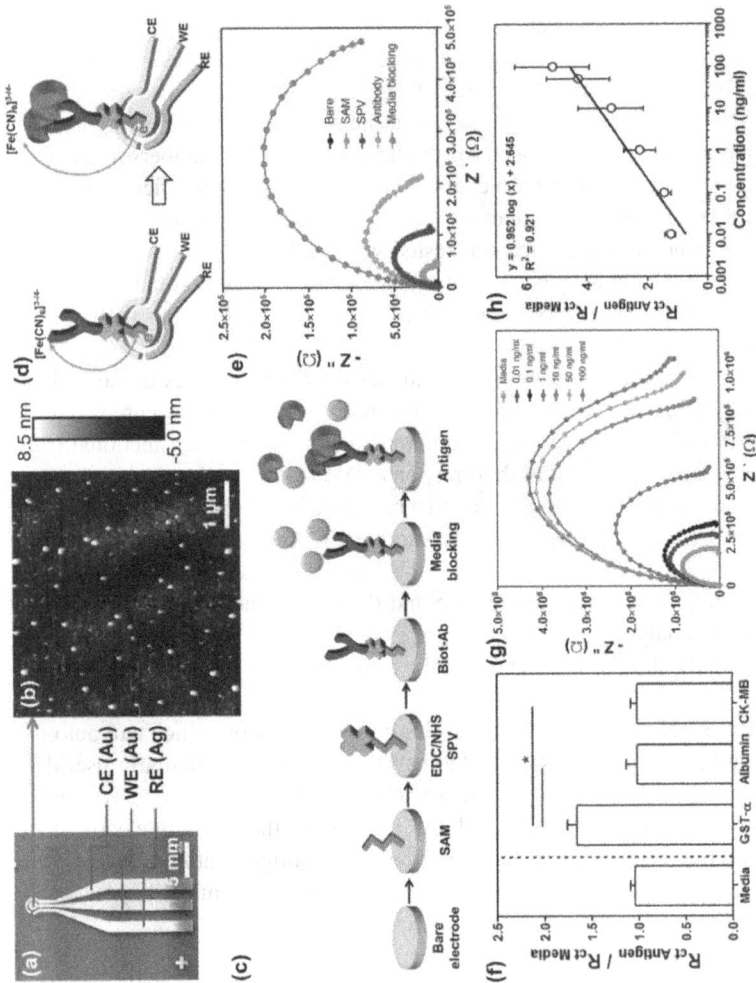

FIGURE 11.6 Detection principle of the label-free EC biosensing system by using microelectrode. a) Image of fabricated microelectrode containing the triple electrode system. b) AFM image of the bare WE surface. c) A schematic illustration for immobilization of antibody using SPV on the surface of the microelectrodes. d) Schematic of charge transfer after antigen binding upon antibody–antigen binding for $[K_3Fe(CN)_6]^{3-/4-}$ redox process. e) Nyquist plots obtained from measurements before and after the deposition of each layer. f) Selectivity study of albumin biosensor showing the effect of media incubation. g) Nyquist plots drawn for different standard human albumin concentrations. h) Calibration curve for human albumin plotted according to the normalized R_{ct} (R_{ct} antigen/R_{ct} media) values. Reprinted from ref. (98) with permission.

TABLE 11.1
Target Analytes, Procedures, and Some Analytical Parameters of Microfluidic-based Electrochemical Biosensors

Microfabrication process	Material	Biosensor Modification	Electrochemical Detection Type	Analyte	Determination Range	LOD	Ref
3D-printer	Acrylonitrile-butadiene-styrene) polymer	carbon nanotubes/gold nanoparticles followed by covalent binding of tyrosinase	CV	phenolic compound	10.0 to 200 nmol L−1	2.94 nmol L−1	(99)
UV-lithography	PDMS SU-8	CNTs/Au	EIS	viral DNA from Hepatitis C and genomic DNA from Mycobacterium tuberculosis	0.1 fM to 1 Pm	2 pM	(100)
photolithography	3D paper-based microfluidic screen-printed electrode	reduced graphene oxide-tetraethylene pentamine	CV	glucose	0.1 mM—25 mM	25 µM	(101)
3D printing mold	PDMS	SPE/GO-Chit	CV, DPV	cystatin C	1–10 mg/L	0.0078 mg/L	(102)
3D printing mold	photosensitive resin chip	β-CD, MWCNTs, NMo2C, MB, and urease	EIS- CV- DPV	Multiplex (MTX, LDH, UA, and urea)	nd	35 nM(MTX) 25 U/L (LDH) 450 nM (UA and Urea)	(103)
inkjection	filter paper	graphene quantum dots	SWV	uric acid and creatinine	0.010–3.0 µmol L−1	8.4 nmol L−1 (uric acid) and 3.7 nmol L−1 (creatinine)	(104)
channel based	PDMS	gold nanoflowers (Au NFs) and DNA tetrahedron structural probes	CV and amperometry	PSA	0.2 ng/mL	1–100 ng/Ml	(105)
photolithography	PDMS, ITO-coated glass	Nd	CV, DPV	B1 (FB1) and deoxynivalenol	97 pg/Ml (FB1) and 35 pg/mL (DON)	0.3–140 ppb (FB1) 0.2–60 ppb (DON)	(106)
soft lithography	Si wafer	Nd	amperometry	dopamine from cerebrospinal fluid and plasma	0.1 – 1000 nM	0.1 nM	(107)
photolithography	PDMS SU-8	ZnO nanorods	amperometry	multi virus (H1N1, H5N1, and H7N9)	1–10 ng/ml	1 pg/ml	(108)

*nd: not determined

simultaneously (110). Surface plasmon resonance (SPR) is a real-time sensing technique that contains a thin gold film in its structure, is highly sensitive, label-free and is suitable for miniaturization (111). Integrating SPR sensing technologies into microfluidics allows the incorporation of superior properties of microfluidics such as increasing the reaction rate, reducing the diffusion time and renewing the surface more efficiently. Thanks to microfluidics, automation allows one to increase the sensitivity and achieve advanced performance (112–113). Rapid and timely detection of infectious agents such as Escherichia coli (E. coli) and Staphylococcus aureus (S. aureus) is critical for the treatment of infectious diseases. In the study (114) with the developed microfluidic-integrated surface plasmon resonance (SPR) platform, a portable, multiplex proposal that rapidly detects and measures both bacteria is presented (Figure 11.7). The platform reliably detects E. coli in phosphate buffered saline (PBS) and peritoneal dialysis (PD) fluid at concentrations ranging from ~105 to 3.2×107 CFUs/mL.

The multiplexing and specificity capability of the platform was also tested with S. aureus samples. During the development of the microfluidic chip, glass wafers were coated with gold. The SPR platform is custom made for the compatibility of microfluidic integration. Prism coupling is used to provide momentum conservation for plasmon excitation by an external light source. As a result, a plasmonic-based microchip has been developed for sensitive pathogen detection with potential use for POC applications.

A limited number of detection methods have been developed for S-layer protein (SLP), a relatively large protein found in the cell wall or cell membrane of many bacteria. Zhuo et al. (115), using a TiO_2-coated, porous silicon (PSi)-based microfluidic biosensor they developed, demonstrated a label-free, highly selective and sensitive

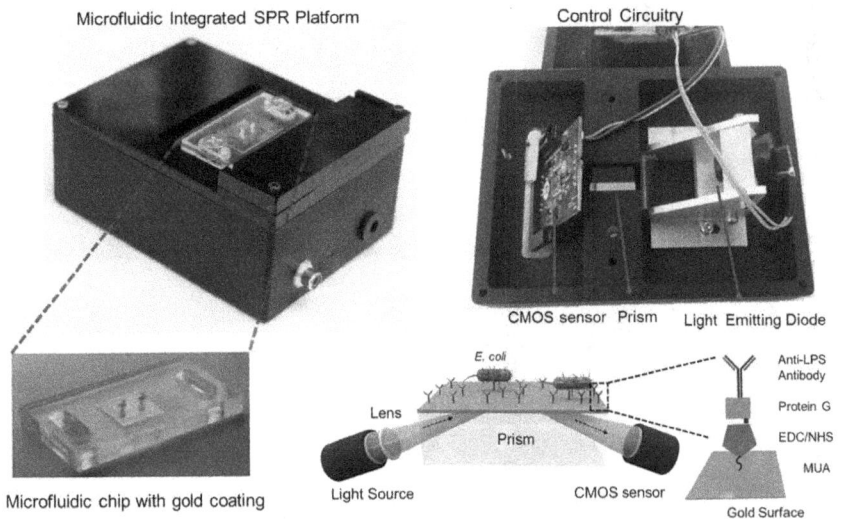

FIGURE 11.7 Portable plasmonic platform for pathogen detection and quantification. Reprinted from ref. (114) with permission.

test for this protein. The fact that the biosensor is unlabeled is because the detection procedure relies on light reflection interference. The monoclonal SLP antibody was non-covalently immobilized to the PSi surface by the creation of specific recognition sites via spacer protein A. In this way, the target protein could be captured on optical basis with very high sensitivity, wide detecti on range (0.001 ~ 10µg/mL) and fast response time (5min).

The tremendously rapid applications in proteomics and drug discovery demand simultaneous screening of many analytes from a single SPR imaging experiment. The SPR imaging sensor is typically integrated with continuous flow-through channels. With this integration, it is possible to detect different concentrations of the same analyte or different concentrations of different analytes (116), to detect a large number of analytes with temperature control (117) or to prevent situations such as signal mixing caused by nonspecific interferences (118). Different microfluidic manufacturing processes can be combined for more analytical characteristics. Karuwan et al. designed an electrowetting-on-dielectric digital microfluidic device in real samples where iodide is the target analyte. This design includes a 3-electrode system, including the Ag reference electrode, Pt counter electrode and Au working electrode, with a T-junction to combine the sample droplets and buffer reagent. The working principle was followed by CV and the electroanalysis was completed within 12 hours (119). Table 11.2 presents information from the electrochemical biosensors integrated into the microfluidic device developed in the last 5 years.

Examining the table, it will be seen that many different biosensing modifications give perspective to microfluidic based devices. While these different analytical modifications create extraordinary differences in the detection range of the target analyte, they also reflect the hope of detecting very different analytes. Microfluidic integration contributes to this window with POC, and LOC applications. Considering that our age is the age of speed, rapid and reliable detection of many assays from agriculture to biomedicine becomes the primary target.

11.6   CONCLUSION AND FUTURE REMARKS

Biosensors are powerful tools that can analyze many different analytes with high sensitivity, from drug discovery, agricultural monitoring, clinical diagnosis and prognosis to biosafety. On the other hand, while performing qualitative and quantitative analyses of many analytes, limited portability and problems from laboratory-based techniques bring many errors. The development of biosensor platforms using microfluidic technology allows the emergence of a compact, small-sized device with a portable and manipulable format and all the necessary instrumentation. All the examples given in this chapter demonstrate that portable, upgradeable high-throughput biosensors can be successfully integrated at acceptable cost, targeting the assay of different analytes. The most attractive aspect of microfluidic integration is that very small amounts of analytes, such as µL or nL, are sufficient for analysis. Microfluidics coupled electrochemical and optical detection systems are particularly advantageous compared to conventional detection systems due to their flexibility, rapid analysis, low production costs, practicality and portability. In terms of POC diagnostics, microfluidic integrated biosensors are particularly promising, although systems produced in this field are

TABLE 11.2
Target Analytes, Procedures, and Some Analytical Parameters of Microfluidic-Based Optical Biosensors

Microfabrication Process	Material	Biosensor Modification	Optical Detection Type	Analyte	Determination Range	LOD	Time	References
SU8 photolithography	PDMS	Nd	fluorescent	Salmonella typhimurium	1.4×10^2 to 1.4×10^6 CFU/mL	58 CFU/mL	15 min	(120)
soft lithography	PDMS, SU-8 negative photoresist	quantum dots aptamer functionalized graphene oxide	fluorescent	peanut allergen	nd	56 ng/mL	10 min	(121)
soft lithography	PDMS	11-MUA SAMs	localized surface plasmon resonance	multiplex (immunoglobulins, cytokines, and C-reactive protein)	nd	nd	4 min	(122)
soft lithography	Square glass substrates	gold film	SPR	HER2	15–75 ng mL−1	nd	nd	(123)
four-channel microfluidic flow cell	SPRi	quantum dot-coupled aptasensor	SPR	insulin	0.8–250 pM	800 fM	nd	(124)
soft lithography	PDMS, paper	gold nanostars (GNSs), and nanorods (NRs)	Surface-enhanced Raman scattering (SERS)	human colorectal cancer cells (SW480) and human peripheral blood mononuclear cells	nd	nd	30 min	(125)
soft lithography	PDMS- glass	Ag NPs	Surface enhanced Raman scattering (SERS)	CA153, CA125 and CEA	nd	LOD of CA153, CA125 and CEA 0.01 U/mL, 0.01 U/mL and 1 pg/mL in serum respectively	nd	(126)
Photolithography	PDMS, paper SU-8	Nd	colorimetric	lactate concentration in real human sweat samples	nd	nd	nd	(127)
3D printing and surface plasma bonding	polystyrene microspheres	combination AuNPs, magnetic nanoparticles, modified with the capture antibodies	colorimetric	E. coli O157:H7	50 CFU/mL	5.0×10^1–5.0×10^4 CFU/mL	20 min	(128)

limited due to the complexity of the system. Personalized POC diagnostic tools based on microfluidics have unique potential for diseases, genetic screening and designing personalized treatments. Especially recent developments in the discovery of biomarkers accelerate this potential. Many research groups mentioned in this chapter produce and characterized integrated microfluidic technologies for simultaneous evaluation of multi-analytes. However, the speed of industrialization of the developed products is not as high as the speed of academic studies. Building a bridge between academic studies and converting to industrial production will give a great impetus to personal health care rates. The society's shift from traditional biochemical assays to microfluidic-based clinical diagnostic devices can be considered an excellent advance in POC and effective disease diagnosis and treatment. In addition, extensive collection of large amounts of data rather than a single patient should be the first step in overcoming the difficulties encountered.

REFERENCES

1. Convery N, Gadegaard N. 2019. 30 years of microfluidics. *Micro and Nano Engineering*. 2: 76–91
2. Whitesides GM. 2006. The origins and the future of microfluidics. *Nature*. 442(7101): 368–373
3. Zimmerman WB (Ed.). 2006. Microfluidics: History, theory and applications. *Springer Science & Business Media*. 466
4. Mark D, Haeberle S, Roth G, Von Stetten F, Zengerle R. 2010. Microfluidic lab-on-a-chip platforms: Requirements, characteristics and applications. *Microfluidics Based Microsystems*. 305–376
5. Garg S, Heuck G, Ip S, Ramsay E. 2016. Microfluidics: A transformational tool for nanomedicine development and production. *Journal of Drug Targeting*. 24(9): 821–835
6. Streets AM, Huang Y. 2013. Chip in a lab: Microfluidics for next generation life science research. *Biomicrofluidics*. 7(1): 011302
7. Chiu DT, Demello AJ, Di Carlo D, Doyle PS, Hansen C, Maceiczyk RM, Wootton RC. 2017. Small but perfectly formed? Successes, challenges, and opportunities for microfluidics in the chemical and biological sciences. *Chem*. 2(2): 201–223
8. Prakash S, Pinti M, Bhushan B. 2012. Theory, fabrication and applications of microfluidic and nanofluidic biosensors. *Philosophical Transactions of the Royal Society A: Mathematical, Physical and Engineering Sciences*. 370(1967): 2269–2303
9. Nikoleli GP, Siontorou CG, Nikolelis DP, Bratakou S, Karapetis S, Tzamtzis N. 2018. Biosensors based on microfluidic devices lab-on-a-chip and microfluidic technology. *Nanotechnology and Biosensors*. 375–394
10. Choi S, Goryll M, Sin LYM, Wong PK, Chae J. 2011. Microfluidic-based biosensors toward point-of-care detection of nucleic acids and proteins. *Microfluidics and Nanofluidics*. 10(2): 231–247
11. Rivet C, Lee H, Hirsch A, Hamilton S, Lu H. 2011. Microfluidics for medical diagnostics and biosensors. *Chemical Engineering Science*. 66(7): 1490–1507
12. Nikhil B, Pawan J, Nello F, Pedro E. 2016. Introduction to biosensors. *Essays Biochem*. 60(1): 1–8
13. Wang J, Ren Y, Zhang B. 2020. Application of microfluidics in biosensors. In *Advances in Microfluidic Technologies for Energy and Environmental Applications*. IntechOpen
14. Alam MK, Koomson E, Zou H, Yi C, Li CW, Xu T, Yang M. 2018. Recent advances in microfluidic technology for manipulation and analysis of biological cells (2007–2017). *Analytica Chimica Acta*. 1044: 29–65

15. Luka G, Ahmadi A, Najjaran H, Alocilja E, DeRosa M, Wolthers K, Hoorfar M. 2015. Microfluidics integrated biosensors: A leading technology towards lab-on-a-chip and sensing applications. *Sensors*. 15(12): 30011–30031

16. Shen MY, Li BR, Li, YK. 2014. Silicon nanowire field-effect-transistor based biosensors: From sensitive to ultra-sensitive. *Biosensors and Bioelectronics*. 60: 101–111

17. Halldorsson S, Lucumi E, Gómez-Sjöberg R, Fleming RM. 2015. Advantages and challenges of microfluidic cell culture in polydimethylsiloxane devices. *Biosensors and Bioelectronics*. 63: 218–231

18. Burger R, Amato L, Boisen A. 2016. Detection methods for centrifugal microfluidic platforms. *Biosensors and Bioelectronics*. 76: 54–67

19. Delgado SMT, Kinahan DJ, Julius LAN, Mallette A, Ardila DS, Mishra R, Mager D. 2018. Wirelessly powered and remotely controlled valve-array for highly multiplexed analytical assay automation on a centrifugal microfluidic platform. *Biosensors and Bioelectronics*. 109: 214–223

20. Reyes DR, Iossifidis D, Auroux PA, Manz A. 2002. Micro total analysis systems. 1. Introduction, theory, and technology. *Analytical Chemistry*. 74(12): 2623–2636

21. Becker H, Locascio LE. 2002. Polymer microfluidic devices. *Talanta*. 56(2): 267–287

22. Iliescu C, Taylor H, Avram M, Miao J, Franssila S. 2012. A practical guide for the fabrication of microfluidic devices using glass and silicon. *Biomicrofluidics*. 6(1): 016505

23. Hwang J, Cho YH, Park MS, Kim BH. 2019. Microchannel fabrication on glass materials for microfluidic devices. *International Journal of Precision Engineering and Manufacturing*. 20(3): 479–495

24. Sia SK, Whitesides GM. 2003. Microfluidic devices fabricated in poly (dimethylsiloxane) for biological studies. *Electrophoresis*. 24(21): 3563–3576

25. Rodrigues RO, Lima R, Gomes HT, Silva AM. 2015. Polymer microfluidic devices: An overview of fabrication methods. *U. Porto Journal of Engineering*. 1(1): 67–79

26. Liu KK, Wu RG, Chuang YJ, Khoo HS, Huang SH, Tseng FG. 2010. Microfluidic systems for biosensing. *Sensors*. 10(7): 6623–6661

27. Dixon C, Lamanna J, Wheeler AR. 2017. Printed microfluidics. *Advanced Functional Materials*. 27(11): 1604824

28. Qin D, Xia Y, Whiteside GM. 2010. Soft lithography for micro-and nanoscale patterning. *Nature Protocols*. 5(3): 491

29. Sohrabi S, Moraveji MK. 2020. Droplet microfluidics: Fundamentals and its advanced applications. *RSC Advances*. 10(46): 27560–27574

30. Sista R, Hua Z, Thwar P, Sudarsan A, Srinivasan V, Eckhardt A, Pamula V. 2008. Development of a digital microfluidic platform for point of care testing. *Lab on a Chip*. 8(12): 2091–2104

31. Fair RB, Khlystov A, Tailor TD, Ivanov V, Evans RD, Srinivasan V, Zhou J. 2007. Chemical and biological applications of digital-microfluidic devices. *IEEE Design & Test of Computers*. 24(1): 10–24

32. Su F, Hwang W, Chakrabarty K. 2006. Droplet routing in the synthesis of digital microfluidic biochips. *Proceedings of the Design Automation & Test in Europe Conference*. 1: 1–6

33. Xu T, Chakrabarty K. 2008. A droplet-manipulation method for achieving high-throughput in cross-referencing-based digital microfluidic biochips. *IEEE Transactions on Computer-Aided Design of Integrated Circuits and Systems*. 27(11): 1905–1917

34. Jain V, Raj TP, Deshmukh R, Patrikar R. 2017. Design, fabrication and characterization of low cost printed circuit board based EWOD device for digital microfluidics applications. *Microsystem Technologies*. 23(2): 389–397

35. Wei Q, Yao W, Gu L, Fan B, Gao Y, Yang L, Che C. 2021. Modeling, simulation, and optimization of electrowetting-on-dielectric (EWOD) devices. *Biomicrofluidics*. 15(1): 014107

36. Shen HH, Fan SK, Kim CJ, Yao DJ. 2014. EWOD microfluidic systems for biomedical applications. *Microfluidics and Nanofluidics*. 16(5): 965–987

37. Li HF, Lin JM. 2009. Applications of microfluidic systems in environmental analysis. *Analytical and Bioanalytical Chemistry*. 393(2): 555–567

38. Jokerst JC, Emory JM, Henry CS. 2012. Advances in microfluidics for environmental analysis. *Analyst*. 137(1): 24–34

39. Vasudev A, Kaushik A, Jones K, Bhansali S. 2013. Prospects of low temperature co-fired ceramic (LTCC) based microfluidic systems for point-of-care biosensing and environmental sensing. *Microfluidics and Nanofluidics*. 14(3–4): 683–702

40. Pol R, Céspedes F, Gabriel D, Baeza M. 2017. Microfluidic lab-on-a-chip platforms for environmental monitoring. *TrAC Trends in Analytical Chemistry*. 95: 62–68

41. Song K, Li G, Zu X, Du Z, Liu L, Hu, Z. 2020. The fabrication and application mechanism of microfluidic systems for high throughput biomedical screening: A review. *Micromachines*. 11(3): 297

42. Faustino V, Catarino SO, Lima R, Minas G. 2016. Biomedical microfluidic devices by using low-cost fabrication techniques: A review. *Journal of Biomechanics*. 49(11): 2280–2292

43. Pihl J, Karlsson M, Chiu DT. 2005. Microfluidic technologies in drug discovery. *Drug Discovery Today*. 10(20): 1377–1383

44. Kang L, Chung BG, Langer R, Khademhosseini A. 2008. Microfluidics for drug discovery and development: From target selection to product lifecycle management. *Drug Discovery Today*. 13(1–2): 1–13

45. Lombardi D, Dittrich PS. 2010. Advances in microfluidics for drug discovery. *Expert Opinion on Drug Discovery*. 5(11): 1081–1094

46. Mairhofer J, Roppert K, Ertl P. 2009. Microfluidic systems for pathogen sensing: A review. *Sensors*. 9(6): 4804–4823

47. Lei KF. 2012. Microfluidic systems for diagnostic applications: A review. *Journal of Laboratory Automation*. 17(5): 330–347

48. Rabiee N, Ahmadi S, Fatahi Y, Rabiee M, Bagherzadeh M, Dinarvand R, Webster TJ. 2020. Nanotechnology-assisted microfluidic systems: From bench to bedside. *Nanomedicine*. 16(3): 237–258

49. Damit B. 2017. Droplet-based microfluidics detector for bioaerosol detection. *Aerosol Science and Technology*. 51(4): 488–500

50. Woronoff G, El Harrak A, Mayot E, Schicke O, Miller OJ, Soumillion P, Ryckelynck M. 2011. New generation of amino coumarin methyl sulfonate-based fluorogenic substrates for amidase assays in droplet-based microfluidic applications. *Analytical Chemistry*. 83(8): 2852–2857

51. Itoh D, Sassa F, Nishi T, Kani Y, Murata M, Suzuki H. 2012. Droplet-based microfluidic sensing system for rapid fish freshness determination. *Sensors and Actuators B: Chemical*. 171: 619–626

52. Yu Z, Boehm CR, Hibberd JM, Abell C, Haseloff J, Burgess SJ, Reyna-Llorens I. 2018. Droplet-based microfluidic analysis and screening of single plant cells. *PLoS One*. 13(5):e0196810

53. Cao J, Köhler JM. 2015. Droplet-based microfluidics for microtoxicological studies. *Engineering in Life Sciences*. 15(3): 306–317

54. Cedillo-Alcantar DF, Han YD, Choi J, Garcia-Cordero JL, Revzin A. 2019. Automated droplet-based microfluidic platform for multiplexed analysis of biochemical markers in small volumes. *Analytical Chemistry*. 91(8): 5133–5141

55. Zhao-Miao LIU, Yang Y, Yu DU, Yan PANG. 2017. Advances in droplet-based microfluidic technology and its applications. *Chinese Journal of Analytical Chemistry*. 45(2): 282–296

56. Feng H, Zheng T, Li M, Wu J, Ji H, Zhang J, Guo J. 2019. Droplet-based microfluidics systems in biomedical applications. *Electrophoresis*. 40(11): 580–1590

57. Sánchez Barea J, Lee J, Kang DK. 2019. Recent advances in droplet-based microfluidic technologies for biochemistry and molecular biology. *Micromachines*. 10(6): 412

58. Mok J, Mindrinos MN, Davis RW, Javanmard M. 2014. Digital microfluidic assay for protein detection. *Proceedings of the National Academy of Sciences*. 111(6): 2110–2115

59. Luan L, Evans RD, Jokerst NM, Fair RB. 2008. Integrated optical sensor in a digital microfluidic platform. *IEEE Sensors Journal*. 8(5): 628–635

60. Farzbod A, Moon H. 2018. Integration of reconfigurable potentiometric electrochemical sensors into a digital microfluidic platform. *Biosensors and Bioelectronics*. 106: 37–42

61. Eicher D, Merten CA. 2011. Microfluidic devices for diagnostic applications. *Expert Review of Molecular Diagnostics*. 11(5): 505–519

62. Kumar S, Kumar S, Ali MA, Anand P, Agrawal VV, John R, Malhotra BD. 2013. Microfluidic-integrated biosensors: Prospects for point-of-care diagnostics. *Biotechnology Journal*. 8(11): 1267–1279

63. Morales MA, Halpern JM. 2018. Guide to selecting a biorecognition element for biosensors. *Bioconjugate Chemistry*. 29(10): 3231–3239

64. Yu J, Le Roux R, Gu Y, Yunus K, Matthews S, Shapter JG, Fisher AC. 2008. Integration of enzyme immobilised single-walled carbon nanotube arrays into microfluidic devices for glucose detection. In *2008 International Conference on Nanoscience and Nanotechnology*, pp. 137–140. IEEE

65. Ali MA, Srivastava S, Solanki PR, Reddy V, Agrawal VV, Kim C, Malhotra BD. 2013. Highly efficient bienzyme functionalized nanocomposite-based microfluidics biosensor platform for biomedical application. *Scientific Reports*. 3(1): 1–9

66. Li X, Ballerini DR, Shen W. 2012. A perspective on paper-based microfluidics: Current status and future trends. *Biomicrofluidics*. 6(1): 011301

67. Zhao C, Thuo, MM, Liu X. 2013. A microfluidic paper-based electrochemical biosensor array for multiplexed detection of metabolic biomarkers. *Science and Technology of Advanced Materials*. doi:10.1088/1468-6996/14/5/054402

68. Chambers JP, Arulanandam BP, Matta LL, Weis A, Valdes JJ. 2008. Biosensor recognition elements. *Current Issues in Molecular Biology*. 10(1–2): 1–12

69. Pinto V, Sousa P, Catarino SO, Correia-Neves M, Minas G. 2017. Microfluidic immunosensor for rapid and highly-sensitive salivary cortisol quantification. *Biosensors and Bioelectronics*. 90: 308–313

70. Ekins R. 1999. Immunoassay and other ligand assays: From isotopes to luminescence. *Journal of Clinical Ligand Assay*. 22(1): 61–77

71. Wang Y, Liu Q, Men T, Liang Y, Niu H, Wang J. 2020. A microfluidic system based on the monoclonal antibody BCMab1 specifically captures circulating tumor cells from bladder cancer patients. *Journal of Biomaterials Science, Polymer Edition*. 31(9): 1199–1210

72. Grochowska K, Siuzdak K, Śliwiński G. 2015. Properties of an indium tin oxide electrode modified by a laser nanostructured thin Au film for biosensing. *European Journal of Inorganic Chemistry*. 7: 1275–1281

73. Bouden S, Dahi A, Hauquier F, Randriamahazaka H, Ghilane J. 2016. Multifunctional indium tin oxide electrode generated by unusual surface modification. *Scientific Reports*. 6(1): 1–9

74. Seenivasan R, Singh CK, Warrick JW, Ahmad N, Gunasekaran S. 2017. Microfluidic-integrated patterned ITO immunosensor for rapid detection of prostate-specific membrane antigen biomarker in prostate cancer. *Biosensors and Bioelectronics*. 95: 160–167

75. MacKay S, Wishart D, Xing JZ, Chen J. 2014. Developing trends in aptamer-based biosensor devices and their applications. *IEEE Transactions on Biomedical Circuits and Systems*. 8(1): 4–14

76. Rozenblum GT, Lopez VG, Vitullo AD, Radrizzani M. 2016. Aptamers: Current challenges and future prospects. *Expert Opinion on Drug Discovery*. 11(2): 127–135

77. Lin Q, Nguyen T. 2009. Aptamer-based microfluidic biosensors. In *2009 9th IEEE Conference on Nanotechnology (IEEE-Nano)*, pp. 812–814. IEEE

78. Nguyen NV, Yang CH, Liu CJ, Kuo CH, Wu DC, Jen CP. 2018. An aptamer-based capacitive sensing platform for specific detection of lung carcinoma cells in the microfluidic chip. *Biosensors*. 8(4): 98

79. Fakhri N, Hosseini M, Tavakoli O. 2018. Aptamer-based colorimetric determination of Pb 2+ using a paper-based microfluidic platform. *Analytical Methods*. 10(36): 4438–4444

80. Shin SR, Zhang YS, Kim DJ, Manbohi A, Avci H, Silvestri A, Khademhosseini A. 2016. Aptamer-based microfluidic electrochemical biosensor for monitoring cell-secreted trace cardiac biomarkers. *Analytical Chemistry*. 88(20): 10019–10027

81. Weng X, Neethirajan S. 2017. Aptamer-based fluorometric determination of norovirus using a paper-based microfluidic device. *Microchimica Acta*. 184(11): 4545–4552

82. Damborský P, Švitel J, Katrlík J. 2016. Optical biosensors. *Essays in Biochemistry*. 60(1): 91–100

83. Ronkainen NJ, Halsall HB, Heineman WR. 2010. Electrochemical biosensors. *Chemical Society Reviews*. 39(5): 1747–1763

84. Rackus DG, Shamsi MH, Wheeler AR. 2015. Electrochemistry, biosensors and microfluidics: A convergence of fields. *Chemical Society Reviews*. 44(15): 5320–5340

85. Lee G, Lee J, Kim J, Choi HS, Kim J, Lee S, Lee H. 2017. Single microfluidic electrochemical sensor system for simultaneous multi-pulmonary hypertension biomarker analyses. *Scientific Reports*. 7(1): 1–8

86. Elgrishi N, Rountree KJ, McCarthy BD, Rountree ES, Eisenhart TT, Dempsey JL. 2018. A practical beginner's guide to cyclic voltammetry. *Journal of Chemical Education*. 95(2): 197–206

87. Lovrić M. 2010. Square-wave voltammetry. In *Electroanalytical Methods*, pp. 121–145. Springer

88. Chen A, Shah B. 2013. Electrochemical sensing and biosensing based on square wave voltammetry. *Analytical Methods*. 5(9): 2158–2173

89. Srikanth S, Mohan JM, Raut S, Dubey SK, Ishii I, Javed A, Goel S. 2021. Droplet based microfluidic device integrated with ink jet printed three electrode system for electrochemical detection of ascorbic acid. *Sensors and Actuators A: Physical*. 325: 112685

90. Suea-Ngam A, Rattanarat P, Chailapakul O, Srisa-Art M. 2015. Electrochemical droplet-based microfluidics using chip-based carbon paste electrodes for high-throughput analysis in pharmaceutical applications. *Analytica Chimica Acta*. 883: 45–54

91. Rattanarat P, Suea-Ngam A, Ruecha N, Siangproh W, Henry CS, Srisa-Art M, Chailapakul O. 2016. Graphene-polyaniline modified electrochemical droplet-based microfluidic sensor for high-throughput determination of 4-aminophenol. *Analytica Chimica Acta*. 925: 51–60

92. Carvalhal RF, Simão Kfouri M, de Oliveira Piazetta MH, Gobbi AL, Kubota LT. 2010. Electrochemical detection in a paper-based separation device. *Analytical Chemistry*. 82(3): 1162–1165

93. Wu Y, Xue P, Hui KM, Kang Y. 2014. A paper-based microfluidic electrochemical immunodevice integrated with amplification-by-polymerization for the ultrasensitive multiplexed detection of cancer biomarkers. *Biosensors and Bioelectronics*. 52: 180–187

94. Loo JF, Ho AH, Turner AP, Mak WC. 2019. Integrated printed microfluidic biosensors. *Trends in Biotechnology*. 37(10): 1104–1120

95. Cate DM, Adkins JA., Mettakoonpitak J, Henry CS. 2015. Recent developments in paper-based microfluidic devices. *Analytical Chemistry*. 87(1): 19–41

96. Delaney JL, Hogan CF, Tian J, Shen W. 2011. Electrogenerated chemiluminescence detection in paper-based microfluidic sensors. *Analytical Chemistry*. 83(4): 1300–1306

97. Fava EL, Silva TA, do Prado TM, de Moraes FC, Faria RC, Fatibello-Filho O. 2019. Electrochemical paper-based microfluidic device for high throughput multiplexed analysis. *Talanta*. 203: 280–286

98. Shin SR, Kilic T, Zhang YS, Avci H, Hu N, Kim D, Khademhosseini A. 2017. Label-free and regenerative electrochemical microfluidic biosensors for continual monitoring of cell secretomes. *Advanced Science*. 4(5): 1600522

99. Caetano FR, Carneiro EA, Agustini D, Figueiredo-Filho LCS, Banks CE, Bergamini MF, Marcolino-Junior LH. 2018. Combination of electrochemical biosensor and textile threads: A microfluidic device for phenol determination in tap water. *Biosensors and Bioelectronics*. 99: 382–388

100. Zribi B, Roy E, Pallandre A, Chebil S, Koubaa M, Mejri N, Haghiri-Gosnet AM. 2016. A microfluidic electrochemical biosensor based on multiwall carbon nanotube/ferrocene for genomic DNA detection of Mycobacterium tuberculosis in clinical isolates. *Biomicrofluidics*. 10(1): 014115

101. Cao L, Han GC, Xiao, H., Chen Z, Fang C. 2020. A novel 3D paper-based microfluidic electrochemical glucose biosensor based on rGO-TEPA/PB sensitive film. *Analytica Chimica Acta*. 1096: 34–43

102. Devi KS, Krishnan UM. 2020. Microfluidic electrochemical immunosensor for the determination of cystatin C in human serum. *Microchimica Acta*. 187(10): 1–12

103. Zhu L, Liu X, Yang J, He Y, Li Y. 2020. Application of multiplex microfluidic electrochemical sensors in monitoring hematological tumor biomarkers. *Analytical Chemistry*. 92(17): 11981–11986

104. Cincotto FH, Fav EL, Moraes FC, Fatibello-Filho O, Faria RC. 2019. A new disposable microfluidic electrochemical paper-based device for the simultaneous determination of clinical biomarkers. *Talanta*. 195: 62–68

105. Feng D, Su J, Xu Y, He G, Wang C, Wang X, Mi, X. 2021. DNA tetrahedron-mediated immune-sandwich assay for rapid and sensitive detection of PSA through a microfluidic electrochemical detection system. *Microsystems & Nanoengineering*. 7(1): 1–10

106. Lu L, Gunasekaran S. 2019. Dual-channel ITO-microfluidic electrochemical immunosensor for simultaneous detection of two mycotoxins. *Talanta*. 194: 709–716

107. Senel M, Dervisevic E, Alhassen S, Dervisevic M, Alachkar A, Cadarso VJ, Voelcker NH. 2020. Microfluidic electrochemical sensor for cerebrospinal fluid and blood dopamine detection in a mouse model of Parkinson's disease. *Analytical Chemistry*. 92(18): 12347–12355

108. Han JH, Lee D, Chew CHC, Kim T, Pak JJ. 2016. A multi-virus detectable microfluidic electrochemical immunosensor for simultaneous detection of H1N1, H5N1, and H7N9 virus using ZnO nanorods for sensitivity enhancement. *Sensors and Actuators B: Chemical*. 228: 36–42

109. Ligler FS. 2009. Perspective on optical biosensors and integrated sensor systems. *Analytical Chemistry*. 81(2): 519–526

110. Fan X, White IM, Shopova SI, Zhu H, Suter JD, Sun Y. 2008. Sensitive optical biosensors for unlabeled targets: A review. *Analytica Chimica Acta*. 620(1–2): 8–26

111. Wang DS, Fan SK. 2016. Microfluidic surface plasmon resonance sensors: From principles to point-of-care applications. *Sensors*. 16(8): 1175

112. Hassani A, Skorobogatiy M. 2006. Design of the microstructured optical fiber-based surface plasmon resonance sensors with enhanced microfluidics. *Optics Express*. 14(24): 11616–11621

113. Lee KH, Su YD, Chen SJ, Tseng FG, Lee GB. 2007. Microfluidic systems integrated with two-dimensional surface plasmon resonance phase imaging systems for microarray immunoassay. *Biosensors and Bioelectronics*. 23(4): 466–472

114. Tokel O, Yildiz UH, Inci F, Durmus NG, Ekiz OO, Turker B. Demirci U. 2015. Portable microfluidic integrated plasmonic platform for pathogen detection. *Scientific Reports*. 5(1): 1–9

115. Zhuo S, Xun M, Li M, Kong X, Shao R, Zheng T, Li Q. 2020. Rapid and label-free optical assay of S-layer protein with high sensitivity using TiO2-coated porous silicon-based microfluidic biosensor. *Sensors and Actuators B: Chemical*. 321: 128524

116. Ouellet E, Lausted C, Lin T, Yang CWT, Hood L, Lagally ET. 2010. Parallel microfluidic surface plasmon resonance imaging arrays. *Lab on a Chip*. 10(5): 581–588

117. Ameen A, Gartia MR, Hsiao A, Chang TW, Xu Z, Liu GL. 2015. Ultra-sensitive colorimetric plasmonic sensing and microfluidics for biofluid diagnostics using nanohole array. *Journal of Nanomaterials*: 460895

118. Grasso G, D'Agata R, Zanoli L, Spoto G. 2009. Microfluidic networks for surface plasmon resonance imaging real-time kinetics experiments. *Microchemical Journal*. 93(1): 82–86

119. Karuwan C, Sukthang K, Wisitsoraat A, Phokharatkul D, Patthanasettakul V, Wechsatol W, Tuantranont A. 2011. Electrochemical detection on electrowetting-on-dielectric digital microfluidic chip. *Talanta*. 84(5): 1384–1389

120. Wang S, Zheng L, Cai G, Liu N, Liao M, Li Y, Lin J. 2019. A microfluidic biosensor for online and sensitive detection of Salmonella typhimurium using fluorescence labeling and smartphone video processing. *Biosensors and Bioelectronics*. 140: 111333

121. Weng X, Neethirajan S. 2016. A microfluidic biosensor using graphene oxide and aptamer-functionalized quantum dots for peanut allergen detection. *Biosensors and Bioelectronics*. 85: 649–656

122. Chen JS, Chen PF, Lin HTH, Huang NT. 2020. A Localized surface plasmon resonance (LSPR) sensor integrated automated microfluidic system for multiplex inflammatory biomarker detection. *Analyst*. 145(23): 7654–7661

123. Monteiro JP, de Oliveira JH, Radovanovic E, Brolo AG, Girotto EM. 2016. Microfluidic plasmonic biosensor for breast cancer antigen detection. *Plasmonics*. 11(1): 45–51

124. Singh V. 2020. Ultrasensitive quantum dot-coupled-surface plasmon microfluidic aptasensor array for serum insulin detection. *Talanta*. 219: 121314

125. Teixeira A, Hernández-Rodríguez JF, Wu L, Oliveira K, Kant K, Piairo P, Abalde-Cela S. 2019. Microfluidics-driven fabrication of a low cost and ultrasensitive SERS-based paper biosensor. *Applied Sciences*. 9(7): 1387

126. Zheng Z, Wu L, Li L, Zong S, Wang Z, Cui Y. 2018. Simultaneous and highly sensitive detection of multiple breast cancer biomarkers in real samples using a SERS microfluidic chip. *Talanta*. 188: 507–515

127. Kuşbaz A, Göcek İ, Baysal G, Kök FN, Trabzon L, Kizil H, Karagüzel Kayaoğlu B. 2019. Lactate detection by colorimetric measurement in real human sweat by microfluidic-based biosensor on flexible substrate. *The Journal of the Textile Institute*. 110(12): 1725–1732

128. Zheng L, Cai G, Wang S, Liao M, Li Y, Lin J. 2019. A microfluidic colorimetric biosensor for rapid detection of Escherichia coli O157: H7 using gold nanoparticle aggregation and smart phone imaging. *Biosensors and Bioelectronics*. 124: 143–149

12 Aptamer Based Sensing Approaches Towards Food and Biomedical Applications

Atul Sharma[1], Rupesh K. Mishra[2,3], and Jean Louis Marty[4]

[1]SGT University
Department of Pharmaceutical Chemistry,
SGT College of Pharmacy
Gurugram Harayana, India

[2]Purdue University
School of Materials Engineering
West Lafayette, IN, USA

[3]Purdue University
Birck Nanotechnology Center
West Lafayette, IN, USA

[4]Université De Perpignan Via Domitia
Laboratoire B.A.E.
Perpignan, France

CONTENTS

DOI: 10.1201/9781003189435-16

12.1 INTRODUCTION

As sensor technology advances, biosensors are becoming increasingly important tools in food safety, environmental monitoring, clinical diagnostics, and other fields (1–2). The fast proliferation and diversity of biosensors have resulted in a lack of discipline in setting performance requirements (3). While each biosensor is evaluated for a unique application, it is still necessary to look at how standard methods for biosensor performance criteria can be defined in compliance with IUPAC principles. "A biosensor is a device that uses specific biochemical reactions mediated by isolated enzymes, immunosystems, tissues, organelles, or whole cells to detect chemical compounds, usually by electrical, thermal, or optical signals (Compendium of Chemical Terminology, 2nd edition (the "Gold Book")," according to the IUPAC definition (4). The signal of a specific biochemical reaction is converted into a quantifiable and detectable signal by the transducer in a standard biosensor design, proportional to the concentration of cognate molecule or group of analyte present. Bio-recognition components distinguish biosensors with excellent specificity and selectivity, which react only with a specific target molecule. Figure 12.1 depicts the many components of a biosensor. The unique and unrivalled features of biosensors that can detect a variety of analytes, even when they are present in complex matrices, are due to the combination of this specificity and sensitive transducer. Even though biosensors, particularly immunosensors, have been

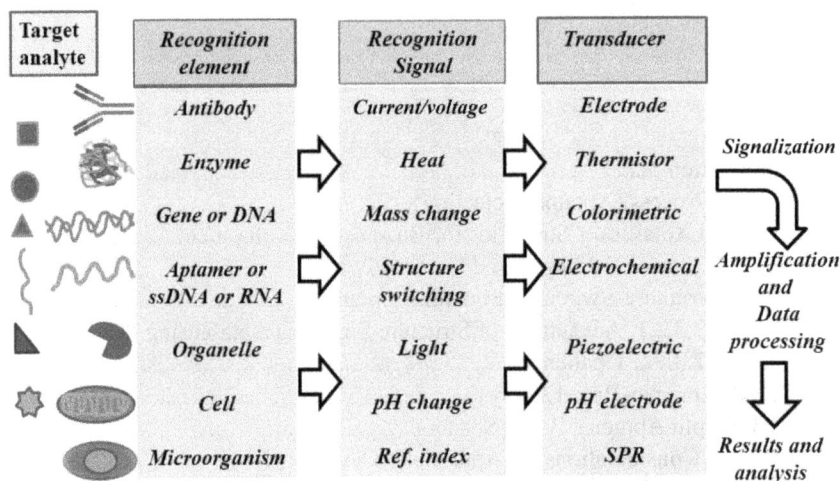

EIS: Electrochemical Impedance Spectroscopy; CV: Cyclic Voltammetry; DPV: Differential Pulse Voltammetry; Ref. index: Refractive Index; SPR: Surface Plasmon Resonance

FIGURE 12.1 Schematic representation of components of biosensors (Adopted from ref. (6), 2016 Copyright 2016 Smith and Fanklin).

effectively deployed in various analytical and diagnostic laboratories, as measured by the sheer quantity of publications and patent applications, general commercialization of biosensor technology has lagged behind research output (5). The high development expenses and technological constraints may be to blame for the commercialization delays. The science of analytical chemistry has matured sufficiently in recent decades, and the focus has switched to miniaturization. Biosensors, typically designed to detect a single target molecule, face a more significant barrier due to these needs.

A biosensor comprises a biological recognition element that detects the target molecule and a signaling fragment that turns the biological recognition into physically detectable signals. Biosensors rely on recognition molecules because their binding affinity and specificity determine sensor performance. So far, a broad spectrum of recognition molecules has been discovered. Because antibodies may selectively bind a wide range of analytes with high affinity, they have become the most often utilized probes. The commercial availability of antibodies also aids antibody research and development. Antibodies, on the other hand, have some restrictions. Antibodies, for example, are sensitive to temperature and prone to irreversible denaturation (7). Animals are frequently subjected to batch-to-batch variance when producing antibodies. Furthermore, antibodies have a considerable molecular weight, making site-specific tagging problematic. Finally, antibody-based tests frequently necessitate immobilization and prolonged washing, and homogenous experiments are challenging to perform. As a result, searching for additional ligands as a novel platform enabling biological diagnosis is highly desirable (8). A characteristics comparison of biorecognition elements is provided in Table 12.1.

Since their discovery in the early 1990s, aptamers have gained tremendous interest in medicines and bio-analytical applications (10). Aptamers are single-stranded (ss)

TABLE 12.1
Comparison of Inherent Characteristics of Commonly Employed Biorecognition Elements in Biosensor Development (9)

Receptor	Merits	Demerits
Antibody	1. Very selective 2. Ultra-sensitive 3. High binding (Ka = 10^6)	1. No catalytic effect 2. Due to high binding, extreme & harsh conditions are needed to reverse the reaction.
Aptamer	1. High stability and specificity 2. High selectivity and sensitivity 3. No immunogenicity	1. Need time to optimize the in-vitro process. But once optimized, ease for modification with improved stability and shelf life
Whole-cell/ Microbes	1. Cheaper source of enzymes than isolated enzymes 2. Less sensitive to inhibition by solutes. Tolerance to pH & temperature changes that leads to longer lifetimes	1. More extended response and regeneration time due to diffusion of the substrate through/ be transported into the cytoplasm 2. Less selective (contain many enzymes like tissues
Enzyme	1. Highly selective and fast-acting 2. Have catalytic activity, thus improving sensitivity	1. Expensive (cost of source, isolation, and purification) 2. Loss of activity on immobilization 3. Loss of activity over time unless stored under appropriate condition

oligonucleotides (DNA/RNA/XNA) or peptides synthesized from an in-vitro method known as Systemic Evolution of Ligands by Exponential Enrichment (SELEX). SELEX selects and amplifies an aptamer sequence from a large and random pool of nucleic acid libraries (11–12). When the targeting specific aptamer binds to the target, it folds into one particular G-quadruplex aptamer-target complex structure as a recognition mechanism. Attributed to the prevalence of nuclease, RNA aptamers are unstable and have a reduced life expectancy in biological fluids or real sample analysis. The ssDNA aptamer has inherent stability and resistance to degradation, making it a suitable option for biosensing applications (13–14). As a result of SELEX, many aptamers that can bind to a range of targets have been produced, including metal ions, small compounds, peptides, proteins, and even complex targets like the whole cell and materials surfaces (15–20). Aptamers appeal to biosensor development because of their small size, high stability (especially DNA aptamers), increased binding affinity and specificity, and ease of modification (21–22).

To begin with, aptamers can theoretically be used to target any target, including those that do not trigger immune responses or are highly hazardous. Second, aptamer chemistry has advanced dramatically, resulting in highly repeatable reagents that can be labeled with a wide range of functional groups at any position. Third, because DNA aptamers are more stable than proteins, they can be reused without losing their ability to bind. Aptamers are also non-immunogenic and non-toxic, which is the fourth advantage. Finally, aptamers can achieve higher immobilization densities and bind to epitopes that antibodies cannot reach because of their small size.

Recently, based on the numerous advantages, promising achievements have been made on developing aptamer-based sensors. For proof, in a search of Web of Science, more than 10k papers (including research and review) have been published under the keywords of "aptamer," "biosensor," and "biomedical applications." Since the last decades, a plethora of literature have been published under the keywords of "aptasensor" and "food applications," and still, the number is growing exponentially. Given the wealth of knowledge gained over the last two decades, this is an excellent opportunity to critically review the profession and identify fresh ideas for moving forward. In this work, we do not attempt to present a thorough assessment of the aptamer biosensor area because numerous review papers have already been published on various parts of the subject at varying stages of development. Given the amount of knowledge accumulated in the past two decades, this is a good time to critically summarize this field and identify new solutions to move forward. Numerous papers have already been published on various aspects of the aptamer biosensor field at different stages of development (14, 18–20, 23–29), and we do not intend to provide a comprehensive review here. First, we'll go over aptamer introduction, selection, and general aptasensor design methodologies in this chapter with their utility in food and biomedical applications. Following that, specific designs that have the potential to be used in clinical settings were examined.

12.2   SELEX PROCESS (APTAMER SELECTION)

Since 1990, the SELEX has isolated the vast majority of aptamers (30–33). The development of the in-vitro SELEX technique provides a platform for isolating oligonucleotide sequences with high affinity and specificity for distinct target molecules. A SELEX method is broken down into five essential steps: binding, partitioning,

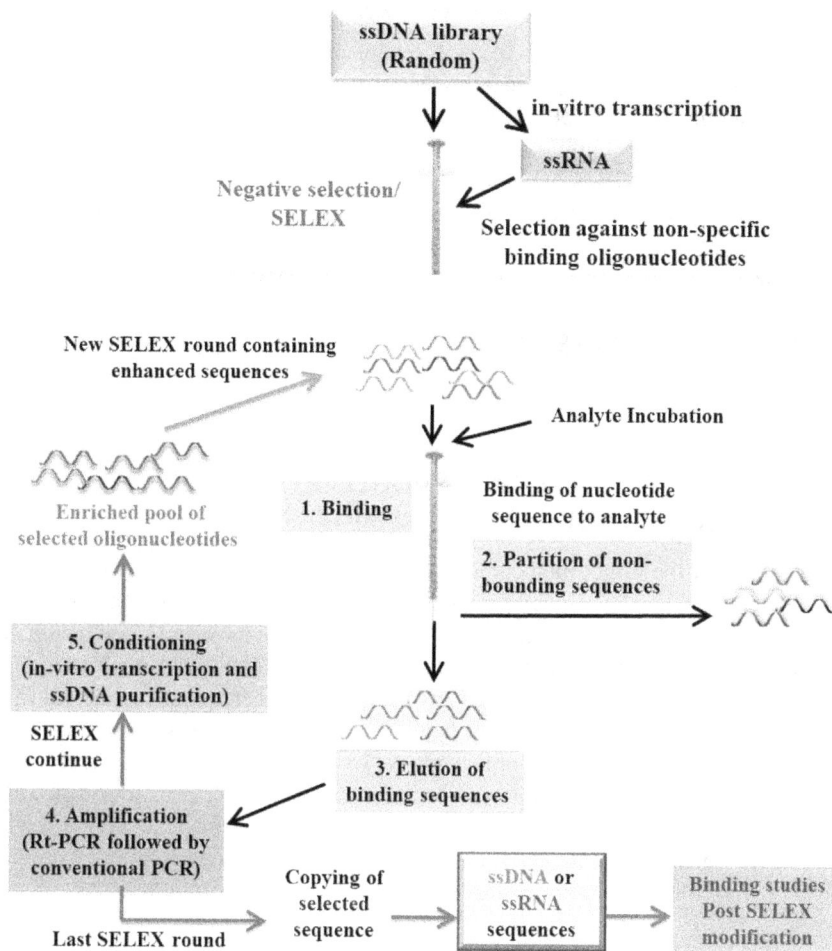

FIGURE 12.2 Schematic representation of the SELEX Process (demonstrating all main five steps): Steps in solid arrow line are specific for ssDNA aptamer selection, and actions in dotted arrow line are exact for RNA aptamer selection (illustrated from ref. (34)).

elution, amplification, and conditioning (Figure 12.2). The sequence library of synthetic oligonucleotides (ssDNA/ssRNA) is randomized from hundreds to thousands at first under a specific set of experimental conditions (34–35). Before binding, the oligonucleotides (ssDNA or ssRNA) library is usually incubated with a target-free selection matrix (negative SELEX) to remove any non-specific oligonucleotides. The randomized oligonucleotide sequences are set with the specific target analyte of interest in the first step of the SELEX cycle, binding. The second and most crucial stage is to use various separation techniques to separate the bound target sequences from unbound or weakly bound oligonucleotides (chromatographic, flow cytometry, electrophoresis, etc.) (36). The determined sequences are then eluted and amplified further using PCR (DNA SELEX) or RT-PCR (RNA SELEX). A pool of enhanced oligonucleotides was reintroduced into the following SELEX cycle. Each cycle's iterative selection and amplification round reduces the nucleotide pool to a smaller number of sequences that have the highest affinity and specificity for cognate molecules. The total number of SELEX cycles done varies significantly depending on the target molecule structure, selection conditions (such as pH, buffer strength, temperature, and so on), and partition step efficiency. However, optimizing SELEX parameters takes time, but once done, it represents the high efficiency conditions of the aptamer towards the target.

The molecular shape conformational, electrostatic/van der Waals, and hydrogen bonding interactions, as well as the stacking of aromatic rings, characterize the event of aptasensor identification and binding to their cognate molecules. Aptasensors provide remarkable selectivity, specificity, speed, stability, and the ability to use various measuring methods for an extensive range of analytes. Researchers are particularly interested in using aptasensors to analyze toxins in food and drinks to create reliable and economic approaches that require little or no sample preparation.

12.3 DESIGNED APTASENSOR STRATEGIES FOR BIOSENSOR DEVELOPMENT

An assay to test the target molecule can be designed after an aptamer sequence is known. Aptamers can be utilized for simple binding in the same way that antibodies can. Aside from that, several more aptamer-specific tests have been demonstrated. The majority of them make use of nucleic acids' programmability. A conformational shift into a well-defined binding structure, for example, is frequently observed (37). To date, numerous pieces of literature have been published on this subject adopting various designs (6, 20, 38–40). In this work, we have summarized some general strategies for the construction of aptamer-based sensors concerning the fluorescence and electrochemical-based platforms; however, each strategy can further conjugated with a variety of signaling techniques such as SPR, colorimetric, and Raman technique, already reported in the literature (18–19, 23, 41–44) as shown in Figure 12.3.

12.3.1 APTAMER FOLDING BASED SENSORS

While antibody binding to its target does not need a substantial conformational change, aptamers have been extensively characterized for their adaptive binding (Figure 12.4). Non-specific binding artifacts can be avoided by conformational modifications like these (46). For illustration, a change in aptamer confirmation, for example, might

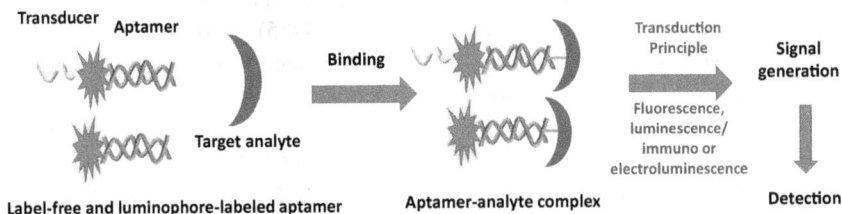

FIGURE 12.3 Aptamer-mediated biosensors for analysis using fluorescence, chemiluminescence, electrochemistry, or immunoluminescence (illustrated from ref. (45), copyright 2020, published and licensed by Dove Medical Press Limited).

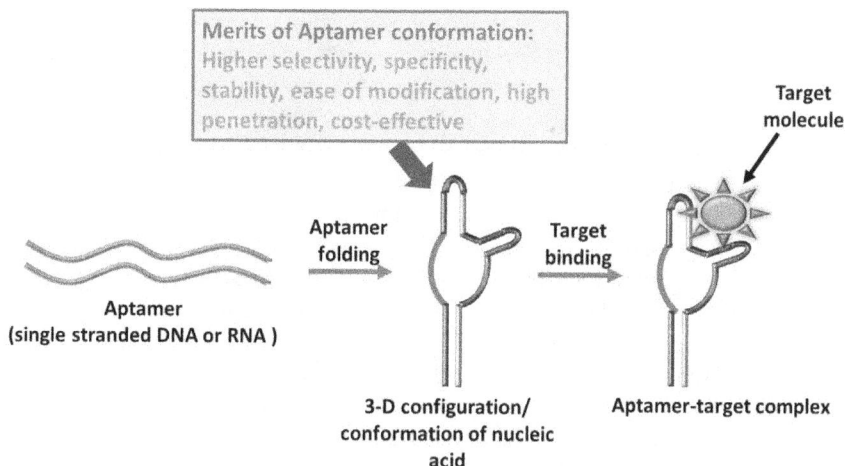

FIGURE 12.4 Aptamer structure, characteristics, and functional illustration for biomedical applications (illustrated from ref. (45), copyright 2020, published and licensed by Dove Medical Press Limited).

affect the local environment of a fluorophore linked to it. Jhaveri and co-workers (2000) (47) investigated single fluorophore tagged aptamers based on this hypothesis. However, selecting the most efficient labeling location is not easy, and the amount of signal change is frequently small (48). The use of an external quencher improves sensor performance substantially. The two ends of an aptamer are usually labeled with a fluorophore and a quencher, respectively, in most designs. As a result, the fluorophore emission is sensitive to changes in end-to-end distance. These methods have been employed to detect various analytes of interest (43, 49–51). Aside from fluorophores, nanoparticles (such as AuNPs) may also be used to detect aptamer folding. Folded aptamers, for example, cannot be readily adsorbed by AuNPs, but unfolded aptamers can be absorbed to improve AuNP colloidal stability (52).

In electrochemical-based folding aptasensors, when the analyte is present, a variation in the aptamer's confirmation is required (53), allowing changes in the efficiency of electron transfer to be detected, which is dependent on the distance that occurs between the electrode's surface and the redox probe. The differences in the structure of the recognition aptamer adsorbed on the electrode surface are advantageous in this strategy (54). This aptamer is commonly modified at the 5' end with a molecule containing a group capable of linking to the electrode surface (for example, a thiol) and tagged at the opposite end (3' end) with a redox probe, such as methylene blue (MB) (55–56). The use of structure-switching aptamers in electrochemical aptasensors is a promising technology since it allows for quick, sensitive, and low-cost determination of various target analytes (57).

12.3.2 STRUCTURE-SWITCHING SIGNALING APTAMERS

In principle, the recognition of structure switching aptaswitches is based on converting an aptamer-target contact event into a quantifiable signal. In structure switching aptaswitches, monochromophoric and bischromophoric techniques are commonly adopted. The aptamer molecule is first labeled with a single fluorophore in the monochromophoric approach. The attachment of the target molecule to the aptamer then causes structural conformational changes in the attached fluorophore, altering its spectroscopic characteristics and generating a signal. In a bischromophoric approach, target-induced conformational changes in the aptamer reduce the binding between the fluorophore and quencher-labeled complementary sequences, altering the fluorescence properties of the attached fluorophore and resulting in an increase in the fluorescence signal based on target analyte affinity (44). Successful fluorescence dequenching is required for effective signal production. The signal amplitude of the bischromophoric method (aptamer beacons) is often significantly higher than that of a single fluorophore-labeled aptamer (58). Other signaling moieties, such as electrode surfaces, quantum dots, or metal nanoparticles (42, 59–61), can be used in place of the fluorophore/quencher for different types of signal transduction. An inorganic surface can be used to adsorb the DNA probe instead of utilizing cDNA to lock the first aptamer conformation. Graphene oxide (GO), which efficiently adsorbs water, is one of the most well-known examples (41).

For electrochemical aptasensors, two primary detection methodologies have been used: labels (enzymes, metal nanoparticles (NPs), and redox chemicals) covalently or noncovalently attached to aptamers or aptamer label-free detection systems (62). The target concentration might then be correlated by measuring changes in electrochemical characteristics following target binding. -Signal-on- (positive readout signal) and -Signal-off- (reduced sensitive readout signal) (negative readout signal) aptasensors based on target binding-induced conformational change of aptamers, target binding-induced strand displacement, or both processes have been described (63–64).

12.3.2.1   Advantages of Structure Signaling Aptasensing Platforms

Aptasensing technologies are well suited for the study of emerging contaminants due to remarkable advancements in the aptamer selection and modification process. Aptamer/DNA/RNA molecules that reversibly change between two or more conformations in response to binding a specific ligand molecule are known as bimolecular aptaswitches. A combination of fluorophore and quencher labeled complementary aptamer sequences based on the universal phenomena of adopting two distinct conformations provides advantages in fluorescence quenching is driven structure-switching aptasensors (65):

(i) Non-specific analytes cannot easily replicate specific-binding conformational changes that transduce signals based on hydrogen bonding and electrostatic interaction.

(ii) Signal transduction is fast, reversible, and reagent-free, allowing for continuous and real-time monitoring.

(iii) Approach versatility allows for customizing the length of aptamer sequences.

12.3.2.2   Challenges

The amount of signal growth is considerable with an external quencher and a significant change in distance upon target attachment (e.g., from zero to infinite). The disadvantage of the structural switching design is that aptamer binding affinity is reduced due to competition with cDNA or inorganic surfaces (66). Furthermore, because the released probe is unlikely to re-hybridize in a short period, it isn't easy to achieve continuous target monitoring when the sensor system consists of two components. Overall, this commonly utilized signaling mechanism is unique to aptamers, and incorporating such sensible structural alterations into an antibody is very hard.

12.3.3 Aptamer-Based Assays

Because aptamers are affinity ligands, they can be used instead of antibodies in ELISA tests (enzyme-linked immunosorbent assay) (67). The mass or refractive index at the surface may be increased due to target binding to immobilized aptamers (68–69). Simple binding tests, on the other hand, necessitate a lot of washing. This is especially crucial for biomedical samples, which frequently contain a high concentration of contaminants. Immunoprecipitation is another antibody-based test (70). The method can crosslink to precipitate because each antibody has numerous Fab arms, and a target protein often has a few epitopes. Similar precipitants can be formed with protein targets by connecting numerous aptamers to a nanoparticle (71). For example, the association of the gold surface plasmophore with AuNPs, the color may shift to blue or purple (72). Enzyme-linked aptamer assay (ELAA), a variant of the classic enzyme-linked immunosorbent assay (ELISA) that uses aptamers instead of antibodies, is appealing for the development of electrochemical aptasensors because it generates an amplified signal via the electrocatalytic activity of the enzyme (73). Two main configurations for ELAA have been developed: an antibody/target/aptamer sandwich and an aptamer/target/aptamer type. In the first configuration, Drolet et al. (74) set the first ELAA-based human vascular endothelial growth factor detection. Secondly, Yu and colleagues (75) designed an antibody/IgE/biotinylated aptamer/streptavidin-alkaline phosphatase sandwich to detect IgE. To create the sandwich model, however, this structure still requires antibodies.

12.3.4 Split aptamer-Based Sensors

Splitting an aptamer into two pieces is also feasible. The two fragments can self-assemble in the presence of target molecules, changing the output signal (8, 58, 76). Stojanovic et al. (77) developed a split-aptamer-based method in which nucleic acid aptamers were divided into two fragments in the presence of ligand to produce a ternary complex. Due to the lack of secondary structures in the two oligonucleotides, false-positive or non-specific signals are not made. The two divided aptamer fragments must be brought close to each other to obtain positive signs. Diverse transduction approaches, including colorimetric (78), fluorescence (79), electrochemical techniques (80), and SPR (81), have been used to detect various targets, tiny molecular targets, using this straightforward methodology.

In the creation of sandwich aptasensors, split aptamers offer a lot of promise. However, there are presently few viable strategies for separating aptamers into fragments. Split aptamers can theoretically be made by dividing an aptamer sequence in half. Because it is challenging to design aptamers into split aptamers, there is a lack of understanding of the aptamer-target complex structures and nucleotide-binding proteins.

12.4 PRACTICAL CONSIDERATIONS FOR APTASENSOR DEVELOPMENT

It's worth noting that, even though hundreds of aptamers have been chosen and over a thousand sensing-related publications have been published, only a few targets have been used. Furthermore, only a few of these targets are of therapeutic significance. This pattern highlights a couple of the field's challenges. Many of the aptamer selections were made in laboratories, and the detailed sequence information isn't available. Furthermore, functional aptamers must be of high enough quality to detect target analytes at physiologically relevant concentrations. Due to a scarcity of high-quality, high-impact aptamers, this field's influence has been limited to academic study. Second, many previously described aptamers lack thorough characterization, making further development difficult for analytical chemists. The secondary structure of aptamers needs to be known for rational biosensor design. The frequent employment of the same model aptamers is most likely due to this. The third challenge is the inaccessibility of target molecules, particularly proteins and species. This has made it difficult for a broader community to research those aptamers, and the labs that produce these reagents may be uninterested in analytical advancement. More collaboration is required in this case. Finally, the sensors and tests must function in a complicated sample matrix.

Fortunately, aptamers have been shown to detect their targets still selectively and are relatively immune to non-specific interactions in serum on numerous occasions. Natural riboswitches function well in the cellular context, so this is not surprising. While many of the sensors are exhibited in simple buffers, the most typical biological samples are blood, serum, urine, and saliva, which are complicated and have a lot of optical interference. Blood, for example, is dense and dark crimson in hue. Because most visible light is scattered or absorbed, optical detection has an extremely high background.

Furthermore, non-specific binding occurs frequently indirect binding experiments as a result of other proteins in the sample. The sample procedures must also be examined in the fifth place. The sensor must work with a small sample volume, and blood samples must be acquired with minimal invasiveness. Finally, existing tests, such as those based on antibodies present a barrier and competition. Aptamer-based assays must be carefully evaluated to verify their accuracy, repeatability, and user-friendliness, eliminating numerous stages of liquid sample transfer and mixing to be generally accepted by the industry. The price must be less than that of competing technology. Many efforts have been undertaken to resolve these issues to date. Numerous literature based on fluorescence-based aptasensors for biomedical and food application is available (6, 8, 20, 82), therefore to gather the readability, in this chapter, we will be reviewing some of the essential and different types of electrochemical-based aptasensors that might be eventually useful for biomedical diagnosis and food applications.

12.5 ELECTROCHEMICAL-BASED APTASENSOR

12.5.1 APTASENSOR FOR BIOMEDICAL APPLICATIONS

The advantages of electrochemical aptasensor-based detection are high sensitivity, rapid response, robustness, low cost, and miniaturization potential (83–84). In a report, the detection of thrombin was reported using an electrochemical aptasensor in which the thrombin was immobilized on the surface and detected using horseradish peroxidase (HRP)- or biotin-labeled aptamer after reaction with streptavidin-HRP (85). Immobilization of the aptamer on the surface by biotin or a thiol group is two more configurations that have been studied. A chromogenic substrate or HRP labeling was used to identify the thrombin once it had attached to the aptamer. A sandwich configuration with a second tagged aptamer was also used to detect thrombin, as was a reagentless electrochemical aptabeacon. Each time the aptamer changed from a duplex structure to a 3-D quadruplex structure after binding to thrombin, the conformational shift was detected electrochemically.

Conceptualizing the "signal-on or off" mechanism, Xiao et al. (86) developed a sensitive, selective, and reusable signal-off sensor for detection of thrombin. The sensor is made by covalently attaching a MB label to a gold electrode via a thiol group on the thrombin-binding aptamer. The MB label is close to the electrode surface because the aptamer chain is initially flexible without thrombin, allowing for excellent electron-transfer efficiency. When the aptamer binds to its target, it folds into its natural binding form, preventing electron transport. The sensor can detect thrombin at concentrations as low as 6.4 nM. The signal-off feature is undesirable because it is more susceptible to background variation and has limited room for signal alteration. The discovery of direct platelet-derived growth factor (PDGF) in blood serum was reported using a "signal-on" technique (87). The 30-end of the PDGF aptamer was tagged with an MB molecule, and the 50-end of the aptamer was modified with a thiol group to allow it to connect to the gold electrode. The MB label is relatively far away from the electrode surface in the absence of the target. The aptamer undergoes a conformational change in response to PDGF binding, forming a stable three-way junction that keeps the MB label close to the electrode and improves electron transport. With a detection limit of 50 pM, this approach could detect up to 10 nM PDGF.

Microfluidic sensor chips are being used in this industry to accomplish continuous, real-time, and high-throughput automated detection (88–90). Rowe and colleagues (91) created a reusable aptamer-based electrochemical array device that can swiftly detect aminoglycoside drugs in blood samples (in less than 10 seconds) to avoid overdosing or side effects. To boost RNA stability, ultrafiltration was utilized to eliminate nuclease. Even though this pre-treatment phase reduces the sensor's total work-flow, it detects aminoglycoside at clinically relevant concentrations (2–6 mM) in less than 30 minutes and requires just a low-cost desktop centrifuge. Electrochemical detection is likely to be a viable method for biological diagnostics based on the examples provided. The use of personal glucose meters to collaborate with aptamer-based reagents is a fascinating recent development (92–93).

Through direct electrochemistry and electrocatalysis in a sandwich-type electrochemical aptasensor for ultrasensitive detection of thrombin, a glucose oxidase-functionalized bioconjugate was synthesized and utilized as a novel trace label for the first time (94). The Fe_3O_4 magnetic nanoparticles/reduced graphene oxide nanocomposite modified glassy carbon electrode (Fe_3O_4/r-GO/GCE) was used to develop a high-performance electrochemical biosensing platform (95). The biocatalytic activity of glucose oxidase(GOx) mounted on Fe_3O_4/r-GO/GCE, and its direct electrochemistry toward glucose oxidation were examined first. Further, the electrochemical oxidation behavior of four DNA-free bases, guanine (G), adenine (A), thymine (T), and cytosine (C) at Fe_3O_4/r-GO/GCE, was then investigated using differential pulse voltammetry (DPV). Additionally, for Immunoglobulin E (IgE), a label-free electrochemical impedance spectroscopy (EIS) technique of detection was devised, in which IgE-specific aptamers immobilized on Fe_3O_4/r-GO/GCE cause impedance changes associated with IgE binding events.

Based on bifunctional Fe_3O_4@Au nanocomposites, a label-free electrochemical aptasensor was reported to detect and quantify adenosine triphosphate (ATP) (96). The obtained bifunctional Fe_3O_4@Au nanocomposites may be easily modified with ATP-specific aptamer molecules through Au–S interaction to produce the nanoprobe and used to immobilize on a glassy carbon electrode with the help of a magnet. The current signal recorded at the nanoprobe-containing magnetic glassy carbon electrode instantly drops when exposed to the target ATP. The drop-in sensor current is because the ATP-specific aptamer prefers to bond with ATP rather than methylene blue, resulting in less methylene blue adsorbed on the aptamer. The drop-in sensor current can be utilized to electrochemical ATP determinations, such as in human serum samples.

Ravalli et al. (2015) developed an electrochemical single-use aptasensor for detecting and analyzing vascular endothelial growth factor (VEGF) (97). First, a mixed monolayer of a primary thiolated DNA aptamer and a spacer thiol, 6-mercapto-1-hexanol, was applied on gold nanostructured graphite screen-printed electrodes. The aptasensor was then treated with the VEGF protein. The coupling of a streptavidin-alkaline phosphatase conjugate with a secondary biotinylated aptamer was then used to create an enzyme-amplified detection method. The enzyme converted the electroactive 1-naphthyl phosphate to 1-naphthol hydrolysis; this electroactive product was identified by differential pulse voltammetry (DPV). The response of the aptasensor was found to be linearly related to the target concentration between 0 to 250 nmol L^{-1} with a detection limit of 30 nmol L^{-1}. The immunoassay's repeatability and selectivity were also investigated.

A novel sandwich-type electrochemical aptasensor for detecting Mycobacterium TB MPT64 antigen quickly and sensitively was reported by Chen et al. (2019) (98). The first time, they reported the synthesis of a novel carbon nanocomposite composed of fullerene nanoparticles, nitrogen-doped carbon nanotubes, and graphene oxide (C60NPs-N-CNTs/GO) was easily synthesized, which not only had a large specific surface area and excellent conductivity with promising electroactive, nanocarrier and redox properties. AuNPs were then uniformly fixed onto the surface of such nanocomposites via Au–N bonds to bind with MPT64 antigen aptamer II (MAA II), forming the tracer label for electrochemical signal production and amplification. Furthermore, a conductive polyethyleneimine (PEI)-functionalized Fe-based metal-organic framework (P-MOF) was used as a sensing platform to absorb bimetallic core-shell Au–Pt nanoparticles (Au@Pt), which may speed up electron transfer and boost MPT64 antigen aptamer I (MAA I) immobilization. The presence of tetraoctylammonium bromide (TOAB) further enhances the inherent electroactivity of the tracer label, resulting in a well-defined current response. Under ideal conditions, the suggested aptasensor demonstrated a wide linear range for MPT64 detection from 1 fg/mL to 1 ng/mL under a low limit of detection (LOD) of 0.33 fg/mL. More crucially, it was found to be effective in detecting MPT64 antigen in human serum, indicating that it could be utilized in clinical practice to diagnose tuberculosis.

An electrochemical aptasensor constructed by immobilizing the aptamer on a carboxyl graphene/thionin/gold nanoparticle modified glassy-carbon electrode was reported for detection of tau381 (99) (Figure 12.5), which is a critical Alzheimer's disease in human serum. The differential pulse voltammetry signal increment rose linearly with the logarithm of tau381 concentration in the range of 1.0 to 100 pM under ideal conditions (Figure 12.6), with a detection limit of 0.70 pM. The

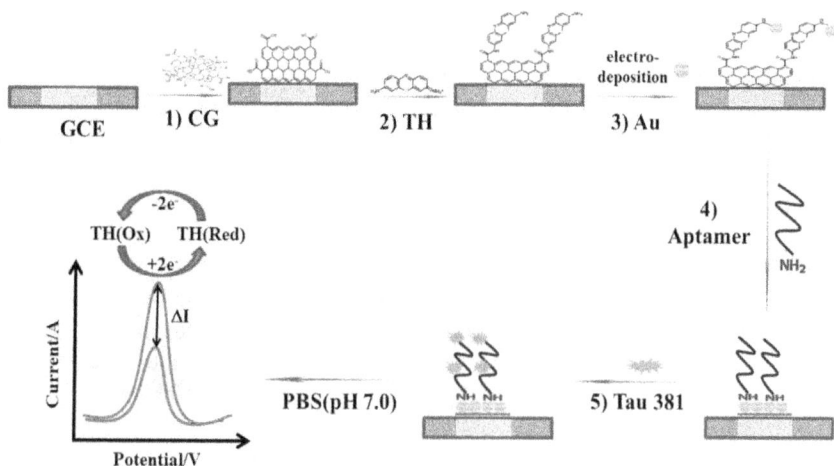

FIGURE 12.5 Preparation of glassy-carbon electrode/carboxyl graphene/thionine/gold nanoparticles/aptamer/tau381 aptasensor. 1) Doped carboxyl graphene on a glassy carbon electrode surface, 2) followed by incubation with thionine, 3) electro-deposition and reduction of $HAuCl_4 \cdot 3H_2O$ to gold nanoparticles, then incubation with aptamer 4) and tau381; 5) finally, differential pulse voltammetry evaluation of tau381 levels in human serum (Reprint with permission from ref. (99), copyrights 2019 MDPI).

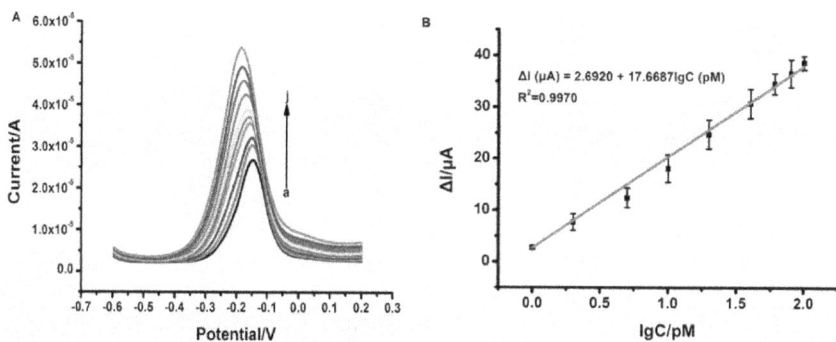

FIGURE 12.6 (A) Differential pulse voltammetry curves of tau381 concentrations including 0, 1, 2, 5, 10, 20, 40, 60, 80, and 100 pM in 0.1 M PBS (pH 7.0); (B) Plot of ΔI vs. Log of tau381 concentration (n > 3) (Reprint with permission from ref. (99), copyrights 2019 MDPI).

selectivity, repeatability, stability, detection limit, and recovery of the aptasensor were all determined. The tau381 aptasensor was found to screen patients with and without Alzheimer's disease based on performance investigation of 10 patients' serum samples. The proposed aptasensor could be useful in early-stage clinical diagnosis of Alzheimer's disease.

An enzyme-linked aptamer assay based on DPV signal generation was reported to detect cocaine (100). The obtained results represent a dynamic range of 0.10 to 50.0 M with a detection limit of 20 nM (S/N = 3). The DPV signal change could be used to detect cocaine sensitively. The proposed aptasensor has the advantages of high sensitivity and low background current. Furthermore, in work, a new ELAA configuration was developed that only requires a single aptamer sequence and can be adapted to detect various targets by cleaving the aptamers into two suitable segments. Exosome-based liquid biopsy is a promising application in which test quality is determined by high sensitivity detection with good selectivity. An electrochemical aptasensor was proposed to detect tumorous exosomes that are both selective and sensitive (101). For signal amplification, it uses the multidirectional hybridization chain reaction (mHCR). The aptamers are initially fixed on a gold electrode surface before interacting with exosomal membrane proteins to capture exosomes. Cholesterol-modified H shape-like DNA unit 1 (cH1) is capable of plunging cholesterol into the exosomal lipid bilayer and attaching it to the exosomal membrane after capturing the target. The mHCR is performed at cH1 in the presence of H shape-like DNA units 1 (H1) and 2 (H2) to generate a large DNA nanostructure that allows many signal molecules to be recruited for electrochemical signal amplification. Under optimized environmental conditions, the designed aptasensor has high sensitivity and a low limit of detection (LOD) of 285 exosomes/L. It also indicates that tumorous and non-tumorous exosomes from cells and serum samples may be distinguished selectively. Overall, the suggested aptasensor has promise for detecting exosomes in cancer early detection and screening

12.5.2 APTASENSOR APPLICATION IN FOOD ANALYSIS

One of the most critical aspects of addressing new food safety concerns worldwide is the development of susceptible testing methods for food pollutants. Electrochemical aptamer-based sensors have been extensively studied as prospective analytical tools, giving the requisite portability, speed, sensitivity, specificity, and lower cost and simplicity than traditional approaches (26, 102).

12.5.2.1 Detection of Antibiotic Residues

Antibiotic residues in environmental samples are becoming more common, signaling increased drug resistance in humans and microbes. As a result, a lot of attention has been paid to detecting antibiotics in environmental commodities early and precisely. In the measurement of signals, electrochemical aptasensors provide significant advantages in terms of sensitivity and selectivity. Recent advances in sensor design have made them compatible with novel microfabrication techniques and allowed for the portability and miniaturization of the sensing platform. As a result, electrochemical aptasensors have piqued the interest of various researchers looking for a simple, quick, and cost-effective way to monitor aptamer recognition events. The increasing occurrence of antibiotic residues in animal-derived foods such as milk and meat has sparked a lot of interest in developing rapid and precise analytical approaches for antibiotic detection using electrochemical aptasensors. Sensitive detection of streptomycin in milk samples was recently reported using an aptasensor (103). The construction of the aptasensor utilized an arch aptamer, and its complementary strand and exonuclease were used in the sensor. Exonuclease works like a digestive enzyme that specifically destroys the ssDNA from its 3′-terminus end. In contrast, streptomycin produces conformational changes in the aptamer-streptomycin combination, resulting in the release of a complementary strand. Exonuclease causes the complementary strand to degrade, resulting in electrochemical signals. The described aptasensor detected streptomycin with a LOD of 11.4 nM and was highly selective. The scientists also used this aptasensor to detect streptomycin in serum, with a LOD of 15.3 nM. For streptomycin in serum (40–1500 nM) and milk (30–1500 nM), the aptasensor provided excellent linear ranges. An aptasensor for label-free kanamycin (KANA) detection in milk was described by Zhou et al. (104). The SWV approach measured kanamycin residues using a current response caused by a conformational change in a specific aptamer connected to an Au-electrode with a remarkable detection range between 10 to 2000 nM. Later, our group reported developing a label-free, and disposable SPCE integrated EIS-based aptasensor for KANA detection (18). The KANA-EIS aptasensor exhibited a dynamic range of 1.2–600 ng mL^{-1} with LOD of 0.11 ng mL^{-1} (S/N = 3). The developed aptasensor showed high selectivity and specificity for KANA, with no interference from competing analogs (streptomycin and gentamicin). The aptasensor performance was tested in spiked milk samples for practical application, and an acceptable recovery percentage of 96.88–100.50% (% RSD = 4.56, n = 3) was obtained in KANA-spiked milk samples. The design of a sandwich-type electrochemical aptasensor for the determination of oxytetracycline residues was reported by Liu et al (105). A graphene-three-dimensional nanostructure gold nanocomposite (GR3DAu) was used to build this aptasensor, which investigated an aptamer—AuNPs—horseradish peroxidase nanoprobe for signal amplification. The proposed aptasensor had a good linear range for detecting oxytetracycline between 5×10^{-10} to 2×10^{-3} g/L, with a LOD at 4.98×10^{-10} g/L.

Using metallic-encoded apoferritin probes and double stirring bars-assisted target recycling for signal amplification, a multiplexed electrochemical aptasensor for antibiotic detection was developed (106). Apoferritin was used to load Cd^{2+} and Pb^{2+} ions into the encoded probes then labeled with duplex DNAs (aptamers corresponding to KANA and AMP hybrid with complementary DNA sequence, respectively). The targets can recurrently react with the probes on the bars in the presence of KANA and AMP and then replace a large number of Apo-Mencoded signal tags in the supernatant. Square wave voltammetry was used to identify the peak currents of Cd2+ and Pb2+ from the labels corresponding to the concentrations of KANA and AMP in one run. As a result, KANA and AMP can be detected simultaneously in the 0.05 pM to 50 nM range. The detection limits (S/N = 3) were 18 fM KANA and 15 fM AMP. An ssDNA aptamer was used as the bioreceptor in a tetracycline-specific aptasensor developed by Kim et al. (107). The binding of biotinylated ssDNA aptamer immobilized on a streptavidin modified SP-Au-electrode and tetracycline was detected using the CV and SWV techniques. The described aptasensor had good tetracycline detection sensitivity, with a range of 0.010–10.0 μM, with an ability to detect tetracycline at low and high levels in food samples.

Very recently, utilizing a Penicillin aptamer as the particular recognition element and an electrospun carbon nanofiber (ECNF) mat electrodeposited gold nanoparticles (AuNPs) as the platform, a novel electrochemical aptasensor for the detection of Penicillin in a milk sample was developed (108). Initially, the ECNF mat electrode was made by electrospinning and heat treatment. Second, to increase the rate of electronic transmission, the prepared ECNF mat electrode was changed with AuNP electro-deposition. Finally, the modified electrode was used to assemble Penicillin aptamer. The CV studies revealed a high selectivity, strong stability, great reproducibility and repeatability, as well as a wide linear range (1–400 ng/mL) with a low detection limit (0.6 ng/mL). Another aptasensor based on a single-stranded DNA-binding protein for the ultrasensitive detection of ciprofloxacin was recently published (109). The created aptasensor used a single-stranded DNA-binding protein (SSB) on an Au-electrode and showed great selectivity for ciprofloxacin (CIP) detection in milk samples LOD of 263 pM. Other aptasensors based on nanoparticles and nanocomposites have been reported for the CIP detection with LOD of 0.63 ng mL^{-1} (110) and 0.50 ng mL^{-1} (111).

12.5.2.2 Detection of Mycotoxin

Food toxins are benign chemical compounds produced by the metabolic processes of many microorganisms. Mycotoxins are fungus secondary metabolites that have severe toxic effects. Mycotoxins are classified based on their chemical and physiological features. Some mycotoxins have been linked to the development of hereditary problems and carcinogenic consequences (20, 112). Mycotoxins can be detected in electrochemical biosensors using aptamers as biorecognition components. Several mycotoxins have been detected using aptasensors, including ochratoxin A (OTA), aflatoxin M1 (AFM1), and aflatoxin B1 (AFB1). Based on the host-guest recognition between ferrocene and -cyclodextrin (-CD), a simple electrochemical aptasensor was described to detect AFB1 (113). AFB1's Fc-labeled aptamer was initially hybridized with its Fc-cDNA complement. Two ferrocene molecules were brought together so closely that they couldn't fit into the cavity of the β-CD modified electrode. The electrochemical aptasensor responded to AFB1 throughout a wide linear range of 0.1 pg/mL to 10 ng/mL, with a low detection limit of 0.049 pg/mL (0.147 pmol/mL) using AC impedance detection. The aptasensor was used to determine AFB1 in genuine peanut oil samples, with recoveries ranging from 94.5 to 106.7 percent and an inter-assay RSD of less than 11.51 percent. The signal response of an aptamer-based biosensor constructed in a multilayer framework using cyclic voltammetry (CV) and EIS was measured using redox indicators: $K(Fe(CN)_6)^{3-/4-}$ (114). This sensor had a detection limit of 0.40 ± 0.03 nM, was regenerable in 0.2 M glycine-HCl, and did not lose its stability after 60 hours of storage at 4 °C. Gaud et al. (115) developed a label-free impedimetric electrochemical aptasensor to detect AFB1 in alcoholic beverages. The authors conducted a comparison study of two aptamer sequences, sequence-A, and sequence-B. The selective recognition of AFB1 was achieved in the study using covalently attached aptamers as a compact monolayer on SPCE via diazonium coupling. EIS reported a quantitative dynamic range of 0.125–16 ng/mL for both sequences, with LODs of 0.12 ng/mL and 0.25 ng/mL for sequence-A and sequence-B, respectively. To illustrate the usefulness of the developed aptasensor, the authors detected AFB1 in beer and wine samples.

Mishra et al. (116) have revealed for the first time a sensitive detection technique for OTA in cocoa beans using a competitive aptasensor developed by DPV. The authors presented an approach in which a free and biotin-labeled OTA competed with a tethered aptamer on an SPCE to attach. After adding avidin-alkaline phosphatase, the detection was done (ALP). The signal was created using an appropriate ALP substrate, 1-naphthyl phosphate (1-NP). The aptasensor demonstrated good linearity between 0.15 and 5 ng/mL, with 0.07 ng/mL LOD. Catanante et al. (57) published a folding mechanism-based aptasensor for OTA detection using MB-tagged anti-OTA aptamers in another paper from our research. Authors have described several aptamer coupling strategies using hexamethylenediamine (HMDA), polyethylene glycol (PEG), and diazonium coupling. With a LOD of 0.01 ng/mL, HMDA coupling on SPCE was shown to be the most effective coupling approach. Nguyen et al. (117) developed an electrochemical aptasensor for AFM1 based on CV and SWV. AFM1 specific aptamers were immobilized on an interdigitated electrode (IDE) polymerized with Fe_3O_4 integrated polyaniline to create the aptasensor. The aptasensor demonstrated good stability, repeatability, and sensitivity (0.00198 g/L) in detecting AFM1. The use of the created aptasensor in real-world sample analysis, however, has yet to be shown.

Saxitoxin (STX) is a type of marine biological toxin that is found in large amounts in seafood. An electrolyte-insulator-semiconductor (EIS) sensor for STX detection was developed using aptamer-modified two-dimensional layered Ti_3C_2Tx nanosheets. MXene's high surface area and abundance of functional groups aided in creating an aptamer that had specialized interactions with STX. The aptasensor was able to detect STX with high sensitivity and specificity, according to the results of capacitance-voltage (C-V) and constant-capacitance (ConCap) measurements. The detection range was from 1.0 nM to 200 nM, with a detection limit of 0.03 nM (118). Fumonisins B1 is the most common member of a toxin family produced primarily in maize, wheat, and other cereals by numerous species of Fusarium molds. To determine fumonisins B1 and B2, Shi et al. (119) designed an aptasensor with dual amplification of Au nanoparticles and graphene/thionine nanocomposites (FB1). Graphene/thionine nanocomposites were released from the electrode surface after a specific combination of the aptamer and its target (FB1) in solution, resulting in a diminished electrochemical signal using the CV technique.

12.6   CONCLUSIONS AND FUTURE PERSPECTIVES

Aptamers have been identified as a unique analytical method for detecting clinically essential biomarkers and dietary contaminants quickly and accurately. Electrochemical-based aptasensors, among the other aptamer-based sensors mentioned, have several advantages, including the ability to miniaturize sensing systems. This chapter summarizes some recent publications on electrochemical aptasensors for clinical diagnostics and food monitoring. Aptamers are helpful as bio-recognition elements because they have a higher binding affinity for the analyte of interest. Even though the presented aptasensors are characterized by their ability to detect low quantities of target analytes, several challenges must be addressed during their development. To begin with, selecting a specific aptamer sequence is a time-consuming process that must be tailored to the target analytes. Second, immobilizing the aptamer probe on a sensing electrode is still a time-consuming process.

To ensure competent binding of the target analyte with specific aptamers, the immobilization of the aptamer sequence should be carefully chosen. Furthermore, the majority of the aptasensors that have been designed work by examining redox probes. As a result, the redox probe must be carefully selected from following the experimental parameters. By enabling multi-analyte and high-throughput sensing platforms for clinical, food, and environmental applications, recent innovations in electrochemical aptasensors have effectively reduced the time consumption associated with traditional lab techniques. The development of electrochemical aptasensors for simultaneous screening of several analytes is tough since immobilizing many aptamers on a single sensing platform will be tricky. However, this worry could be solved by developing an engineered aptamer sequence for specific binding to numerous targets with identical affinity. Furthermore, new structural switching aptamers for the selective recognition of multiple therapeutically significant biomarkers and dangerous substances are expected to be developed in the near future. The combination of lab-on-chip platforms with aptamers and functional nanomaterials will offer up new possibilities for clinical diagnostics, food analysis, and environmental monitoring devices.

ACKNOWLEDGMENTS

A.S. and R.K.M. would like to acknowledge the library search engine of SGT University, Gurugram, India and Purdue University for providing continuous support.

REFERENCES

1. Rodriguez-Mozaz S, Lopez de Alda MJ, Barceló D. 2006. Biosensors as useful tools for environmental analysis and monitoring. *Analytical and Bioanalytical Chemistry*. 386: 1025–1041
2. Sezgintürk MK. 2020.Chapter one—introduction to commercial biosensors. In *Commercial Biosensors and Their Applications*, ed. MK Sezgintürk, pp. 1–28. Elsevier
3. Thévenot DR, Toth K, Durst RA, Wilson GS. 2001. Electrochemical biosensors: Recommended definitions and classification. *Biosensors and Bioelectronics*. 16: 121–131
4. Nic M, Jirat J, Kosata B. 2006. Chemical terminology at your fingertips. *Chemistry International*. 28(6): 28
5. Luong JHT, Male KB, Glennon JD. 2008. Biosensor technology: Technology push versus market pull. *Biotechnology Advances*. 26: 492–500
6. Sharma A, Chandra Singh A, Bacher G, Bhand S. 2016. Recent advances in aptamer-based biosensors for detection of antibiotic residues. *Aptamers Synth Antibodies*. 2: 43–54
7. Schadewaldt P, Bodner-Leidecker A, Hammen HW, Wendel U. 1999. Significance of L-alloisoleucine in plasma for diagnosis of maple syrup urine disease. *Clinical Chemistry*. 45: 1734–1740

8. Zhou W, Huang PJ, Ding J, Liu J. 2014. Aptamer-based biosensors for biomedical diagnostics. *Analyst*. 139: 2627–2640

9. Shruthi GS, Amitha CV, Mathew BB. 2014. Biosensors: A modern day achievement. *Journal of Instrumentation Technology*. 2: 26–39

10. Gold L, Janjic N, Jarvis T, Schneider D, Walker JJ, Wilcox SK, et al. 2012. Aptamers and the RNA world, past and present. *Cold Spring Harbor Perspectives in Biology*. 4

11. Song S, Wang L, Li J, Fan C, Zhao J. 2008. Aptamer-based biosensors. *TrAC Trends in Analytical Chemistry*. 27: 108–117

12. Cox JC, Ellington AD. 2001. Automated selection of anti-Protein aptamers. *Bioorganic & Medicinal Chemistry*. 9: 2525–2531

13. Wang RE, Wu H, Niu Y, Cai J. 2011. Improving the stability of aptamers by chemical modification. *Current Medicinal Chemistry*. 18: 4126–4138

14. Zon G. 2020. Recent advances in aptamer applications for analytical biochemistry. *Analytical Biochemistry*. 113894

15. Fang X, Tan W. 2010. Aptamers generated from cell-SELEX for molecular medicine: A chemical biology approach. *Accounts of Chemical Research*. 43: 48–57

16. Mehlhorn A, Rahimi P, Joseph Y. 2018. Aptamer-based biosensors for antibiotic detection: A review. *Biosensors (Basel)*. 8

17. Huang Z, Qiu L, Zhang T, Tan W. 2021. Integrating DNA nanotechnology with aptamers for biological and biomedical applications. *Matter*. 4: 461–489

18. Sharma A, Istamboulie G, Hayat A, Catanante G, Bhand S, Marty JL. 2017. Disposable and portable aptamer functionalized impedimetric sensor for detection of kanamycin residue in milk sample. *Sensors and Actuators B: Chemical*. 245: 507–515

19. Sharma A, Catanante G, Hayat A, Istamboulie G, Ben Rejeb I, Bhand S, et al. 2016. Development of structure switching aptamer assay for detection of aflatoxin M1 in milk sample. *Talanta*. 158: 35–41

20. Sharma A, Khan R, Catanante G, Sherazi TA, Bhand S, Hayat A, et al. 2018. Designed strategies for fluorescence-based biosensors for the detection of mycotoxins. *Toxins*. 10: 197

21. Jayasena SD. 1999. Aptamers: An emerging class of molecules that rival antibodies in diagnostics. *Clinical Chemistry*. 45: 1628–1650

22. Zhou J, Battig MR, Wang Y. 2010. Aptamer-based molecular recognition for biosensor development. *Analytical and Bioanalytical Chemistry*. 398: 2471–2480

23. Song S-H, Gao Z-F, Guo X, Chen G-H. 2019. Aptamer-based detection methodology studies in food safety. *Food Analytical Methods*. 12: 966–990

24. Kong R-M, Zhang X-B, Chen Z, Tan W. 2011. Aptamer-assembled nanomaterials for biosensing and biomedical applications. *Small*. 7: 2428–2436

25. Ding F, Gao Y, He X. 2017. Recent progresses in biomedical applications of aptamer-functionalized systems. *Bioorganic & Medicinal Chemistry Letters*. 27: 4256–4269

26. Li Z, Mohamed MA, Vinu Mohan AM, Zhu Z, Sharma V, Mishra GK, et al. 2019. Application of electrochemical aptasensors toward clinical diagnostics, food, and environmental monitoring: Review. *Sensors*. 19: 5435

27. Mishra RK, Hayat A, Catanante G, Ocaña C, Marty J-L. 2015. A label free aptasensor for Ochratoxin A detection in cocoa beans: An application to chocolate industries. *Analytica Chimica Acta*. 889: 106–112

28. Smart A, Crew A, Pemberton R, Hughes G, Doran O, Hart JP. 2020. Screen-printed carbon based biosensors and their applications in agri-food safety. *TrAC Trends in Analytical Chemistry*. 127: 115898

29. Foo ME, Gopinath SCB. 2017. Feasibility of graphene in biomedical applications. *Biomedicine & Pharmacotherapy*. 94: 354–361

30. Rowsell S, Stonehouse NJ, Convery MA, Adams CJ, Ellington AD, Hirao I, et al. 1998. Crystal structures of a series of RNA aptamers complexed to the same protein target. *Nature Structural Biology*. 5: 970–975

31. Shaw J-P, Fishback JA, Cundy KC, Lee WA. 1995. A novel oligodeoxynucleotide inhibitor of thrombin. I. In vitrometabolic stability in plasma and serum. *Pharmaceutical Research*. 12: 1937–1942

32. Wilson DS, Szostak JW. 1999. In vitro selection of functional nucleic acids. *Annual Review of Biochemistry*. 68(1): 611–647

33. Kotia RB. 2000. *DNA Aptamers: A Novel Approach to Chemical Separations*. Duke University

34. Stoltenburg R, Reinemann C, Strehlitz B. 2007. SELEX—A (r)evolutionary method to generate high-affinity nucleic acid ligands. *Biomolecular Engineering*. 24: 381–403

35. Meyers RA. 2000. *Encyclopedia of Analytical Chemistry*. Wiley

36. Musheev MU, Krylov SN. 2006. Selection of aptamers by systematic evolution of ligands by exponential enrichment: Addressing the polymerase chain reaction issue. *Analytica Chimica Acta*. 564: 91–96

37. Hermann T, Patel DJ. 2000. Adaptive recognition by nucleic acid aptamers. *Science*. 287: 820–825

38. Stanciu LA, Wei Q, Barui AK, Mohammad N. 2021. Recent advances in aptamer-based biosensors for global health applications. *Annual Review of Biomedical Engineering*. 23: 433–459

39. Muhammad M, Huang Q. 2021. A review of aptamer-based SERS biosensors: Design strategies and applications. *Talanta*. 227: 122188

40. Han K, Liang Z, Zhou N. 2010. Design strategies for aptamer-based biosensors. *Sensors*. 10: 4541–4557

41. Lu CH, Yang HH, Zhu CL, Chen X, Chen GN. 2009. A graphene platform for sensing biomolecules. *Angewandte Chemie International Edition*. 48: 4785–4787

42. Lu Z, Chen X, Hu W. 2017. A fluorescence aptasensor based on semiconductor quantum dots and MoS2 nanosheets for ochratoxin A detection. *Sensors and Actuators B: Chemical*. 246: 61–67

43. Mishra GK, Sharma V, Mishra RK. 2018. Electrochemical aptasensors for food and environmental safeguarding: A review. *Biosensors*. 8: 28

44. Nutiu R, Li Y. 2004. Structure-switching signaling aptamers: Transducing molecular recognition into fluorescence signaling. *Chemistry—A European Journal*. 10: 1868–1876

45. Guan B, Zhang X. 2020. Aptamers as versatile ligands for biomedical and pharmaceutical applications. *International Journal of Nanomedicine*. 1059–1071

46. Lubin AA, Plaxco KW. 2010. Folding-based electrochemical biosensors: The case for responsive nucleic acid architectures. *Accounts of Chemical Research*. 43: 496–505

47. Jhaveri S, Kirby R, Conrad R, Maglott E, Bowser M, Kennedy R, et al. 2000. Designed signaling aptamers that transduce molecular recognition to changes in fluorescence intensity. *Journal of the American Chemical Society*. 122

48. Didenko VV. 2006. *Fluorescent Energy Transfer Nucleic Acid Probes: Designs and Protocols*, p. 335. Springer Science & Business Media

49. Terracciano M, Rea I, Borbone N, Moretta R, Oliviero G, Piccialli G, et al. Porous silicon-based aptasensors: The next generation of label-free devices for health monitoring. *Molecules*. 24: 2216

50. Sun Y, Lu J. 2019. Chemiluminescence-based aptasensors for various target analytes. *Luminescence*. 33: 1298–1305

51. Avino A, Fabrega C, Tintore M, Eritja R. 2012. Thrombin binding aptamer, more than a simple aptamer: Chemically modified derivatives and biomedical applications. *Current Pharmaceutical Design*. 18: 2036–2047

52. Zhang J, Wang L, Pan D, Song S, Boey FYC, Zhang H, et al. 2008. Visual cocaine detection with gold nanoparticles and rationally engineered aptamer structures. *Small*. 4: 1196–1200

53. Ferapontova EE, Olsen EM, Gothelf KV. 2008. An RNA aptamer-based electrochemical biosensor for detection of theophylline in serum. *Journal of the American Chemical Society*. 130: 4256–4258

54. Schoukroun-Barnes LR, Wagan S, White RJ. 2014. Enhancing the analytical performance of electrochemical RNA aptamer-based sensors for sensitive detection of aminoglycoside antibiotics. *Analytical Chemistry*. 86: 1131–1137

55. García-González R, Costa-García A, Fernández-Abedul MT. 2014. Methylene blue covalently attached to single stranded DNA as electroactive label for potential bioassays. *Sensors and Actuators B: Chemical*. 191: 784–790

56. Wu L, Zhang X, Liu W, Xiong E, Chen J. 2013. Sensitive electrochemical aptasensor by coupling "signal-on" and "signal-off" strategies. *Analytical Chemistry*. 85: 8397–8402

57. Catanante G, Mishra RK, Hayat A, Marty J-L. 2016. Sensitive analytical performance of folding based biosensor using methylene blue tagged aptamers. *Talanta*. 153: 138–144

58. Yamamoto R, Baba T, Kumar PK. 2000. Molecular beacon aptamer fluoresces in the presence of Tat protein of HIV-1. *Genes Cells*. 5: 389–396

59. Niazi S, Khan IM, Yu Y, Pasha I, Lv Y, Mohsin A, et al. 2020. A novel fluorescent aptasensor for aflatoxin M1 detection using rolling circle amplification and g-C3N4 as fluorescence quencher. *Sensors and Actuators B: Chemical*. 315: 128049

60. Yang J, Zhang Z, Pang W, Chen H, Yan G. 2019. Graphene oxide based fluorescence super-quencher@QDs composite aptasensor for detection of Ricin B-chain. *Sensors and Actuators B: Chemical*. 301: 127014

61. Wen L, Qiu L, Wu Y, Hu X, Zhang X. 2017. Aptamer-modified semiconductor quantum dots for biosensing applications. *Sensors*. 17(8): 1736

62. Hasanzadeh M, Shadjou N, de la Guardia M. 2017. Aptamer-based assay of biomolecules: Recent advances in electro-analytical approach. *TrAC Trends in Analytical Chemistry*. 89: 119–132

63. Kang K, Sachan A, Nilsen-Hamilton M, Shrotriya P. 2011. Aptamer functionalized microcantilever sensors for cocaine detection. *Langmuir: The ACS Journal of Surfaces and Colloids*. 27: 14696–14702

64. Teengam P, Tuantranont A, Henry C, Vilaivan T, Chailapakul O. 2016. Electrochemical paper-based peptide nucleic acid biosensor for detecting human papillomavirus. *Analytica Chimica Acta*. 952

65. Vallée-Bélisle A, Plaxco KW. 2010. Structure-switching biosensors: Inspired by nature. *Current Opinion in Structural Biology*. 20: 518–526

66. Nutiu R, Li Y. 2003. Structure-switching signaling aptamers. *Journal of the American Chemical Society*. 125: 4771–4778

67. Kirby R, Cho EJ, Gehrke B, Bayer T, Park YS, Neikirk DP, et al. 2004. Aptamer-based sensor arrays for the detection and quantitation of proteins. *Analytical Chemistry*. 76: 4066–4075

68. Sassolas A, Blum LJ, Leca-Bouvier BD. 2011. Optical detection systems using immobilized aptamers. *Biosensors and Bioelectronics*. 26: 3725–3736

69. Nguyen HH, Lee SH, Lee UJ, Fermin CD, Kim M. 2019. Immobilized enzymes in biosensor applications. *Materials (Basel)*. 12: 121

70. Anderson DJ, Blobel G. 1983. Immunoprecipitation of proteins from cell-free translations. *Methods Enzymol*. 96: 111–120

71. Zhou J, Soontornworajit B, Wang Y. 2010. A temperature-responsive antibody-like nanostructure. *Biomacromolecules*. 11: 2087–2093

72. Liu J, Peng Q. 2017. Protein-gold nanoparticle interactions and their possible impact on biomedical applications. *Acta Biomaterialia*. 55: 13–27

73. Zhao J, Zhang Y, Li H, Wen Y, Fan X, Lin F, et al. 2011. Ultrasensitive electrochemical aptasensor for thrombin based on the amplification of aptamer—AuNPs—HRP conjugates. *Biosensors and Bioelectronics*. 26: 2297–2303

74. Drolet DW, Moon-McDermott L, Romig TS. 1996. An enzyme-linked oligonucleotide assay. *Nature Biotechnology.* 14: 1021–1025

75. Feng K, Kang Y, Zhao JJ, Liu YL, Jiang JH, Shen GL, et al. 2008. Electrochemical immunosensor with aptamer-based enzymatic amplification. *Analytical Biochemistry.* 378: 38–42

76. Chen A, Yan M, Yang S. 2016. Split aptamers and their applications in sandwich aptasensors. *TrAC Trends in Analytical Chemistry.* 80: 581–593

77. Stojanovic MN, de Prada P, Landry DW. 2000. Fluorescent sensors based on aptamer self-assembly. *Journal of the American Chemical Society.* 122: 11547–11548

78. Sharma AK, Kent AD, Heemstra JM. 2012. Enzyme-linked small-molecule detection using split aptamer ligation. *Analytical Chemistry.* 84: 6104–6109

79. Kumke MU, Li G, McGown LB, Walker GT, Linn CP. 1995. Hybridization of fluorescein-labeled DNA oligomers detected by fluorescence anisotropy with protein binding enhancement. *Analytical Chemistry.* 67: 3945–3951

80. Sharma AK, Heemstra JM. 2011. Small-molecule-dependent split aptamer ligation. *Journal of the American Chemical Society.* 133: 12426–12429

81. Wang Q, Huang J, Yang X, Wang K, He L, Li X, et al. 2011. Surface plasmon resonance detection of small molecule using split aptamer fragments. *Sensors and Actuators B: Chemical.* 156: 893–898

82. Deisingh AK. 2006. Aptamer-based biosensors: Biomedical applications. *RNA Towards Medicine.* 341–357

83. Yu Z, Cui P, Xiang Y, Li B, Han X, Shi W, et al. 2020. Developing a fast electrochemical aptasensor method for the quantitative detection of penicillin G residue in milk with high sensitivity and good anti-fouling ability. *Microchemical Journal.* 157: 105077

84. Rezaei B, Jamei HR, Ensafi AA. 2018. An ultrasensitive and selective electrochemical aptasensor based on rGO-MWCNTs/Chitosan/carbon quantum dot for the detection of lysozyme. *Biosensors and Bioelectronics.* 115: 37–44

85. Mir M, Katakis I. 2005. Towards a fast-responding, label-free electrochemical DNA biosensor. *Analytical and Bioanalytical Chemistry.* 381: 1033–1035

86. Xiao Y, Lubin AA, Heeger AJ, Plaxco KW. 2005. Label-free electronic detection of thrombin in blood serum by using an aptamer-based sensor. *Angewandte Chemie International Edition.* 44: 5456–5459

87. Lai RY, Plaxco KW, Heeger AJ. 2007. Aptamer-based electrochemical detection of picomolar platelet-derived growth factor directly in blood serum. *Analytical Chemistry.* 79: 229–233

88. Altintas Z, Akgun M, Kokturk G, Uludağ Y. 2017. A fully automated microfluidic-based electrochemical sensor for real-time bacteria detection. *Biosensors and Bioelectronics.* 100

89. Sun Y, Haglund TA, Rogers AJ, Ghanim AF, Sethu P. 2018. Review: Microfluidics technologies for blood-based cancer liquid biopsies. *Analytica Chimica Acta.* 1012: 10–29

90. Wu J, He Z, Chen Q, Lin J-M. 2016. Biochemical analysis on microfluidic chips. *TrAC Trends in Analytical Chemistry.* 80: 213–231

91. Rowe AA, Miller EA, Plaxco KW. 2010. Reagentless measurement of aminoglycoside antibiotics in blood serum via an electrochemical, ribonucleic acid aptamer-based biosensor. *Analytical Chemistry.* 82: 7090–7095

92. Xiang Y, Lu Y. 2011. Using personal glucose meters and functional DNA sensors to quantify a variety of analytical targets. *Nature Chemistry.* 3: 697–703

93. Yan L, Zhu Z, Zou Y, Huang Y, Liu D, Jia S, et al. 2013. Target-responsive "sweet" hydrogel with glucometer readout for portable and quantitative detection of non-glucose targets. *Journal of the American Chemical Society.* 135: 3748–3751

94. Bai L, Yuan R, Chai Y, Yuan Y, Wang Y, Xie S. 2012. Direct electrochemistry and electrocatalysis of a glucose oxidase-functionalized bioconjugate as a trace label for ultrasensitive detection of thrombin. *Chemical Communications*. 48: 10972–10974

95. Teymourian H, Salimi A, Firoozi S. 2014. A high performance electrochemical biosensing platform for glucose detection and IgE aptasensing based on Fe3O4/reduced
graphene oxide nanocomposite. *Electroanalysis*. 26: 129–138

96. Jin X, Lv M, Pan Q, Fang S, Zhu N. 2021. An electrochemical aptasensor based on
bifunctional Fe3O4@Au nanocomposites for adenosine triphosphate assay. *Journal of
Solid State Electrochemistry*. 25: 1073–10781

97. Ravalli A, Rivas L, De La Escosura-Muñiz A, Pons J, Merkoçi A, Marrazza G. 2015.
A DNA aptasensor for electrochemical detection of vascular endothelial growth factor.
Journal of Nanoscience and Nanotechnology. 15: 3411–3416

98. Chen Y, Liu X, Guo S, Cao J, Zhou J, Zuo J, et al. 2019. A sandwich-type electrochemical aptasensor for mycobacterium tuberculosis MPT64 antigen detection using C60NPs
decorated N-CNTs/GO nanocomposite coupled with conductive PEI-functionalized
metal-organic framework. *Biomaterials*. 216: 119253

99. Tao D, Shui B, Gu Y, Cheng J, Zhang W, Jaffrezic-Renault N, et al. 2019. Development
of a label-free electrochemical aptasensor for the detection of Tau381 and its preliminary
application in AD and non-AD patients' sera. *Biosensors*. 9: 84

100. Zhang D-W, Sun C-J, Zhang F-T, Xu L, Zhou Y-L, Zhang X-X. 2012. An electrochemical aptasensor based on enzyme linked aptamer assay. *Biosensors and Bioelectronics*.
31: 363–368

101. Wang L, Zeng L, Wang Y, Chen T, Chen W, Chen G, et al. 2021. Electrochemical
aptasensor based on multidirectional hybridization chain reaction for detection of tumorous exosomes. *Sensors and Actuators B: Chemical*. 332: 129471

102. Vasilescu A, Marty J-L. 2016. Electrochemical aptasensors for the assessment of food
quality and safety. *TrAC Trends in Analytical Chemistry*. 79: 60–70

103. Mohammad Danesh N, Ramezani M, Sarreshtehdar Emrani A, Abnous K, Taghdisi
SM. 2016. A novel electrochemical aptasensor based on arch-shape structure of aptamer-complimentary strand conjugate and exonuclease I for sensitive detection of streptomycin. *Biosensors and Bioelectronics*. 75: 123–128

104. Zhou N, Luo J, Zhang J, You Y, Tian Y. 2015. A label-free electrochemical aptasensor for
the detection of kanamycin in milk. *Analytical Methods*. 7: 1991–1996

105. Liu S, Wang Y, Xu W, Leng X, Wang H, Guo Y, et al. 2017. A novel sandwich-type
electrochemical aptasensor based on GR-3D Au and aptamer-AuNPs-HRP for sensitive
detection of oxytetracycline. *Biosensors and Bioelectronics*. 88: 181–187

106. Shen Z, He L, Cao Y, Hong F, Zhang K, Hu F, et al. 2019. Multiplexed electrochemical aptasensor for antibiotics detection using metallic-encoded apoferritin probes and
double stirring bars-assisted target recycling for signal amplification. *Talanta*. 197:
491–499

107. Kim Y-J, Kim YS, Niazi JH, Gu MB. 2009. Electrochemical aptasensor for tetracycline
detection. *Bioprocess and Biosystems Engineering*. 33: 31

108. Ebrahimi Vafaye S, Rahman A, Safaeian S, Adabi M. 2021. An electrochemical aptasensor based on electrospun carbon nanofiber mat and gold nanoparticles for the sensitive
detection of Penicillin in milk. *Journal of Food Measurement and Characterization*. 15:
876–882

109. Abnous K, Danesh NM, Ramezani M, Taghdisi SM, Emrani AS. 2016. A novel electrochemical aptasensor based on H-shape structure of aptamer-complimentary strands
conjugate for ultrasensitive detection of cocaine. *Sensors and Actuators B: Chemical*.
224: 351–355

110. Hu X, Wei P, Catanante G, Li Z, Marty JL, Zhu Z. 2019. Ultrasensitive ciprofloxacin assay based on the use of a fluorescently labeled aptamer and a nanocomposite prepared from carbon nanotubes and MoSe2. *Microchimica Acta*. 186: 507

111. Hu X, Goud KY, Kumar VS, Catanante G, Li Z, Zhu Z, et al. 2018. Disposable electrochemical aptasensor based on carbon nanotubes- V2O5-chitosan nanocomposite for detection of ciprofloxacin. *Sensors and Actuators B: Chemical*. 268: 278–286

112. Marin S, Ramos AJ, Cano-Sancho G, Sanchis V. 2013. Mycotoxins: Occurrence, toxicology, and exposure assessment. *Food and Chemical Toxicology*. 60: 218–237

113. Wu SS, Wei M, Wei W, Liu Y, Liu S. 2019. Electrochemical aptasensor for aflatoxin B1 based on smart host-guest recognition of β-cyclodextrin polymer. *Biosensors and Bioelectronics*. 129: 58–63

114. Castillo G, Spinella K, Poturnayová A, Šnejdárková M, Mosiello L, Hianik T. 2015. Detection of aflatoxin B1 by aptamer-based biosensor using PAMAM dendrimers as immobilization platform. *Food Control*. 52: 9–18

115. Yugender Goud K, Catanante G, Hayat A, M S, Vengatajalabathy Gobi K, Marty JL. 2016. Disposable and portable electrochemical aptasensor for label free detection of aflatoxin B1 in alcoholic beverages. *Sensors and Actuators B: Chemical*. 235: 466–473

116. Mishra RK, Hayat A, Catanante G, Istamboulie G, Marty J-L. 2016. Sensitive quantitation of ochratoxin A in cocoa beans using differential pulse voltammetry based aptasensor. *Food Chemistry*. 192: 799–804

117. Nguyen BH, Tran LD, Do QP, Nguyen HL, Tran NH, Nguyen PX. 2013. Label-free detection of aflatoxin M1 with electrochemical Fe3O4/polyaniline-based aptasensor. *Materials Science and Engineering: C*. 33: 2229–2234

118. Ullah N, Chen W, Noureen B, Tian Y, Du L, Wu C, et al. 2021. An electrochemical Ti3C2Tx aptasensor for sensitive and label-free detection of marine biological toxins. *Sensors*. 21: 4938

119. Shi Z-Y, Zheng Y-T, Zhang H-B, He C-H, Wu W-D, Zhang H-B. 2015. DNA electrochemical aptasensor for detecting fumonisins B1 based on graphene and thionine nanocomposite. *Electroanalysis*. 27: 1097–1103

13 Validation Requirements in Biosensors

Leyla Karadurmus[1,2], S. Irem Kaya[1,3],
Goksu Ozcelikay[1], Mehmet Gumustas[4],
Bengi Uslu[1], and Sibel A. Ozkan[1]

[1]Ankara University
Faculty of Pharmacy, Department of Analytical Chemistry
Ankara, Turkey

[2]Adıyaman University
Faculty of Pharmacy, Department of Analytical Chemistry
Adıyaman, Turkey

[3]University of Health Sciences
Gulhane Faculty of Pharmacy, Department of Analytical Chemistry
Ankara, Turkey

[4]Ankara University
Institute of Forensic Sciences, Department of Forensic Toxicology
Ankara, Turkey

CONTENTS

DOI: 10.1201/9781003189435-17

13.1 INTRODUCTION

Validation is vital for many aspects of chemistry, biochemistry, clinical, pharmacy, food production, life science, environment, etc. Analytical method validation provides verification of the method with various criteria. Optimized and developed analytical methods should be checked according to required validation parameters before use. New drug applications, bioequivalence, and bioavailability studies should be periodically examined by validation parameters (1). Validation is a tool of analytical parameters that comprise formal, systematic, and documented to provide reliable, accurate, consistent, and reproducible data (2). The design of experiments must improve to obtain considerable value. Analytical method validation aims to confirm its suitability for the proposed purpose (3). Depending on the analyte, analytical method, and/or analyte matrix, parameters can differ from the anticipated laboratory requirements. In other words, all the analytical parameters are not always evaluated and validated that are available for a specific technique or required validation group. Method validation is the process of demonstrating that analytical procedures are employed for a particular test that is suitable for its intended use (4).

Method performance should be validated according to authorities' guidelines: Pharmacopoeias included as part of registration applications submitted within the European Pharmacopeia (E.P.), Japanese Pharmacopeia (J.P.), United States Pharmacopeia (USP), Indian Pharmacopeia (I.P.), Chinese Pharmacopeia (C.P.), Turkish Pharmacopeia (T.P.), etc., On the other hand, Food and Drug Administration (FDA), Food and Agricultural Organization (FAO), European Medicinal Agency (EMA), International Laboratory Accreditation Cooperation (ILAC), Association of Analytical Chemists (AOAC), American Society for Testing and Material (ASTM), Codex Committee on Methods of Analysis and Sampling (CCMAS), Cooperation on International Traceability in Analytical Chemistry (CITAC), Environmental Protection Agency (EPA), European Analytical Chemistry Group (EURACHEM), European Committee for Normalization (CEN), European Commission, European Cooperation for Accreditation (E.A.), International Council for Harmonisation (ICH) guidelines addressing the validation of analytical methods (5). ICH has been taken as preferable because this guideline has been developed for harmonization purposes (6).

The availability of some pharmacopeia and guidelines has contributed significantly to such errors in evaluating the analytical process and correcting these misunderstandings (7). The aim of mentioned guidelines and pharmacopeias that serve as a collection of terms and definitions is to bridge the differences between various compendia and regulators.

Qualitative and quantitative analyses are performed to obtain the best decision. Moreover, the appropriate statistical tools evaluate what is claimed or intended data.

The researchers should be aware of the quality of the data whenever analytical methods are used such as validation parameters, calculations, and specific applications. Although method validation and analytical procedures must be intertwined, validation should not be thought of separately (8).

Method validation is a fundamental step of the measurements to produce reliable analytical data but time-consuming activity for analytical laboratories where the method development is performed (9). Authorized institutions approve the application certificate of the analytical methods that pass the validation parameters (10). The endorsed certificate requires registration in, or export to, the world because it discusses the characteristics for consideration during the validation of the analytical procedures. Analytical validation under Good Manufacturing Practices (GMP) is needed in pharmaceutical companies to obtain scientific evidence that an analytical method provides reliable results (11).

This chapter aims to critically discuss validation for analytical methods in the frame of biosensors, which can be classified under the ligand binding assays by evaluating the analytical performance parameters and their limitations. Thus, it also highlights analytical method validation and part of the regulatory guidelines used as a standard and exhaustive checklist.

13.2 HISTORICAL OVERVIEW OF ANALYTICAL VALIDATION

The word "validation" comes from the Latin term *validus*, and the meaning of this word is strong and worthy, suggesting that something is reliable, useful, and true. Analytical method validation has become a regulatory requirement in the USA for Federal Register after two FDA officials, who are Ted Byers and Bud Loftus, proposed the concept of validation for finding the solution to the sterility of large volume parenteral market and improving the quality of pharmaceuticals in the mid-1970s (12). The first validation attempts start with the processes involved in making these products and then continue the associated steps of the pharmaceutical process. cGMP regulations and definition of process validation did not appear in any part of the literature of FDA till 1978 (13). However, one of the guidelines in The U.S. Food and Drug Administration (FDA) is entitled "Submitting Samples and Analytical Data for Methods Validation". In addition, the International Council for Harmonisation (ICH), formerly known as the International Conference on Harmonisation, published guidelines under the heading "Requirements for Pharmaceuticals for Human Use" (14–15).

As mentioned earlier, the Current Good Manufacturing Practice (cGMP), Agência Nacional de Vigilância Sanitária (ANVISA) (2017), National Drug Administration (NDA), World Health Organization (WHO) (2016), EMA (2016), FDA (2015), USP Convention, AOAC, the American Public Health Association, and other useful protocols have derived from ICH as a worldwide basis for both regulatory authorities and the pharmaceutical industry (16).

ICH Q2(R1) is known as the primary reference for recommendations and definitions on validation characteristics for analytical procedures (8). Moreover, The FDA guidance has validation characteristics of Chromatographic Methods for industry (17).

13.3   ANALYTICAL VALIDATION CHARACTERISTICS
OF BIOSENSORS

The primary goal of an analytical measurement is to provide stable, accurate, and reliable data, and method validation is an important tool in achieving this goal. Before all of these, the selection, design, and development of analytical methods should be performed based on a systematic approach.

Initially, the purpose for the study should be identified: Impurities, safety and characterization, and identification issues with the stability of the analytes should be carefully examined. In the next step, instrumentation, reagents, the risk from analysts, and variability of the method need to be identified. On the other hand, the analyte of interest (in vitro and in vivo metabolism physiochemical properties of the studied compound, protein binding capability, etc.) should be very well understood. Then, the method development process should be realized: Method development has not been a scope of validation activities. However, it is an essential step before validation. Draft of the ICH M10 guideline mainly refers that "The purpose of bioanalytical method development is to define the design, operating conditions, limitations and suitability of the method for its intended purpose and to ensure that the method is ready for validation" (18). In this context, the difference between the terms of qualification and the validation of an assay should be very well known. Qualification of an assay determines whether an assay is suitable for its intended purpose, and assay validation is assuring the assay is suitable for its intended purpose on a routine basis.

Validation is needed for the build-up of a novel analytical method. Validation and evaluation of analytical and/or electroanalytical method productivity are necessary to determine the degree of expected error due to inaccuracy and precision and verify that the degree of error meets expected laboratory or clinical requirements. Recommended techniques for method validation vary depending on the kind of test and the anticipated use. Experiments should be designed in such a way that accurate data is obtained. As mentioned earlier, in the frame of this chapter, the method validation is discussed and explained specifically from the biosensors point of view.

Beyond the outstanding advantages of biosensors, they should be comparable to conventional analytical systems in terms of reliability, sensitivity, selectivity, specificity, and robustness. Therefore, biosensor measurements need to be verified and validated before being accepted. The validation of biosensors can present some challenges as other aspects such as stability of the biological element or the element immobilized in the transducer may need to be evaluated and the general quality parameters (19–22). As mentioned earlier ICH M10 Guideline and specifically section 4.2 (validation of ligand binding assays) can be carefully examined and performed with some modifications for biosensor studies.

Before discussing the analytical method validation, critical reagents for the bioanalytical assays need to be highlighted. One of the most important classes of reagents is reference standard materials. These materials should be well characterized and identified. Furthermore, the documentation of these materials must be complete and classified. Especially, even if it is necessary to change the batch of a reference standard in bioanalysis, the bioanalytical evaluation must be carried out before use. Based on the evaluation, the performance characteristics of the method should be checked within the acceptance limits.

Other critical reagents, including aptamers, antibodies, proteins, enzymes, etc., have significantly affected the results of the experiments. In most cases, critical reagents mentioned earlier interact with the analytes and bind to them. These analyte--reagent complexes give signals based on the concentration of the analytes. Suppose the composition of a reagent (source, purity, etc.) varied from batch to batch. In that case, it also affects the binding, and due to this, experimental data may lose its precision and accuracy. As a minor variation, the source of a reagent can be different: For example, assuming the production method is the same, the source of β-Glucuronidase can be limpets, *E.coli*, bovine liver, *Helix pomatia*, *Helix aspersa*, abalone, etc. In this case, a comparison between the reagents in terms of precision and accuracy is sufficient. But in some cases, major changes are available between the reagents; antibodies are commonly used in biosensor development, and any change in supplier of the monoclonal antibody or production method may significantly result in the experiments and additional validation experiments will be performed. Because of these reasons, critical reagents used in the analytical method should be defined and identified (18).

Before discussing each type of characteristic validation parameters and their contents needs to be understood. There are 3 types of method validation strategies;

- partial validation,
- full validation and
- cross-validation.

Partial validation is required

- When transferring the actual method to another laboratory
- When analysis performed by different analysts
- When the instrumentation changes
- When the software changes
- When the extraction or sample preparation procedure changes
- When the species in the matrix changes
- When the matrix changes within the same analytes

Full validation is performed

- When the method is developed and optimized the first time
- When the metabolites are added for the quantitative purposes
- When the new drug, degradation product and/or impurity entity is being analyzed

The last type of method validation strategy is cross-validation which is used to compare at least 2 bioanalytical methods (reference and test methods). The goal is to determine whether the obtained data are comparable.

- The same analytical methods are being used in different laboratories (at least two of them) for comparing the data from quality control samples in the same analysis
- Different analytical methods can be used for the same quality control samples

TABLE 13.1

Comparison Between the Most Used Guidelines in Terms of Characteristics of the Analytical Method Validation

Parameters	ICH	USP	EP
Specificity	✓	✓	✓
Linearity	✓	✓	Only for quantitative purposes
Range	✓	✓	Only for quantitative purposes
Limit of Detection	✓	Limit tests	Limit tests
Limit of Quantification	✓	Only for quantitative purposes	Limit tests
Repeatability	✓	✓	Only for quantitative purposes under the title of precision
Intermediate Precision	✓	✓	
Reproducibility	✓	✓	
Accuracy	✓	✓	Only for quantitative purposes
Robustness	Suggested	✓	X
Ruggedness	✓	Suggested	✓

The analytical method validation characteristics for biosensor studies are demonstrated in Table 13.1 and described in detail.

13.3.1 Specificity (Selectivity)

Biosensors are widely used to detect the presence of chemical, and pharmaceutical entities in a sample, by using biological materials as binding compounds (antibodies, enzymes, etc.), but their application is subject to limitations such as selectivity. Specificity is considered the most vital property of a biosensor, as it describes the sensor's ability to distinguish between targeted and untargeted biological entities in a sample. Specificity is the ability of an analytical method to differentiate and quantify in a sample the analyte in the presence of endogenous matrix components, metabolites, degradation products, and other components (21–25). While constructing a biosensor, selectivity is the main consideration when choosing a bioreceptor. The substance's ability to be measured separately from other components present in the sample to be analyzed may interfere with the analyzed substance. A good selectivity is required while working on qualitative and quantitative purposes. If the selectivity is not sufficient, doubt arises about the method's accuracy and precision. Depending on the possible interfering compounds, the evaluation of the specificity for the particular methodology can be different:

The selectivity is determined by spiked analytes and compares with the blank matrix from several sources if a compound directly belongs to the sample or matrices. On the other hand, the selectivity was tested by analyzing samples with or without analytes besides their possible interferences. The recommendation with the selectivity is the evaluation of higher analyte concentrations. It will be better to evaluate this parameter using blanks obtained from individual sources (n:10). The levels for spiking can be the lower limit of quantification (LLOQ) concentration with the accuracy of ±25% and high concentration of quality control (Q.C.) level with the accuracy of ±20%. The responses should not be higher than the LLOQ at least 8 out of 10 individual sources (26–30).

13.3.2 CALIBRATION CURVE AND LINEAR RANGE

The definition of linearity refers to the function of the analyte concentration to the response of the analytical methodology. As for bioanalytical purposes, the calibrator solutions should be prepared as fresh even if they will not be applicable, and the frozen standards should be used within their stability period. Calibration standards are prepared by spiking the known amounts of the analytes to the sample matrix. The calibration range is constructed from LLOQ to the upper limit of quantification (ULOQ), which refers to the highest level of the calibration. The LLOQ can be measured with acceptable accuracy and precision. The criteria for LLOQ is that the response should be at least 5 times compared with the blank response, and the accuracy and precision should be maximum ±25%. The calibration range is an indispensable validation parameter for quantification purposes. There must be at least six different concentration values to generate the calibration. As mentioned earlier, it should be contained the analysis of a blank sample (it should not be included in the calculation of the calibration equation), LLOQ, and ULOQ standards. Usually, three repetitive measurements are taken for each concentration level. A minimum of 6 independent runs is recommended to be evaluated over several days to demonstrate between run variations. Linearity is the property that indicates the accuracy of the measured response to a straight line, mathematically represented as $y=mc+n$, where c is the concentration of the analyte, y is the output signal, n is the intercept, and m is the sensitivity of the biosensor. If a linear relationship is determined as a result of the analysis, the study results are shown by making statistical calculations. The slope (m), cut (y), correlation coefficient (r), and determination coefficient (r^2) of the linearity formed are examined in the expression of linearity. The mathematical value to be obtained for the r and r^2 coefficients should be close to 1.00. Also, 0.99 is an acceptable value for the demonstration of the linearity of the responses. To control linearity and related validation parameters, RSD % or standard error (S.E.) values of slope and intersection are calculated and reported. The correlation coefficient (r) shows the numerical expression of the relationship between the analytical response and the concentration. The accuracy of the back-calculated concentrations of each calibration level should be 25% of LLOQ and ULOQ levels while 20% for the other levels. In statistical analyses, slope, intercept, and coefficient of determination are affected by LLOQ and ULOQ limit of linearity. The slope of the calibration curve is also an expression of the precision of the method.

The linearity of the biosensor can be related with the resolution of the biosensor and range of analyte concentrations under testing. The resolution of the biosensor is described as the minimum change in the concentration of an analyte that is needed to bring a changing in the reply of the biosensor. Depending on the application, a good resolution is required as most biosensor applications require not only analyte detection but also measurement of concentrations of analyte over a wide working range (27). Another term associated with linearity is the working range, defined as the range of analyte concentrations at which the biosensor response varies linearly with concentration (31–33). The concentration range should also include the target concentration of the compound to be studied. It should contain at least the expected or required range of analytical results, the latter being directly related to the acceptance limits of the specification or the target test concentration of the analytical procedure style. The working range depends on the analyte, the type of analytical methodology, and the application environment. The range is generally expressed in the same units as the test results obtained by the recommended analytical method (34).

13.3.3 ACCURACY

Accuracy is one of the essential requirements in the analytical method validation of biosensor-based methodologies. It refers to the closeness of mean test results obtained by the proposed methodology to the nominal concentration of the analyte. It is also a measure of the difference between the expectations of the test result and the accepted reference value due to method and experimental error. Finally, accuracy is a qualitative term for the degree to which sensor results converge to the reference value, which may be the actual or expected value, depending on agreement or definition.

Accuracy can be defined as the sum of systematic and random errors. The number of iterations can be increased to eliminate or reduce the random error. Moreover, the systematic error can only be eliminated by eliminating its cause or by corrective calculations involving the application of the recovery function. Even the accuracy and precision appear to be two independent properties. They are a combination of both parameters. However, it can be noticed that accuracy depends on precision if the accuracy studies exist. Preparation of the Q.C. samples become of utmost importance for bioanalytical purposes. The quality control samples are mimicking the working samples prepared in the known concentrations in the matrix. Q.C. series and the calibration series should be different. Q.C. series should be prepared from a single stock, and the selected levels should not be used in the calibration set. 5 concentration levels for the preparation of the Q.C. series are recommended. The analyte should be spiked at:

- The lower limit of quantification (LLOQ)
- Low QC (3 × LLOQ)
- Medium QC (Mean of the calibration)
- High QC (at least 75 % of the upper limit of calibration)
- Upper limit of quantification (ULOQ)

Accuracy is usually reported as the recovery % with the recommended analytical procedure of the known amount of working compound added in the real sample or as the difference between the mean and the true value (Bias%) along with confidence intervals (Figure 13.1). As mentioned earlier Q.C. samples should be used to check the accuracy and precision by determining them within the run and between runs. They are required to be determined by analyzing three repetitions per run for each concentration level for at least two days or more. Remarkably, within run precision and accuracy should be checked for each run. If this data criterion is not met in all runs, overall estimation for within-run values for each Q.C. level should be calculated. Furthermore, accuracy and intermediate precision (between runs) should be calculated by combining the data from all runs. Recovery of an analyte in an assay is the current response from an amount of analyte added to and extracted from the

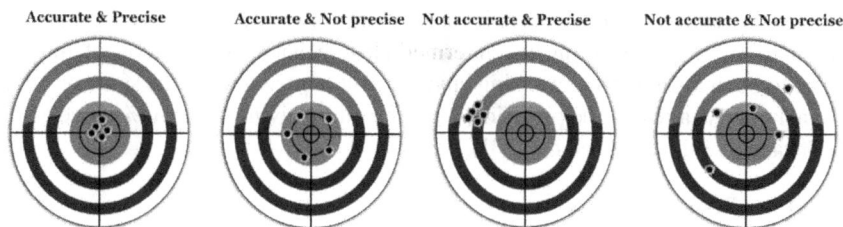

FIGURE 13.1 The schematic representation of accuracy and precision.

matrix compared to the current response for the actual concentration of the pure compound. In accuracy studies, bias is the difference between the expectation of test results and an accepted reference value. It may consist of more than one systematic error component. The deviation can be measured as a percentage deviation from the accepted reference value. Bias indicates how well the measurement is relative to the size of the thing being measured and is also called Systematic (or determinate) or Relative error (RE). It affects the accuracy of an analytical measurement causing all results to be equally erroneous. The bias is the difference between the mean value of multiple test results and the accepted reference value of the compound analyzed. The bias identified is the accuracy of the method. RE associated with biosensor-based data can be identified by bias, which can be positive or negative. The overall accuracy for both between run and within run studies should be within ±20% of the nominal values, while it should be ±25% for LLOQ and ULOQ levels. When examining the precision data for the Q.C. samples, the RSD% values should not exceed 25% for LLOQ and ULOQ and 20% for the other levels (18, 33, 35–37).

13.3.4 PRECISION

The precision is indicated by the slope value of the equation of the calibration line. Precision can be used to detect deviations in linearity. It is hard to obtain good reproducibility and repeatability results of the analytical method under real operating conditions. The precision of the analytical method is expressed as the percentage of the relative standard deviation (RSD%) or percentage of the coefficient of variation (CV). Precision is performed in three stages: repeatability, intermediate precision, and reproducibility (21–22, 31, 33, 38). Precision should be measured using a minimum of 3 concentrations and 5 repetitive analyses per level. As mentioned in the section on accuracy, RSD% values should not exceed 25% for LLOQ and ULOQ and 20% for the other levels.

13.3.4.1 Repeatability

Repeatability refers to the precision under the same operating and application conditions (by the same analyst, the same device, the same laboratory, and in the same period) within a short period of time. As a result, precision is expressed by giving the RSD% and/or CV% of the obtained values (31).

13.3.4.2 Intermediate Precision

Intermediate Precision is a level of precision that indicates within-laboratory experimental differences using different analysts or different instruments on different days.

Intraday repeatability can also be referred to as between-study or between-experiment precision. While determining the intermediate precision, the solution of the substance to be analyzed in the working environment is prepared by the same or different analysts. As a result, the precision is expressed by giving the RSD% value and/or CV% values of the obtained data (31).

13.3.4.3   Reproducibility

Reproducibility is a level of precision used in method standardization, obtained as a result of the practices of different laboratories and indicating experimental differences between one laboratory to the other. It is the realization of the analysis by preparing the solution in the working environment of the substance to be analyzed on different days (with a few days intervals). In some cases, this parameter can be used instead of intermediate precision. If intermediate precision parameters were performed and examined, reproducibility is not required. However, if the method is used in different laboratories, this parameter becomes significant (31).

13.3.5   Limit of Detection (LoD)

The detection limit is defined as the minimum amount of analyte that can be detected. This parameter especially becomes important when the threshold limit is very low. A biosensor is only useful if the detection limit is lower than that threshold limit (39). The terms "sensitivity" and "detection limit" could be used interchangeably. It also defines the lowest concentration of the analyzed sample, which is located under the specified experimental conditions but cannot be quantified accurately, and precisely. The point is LoD value is a variable parameter, it especially has poor reproducibility between days analysis.

For this reason, it should be determined daily basis for the demonstration of actual operating performance. Current efforts are focused on reducing the detection limit of a constructed biosensor while maintaining its sensitivity and reliability. Furthermore, LoD can be varied in the matrix and is also called matrix sensitive. Significantly, that should be determined in a matrix that matches the real sample matrix. The approaches for the calculation of LoD depend on whether the process is performed with or without tools. The most used type of LoD calculation is recommended by official guidelines and pharmacopeias (31, 33, 40–41).

13.3.5.1   Based on Visual LoD Inspection

LoD is defined by analyzing compounds with known concentrations of the analyte and establishing the minimum level at which the analyte can be reliably detected. Hence, this approach can be used mostly for non-instrumental methods, but it can also be used with instrumental methods in general, peak or signal shapes, as well as areas and/or heights, and may be varied between the samples with the same level of concentration.

13.3.5.2   Based on Signal-to-Noise Ratio (S/N)

Determination of the S/N is accomplished by comparing the measured signals from samples known to have low concentrations of the analyte with those of blank samples and establishing the minimum concentration at which the analyte can be reliable. Primarily when the final measurement is based on the instrumental output, the background response will need to be considered. For procedures that exhibit fundamental noise, the LoD may be calculated based on the S/N. Most of the electroanalytical-based biosensor setups have a background noise subtraction option. Due to this, noise cannot be obtained clearly due to the nature of the methods. However, LoD is determined by establishing the lowest level of concentration at which the S/N ratio of 3:1 (Figure 13.2) or between 3:1–2:1 is generally considered acceptable for estimating LoD.

13.3.5.3   Based on the Standard Deviation of the Response

Another calculation strategy that is most commonly used in electroanalytical-based biosensor studies is to calculate the LoD from the standard deviation of the response and the slope of the corresponding calibration curve.

A specific calibration equation is examined using samples containing an analyte close to the LoD level. For example, the following equation can express the calculation for this method:

$$LoD = \frac{3.3.\sigma}{m}$$

where σ is the standard deviation of the signal and m is the slope of the corresponding calibration graph. Thus, σ can be determined in different ways following:

* The standard deviation of intercept
* The standard deviation of blank sample analysis (n:10)
* The standard deviation of the repetitive analysis of calibration point close to the LoD level

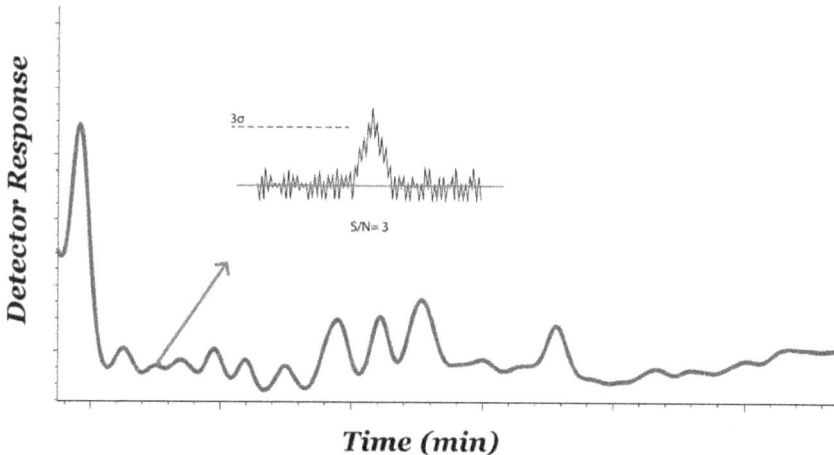

FIGURE 13.2   Schematic demonstration of S/N ratio for LoD.

Therefore, the answer can be converted to concentration units using the previously given equation.

13.3.6 LIMIT OF QUANTIFICATION (LOQ)

Several definitions can be used for this term, such as the limit of quantification, limit of quantitation, lower limit of quantification, limit of determination, quantitation limit, quantification limit, etc. in different guidelines and standards. The LoQ is defined as the lowest amount of analyte in a sample that can be quantified with appropriate precision and trueness. The LoQ is always higher than the LoD and is taken less than or equal to the lower limit of its range. If the result is found between LoD and LoQ, it can be reported as the analyte has been detected but below LoQ. Approaches to determine both LoD and LoQ are similar and are given later as details (31, 33, 41).

13.3.6.1 Based on Visual LoQ Inspection

While using the visual evaluation method, the LoQ is taken as the minimum concentration level of calibration where the analyte can be quantified with acceptable accuracy and precision. In another way, LoQ can be found by multiplying LoD by 3 or 3.3.

13.3.6.2 Based on Signal-to-Noise Ratio

The main difference with the LoQ from LoD in terms of S/N ratio is, the LOQ is the minimum measurement concentration at which the analyte can be reliably measured at 10:1 S/N (Figure 13.3). Using such a calculation method of LoQ, the S/N ratio is performed by comparing the measured signals from samples with known low concentrations of the analyte with blank samples similar to LoD.

13.3.6.3 Based on the Standard Deviation of the Response

The most commonly used and appropriate way of calculating LoQ with the following equation as mentioned in the ICH guidelines:

$$LoQ = \frac{10.\sigma}{m}$$

In the equation, σ is the standard deviation and m is the slope of the corresponding calibration line. The result can be easily converted to concentration units via the equation. Also, the relationship between LoQ and sensitivity is verified using the above equation based on the function of the slope. σ estimation can be obtained using the same steps and different approaches given in the LoD section.

13.3.7 SENSITIVITY

While Eurochem expressed this definition as "the change in the response of a measuring instrument divided by a corresponding change in the stimulus" (42), ICH and

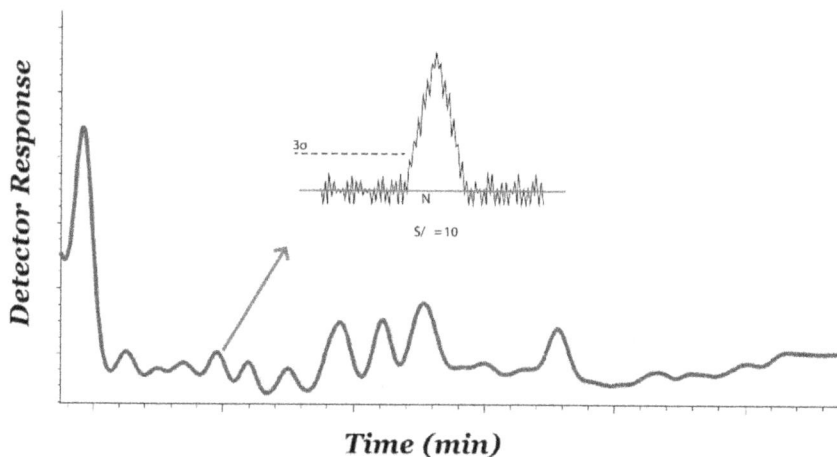

FIGURE 13.3   Schematic demonstration of S/N ratio for LoQ.

EMA do not mention this term in their guidelines. From the viewpoint of FDA's Bioanalytical Method Validation guideline, the definition is so close to the definition of LOQ, which is "the lowest analyte concentration that can be measured with acceptable precision and accuracy".

A biosensor should be capable of detecting very low analyte concentrations to confirm the presence of even a trace amount of analyte in a sample. Furthermore, sensitivity is another significant approach used for the detection of deviations from linearity. It is the rate of change of meter response with change in concentration. The sensitivity of a method is the gradient of its response curve. It may also be known as the slope of the calibration line and can be measured during linearity experiments. Sensitivity is often used in conjunction with the method's LoD and LoQ values. The slope of the calibration equation is used for the calculation of LoD and LoQ. For tests where the response function is linear, the sensitivity is constant concerning concentration and equal to the calibration curve's slope. The relationship between sensitivity and LoD is valid only for determination by linear response function. Such a relationship applies only to the linear response function and not to nonlinear species due to the sensitivity changes with concentration. The higher the sensitivity, the better the method can distinguish between similar concentrations because a small difference in concentrations will result in a large difference in the observed response. The sensitivity depends on the analyte and the selected technique. It can be expressed by the following equation:

$$S = \frac{\Delta x}{\Delta C}$$

where Δx is the change of response and ΔC is the change of concentration.

Sensitivity can be defined in two complementary ways. One of them is detecting and determining a small amount of analyte in the application environment. The other is the power to discriminate between similar amounts of an analyte in the application environment (27, 31, 33, 41).

13.3.8 ROBUSTNESS AND RUGGEDNESS

Robustness and the ruggedness of a method refer to the capability of an analytical method, in our case and electrochemical based biosensor, to remain unaffected by small variations in the parameters: It can be the pH of the buffer system, type of the reference electrode, duration of the surface modification of used working electrode, environmental factors such as temperature, humidity, etc. In these types of studies, two possible variations can be expected. First, while the method parameters varied, the detected amount of the analyte should not be changed significantly. Last, when the method parameters change, the critical performance characteristics such as sensitivity, etc., should not be differed. Within the same laboratory, robustness included the stability of an analyte in a given environment at its intended storage temperature, under the influence of periods of freezing or thawing, at room temperature, or under the influence of other environmental factors mentioned earlier. Thus, robustness tests show the effect of the small changes during the analysis result in different laboratories or under different circumstances. Generally, method performance parameters such as specificity, selectivity, sensitivity, precision, and accuracy are used to evaluate robustness (27, 41).

Although the terms ruggedness and robustness are often treated the same and used interchangeably, there are separate definitions. Robustness is defined as the degree of repeatability of results from the analysis under normal operating conditions and can be determined as part of a collaborative experiment. It can also be defined as the sensitivity of an analytical method to changes in experimental conditions, expressed as sample materials, analytes, analysts, laboratories, storage conditions, instruments, reagent lots, elapsed test times, environment, or sample list. Requirements of preparation under which the method can be applied as presented or with minor modifications are indicated. In practice, these are pH, temperature, stability of reagents, etc. For all experimental conditions that may be subject to fluctuations, such as a short definition, robustness refers to intra-laboratory and interlaboratory work. Robustness tests examine the effects of changing parameters on the accuracy and precision of final results. The individual laboratory typically assesses robustness prior to participating in this joint trial and should be considered at an earlier stage during the development and/or optimization of the electroanalytical method. Robustness is a performance criterion that examines the applicability and transferability of the method. These tests are performed as part of precision studies. The aim is to reveal the effect of small changes in method conditions on the qualitative and quantitative capabilities of the method. The ruggedness is used to evaluate the results when the external factors are interpreted like analyst, the brand of the instrument, laboratory conditions, different brands of reagents, etc. As briefly, ruggedness was used for the demonstration of the method against extraneous influencing factors. Like in robustness, it is generally expressed as RSD% of data obtained with the changed parameter compared to the same data obtained at the beginning. Hence, student t-test and F-test can be used to assure the statistical variations of both results (27, 41, 43).

In the context of this chapter the validation parameters have been discussed as details. Hence, steps for the analytical method validation are described briefly in the following, except the stability, which is given as Figure 13.4:

Checklist after the method optimization

Selectivity	• Analyze at least 10 blank samples • Analyze spiked samples at LLOQ level	Each analyte should be tested to ensure that there are no interferences
Linearity and Range	• Analyze matrix matched solutions at least 6 level (10 is better) at triplicates • LLOQ, Low QC, Medium QC, High QC covered in calibration • 3 calibration equation prepared at least duplicate analysis	r or r^2 should be close to 0.99
LoD and LoQ	• Measurements should be done on 5 different days over a long time period • LoD=3.3 σ/m, LoQ=10 σ/m	LoD and LoQ should not be higher than the LLOQ in the linearity
Precision	• Minimum 3 concentration level should be analyzed at least 5 replicates on the same day for the repeatability • Intermediate precision measurements should be performed at least 3 consecutive days same procedure as the repeatability	Given as RSD% or CV% and test it for "fit for purpose"
Accuracy	• Minimum 3 concentrations of 5 replicates of the samples containing known amounts of the analyte	Calculate RSD% or Bias and compare to legal requirements if available
Robustness	• Select method parameters having the strongest influence to your performance. Make a small change in selected parameters to test the method performance	Obtained results should be in the limits after small variations on the method

---

13.3.9 STABILITY

This parameter is not essential and not included in all guidelines as well as pharmacopeias. It is a recommendation to obtain reproducible and reliable results. As it is known, reproducible, accurate, and precise results are obtained with stock solutions, samples, reference standards, reagents, and solvents that can maintain their stability for a certain period of time. FDA recommended the following parameters to ensure stability:

- Freeze & thaw stability (three thawing cycles)
- Short term stability or Bench-top stability (room temperature or sample processing temperature)
- Long term stability

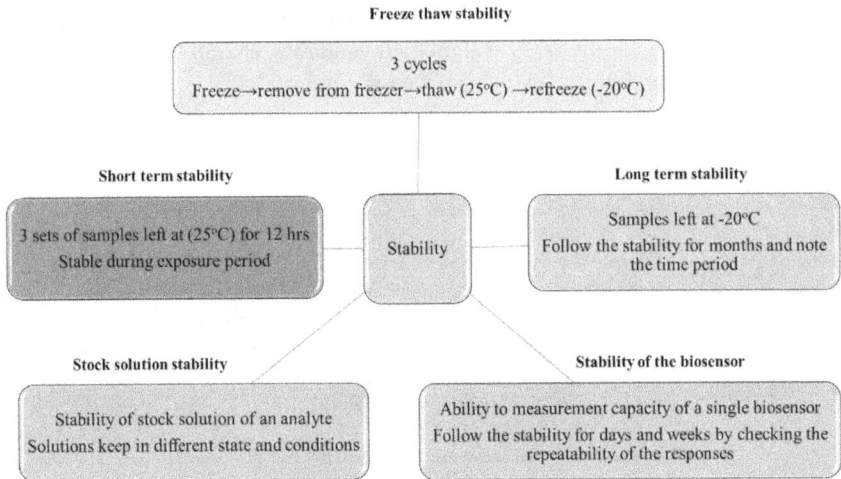

Freeze thaw stability

3 cycles
Freeze→remove from freezer→thaw (25°C) →refreeze (-20°C)

Short term stability

3 sets of samples left at (25°C) for 12 hrs
Stable during exposure period

Stability

Long term stability

Samples left at -20°C
Follow the stability for months and note the time period

Stock solution stability

Stability of stock solution of an analyte
Solutions keep in different state and conditions

Stability of the biosensor

Ability to measurement capacity of a single biosensor
Follow the stability for days and weeks by checking the repeatability of the responses

FIGURE 13.4   Scheme for stability testing.

- Stock solution stability
- Sample stability (e.g., stability in the matrix)

To demonstrate stability, at least two concentration levels (low and high) for the raw analyte and the matrix-matched samples should be studied. For this reason, samples were analyzed at different time points at least 6 replicates. The average percentage of the analyte under the specific condition compared by the freshly prepared standard solution (it is accepted as 100%). The acceptance criteria for this parameter is the tested samples should be in the range between 80%–120%. When reporting the results related to stability, the concentrations of analyzed samples and storage/analysis times, pH of the buffer solution, and instrumental parameters used in the analyses should be carefully noted. From the viewpoint of the biosensor side, it is crucial for biomonitoring. Stability is the degree of sensitivity to environmental disturbances in and around the biosensing system. These disturbances can cause a shift in the output signals of a biosensor under measurement. This may cause an error in the measured concentration and affect the precision and accuracy of the biosensor. In applications where a biosensor requires long incubation steps or continuous monitoring, stability is primarily important. The response of transducers and electronics can be temperature sensitive, which can affect the stability of a biosensor. Therefore, the electronics must be appropriately adjusted to ensure a stable response to the sensor. Another factor that can affect stability is the affinity of the bioreceptor, the degree to which the analyte binds to the bioreceptor. High-affinity bioreceptors promote either strong electrostatic coupling or covalent coupling of the analyte, which enhances its stability. Another factor affecting the stability of measurement is the degradation of the bioreceptor over a period of time. Also, the aging or product stability of a biosensor is a major limitation for commercial success. The aging mechanisms or reduced sensitivity of a biosensor are complex and affect every layer and reagent used. Biosensor aging is, therefore, the total of changes in the functionality of the complex, whether it is an enzyme or antibody layer, a signaling medium or a protective membrane, a biological component. Many biosensor stability studies have focused only on single-use and shelf life estimates (44–48).

13.4   SELECTED APPLICATIONS FOR THE DEMONSTRATION OF ANALYTICAL METHOD VALIDATION

Biosensors have become a topic that has been studied in an accelerated manner for years (Figure 13.5). Analysis of biomarkers (cancer, Parkinson's disease, Alzheimer's disease, etc.), biomedical and clinical applications, wearable sensors, etc., can be classified as the most featured applications of biosensor studies (27, 49). Therefore, it is also important to pay attention to the analytical method validation parameters during the development processes of biosensors. Analytical method validation parameters prove the reliability and applicability of any biosensor. Therefore, each parameter should be elaborately evaluated in the development and construction of biosensor-based research. Each parameter demonstrates the different characteristics point by point but examining all the parameters can help identify more substantive results (47, 50).

Some basic examples for this situation are the results for the sensitivity given by using only the LoD results, and there is not any value mentioned for the LoQ data (51–53). As another example, precision may be demonstrated by only repeatability and reproducibility, without performing intermediate precision studies (54). In addition to these, determination of any kind of analyte presented with partial validation such as only calibration and LoD and/or LoQ results can be explained as fully analytical method validation, and this may cause a misunderstanding (53). In this case, explaining the results properly is as much important as performing the validation analysis. Tables 13.2 and 13.3 evaluate the selected latest biosensor studies in terms

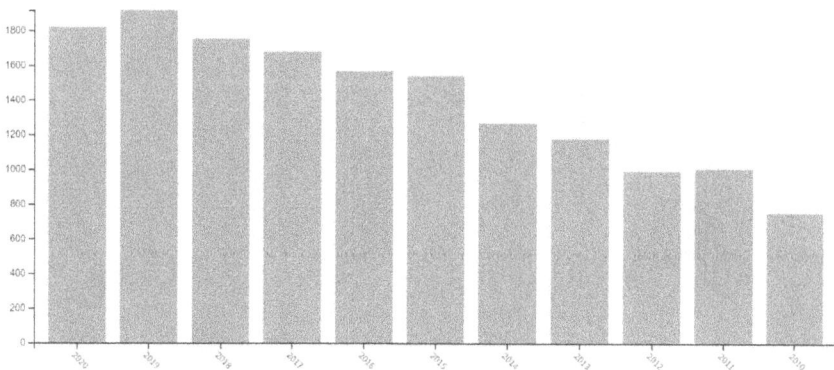

FIGURE 13.5   Illustration of the Web of Science-based published research papers data by using biosensor and electrochemical as the keywords between 2020–2010. When the studies on biosensors in recent years (Tables 13.2 and 13.3) are examined, it is seen that the researchers did not evaluate analytical figures of merits as details and/or the results of the validation studies may not be adequately explained.

TABLE 13.2
Evaluation of the Selected Latest Biosensor Studies in Terms of Essential Analytical Method Validation Characteristics

Measurement Type—Analyte	Analytical Validation Characteristics							Ref.
	Selectivity	Linearity	Range	Accuracy (%)	LoD	LoQ	Sensitivity	
Amperometric—Hydrogen peroxide and glucose	glucose (2.5 mM), uric acid (1.0 mM), dopamine (1.0 mM), glucose (1.0 mM), ascorbic acid (1.0 mM), sucrose (1.0 mM), fructose (1.0 mM), maltose (1.0 mM), L-cysteine (1.0 mM), glutathione (1.0 mM), acetaminophen (1.0 mM)	r=0.997 and r=0.998	0.01–29 nM and 0.01–31.5 mM	99.47, 102.24, 101.25	3.8 µM and 0.3 µM	-	575.75 $\mu A.mM^{-1}.cm^{-2}$ and 241.82 $\mu A.mM^{-1}\cdot cm^{-2}$	(54)
Electrochemical—Adiponectin	-	r=0.99	25–2500 $pg.mL^{-1}$	99.10–103.36	148 $pg.mL^{-1}$	448 $pg.mL^{-1}$	6.6546 $nA.mL.pg^{-1}$	(55)
Electrochemical—Sulfamethazine	Selective at 100-fold concentrations of kanamycin and tetracycline	r=0.99	0.0001–100 $ng.mL^{-1}$	95.24–104.4	0.069 $pg.mL^{-1}$	-	1.99 $\mu A.mL.ng^{-1}$	(56)
Quartz crystal microbalance—L-tryptophan	Selectivity coefficient (k)=12 (ascorbic acid) and 7.3 (D-tryptophan)	r=0.97	1 $\mu g.mL^{-1}$–5.0 $ng.mL^{-1}$	97–104	0.73 $ng.mL^{-1}$	2.0 $ng.mL^{-1}$	-	(57)
Electrochemical—Doxorubicin hydrochloride	Selective in the presence of ascorbic acid, uric acid, L-cystine, glucose, and tyrosine	r=0.99	1×10^{-11}–1×10^{-6} M	95.27–104.2	6.5 pM	-	272 $\mu A.cm^{-2}.M^{-1}$	(58)
Electrochemical—Microalbuminuria	-	r=0.99 and r=0.99	0.1–10 $mg.dL^{-1}$ and 10–40 $mg.dL^{-1}$	102.6	0.1 $mg.dL^{-1}$	-	0.113 $\mu A.mg^{-1}.dL$ and 0.041 $\mu A.mg^{-1}.dL$	(59)
Potentiometric—Formaldehyde	-	r=0.99	0.5–220.0 mM	-	0.1 mM	-	59.23 ± 0.85 mV/decade	(60)

(Continued)

TABLE 13.2
(Continued)

Analytical Validation Characteristics

Measurement Type—Analyte	Selectivity	Linearity	Range	Accuracy (%)	LoD	LoQ	Sensitivity	Ref.
Chronoamperometric—Glucose	−	r=0.99	1.0–25.0 mM	−	0.32 mM	−	11.2 μA.mM⁻¹.cm⁻²	(61)
Colorimetric—Epinephrine	+	r=0.99	1–400 μM	−	0.6 μM	−	0.000655 pixel value. μM⁻¹	(62)
Electrochemical—Oxytetracycline	+	r=0.99	1 – 200 ng.mL⁻¹	−	0.33 ng.mL⁻¹	1.1 ng.mL⁻¹	0.1378 μA.mL.ng⁻¹	(63)
Electrochemical—Hydrogen peroxide	+	r=0.99	6×10^{-7}–4.5×10^{-6} M and 4.5×10^{-6}–1.7×10^{-3} M	98–103	5×10^{-7} M	−	0.43 A.M⁻¹.cm⁻² and 0.027 A.M⁻¹.cm⁻²	(64)
Amperometric—Hydrogen peroxide	No current change in the presence of ascorbic acid, leucine, dopamine, and glucose	r=0.99	50–300 μM	95–103.4	4.4 μM	−	−	(65)
Interferometric reflectance spectroscopy—Cytochrome c	+	r=0.98	1–100 nM	97.6	0.5 nM	−	+	(66)
Electrochemical—circulating miRNAs	~4% of the response provided by the target miRNA	r=0.99	10 fM–2 pM	−	5 fM	−	−	(67)
Surface plasmon resonance imaging—CA125/MUC16	Albumin, leptin, interleukin 6 and metaloproteinase-2 have no interfering effect	r=0.99	2.2–150 U.ml⁻¹	97–128	−	−	SPRI signal (Arbitrary Units).mL.U⁻¹	(68)

(Continued)

TABLE 13.2
(Continued)

Measurement Type—Analyte	Analytical Validation Characteristics							
	Selectivity	Linearity	Range	Accuracy (%)	LoD	LoQ	Sensitivity	Ref.
Amperometric—Tyramine	<2% change of the response in the presence of xanthine, hypoxanthine, and L-tyrosine	r=0.99	10–120 μM	93–97	0.71 μM	-	10 nA.cm^{-2}.μM^{-1}	(69)
Electrochemical—Mycoplasma ovipneumonia	Responses of M1 DNA, M2 DNA, M3 DNA are around 14.2–21.4% of the target analyte response	r=0.99	10 aM—0.1 nM	102–106	3.3 aM	-	+	(70)
Electrochemical—Dopamine	Selectivity coefficient <5%	r=0.99	0.1–6.0 μmol.dm^{-3}	94.6–98	-	-	22.344 μA.dm^{3}μmol^{-1}	(71)
Electrochemical—Choline Acetylcholine	+	r=0.99	5–1000 nmol.L^{-1}	100.18–103.08 and 95.3–106.25	0.885 nmol.L^{-1} and 1.352 nmol.L^{-1}	3.202 nmol.L^{-1} and 4.432 nmol.L^{-1}	0.664 μA.L.nmol^{-1}	(72)
Electrochemical—Pyruvate	150 μM ascorbic acid, glucose and citric acid have no interfering effect.	r=0.99	5–140 pM	93–104	8.69 nM	20.90 μM	9.92×10^{-5} μA.μM^{-1}	(73)
Electrochemical—Temodal	1000- fold Mg^{2+}, Li$^+$, Br$^-$, glucose, and alanine and 400-fold of uric acid have no interfering effects.	r=0.99	0.005–45 μM	98.45–105.2	0.001 μM	-	0.1652 μA.μM^{-1}	(74)
Electrochemical—Albumin	+	r=0.99	1×10^{-10}–1×10^{-4} g.L^{-1}	90–105	3×10^{-11} g.L^{-1}	-	2.126 μA.g^{-1}.L	(75)
Amperometric—Glycerol	+	r=0.99	0.05–1 mM	92–106.67	18 μM	-	29.2 ± 0.9 μA.mM^{-1}.cm^{-2}	(76)

(Continued)

TABLE 13.2
(Continued)

Analytical Validation Characteristics

Measurement Type—Analyte	Selectivity	Linearity	Range	Accuracy (%)	LoD	LoQ	Sensitivity	Ref.
Electrochemical—Phenol	Signal change (μM)=0.003–0.017	r=0.99	5.0–40.0 μM	97.15–104.35	0.082 μM	-	0.078	(77)
Electrochemical—3-chloro-1,2-propandiol	-	r=0.99	1–160 mg.L^{-1}	97.4–99.6	0.25 mg.L^{-1}	-	0.0178 μA.mg^{-1}.L	(78)
Electrochemical—Topotecan	-	r=0.99	0.35–100 μM	97.3–104	0.1 μM	-	0.1431 μA.μM^{-1}	(79)
Fluorescence—Dopamine	Ca^{2+}, Zn^{2+}, uric acid, sucrose, fructose and glucose have no interfering effects.	r=0.99	0.5–10 μM	92–102	0.35 μM	-	+	(80)
Amperometric—Paraoxon	Citric acid, glucose, carbamide, Mg^{2+}, Fe^{3+}, SO_4^{2-} and NO– have no interfering effects.	r=0.99	3.6 pM—100 nM	98.6–105.2	3.6 pM	-	7.8 μA.mM^{-1}	(81)
Amperometric—Sarcosine	Interference effect of citric acid, glutamic acid, uric acid, ascorbic acid, and urea: 4.20%, 2.70%, 1.50%, 3.80%, and 2.80%	-	0.1–100 μM	94.22–97.81	0.1 pM	-	-	(82)
Electrochemical—Didanosine	No interference effect at 750-fold concentrations of glucose, sucrose and uric acid; 1000-fold concentrations of K^+, Cl^-, Na^+, and Br^-	-	0.02–50 μM	95.1–102	8 nM	-	-	(83)

+: Numeric data not provided.

—: Data not available.

TABLE 13.3
Evaluation of the Selected Latest Biosensor Studies in Terms of Essential Analytical Method Validation Characteristics

Measurement Type—Analyte	Analytical Validation Characteristics					Stability of the Biosensor	Ref.
	Precision		Reproducibility (RSD %)	Robustness	Ruggedness (RSD %)		
	Repeatability (RSD %)	Intermediate Precision (RSD %)					
Amperometric—Hydrogen peroxide and glucose	–	–	7.37	–	✓	RSD=10.6% over two weeks	(54)
Electrochemical—Adiponectin	4.34	–	+	–	✓	6 weeks	(55)
Electrochemical—Sulfamethazine	–	–	2.3	–	✓	–	(56)
Quartz crystal microbalance—L-tryptophan	2.85	–	+	–	+	–	(57)
Electrochemical—Doxorubicin hydrochloride	+	–	3.4	–	✓	94.7% peak current density/2 weeks	(58)
Electrochemical—microalbuminuria	–	–	–	–	–	–	(59)
Potentiometric—Formaldehyde	7.8	–	1.4	–	✓	80 days	(60)
Chronoamperometric—Glucose	4.8	–	6.1	–	✓	4.9% decrease in 5 days	(61)
Colorimetric—Epinephrine	–	–	–	–	–	60 min	(62)
Electrochemical—Oxytetracycline	–	–	5.7–9.81	–	✓	2 weeks	(63)
Electrochemical—Hydrogen peroxide	4–5	–	+	–	+	91%/10 days	(64)
Amperometric—Hydrogen peroxide	–	–	–	–	–	Biosensor can be reused after 30 cycles of CV scan.	(65)
Interferometric reflectance spectroscopy—Cytochrome C	+	–	5.2	–	✓	7.3% decrease in the response in 14 days	(66)
Electrochemical—circulating miRNAs	+	–	3.9	–	✓	≥ 2 months	(67)
Surface plasmon resonance imaging—CA125/MUC16	–	–	–	–	–	–	(68)
Amperometric— Tyramine	4.3	–	–	–	–	RSD%=4.3% in 20 days	(69)

(Continued)

TABLE 13.3
(Continued)

Measurement Type—Analyte	Analytical Validation Characteristics						Ref.
	Precision			Robustness	Ruggedness (RSD %)	Stability of the Biosensor	
	Repeatability (RSD %)	Intermediate Precision (RSD %)	Reproducibility (RSD %)				
Electrochemical—Mycoplasma ovipneumonia	—	—	4.82	—	✓	Signal decreases to 93.2% after 20 days.	(70)
Electrochemical—Dopamine	—	—	—	—	—	25 days	(71)
Electrochemical—Choline Acetylcholine	3.12, 2.66 and 2.18, 4.72, 2.51, 3.81	—	3.12, 2.66 and 2.18 4.72, 2.51, 3.81	—	✓	Signal decreases to 93.2% after 90 days.	(72)
Electrochemical—Pyruvate	3.6	—	4.3	—	✓	Signal decreases to 94% after 2 weeks.	(73)
Electrochemical—Temodal	—	—	~3.9	—	✓	—	(74)
Electrochemical—Albumin	—	—	4.4	—	✓	Signal decreases to 95.3% after 25 days.	(75)
Amperometric—Glycerol	—	—	—	—	—	The remaining activity is 20% after 30 days.	(76)
Electrochemical—Phenol	6.75	—	2.69	—	✓	Signal decreases to 75% after 5 weeks.	(77)
Electrochemical—3-chloro-1,2-propandiol	—	—	2.9	—	✓	Signal decreases by 4.7% after 7 days.	(78)
Electrochemical—Topotecan	—	—	—	—	—	—	(79)
Fluorescence—Dopamine	+	—	—	—	+	—	(80)
Amperometric—Paraoxon	3.3	—	4.6	—	✓	Signal decreases to 80% after 30 days	(81)
Amperometric—Sarcosine	—	—	—	—	—	180 days	(82)
Electrochemical—Didanosine	Guanine: 1.9 Adenine: 2.1	—	—	—	—	—	(83)

+: Numeric data not provided.

—: Data not available.

✓: Stated with reproducibility values.

of essential analytical method validation characteristics. Furthermore, some selected applications from the literature are discussed later in detail.

Ince et al. (55) fabricated a new electrochemical biosensor for the determination of adiponectin, a cytokine associated with type 2 diabetes and obesity, using indium tin oxide covered poly ethylene terephthalate sheets as a working electrode. For the analytical validation of the biosensor, various concentrations of adiponectin in a range between 25 pg.mL^{-1} and 2500 pg.mL^{-1} were measured by EIS to acquire a calibration graph. The regression equation was found as y = 6.6546[Adiponectin] + 480.9 (r=0.99) with LoD and LoQ values of 148 pg.mL^{-1} and 448 pg.mL^{-1}, respectively. The correlation coefficient of 0.99 shows the linearity, and the slope of the regression equation gives the sensitivity of the biosensor. Real serum samples were analyzed by spiking, and the recovery results were found satisfactory (99.10%–103.36%). In order to evaluate the repeatability of the constructed sensor, the researchers tested 20 different electrodes under the same experimental conditions, and the standard deviation was found as 4.34%.

Furthermore, reproducibility of the developed method was performed by using 80 different electrodes. 10 calibration graphs in the same linearity range were obtained, and these results proved the precision and ruggedness of the biosensor. Stability experiments were performed for ten weeks period resulting in stability for the first 6 weeks.

Prabakaran and co-workers (57) developed a quartz crystal microbalance biosensor based on molecularly imprinted polymer (MIP) for the determination of an essential amino acid L-tryptophan. To validate the selectivity of the MIP biosensor, D-tryptophan, ascorbic acid, and their mixture were studied. The selectivity constants were calculated for each compound. According to the results, the biosensor has selectivity toward L-tryptophan. LoD and LoQ values were found as 0.73 ng.mL^{-1} and 2 ng.mL^{-1}. A developed biosensor was also applied for food products and human urine with acceptable recovery results in the range of 97.6% and 104%. To evaluate the repeatability and reproducibility of the biosensor the measurements were performed five times, and the RSD% value was found lower than 5% and 2.82%, respectively.

Jampasa et al. (63) fabricated an electrochemical biosensor for oxytetracycline determination. Selectivity of the biosensor was evaluated using similar structured compounds tetracycline, chlortetracycline, doxytetracycline, chloramphenicol, and norfloxacin. Acquired RSD% values were lower than 10%, indicating the selective properties of the biosensor. The calibration curve was obtained in a concentration range of oxytetracycline between 1 and 275 ng.mL^{-1}. The regression equation was found as y=1378x+1.5708 (r^2=0.99) with LoD and LoQ values of 0.33 ng.mL^{-1} and 1.1 ng.mL^{-1}, respectively. These data demonstrate the linearity, range, and sensitivity of the biosensor. When the proposed biosensor was applied to milk, shrimp, and honey samples, the obtained results were compared with the standard HPLC methods' results. It was seen that the experiments gave accurate results compared to standard methods. Reproducibility of the biosensor was tested using 10, 50, and 100 ng.mL^{-1} of oxytetracycline. The results (RSD%<10) indicate that the biosensor is reproducible enough. Stability experiments were carried out to evaluate the long-term stability of the sensor for four weeks. RSD% value of 9.67% in the first two weeks demonstrated the stability of the biosensor.

Fabrication of an Interferometric reflectance spectroscopy-based biosensor for the cancer marker cytochrome C (Cyt C) was studied by Tabrizi et al. (66). Interfering species glucose, urea, dopamine, human IgG antibody, human serum albumin, insulin, and amyloid-beta were used to test the selectivity of the biosensor, and it was found that they have no interfering effect on Cyt C determination. The logarithm of Cyt C concentration showed a linear response between 1 nM and 100 nM. The calibration curve was obtained by repeating each measurement five times. The regression equation was found as peak area (a.u)= $- 399048$ log $(C_{Cyt C})$ (nM) $+ 2 \times 10^6$ (r=0.98) demonstrating the linearity, range and sensitivity of the biosensor. LoD value was calculated as 0.5 nM. A human serum sample was used for the biosensor's practical application, and the recovery value was calculated as 97.6%. Reproducibility of the proposed biosensor was demonstrated with an RSD% value of 5.2%. Stability experiments were performed for 14 days at 4°C. The decrease of the analytical response was found as 7.3%.

Pourghobadi et al. (80) reported a fluorescence-based biosensor using thioglycolic acid-capped cadmium telluride quantum dots for the determination of dopamine. The selectivity of the biosensor toward dopamine was investigated in the presence of Ca^{2+}, Zn^{2+}, fructose, glucose, sucrose, and uric acid. The results showed that the biosensor is able to determine dopamine, selectively. The developed biosensor showed a linear response in the concentration range of dopamine between 0.5 and 1 μM. The correlation coefficient was found as 0.991 and LoD was calculated as 0.35 μM. The repeatability of the biosensor was demonstrated by 10 repetitive analyses resulting in the RSD% of 1.45% and 1.82% for 0.5 μM (low concentration) and 10 μM (high concentration), respectively. Plasma samples were used to test applicability, thus proving the accuracy of the biosensor with acceptable recovery values (92–106%).

13.5   CONCLUSION

This chapter overviewed the main regulatory and method validation issues affecting the overall performance of biosensors. In accordance with the guidelines and pharmacopeias, the essential validation parameters are essential such as accuracy, precision, selectivity, sensitivity, LoD, LoQ, and linearity range. These validation parameters help to construct powerful biosensors for in vitro diagnostics (84). The electroanalytical method validation process is required to assess the degree of error expected to lower the detection limit, increase specificity, and confirm the degree of error (85). The misestimate due to a false-positive or false-negative result should be corrected with the appropriate statistical tools. The validation process of the analytical method is that the required validation parameters are firstly identified, then the experiments are designed, and acceptance criteria are determined. Standardization organizations and regulatory agencies developed guidelines, standards, performance criteria, and operating procedures for reducing the significant issues of biosensors manufacturing and their use. The quality system requirements should be done to enable the development of a safe and effective diagnostics device.

REFERENCES

1. Ozkan SA, Kauffmann J-M, Zuman P. 2015. *Electroanalysis in Biomedical and Pharmaceutical Sciences*, pp. 83–114. Springer
2. Barnett KL, Harrington B, Graul TW. 2013. *Validation of Liquid Chromatographic Methods*, pp. 57–73. Elsevier Inc
3. European Medicines Agency. 2020. European medicines agency. *Definitions*. 2(June 1995): 1–15
4. Krause SO. 2002. Good analytical method validation practice deriving acceptance criteria for the AMV protocol : Part II. *Valid Analytical Methods and Procedures*. 31–47
5. Bliesner DM. 2006. Appendix I: Glossary of methods validation terms. *Validation of Chromatographic Methods of Analysis*. 57–71
6. Konieczka P. 2007. The role of and the place of method validation in the quality assurance and quality control (QA/QC) system. *Critical Reviews in Analytical Chemistry*. 37(3): 173–190
7. Imre S, Vlase L, Muntean DL. 2008. Bioanalytical method validation. *Revista Romana de Medicina de Laborator*. 10(1): 13–21
8. Ich Harmonised Tripartite Guideline. 1996. Guidance for industry. *Q2B Validation of Analytical Procedures: Methodology*. 20857(November): 301–827
9. Chan CC. 2011. Principles and practices of analytical method validation: Validation of analytical methods is time-consuming but essential. *Quality Assurance Journal*. 14(3–4): 61–64
10. Dualde P, Pardo O, Fernández SF, Pastor A, Yusà V. 2019. Determination of four parabens and bisphenols A, F and S in human breast milk using QuEChERS and liquid chromatography coupled to mass spectrometry. *Journal of Chromatography. B, Analytical Technologies in the Biomedical and Life Sciences*. 1114–1115(December 2018): 154–166
11. U.S. Department of Health and Human Services. 2015. *Analytical Procedures and Methods Validation for Drugs and Biologics Guidance for Industry*. U.S. Department of Health and Human Services
12. da Silva Guedes J, da Silva Guedes ML. 2006. Step-by-step analytical protocol in the quality system methods validation and compliance industry. *Revista de Saúde Pública*. 40(6): 951–961
13. Ahir K, Singh K, Yadav S, Patel H, Poyahari C. 2014. Overview of validation and basic concepts of process validation. *Scholars Academic Journal of Pharmacy*. 3(2): 178–190
14. Shah VP. 2007. The history of bioanalytical method validation and regulation: Evolution of a guidance document on bioanalytical methods validation. *AAPS Journal*. 9(1)
15. International conference on harmonisation of technical requirements for registration of pharmaceuticals for human use. 1994. *ICH Harmonised Tripartite Guideline Pharmacokinetics: S3B*, October. https://pubmed.ncbi.nlm.nih.gov/11590294/
16. World Health Organization (WHO). 2018. *Guidelines on Validation—Appendix 4 — Analytical Method Validation*, pp. 1–11. WHO
17. Center for Drug Evaluation and Research (CDER). 1994. *Reviewer Guidance' Validation of Chromatographic Methods*. CDER
18. European Medicines Agency. *ICH M10 on bioanalytical method validation*. www.ema.europa.eu/en/ich-m10-bioanalytical-method-validation
19. Rodriguez-Mozaz S, Alda MJL de, Barceló D. 2006. Biosensors as useful tools for environmental analysis and monitoring. *Analytical and Bioanalytical Chemistry*. 386(4): 1025–1041
20. Bhattarai P, Hameed S. 2020. Basics of biosensors and nanobiosensors. *Nanobiosensors: From Design to Applications*. 10(4)

21. Cárdenas S, Valcárcel M. 2005. Analytical features in qualitative analysis. *TrAC Trends in Analytical Chemistry*. 24(6): 477–487
22. Trullols E, Ruisánchez I, Rius FX. 2004. Validation of qualitative analytical methods. *TrAC Trends in Analytical Chemistry*. 23(2): 137–145
23. Persson BA, Vessman J. 2001. The use of selectivity in analytical chemistry—some considerations. *TrAC Trends in Analytical Chemistry*. 20(10): 526–532
24. Vessman J. 2001. Selectivity—the hallmark of an analytical chemist: The current situation in the analytical sciences. *Accreditation and Quality Assurance*. 6(12): 522–527
25. Valcárcel M, Gómez-Hens A, Rubio S. 2001. Selectivity in analytical chemistry revisited. *TrAC Trends in Analytical Chemistry*. 20(8): 386–393
26. Peveler WJ, Yazdani M, Rotello VM. 2016. Selectivity and specificity: Pros and cons in sensing. *ACS Sensors*. 1(11): 1282–1285
27. Bhalla N, Jolly P, Formisano N, Estrela P. 2016. Introduction to biosensors. *Essays in Biochemistry*. 60(1): 1
28. Ahmed S, Shaikh N, Pathak N, Sonawane A, Pandey V, Maratkar S. 2019. An overview of sensitivity and selectivity of biosensors for environmental applications. In *Tools, Techniques and Protocols for Monitoring Environmental Contaminants*, pp. 53–73. Elsevier
29. Yu C, Irudayaraj J. 2007. Quantitative evaluation of sensitivity and selectivity of multiplex nanoSPR biosensor assays. *Biophysical Journal*. 93(10): 3684–3692
30. Rozet E, Marini RD, Ziemons E, Boulanger B, Hubert P. 2011. Advances in validation, risk and uncertainty assessment of bioanalytical methods. *Journal of Pharmaceutical and Biomedical Analysis*. 55(4): 848–858
31. European Medicines Agency. *ICH Q2 (R1) Validation of Analytical Procedures: Text and Methodology*. www.ema.europa.eu/en/ich-q2-r1-validation-analytical-procedures-text-methodology
32. Krull IS, Swartz M. 1999. Analytical method development and validation for the academic researcher. *Analytical Letters*. 1067–1080
33. Riley CM, Rosanske TW. 1996. *Development and Validation of Analytical Methods*, p. 352. Pergamon
34. Ozkan SA, Kauffmann J-M, Zuman P. 2015. Electroanalytical techniques most frequently used in drug analysis. *Monographs in Electrochemistry*. 45–81
35. Ozkan SA, Kauffmann J-M, Zuman P. 2015. *Electroanalytical Method Validationmethod validation in Pharmaceutical Analysis and Their Applications*, pp. 235–266. Springer
36. Ozkan SA, Kauffmann J-M, Zuman P. 2015. Introduction. In *Electroanalytical Method Validationmethod validation in Pharmaceutical Analysis and Their Applications*, pp. 1–5. Springer
37. Park S. 2019. Improvement of biosensor accuracy using an interference index detection system to minimize the interference effects caused by icterus and hemolysis in blood samples. *Analyst*. 144(17): 5223–5231
38. Ermer J, John JH. 2005. Method validation in pharmaceutical analysis: A guide to best practice. *Method Validation in Pharmaceutical Analysis: A Guide to Best Practice*. 1–403
39. Varshney M, Mallikarjunan K. 2009. Challenges in biosensor development-detection limit, detection time, and specificity. *Resources Magazine*. 16(7): 18–21
40. Ozkan SA. 2018. Analytical method validation: The importance for pharmaceutical analysis. *Journal of Pharmaceutical Sciences*. 24(1): 1–2
41. Gumustas M, A. Ozkan S. 2014. The role of and the place of method validation in drug analysis using electroanalytical techniques. *The Open Analytical Chemistry Journal*. 5(1): 1–21
42. Barwick V, Bravo PPM, Ellison SLR, Engman J, Gjengedal E, et al. 2014. *The Fitness for Purpose of Analytical Methods A Laboratory Guide to Method Validation and Related Topics Second Edition*. undefined

43. Vander Heyden Y, Massart DL. 1996. Chapter 3 review of the use of robustness and ruggedness in analytical chemistry. *Data Handling in Science and Technology.* 19(C): 79–147

44. Dadgar D, Burnett PE, Gerry Choc M, Gallicano K, Hooper JW. 1995. Application issues in bioanalytical method validation, sample analysis and data reporting. *Journal of Pharmaceutical and Biomedical Analysis.* 13(2): 89–97

45. Vander Heyden Y, Nijhuis A, Smeyers-Verbeke J, Vandeginste BGM, Massart DL. 2001. Guidance for robustness/ruggedness tests in method validation. *Journal of Pharmaceutical and Biomedical Analysis.* 24(5–6): 723–753

46. Mulholland M. 1988. Ruggedness testing in analytical chemistry. *Trends in Analytical Chemistry.* 7(10): 383–389

47. Reyes-De-Corcuera JI, Olstad HE, García-Torres R. 2018. Stability and stabilization of enzyme biosensors: The key to successful application and commercialization. *Annual Review of Food Science and Technology.* 9: 293–322

48. Gavalas VG, Chaniotakis NA, Gibson TD. 1998. Improved operational stability of biosensors based on enzyme-polyelectrolyte complex adsorbed into a porous carbon electrode. *Biosensors and Bioelectronics.* 13(11): 1205–1211

49. Yabuki S. 2014. Supporting materials that improve the stability of enzyme membranes. *Analytical Sciences.* 30(2): 213–217

50. Bettazzi F, Marrazza G, Minunni M, Palchetti I, Scarano S. 2017. Biosensors and Related Bioanalytical Tools. In *Comprehensive Analytical Chemistry*, vol. 77, pp. 1–33. Elsevier

51. Lee YH, Mutharasan R. 2005. Biosensors. In *Sensor Technology Handbook*, pp. 161–180. Elsevier

52. Ermer J, Nethercote P (Eds.). 2005. *Method Validation in Pharmaceutical Analysis*, 2nd edition. Wiley

53. Raposo F, Ibelli-Bianco C. 2020. Performance parameters for analytical method validation: Controversies and discrepancies among numerous guidelines. *TrAC Trends in Analytical Chemistry.* 129: 115913

54. Lu Z, Wu L, Dai X, Wang Y, Sun M, et al. 2021. Novel flexible bifunctional amperometric biosensor based on laser engraved porous graphene array electrodes: Highly sensitive electrochemical determination of hydrogen peroxide and glucose. *Journal of Hazardous Materials.* 402(August 2020): 123774

55. İnce B, Sezgintürk MK. 2021. A high sensitive and cost-effective disposable biosensor for adiponectin determination in real human serum samples. *Sensors & Actuators, B: Chemical.* 328(October 2020)

56. Li M, He B. 2021. Ultrasensitive sandwich-type electrochemical biosensor based on octahedral gold nanoparticles modified poly (ethylenimine) functionalized graphitic carbon nitride nanosheets for the determination of sulfamethazine. *Sensors & Actuators, B: Chemical.* 329(November 2020): 129158

57. Prabakaran K, Jandas PJ, Luo J, Fu C, Wei Q. 2021. Molecularly imprinted poly(methacrylic acid) based QCM biosensor for selective determination of L-tryptophan. *Colloids and Surfaces A: Physicochemical and Engineering Aspects.* 611(November 2020)

58. Sharifi J, Fayazfar H. 2021. Highly sensitive determination of doxorubicin hydrochloride antitumor agent via a carbon nanotube/gold nanoparticle based nanocomposite biosensor. *Bioelectrochemistry.* 139: 107741

59. Tseng CC, Ko CH, Lu SY, Yang CE, Fu LM, Li CY. 2021. Rapid electrochemical-biosensor microchip platform for determination of microalbuminuria in CKD patients. *Analytica Chimica Acta.* 1146: 70–76

60. Nurlely, Ahmad M, Yook Heng L, Ling Tan L. 2021. Potentiometric enzyme biosensor for rapid determination of formaldehyde based on succinimide-functionalized polyacrylate ion-selective membrane. *Journal of the International Measurement Confederation.* 175(August 2020): 109112

61. Soganci K, Bingol H, Zor E. 2021. Simply patterned reduced graphene oxide as an effective biosensor platform for glucose determination. *Journal of Electroanalytical Chemistry*. 880: 114801

62. Zhang M, Zhang Y, Yang C, Ma C, Tang J. 2021. A smartphone-assisted portable biosensor using laccase-mineral hybrid microflowers for colorimetric determination of epinephrine. *Talanta*. 224(October 2020): 121840

63. Jampasa S, Pummoree J, Siangproh W, Khongchareonporn N, Ngamrojanavanich N, et al. 2020. "Signal-On" electrochemical biosensor based on a competitive immunoassay format for the sensitive determination of oxytetracycline. *Sensors & Actuators, B: Chemical*. 320(May): 128389

64. Peng R, Offenhäusser A, Ermolenko Y, Mourzina Y. 2020. Biomimetic sensor based on Mn(III) meso-tetra(N-methyl-4-pyridyl) porphyrin for non-enzymatic electrocatalytic determination of hydrogen peroxide and as an electrochemical transducer in oxidase biosensor for analysis of biological media. *Sensors & Actuators, B: Chemical*. 321(May)

65. Fatima B, Hussain D, Bashir S, Hussain HT, Aslam R, et al. 2020. Catalase immobilized antimonene quantum dots used as an electrochemical biosensor for quantitative determination of H_2O_2 from CA-125 diagnosed ovarian cancer samples. *Materials Science and Engineering C*. 117(July)

66. Amouzadeh Tabrizi M, Ferré-Borrull J, Marsal LF. 2020. Highly sensitive IRS based biosensor for the determination of cytochrome c as a cancer marker by using nanoporous anodic alumina modified with trypsin. *Biosensors and Bioelectronics*. 149(September 2019)

67. Zouari M, Campuzano S, Pingarrón JM, Raouafi N. 2020. Femtomolar direct voltammetric determination of circulating miRNAs in sera of cancer patients using an enzymeless biosensor. *Analytica Chimica Acta*. 1104: 188–198

68. Szymańska B, Lukaszewski Z, Hermanowicz-Szamatowicz K, Gorodkiewicz E. 2020. A biosensor for determination of the circulating biomarker CA125/MUC16 by surface plasmon resonance imaging. *Talanta*. 206(March 2019): 120187

69. da Silva W, Ghica ME, Ajayi RF, Iwuoha EI, Brett CMA. 2019. Tyrosinase based amperometric biosensor for determination of tyramine in fermented food and beverages with gold nanoparticle doped poly(8-anilino-1-naphthalene sulphonic acid) modified electrode. *Food Chemistry*. 282(July 2018): 18–26

70. Zhao S, Zhou Y, Wei L, Chen L. 2020. Low fouling strategy of electrochemical biosensor based on chondroitin sulfate functionalized gold magnetic particle for voltammetric determination of mycoplasma ovipneumonia in whole serum. *Analytica Chimica Acta*. 1126: 91–99

71. Wardak C, Paczosa-Bator B, Malinowski S. 2020. Application of cold plasma corona discharge in preparation of laccase-based biosensors for dopamine determination. *Materials Science and Engineering C*. 116(June): 111199

72. Albishri HM, Abd El-Hady D. 2019. Hyphenation of enzyme/graphene oxide-ionic liquid/glassy carbon biosensors with anodic differential pulse stripping voltammetry for reliable determination of choline and acetylcholine in human serum. *Talanta*. 200(March): 107–114

73. Mirzaei F, Mirzaei M, Torkzadeh-Mahani M. 2019. A hydrophobin-based-biosensor layered by an immobilized lactate dehydrogenase enzyme for electrochemical determination of pyruvate. *Bioelectrochemistry*. 130: 107323

74. Jahandari S, Taher MA, Karimi-Maleh H, Khodadadi A, Faghih-Mirzaei E. 2019. A powerful DNA-based voltammetric biosensor modified with Au nanoparticles, for the determination of Temodal; an electrochemical and docking investigation. *Journal of Electroanalytical Chemistry*. 840(March): 313–318

75. Zhang G, Yu Y, Guo M, Lin B, Zhang L. 2019. A sensitive determination of albumin in urine by molecularly imprinted electrochemical biosensor based on dual-signal strategy. *Sensors & Actuators, B: Chemical*. 288(378): 564–570

76. Ramonas E, Ratautas D, Dagys M, Meškys R, Kulys J. 2019. Highly sensitive amperometric biosensor based on alcohol dehydrogenase for determination of glycerol in human urine. *Talanta*. 200(March): 333–339

77. Manan FAA, Hong WW, Abdullah J, Yusof NA, Ahmad I. 2019. Nanocrystalline cellulose decorated quantum dots based tyrosinase biosensor for phenol determination. *Materials Science and Engineering C*. 99(January): 37–46

78. Yuan Y, Wang J, Ni X, Cao Y. 2019. A biosensor based on hemoglobin immobilized with magnetic molecularly imprinted nanoparticles and modified on a magnetic electrode for direct electrochemical determination of 3-chloro-1, 2-propandiol. *Journal of Electroanalytical Chemistry*. 834(September 2018): 233–240

79. Mahmoudi-Moghaddam H, Tajik S, Beitollahi H. 2019. A new electrochemical DNA biosensor based on modified carbon paste electrode using graphene quantum dots and ionic liquid for determination of topotecan. *Microchemical Journal*. 150(July): 104085

80. Pourghobadi Z, Mirahmadpour P, Zare H. 2018. Fluorescent biosensor for the selective determination of dopamine by TGA-capped CdTe quantum dots in human plasma samples. *Optical Materials (Amst)*. 84(August): 757–762

81. Lu X, Tao L, Song D, Li Y, Gao F. 2018. Bimetallic Pd@Au nanorods based ultrasensitive acetylcholinesterase biosensor for determination of organophosphate pesticides. *Sensors & Actuators, B: Chemical*. 255: 2575–2581

82. Kumar P, Narwal V, Jaiwal R, Pundir CS. 2018. Construction and application of amperometric sarcosine biosensor based on SOxNPs/AuE for determination of prostate cancer. *Biosensors and Bioelectronics*. 122(September): 140–146

83. Karimi-Maleh H, Bananezhad A, Ganjali MR, Norouzi P, Sadrnia A. 2018. Surface amplification of pencil graphite electrode with polypyrrole and reduced graphene oxide for fabrication of a guanine/adenine DNA based electrochemical biosensors for determination of didanosine anticancer drug. *Applied Surface Science*. 441: 55–60

84. Naresh V, Lee N. 2021. A review on biosensors and recent development of nanostructured materials-enabled biosensors. *Sensors (Switzerland)*. 21(4): 1–35

85. Vo-Dinh T, Cullum B. 2000. Biosensors and biochips: Advances in biological and medical diagnostics. *Fresenius' Journal of Analytical Chemistry*. 366(6–7): 540–551

Part V

Challenges and Future Prospects

14 Nanomaterials-Based Wearable Biosensors for Healthcare

Jose Marrugo-Ramírez[+1], L. Karadurmus[+2,3], Miguel Angel Aroca[+4], Emily P. Nguyen[1], Cecilia de Carvalho Castro e Silva[1,5], Giulio Rosati[1], Johann F. Osma[4], Sibel A. Ozkan[2], and Arben Merkoçi[1,6]

[1]University of Barcelona
Nanobioelectronics & Biosensors Group, Catalan
Institute of Nanoscience and Nanotechnology (ICN2)
Barcelona, Spain

[2]Ankara University
Faculty of Pharmacy, Department of Analytical Chemistry
Ankara, Turkey

[3]Adıyaman University
Department of Analytical Chemistry, Faculty of Pharmacy
Adıyaman, Turkey

[4]Universidad de los Andes
Department of Electrical and Electronic Engineering
Bogotá, Colombia

[5]Mackenzie Presbyterian University
MackGraphe – Mackenzie Institute for Research
in Graphene and Nanotechnologies
São Paulo, Brazil

[6]ICREA – Institució Catalana de Recerca i Estudis Avançats
Barcelona, Spain

CONTENTS

DOI: 10.1201/9781003189435-19

14.1 INTRODUCTION

With the advent of smartphones and smart watches, wearable sensors have received great attention from the research community, even more in the healthcare field. Most of the wearable sensors are based on exploring the physical parameters of the users, correlated to their life quality and health. These devices can monitor the mobility and vital signs of the individuals, such as the number of steps, heart and respiration rate, oxygen saturation and calories burned (1). However, to improve patients' quality of life and allow constant assessment of their responses to treatments, it is necessary to develop wearable sensors capable of detecting metabolites and biomarkers in biological fluids in real-time and in a non-invasive way. In this scenario, they highlight wearable biosensors.

A wearable biosensor can be defined by the incorporation of a biological recognition element into a sensor device (for example, enzyme, antibody, cell receptor or organelle) (2). These sensors offer unique advantages over other traditional biosensors: continuous, real-time data acquisition, saving time and money, reducing the waste of disposable medical materials, more efficient and high-quality measurements in a less invasive way. Wearable biosensors, which can also be combined with smartphones, can assess the measured data and send it to emergency services or patient relatives in critical situations. These devices are used in cardiovascular diseases to detect heart rate, electrocardiography and blood pressure.

Wearable biosensors, which serve to interface with a specific body part and thus provide reliable long-term recording of clinically useful bio-signals, have attracted great interest for various health monitoring applications. Thanks to these sensors, physiological conditions in a human body related to a particular disease can be monitored because of biochemical or/and biophysical changes. Biomarkers are biological materials found in body fluids, and they specify the normal or pathological processes or responses of the body. Biomarkers are critical tools for the diagnosis and following of many diseases. Body fluids such as sweat, saliva and tears are a rich resource of many important biomarkers and therefore provide clinically useful information for monitoring many metabolic disorders such as diabetes, gout, and Parkinson's disease (1–5).

The development of wearable biosensors demands interdisciplinary work. It is necessary to develop an efficient sampling process capable of performing the sample collection (i.e. sweat, interstitial fluid and tears), extraction and storage; a flexible, biocompatible and highly sensitive sensor platform, allowing good adhesion and conformation into human skin and highly sensitive detection biomarkers; and finally a signal readout with wireless transmission. This chapter will give an overview of the main characteristics of wearable biosensors for healthcare applications, from the sampling process to the signal readouts. We will discuss the open issues regarding their development and application. In particular, we will explore the role of nanomaterials in outcomes of the issues related to mechanical features and improving the sensitivity in the wearable biosensors. Finally, we will discuss the most advanced technologies for signal readout and wireless data transmission. This chapter proposes to give a critical overview with respect to the nanomaterials-based wearable biosensors for healthcare applications.

14.2 SAMPLING AND APPLICATIONS

14.2.1 PATHOLOGIES

Wearable biosensors offer the possibility of real-time monitoring of biomarkers and physical conditions of patients related to various diseases. Monitoring these biomarkers in real time allows physicians to better understand the patient's condition and response to treatment, promoting an improvement in patients' quality of life. In this section, we will discuss the most relevant pathologies, where the use of wearable biosensors will be extremely relevant.

Alcoholic hepatitis: Alcoholic hepatitis (AH) is an alcohol-induced liver failure syndrome that is usually seen in patients who have consumed heavy alcohol for decades (6–8). Alcoholic liver disease depends on many factors such as gender, alcohol consumption and the risk increase with the amount and duration of alcohol taken. Excessive alcohol consumption is the main cause of many diseases and socioeconomic problems such as liver damage, cancer, accidents at work or driving and deterioration of social and family relationships. For these reasons, the rapid, periodic monitoring and detection of alcohol after consumption with a non-invasive measurement has been an interesting area of research in recent years (9–16).

Hyponatremia: Wearable sensors can be used to record, monitor and prevent the onset of conditions such as dehydration and hyponatremia. These are often caused by under or overhydration and can lead to serious, even fatal health complications. Hyponatremia is the most common fluid electrolyte disorder in clinical practice and serum sodium concentration is lower than 130 mEq/L. It mostly shows an asymptomatic clinical course. Acute and severe hyponatremia is an important cause of morbidity and mortality (17–20).

Gout: Gout is an inflammation of the joints as a result of an excessive increase in uric acid in the blood. The height of uric acid alone is not enough for disease to occur. If gout is not diagnosed early and the increase in the amount of uric acid is not reduced, unfortunately, uric acid accumulates in the joints, soft tissues adjacent to the joint, bone protrusions and cartilage surfaces and causes swelling and damage in those areas. So, it is necessary to reduce the increased uric acid in these patients. Because this is the essential point in this disease, patients stop coming to the doctor again after the inflamed joint heals and think that it has recovered until the next attack. However, keeping uric acid at reasonable levels in the blood reduces subsequent attacks. For this reason, the amount of uric acid should be monitored continuously in these patients. Wearable sensors are very convenient tools for monitoring the amount of uric acid (21–25).

Tyrosinemia: Tyrosinemia is a metabolic disease characterized by an increase in the level of tyrosine in the blood due to the deficiency of the tyrosine aminotransferase enzyme and the presence of tyrosine in the urine. There are 3 types of tyrosinemia, each of which is genetically inherited, differing from each other in terms of cause and symptoms. In tyrosinemia, regular monitoring of some laboratory results is part of the treatment and enables healthcare professionals to establish appropriate treatment and follow-up (26–27).

Diabetes: Diabetes is a chronic and metabolic disease that causes severe damage to the heart, blood vessels, eyes, kidneys and nerves over time, characterized by high blood glucose levels. Approximately 422 million people suffer from diabetes worldwide, and 1.6 million deaths each year are directly attributed to diabetes. Diabetes can be prevented or delayed with regular screening and treatment. Abnormal concentrations of metabolites are related to health problems. For instance, chloride concentration is the gold standard to determine cystic fibrosis, and glucose concentration is being examined for diabetes management. With the use of wearable devices and sensor technologies, it has the power to transform the diabetes field by monitoring patients' symptoms, physiological data and behaviors continuously and without load. Wearable sensors have a strong potential to identify new digital markers and risk patterns that can prevent diabetes-related complications thanks to the amount and variety of information they represent. In addition, they have the potential to improve diabetes management and quality of life when combined with clinical data (28–32).

Cardiovascular diseases: Cardiovascular diseases are a few of the many life-threatening statuses that can be detected and efficiently cured using the wearable sensor. Wearable sensors are distinctively used in the treatment of cardiovascular diseases. Cardiovascular disease (CVD) is a general nomenclature given to the group that includes diseases of the heart or blood vessels. Cardiovascular disease describes any disease that affects the circulatory system. These include numerous heart-related complications such as cardiac arrest, arrhythmia, congestive heart failure, coronary artery disease, etc. More than 17 million people die from CVDs each year, which is around 31% of all deaths worldwide. If the current situation is allowed to persist, it is estimated that by 2030, an estimated 23.6 million people will die from cardiovascular disease (33–39).

Parkinson's disease: Parkinson's disease is a slowly progressive neurodegenerative disease, and its onset is characterized by signs of movement system dysfunction. The classic findings are tremor, rigidity, bradykinesia, slowed movements and standing posture. Parkinson's disease often develops gradually and its manifestations throughout the body are often asymmetrical. Gradually, the condition of the disease progresses. There is a response to dopaminergic drugs. The digital age in healthcare has opened new hopes for Parkinson's disease with the use of new wearable biosensors that enable quantitative, reliable, reproducible and multidimensional measurements to be made with minimal disturbance (40–43).

Alzheimer's disease: Alzheimer's disease is a medical condition that develops in the form of memory loss, dementia, and a general decrease in cognitive functions due to the death of brain cells over time. In this disease, which causes a decrease in thought, memory and behavioral functions, the symptoms appear gradually with age. This disease is becoming more common worldwide, especially in ageing populations. Alzheimer's, a neurological disease, is also the most common type of dementia. The use of wearable sensors could help clinician-researchers eventually enable, noninvasively and cost-effectively, to detect sleep defects and autonomic dysregulation associated with early AD pathology, in a possible effort to tailor risk reduction interventions that could help retard cognitive decline (44–45).

Schizophrenia: Schizophrenia is a psychiatric disorder that distorts an individual's behaviors, movements, perception of reality and thoughts, and disrupts his relationships with his family and social environment. In schizophrenia, which is a serious and chronic disease, patients tend to lose their connection with reality, exhibit different behaviors, believe in unreal events and change their personalities. It is a lifelong disease and therefore requires constant treatment. With the right treatment, the disease can be brought under control in patients with schizophrenia (46–51).

Cystic fibrosis (CF): CF is a life-limiting recessive genetic disease that affects the cells that produce mucus, sweat and digestive juices and causes serious damage to the lungs, digestive tract and other organs in the body. Early diagnosing of cystic fibrosis means that treatment can begin right away. Diagnosis is possible with some screening tests. A baby's screening test checks for higher-than-normal levels of a chemical called immunoreactive trypsinogen (IRT), released by the pancreas, in a blood sample. However, IRT levels alone are not sufficient to confirm the diagnosis of cystic fibrosis and other tests may be required. In most cases, the diagnosis of CF is made because of the presence of one or more typical clinical features and then confirmed by demonstrating a high (>60 mmol/L) sweat chloride concentration (52–55).

14.2.2 SAMPLES AND CONDITIONS MONITORING

Wearable biosensors enable the monitoring of body fluids such as tears, sweat, saliva and interstitial fluid (ISF) and have been an exceptionally good alternative to traditional complex analytical methods due to their ease of use and portability.

Sweat Monitoring: Sweat is a form of thermoregulation in which between 500 and 700 ml of hypotonic fluid is naturally secreted per day in the average adult human body at rates that may reflect underlying health conditions. It is also rich in physiological data, including sweat easily accessible from the skin surface of the human body, electrolytes such as sodium and potassium ions, and metabolites such as lactate and glucose (56). Sweat contains several biomarkers of different diseases, such as cystic fibrosis and diabetes. The majority of cystic fibrosis cases are diagnosed by measuring the chloride concentration in sweat. It may also have potential use in continuous monitoring for diabetes; lactate can be measured to detect ischemia, and the temperature of the skin surface can be used to detect various skin injuries and diseases. Unlike other biofluids such as blood, its easy accessibility makes sweat a particularly useful biofluid as it can be removed by non-invasive methods (57–58). Wearable sensors for sweat analysis are one of the most potentially used sensors to monitor continuous measurements of useful noninvasive biomarkers. Wearable sweat sensors have the potential to monitor changes occurring in the body quickly, continuously and non-intrusively. However, sweat-based sensing has several challenges to overcome, such as the easy degradation of biomaterials by repeated testing, the limited detection range and sensitivity of enzyme-based biosensors caused by the lack of oxygen in sweat, and the poor shelf life of all-in-one sensors, low secretion rates and evaporation (59–63).

Tear Monitoring: The tear, secreted by the lacrimal gland as a protective film covering the eye, contains many compounds such as metabolites, proteins, lipids, electrolytes, and peptides. One of the biological fluids that can be used to monitor the physiological state of people with wearable sensors is tears. Biomarker molecules in tears diffuse directly from the blood and exhibit close tear-blood concentration correlations, and tear analysis also offers opportunities for the diagnosis of ocular disease. Tears are also less complex than blood and are part of the eye's toxic action mechanism. These properties make human tears an attractive diagnostic biofluid for healthcare monitoring applications that can be sampled without blood contact. Dopamine concentration in tears was linked to myopia. Produced by nerve cells, dopamine is an important hormone and neurotransmitter and is considered an important neuronal active substance in the retina of vertebrates. Significant efforts have been devoted to the development of wearable sensors that can monitor glucose concentration in the human body in real time (2, 64–68).

Saliva Monitoring: In recent years, wearable sensor technology has focused on disease prevention through early detection using bodily fluids. Non-invasive in vitro diagnosis has several advantages in terms of being easier and simpler than existing invasive tests. Saliva is an excellent body fluid that plays an important role in overall health, providing an alternative to blood measurements that are non-invasive and provide continuous real-time analysis. Some studies developed using saliva have reported the usability of saliva to prevent and diagnose systemic diseases. Saliva pH measurement can give information about the dental health of the patients. Detection systems that analyze saliva in the oral cavity have been developed to measure uric acid, lactate, glucose, cholesterol, and bacteria. For example, salivary pH in patients with chronic gingivitis is more alkaline than in clinically healthy gums. Many pandemics, including COVID-19, have greatly affected the whole world and continue to do so. Experts predict that these epidemics may continue in the future. Therefore, instant diagnosis systems and wearable devices are particularly important in this field. For example, most viruses infiltrate the mouth, so it is possible to actively monitor the virus using saliva sensors. A thin and special wearable device that is not noticed by others will be especially important for virus detection (69–74).

Body Temperature Monitoring: One of the vital signs of the body is body temperature. Body temperature is an indicator of the balance between heat production and heat loss in the body. It is important to keep body temperature within normal limits, as irregular intervals can cause bad conditions. Continuous measurements of body temperature are essential for many applications such as monitoring athletes, soldiers and emergency personnel. Other important applications include monitoring core temperature during sleep studies, monitoring patients under recovery, and monitoring the Anuar Leowe temperature of tissues during hyperthermia and other treatments (75–79).

Breath Monitoring: Breath can contain a large amount of biochemical and physiological information about one's health and can be used for non-invasive medical diagnosis of diseases. In the case of lung diseases, hydrogen peroxide is an important biomarker associated with asthma, chronic obstructive pulmonary disease and lung cancer and can be detected in exhaled breath. The current method of breath analysis involves concentration of exhaled breath, is not continuous or real-time, and requires two separate and bulky devices, making periodic or long-term monitoring of a patient difficult (80–84).

Motion Monitoring: Continuous, long-term and accurate daily physiological monitoring function can be provided to users through wearable sensors without interfering with the daily activities of the users. Monitoring the user's physiological parameters and "knowing" the user's movement state can distinguish the human body is a physiological abnormality or a pathological physiological abnormality caused by the disease caused by strenuous exercise. To facilitate the expansion and reconfiguration of the system, a wearable health monitoring system architecture based on human motion state recognition was created. According to the characteristics of the short-term persistence of human movement, the state of movement is divided into a steady state and an unstable state (85–90).

Blood Pressure Monitoring: Blood pressure is considered an indicator for evaluating a person's health or condition. Blood pressure is a good indicator of potential cardiovascular disorders and circulatory system diseases, requiring constant attention and an effective monitoring system. Especially, high blood pressure can induce major illnesses and ailments such as stroke and heart and kidney disease. The blood pressure also contains numerous information about cardiovascular situations. A wearable device is being developed that can record physiological signals, immediately process them to estimate pulse-to-beat blood pressure, transform it into understandable information adapted to context, and provide feedback to the user (91–92).

One of the key steps in using a wearable biosensor for real-time monitoring is the sampling process. The accuracy of wearable biosensors also depends on the sampling process. The successful detection of biomarkers in body fluids, such as sweat and interstitial fluid, is correlated with the process of sample collection, extraction and storage. Thus, several strategies have been developed to carry out the sampling process; one of the most explored is the use of microfluidic devices.

14.2.3 MICROFLUIDICS

Microfluidic devices with the ability to control liquid samples in micrometer-wide channels are excellent candidates for wearable device design and are already widely used due to their advantages such as detailed microstructure, low sample consumption, fast analysis, high integration of multiple functions and low cost. Compared to traditional techniques, microfluidic techniques are considered important tools in analytical chemistry thanks to these advantages. Microfluidic devices designed for wearables can perform excellent multiple functions, including sample analysis, sample collection, transport and storage, signal conversion and amplification, mechanical sensing and power supply. Moreover, microfluidic wearables with greater integration of wireless modules have demonstrated potential applications in health monitoring, clinical assessment, and human and smart device interaction. The next section primarily focuses on the latest developments in microfluidic wearables, including the overall functions and designs of microfluidic wearables, and their specific applications in physiological signal monitoring, clinical diagnosis and therapeutics and healthcare.

14.2.3.1 Fabrication

Selecting the appropriate soft material for microfluidic sensors is important for its sensing performance, wearability, and scope of application. Most work with microfluidic devices is based on the analysis of microvolume liquid samples using low-cost, easy-to-use, disposable, and miniature devices made of polymers, paper and fabric. Various substrate materials, including fabrics and polymers, are widely used because of their inherent flexibility, stretchability and excellent biocompatibility. In this section, classification of microfluidic sensors according to substrate materials and manufacturing methods are explained.

Fabric-based devices: The emerging fabric as an alternative to paper for the manufacture of microfluidic devices can be easily produced using a variety of natural and synthetic materials containing a wide variety of functional groups that can participate in binding to different types of molecules without further functionalization. Yarns, which are the raw materials of fabric microfluidic substrates, have wicking properties, excellent flexibility, stretchability and durability, and can be obtained at low cost from many natural or artificial materials. Fabric-based microchannels are produced by varying the hydrophobicity and wettability of the fabric to control fluid flow. Due to their capillary action and flexibility, wearable fabric-based microfluidics can capture sweat from the skin surface and deliver it to sensing sites via their defined microchannels. To realize the electrical connection, conductive threads are produced by varying the material composition and grafting various conductive nanomaterials such as conductive polymers. By combining multiple types of yarns, wearable fabric-based microfluidic sensors can be obtained for physical or chemical analysis (93–96).

Polymer-based devices: Polymeric microfluidic substrates are generally based on elastomers such as Ecoflex, poly(dimethylsiloxane), poly(styrenisoprene-styrene). These polymeric materials are widely used in wearable sensors due to their unique properties such as biocompatibility, optical transparency, low modulus, excellent flexibility and stretchability. Thanks to these unique features, it can be used on curved surfaces such as forearms, thighs and joints without tearing or deforming, and does not restrict the user in daily life. Poly(dimethylsiloxane) is the most widely used in microfluidic sensors due to its low cost and simple fabrication methods. Poly(styreneisoprene-styrene) has excellent properties such as high flexibility, greater physical strength, low water permeability and absorbency. Due to their polymeric nature, these polymeric substrate materials also typically exhibit a hydrophobic property (5, 97–100).

Paper-based devices: Paper-based microfluidic sensors have numerous advantages such as environmental friendliness, low cost, fast and sensitive response, disposable and easy to manufacture. Because of these advantages, many microfluidic paper-based analytical instruments have been proposed by researchers to measure the analyte concentration in an aqueous solution. Fabrication of microfluidic paper-based analytical instruments includes inkjet etching, wax printing, and screen printing, which are fast, cost-effective, simple and suitable for use in developing countries. Fabrication of microfluidic paper-based analytical instruments includes inkjet etching, wax printing and screen printing, which are fast, cost-effective, simple and suitable for use in developing countries. Paper is generally produced using natural cellulose fibers with different surface chemistry, porosity, and optical properties to allow for a variety of applications such as filtration and biomolecule immobilization (1, 93, 101–103).

The different microfluidic platforms offer many advantages for the sampling process in wearable biosensors. In the next section, we will highlight the latest application of microfluidic devices in wearable biosensors.

14.2.4 Recent Applications of Wearable Microfluidic Sensors

In recent years, many studies have been conducted into multifunctional wearable device research and even these products have been applied in clinical diagnosis and therapeutics. Wearable devices compatible with the skin, developed for effective and real-time monitoring of physiological changes in the human body, are becoming widespread with the development of modern technology (1–3, 5). This section focuses on the latest advances in microfluidic-based wearables and covers specific applications in the healthcare field.

Blood is the most understood standard sample for diagnostic measurements. Ion et al. designed a wearable, low-cost pressure sensor made of flexible, biocompatible materials that can be integrated into a blood pressure monitor (104). The developed sensor is made of an encapsulated microfluidic channel between two polymer layers; one layer is coated with metal converters and the other consists of a flexible membrane containing a microfluidic channel that acts as a seal for the structure. The developed sensor has a high sensitivity for pressure values between 0 and 150 mm Hg and will provide low production cost and ease of use. The polydimethylsiloxane membrane proved to have very good mechanical features for the designed sensor. Techniques of functionalization of substrates used by 3-aminopropyltrimethoxysilane)/polydimethylsiloxane and 3-glycidoxypropyltriethoxysilane/polyethylene terephthalate have been successfully applied. A pulse pressure generator and an artificial arm were used to simulate the measurement conditions to obtain the wave function close to real human blood pressure. The pressure sensor designed can greatly increase access to hemodynamic research for the diagnosis and prevention of heart disease, thanks to its advantages such as being accessible and easy to use.

Cleanroom-free fabrication of wearable microfluidic sensors using a screen-printed carbon master for electrochemical monitoring of sweat biomarkers during exercise activities was proposed by Vinoth et al. (105). Simultaneous potentiometric Na^+, K^+, and pH ions and amperometric lactate detection were performed by a miniature circuit board with signal acquisition and wireless signal transmission. With the designed sensors, sensitive, selective, stable and reproducible results were obtained. The epidermis-mounted fully integrated pumpless microfluidic device is used for real-time multi-decoding of sweat during stationary cycling. The proposed technique provides a large-scale production of economic, highly efficient wearable sensors for personalized point-of-care and athletic applications.

In the Table 14.1 it is highlighted the latest wearable biosensors based on microfluidic devices, exploring the different types of transduction mechanisms, sample matrix and their limit of detection (LOD).

TABLE 14.1

Evaluation of the Latest Wearable Biosensor Based on Microfluidic Devices Selected for Their Analytical Properties

Target Body Fluid or Body Area	Detection Method	Target Biomarker	Related Human Disease	LOD	Ref.
Blood	-	Blood pressure	Cardiovascular disease	-	(104)
Epidermal	Colorimetric/ Recombinase Polymerase Amplification (RPA)	Nucleic acid fragments of zika virus	Pathogen Detection	10 copies/µL	(106)
Epidermal	Fluorescence	IgG AD7c-NTP	Neurodegenerative disease	12 ng mL^{-1} 4.5 ng mL^{-1}	(107)
Epidermal	Optical	Pressure	-	≈98 Pa	(108)
Interstitial fluid	Colorimetric	Glucose	Diabetes	68 µM	(109)
		Lactate	Ischemia	17 µM	
Ophthalmic	-	Intraocular pressure	Glaucoma	<0.06% for uniaxial and <0.004% for biaxial strain	(110)
Sweat	Fluorimetric	pH	Acid—base disorders	-	(111)
		Cl$^-$	Cystic fibrosis	5 mM	
		Glucose	Diabetes	7 µM	
		Lactate	Ischemia	0.4 mM	
Sweat	Amperometric/ Potentiometric	Lactate	Ischemia	0.2 mM	(105)
		Na$^+$	Hyponatremia, Cystic fibrosis	-	
		Cl$^-$	Cystic fibrosis	-	
Sweat	Colorimetric	Glucose	Diabetes	0.03 mM	(112)
Sweat	Colorimetric	Lactate	Ischemia	0.06 mM	(113)
Sweat	Electrochemical	Glucose	Diabetes	17.05 µM	(114)
		Lactate	Ischemia	3.73 µM	
Sweat	Electrochemical	Glucose	Diabetes	5 µM	(115)
Sweat	Electrochemical	K$^+$	Dehydration	-	(116)
Sweat	Electrochemical	Uric acid	Gout	0.74 µM	(117)
		Tyrosine	Metabolic disorders	3.6 µM	
Sweat	Optical	K$^+$	Dehydration	-	(118)
Saliva	Colorimetric	Glucose	Diabetes	27 µM	(119)

14.3 NANOMATERIALS IN WEARABLE APPLICATIONS

Conventional approaches in wearable biosensing have been metal-, silicon- and/or organic-based (120–121), which has contributed to limitations regarding biocompatibility, sample collection, on-body measurements, conformability and flexibility, due to the rigid or brittle nature of the substrate. Therefore, addressing material technology, and tunning their properties into more flexible and comfortable miniaturized approaches, would considerably enhance the performance of wearable biosensors. More than that, the search for low-level detection of biomarkers in patient samples requires exploring new classes of materials that promote amplification of the detection signal.

The development of new wearable sensors has been focused on seeking materials that can be easily integrated into the human body. In order to minimize discomfort or constraints during daily activities, a wearable sensor must have a low Young's modulus, but a high stretching capability, since the epidermis modulus ranges from 140–600 kPa, while the dermis modulus ranges from 2–80 kPa (122–123). Additionally, it is possible to extend skin to 15% and up to 100%, including wrinkles and creases. Another characteristic that is required in wearable biosensors is practicality. Not only should they be comfortable to use, but they must also be light, highly sensitive and energy-efficient (124–125). For example, significant efforts have been undertaken to make gel-free (dry) wearable electrodes, which allow secure and comfortable contact with the skin (126). In definitive agreement with this proposal, nanomaterial-based electrodes have the properties of high conformability, allowing better skin contact with the electrodes, and presenting exceptional physicochemical properties, amplifying the sensitivity of these sensors (124, 127).

14.3.1 TYPES OF NANOMATERIALS

In order to highlight their significance in wearables biosensing platforms, the use of nanomaterials must be defined. According to the ISO/TS 80004, a nanomaterial is a material that has any of its dimensions in the nanoscale (10^{-9} m), i.e., 10,000 times smaller than the diameter of a human hair. For many years, this particular characteristic has allowed scientists to use their size-dependent and naturally occurring unique properties in terms of electrical and thermal conductivity, chemical reactivity, surface area and flexibility, among others.

Considering their outstanding electrical and mechanical properties and their biocompatibility, the use of nanomaterials is a promising candidate for further improvement of wearable biosensors.

The nanomaterials can be synthesized in several ways and depending on the final application, from zero-dimensional (0D), such as metallic and polymeric nanoparticles (NPs); one-dimensional (1D), like carbon-based nanowires and nanotubes, to two-dimensional (2D) like graphene and MXenes (128). Due to their surface chemistry and layer-dependent structure, these materials demonstrated outstanding capabilities as an emerging approach for improving wearable biosensors, being more prone to modifications by incorporating "defects" or direct immobilization of bioreceptors in their surface (126, 129).

The widespread use of highly conductive metallic nanostructures is also due to the ease and cost-effective development method. Though silver nanostructures have received the most attention over the last decades, copper-based nanobiosensors are considerably rising due to their lower fabrication cost and higher availability than Ag-based ones (123, 127). Owing to the cost-ineffective and complex synthesis of high-quality gold nanostructures, which have greater oxidation resistance and biocompatibility than silver and copper, the first ones are less often used in wearable biosensors. On the other hand, carbon-based nanomaterials are less expensive, and they are known to be doped with conductive or semiconducting properties, so they could be suitable and potential alternatives for green wearable biosensors using biodegradable materials (130–132).

Traditionally, wearable nanobiosensors have consisted of nanomaterials embedded or chemically functionalized on the surface of polymeric matrices, such as fabrics or plastics (3, 125, 133–134). Often, a synergy between nanomaterials and materials engineering might increase the inherent stretchability of the system, such as introducing highly deformable designs to enhance their mechanical properties, as well as generating a better strain absorption, e.g., serpetine-based and fractal structures (135–137).

Research into wearable biosensors has been very active in recent years. Various fabrication procedures are being used, from the drop and/or dip coating to directly print into the substrate (127, 132). Nevertheless, long-term biocompatibility studies of nanomaterials are still in the early stages and must be remedied to guarantee a practical application in wearable biosensors (126, 132, 138–139).

Notably, 1D nanomaterials, such as nanowires, nanofibers or carbon nanotubes (CNTs), have the compelling advantages of a high aspect ratio, exceptional conductivity and their significantly higher mean free path for the charge carriers (136, 129). Although few sensor designs may retain adequate sensitivity under stress, 1D nanomaterials are suitable for developing stretchable, conductive sensors with transparent networks. One example is shown in Figure 14.1 as a stretchable temperature-responsive matrix was created by combining polyaniline nanofibers and SWCNT thin-film transistors (140).

A patterned conductive carbon-based nanocomposite can have advantageous features by integrating CNTs and graphene nanofillers to mimic gecko-inspired characteristics (Figure 14.2). Due to their extremely high flexibility and conductivity (~100 Ω cm), the nanocomposite achieved a significant electrocardiogram accuracy (123, 141). Notably, the introduction of direct printing techniques such as screen and inkjet printing, aided by the outstanding properties provided by nanomaterials and chosen substrate, also allowed for the development of highly accurate and complex wearable biosensors (121, 142–143).

14.3.2 TRANSDUCTION MECHANISM

Temperature: Although there would be temporal and spatial variations due to heat transfer among biological tissues/organs, the importance of having a highly sensitive and accurate body temperature reading is highlighted in several disease conditions, including viral/bacterial infection (3) and cancer (144–145). Current temperature biosensors must have demonstrated long-term stability and high sensitivity, even under environmental alterations and temperature changes ranging from 25 to 40°C (129, 146–147). In this way, incorporating nanomaterials in current wearable biosensors plays a significant role in enhancing their temperature requirements. More specifically, carbon-based nanomaterials have demonstrated outstanding thermal conductivity due to their existent acoustic phonons, main heat carriers in this type of 2D materials (128, 148–150). Researchers have employed several nanomaterials, such as carbon-based (graphene, CNT/CNW), metal NPs/NWs and conductive polymers, as thermal sensing layers.

Pressure and strain: For wearable pressure and strain sensors, the obtained signal and the proper sensitivity of the device are mainly relative to the deformation of the layer in contact with the skin, usually being based on a dielectric material. Hence, low Young's modulus nanomaterials are preferred as an alternative approach, incorporating crack propagation theory and microstructures.

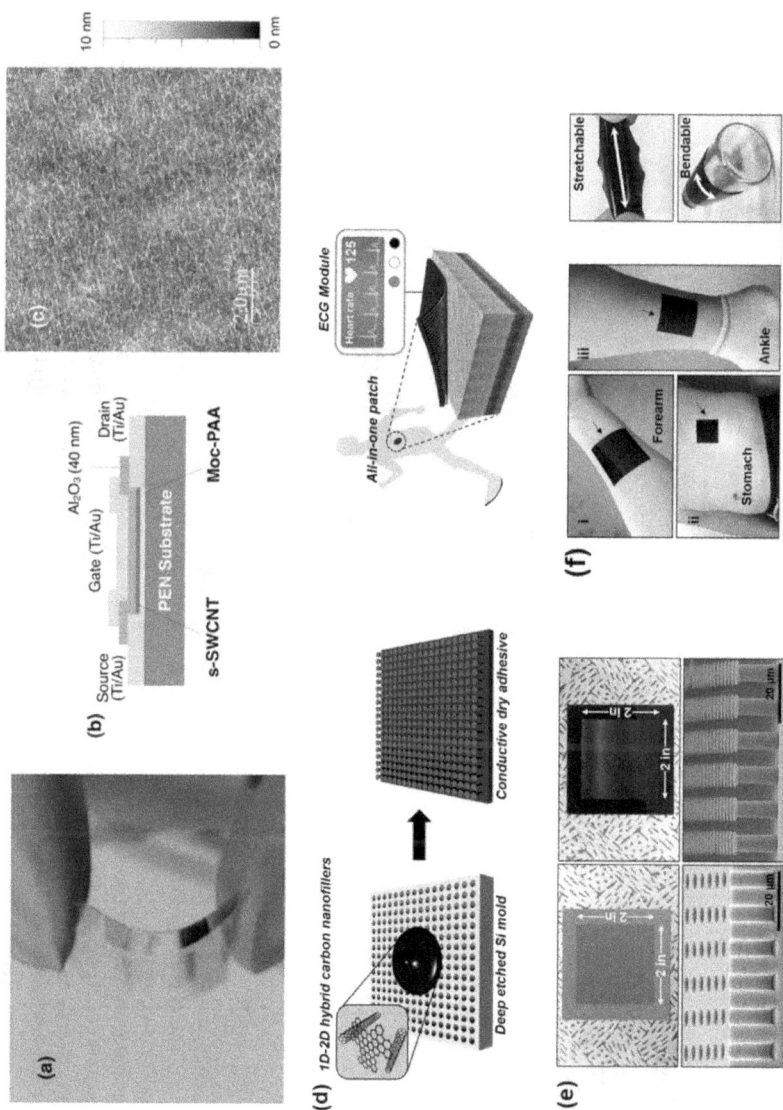

FIGURE 14.1 (a) Photograph and (b) graphic representation of single-walled carbon nanotubes-based (SWCNT) thin film transistors (TFT) deposited on a flexible polyethylene naphthalate (PEN) substrate. (c) AFM image of the modified PEN substrate. Reprinted from ref. (140) with permission. (d) Schematic representation of the carbon-based conductive dry adhesives fabrication, along with using it as ECG electrodes. (e) Images of silicon-based molds with gecko-inspired micropatterns and its replication using the material in (d). (f) Images of the modified adhesive conformally attached on different body parts. Reprinted from ref. (141) with permission.

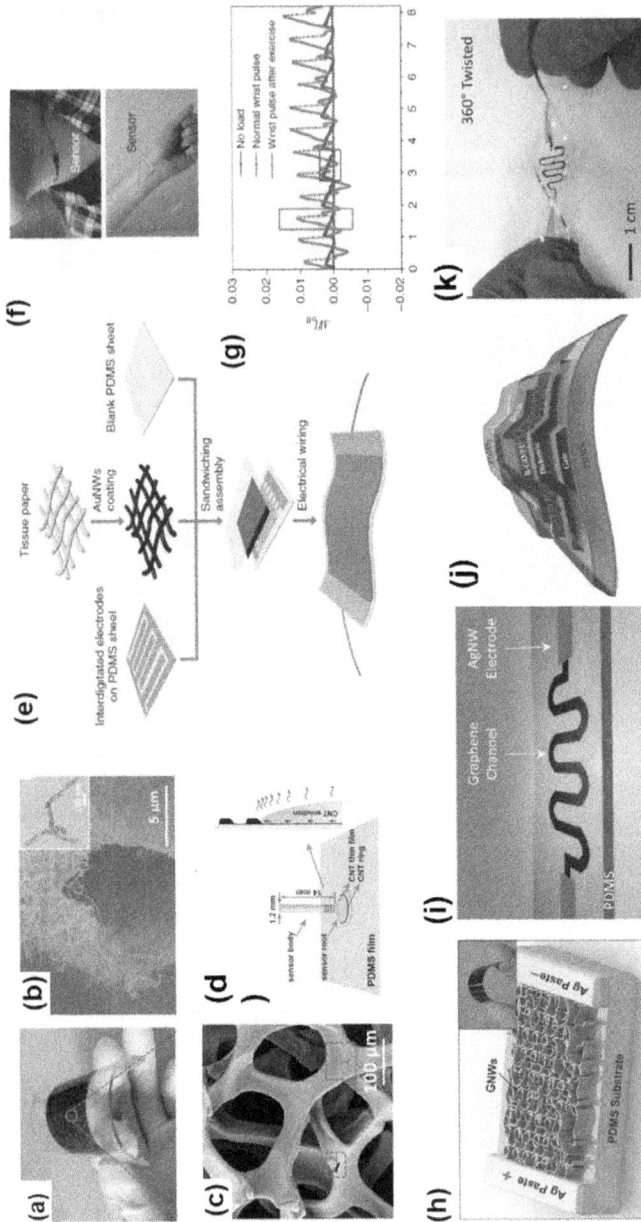

FIGURE 14.2 (a) Photograph of the strain sensor, based on nanocomposite of 3-D Graphene Foam and CNTs, under bending conditions. (b) SEM image of a said material's percolation network, along with (c) an enlarged image showing the graphene foam. (d) Graphical representation of the Graphene nanowall/PDMS array connected to Ag paste for temperature sensing purposes. Reprinted from ref. (123) with permission. (e) Graphical description of the fabrication of flexible AuNW-based pressure sensor. (f) Nanomaterial- and crack-junction-based biosensor attached on the neck and wrist for several applications. (g) Response of (f) against the heart-beat rate resting and during exercise Reprinted from ref. (151) with permission. (h) Schematic of a graphene nanowires network-based temperature sensor with silver paste contacts. (i) Schematic illustration of a nanomaterial-based thermistor PDMS array molding a stretchable graphene thermoresistive sensing channel and AgNW electrodes. (j) Cross-section of a FET based flexible device on PDMS. (k) Image of (i) being twisted. Reprinted from ref. (147) with permission.

Wearable pressure sensors experience a wide range of possible uses due to their outstanding flexibility and ability to mimic human skin's electrical signal. Related electrical transduction concepts are used in pressure sensors as strain sensors, such as triboelectric sensors, which have also gained interest due to their power generation capabilities in response to pressure stimuli (151). Conventional sensors have a planar structure, resulting in a diminishment of the performance. Nanomaterials and nanocomposites have also played a key role as well, where research has been done to develop pressure wearable sensors of excellent efficiency due to incorporation of innovative geometric microstructures, such as μ-pillars, porous membranes (120, 152–153).

Wearable strain sensors with a high stretchability (e.g. torsion and limb bending), sensitivity (changes caused by swallowing, heartbeat, pulses etc.) (121, 123, 138) as well as mechanical/chemical conformability (127, 151), enable measuring critical parameters for a wide variety of applications, including personalized medicine and daily healthcare continuous monitoring, also characterized by their fast response. Although the conventional approach involves using fragile and rigid films, including metals and semiconductors, these sensors possess minimal stretchability and cannot be appropriately integrated to different textiles (120, 131, 147).

Seizing the outstanding intrinsic mechanical, thermal, and electrical properties of nanomaterials, in addition to their enhanced surface area and crack propagation mechanism (147, 154–156), numerous studies have demonstrated the use of various types, including graphene, carbon nanotubes, polymer nanofibers, as well as metal NWs, NPs, in order to overcome current limitations (131, 154, 157).

Structural diversity also allows diverse density distributions of cracks to emerge through stretching, thereby accomplishing the objective of regulating susceptibility (124, 132).

Electrochemical: As aforementioned, by shifting from macro- to nanoscale structures, biosensing performance would be considerably increased. The size and shape effects in nanomaterials play a key role in enhancing numerous important parameters, such as electrochemical and electrocatalytic activity, number of binding sites, specific interactions, and thermodynamics. Proper and selective interaction with the biomarker is critical for the development of highly sensitive electrochemical sensors, which are being aided by the incorporation of nanomaterials with extraordinary electrocatalytic properties (158–159).

Being an inherent and excellent source of metabolites and electrolytes, non-invasive chemical processing and analysis of biological fluids (including sweat, saliva, tears, among others) has shown to be commonly adopted in a variety of application sites, as they minimize risk of infection/harm (122, 160–161). Biosensors covering a broad array of relevant analytes have been used in these biofluids in order to study their dynamic biochemical processes. A label-free graphene-based wearable biosensors in tooth enamel for single-cell bacteria detection was developed by (162), employing a battery-free and wireless approach, as well as a antimicrobial peptide as biorecognition element.

14.3.3 APPLICATIONS

Due to their biocompatibility, stretchability, and human skin-like mimicking properties of ideal wearable biosensors, their placement should be adjustable to clothes or body parts and even incorporated into daily accessories, such as watches, clocks, or necklaces (121, 132, 163). Bringing together the strengths of wearable and electrochemical technologies opens the door to a multitude of applications.

Carbon-based nanomaterials have been extensively incorporated into wearable sensing platforms, especially for noninvasive and on-site approaches, like fitness and wellness (121, 164–165). This is due to their large specific surface area, presence of structural defects, high electric conductivity, high mechanical strength, and excellent chemical and thermal stability (123, 128) have demonstrated the viability of embedding CNTs into textiles for electrochemical monitoring of electrolytes in sweat.

Graphene, as well as its derivatives, have also attracted significant attention in the last decades. It has been employed for several applications, including a multiplexed detection of glucose in sweat (166), single-cell bacteria detection in saliva (162), thermoresponsive μ-needles for diabetes treatment and monitoring (129).

For on-body continuous monitoring of thermal properties, along with enabling novel thermotherapy and drug delivery approach, Tabish et al. (167) have developed a graphene-based strain wearable sensor for transdermal drug delivery. Additionally, a fully printed smart bandage has also been studied for interactive health monitoring (168). Drug-delivery platforms can be integrated to enable therapeutic treatments in response to the information collected by the wearable sensors (138, 169).

As mentioned before, embedding flexible pressure/strain sensors into daily accessories could help obtain important information about the wearer's condition, including breathing, talking, exercising etc. Graphene, CNTs and AuNP-based approaches have been employed for detecting changes in resistance due to heart motions for fitness purposes (121, 170), as shown in Figure 14.1.

Developing wearables is an active field owing to a variety of reasons. Actuators and sensors are critical wearable electronics components since they can provide an interactive interface with the wearer and its surroundings. Beyond the wearable principle, advanced functionalities could be obtained if various other modules can be integrated. Among them are drug delivery, portable power supplies and data acquisition equipment, along with wireless data transmission. Adequately addressing these challenges would encourage further research and development in the field of nanomaterial-based wearable (bio)sensors.

On the other hand, due to the need for continuous biochemical monitoring involving a highly sensitive and specific detection, revenue growth for nanomaterial-enabled wearable biosensors is considerably more difficult. Although it is essential to produce accurate and robust readings, this can only be achieved by addressing a variety of other possible limitations, such as accumulation (biofouling) of metabolic byproducts on the sensor's surface, inadequate sample flow, instability of the target binding under external perturbations as well as the complexity of establishing a valid working range for the on-body type ones.

Besides the sampling process and the amplification of signal detection by using nanomaterials, signal readout systems play a critical role in developing wearable nanobiosensors. In the next section, we will discuss the main strategies to build readout systems that meet the needs of wearable nanobiosensors.

14.4   SIGNAL READOUT

Continuous wireless monitoring of physiological variables in biomedical, sports and healthcare applications is one of the increasingly used technologies (171–173). This technology combines the biorecognition and sensing part with compact electronics that allow the communication of the collected data and optimize electricity consumption. (174), or even harvesting the required energy from the environment in which the biosensor works (175).

Notable advances in the ecosystem called the Internet of Things (IoT) have made it possible to strengthen communication technologies and protocols for the interaction, collection and exchange of more significant amounts of data between humans and a variety of physical and virtual devices, with the lowest energy consumption (176–177).

14.4.1   Wireless Communication

Bluetooth Low Energy (BLE), Wi-Fi, Near-Field Communication (NFC), Radio-Frequency Identification (RFID) and ZigBee, are the most widely adopted wireless technologies in the market for the transmission of information between biosensors and platforms based on smart terminals (178). These technologies can help reduce the size, cost of implementation, energy consumption and processing resources (178), by replacing the electronic components embedded in the system, such as screens, buttons and control interface, by the resources of hardware and software of the new interconnected smart devices (179).

14.4.1.1   Bluetooth Low Energy (BLE)

Bluetooth Low Energy (BLE) is a subset of the Bluetooth v4.x standard that operates in the 2.4 GHz ISM band developed for applications in the areas such as health care, fitness, beacons, security, home, agriculture, and entertainment industries (180–182).

Unlike traditional Bluetooth, this communication protocol has very low power requirements, an acceptable range, and considerable miniaturization of the electronics implemented, which allows its use in devices with energy restrictions such as wearables, sensors, etc. (182–183). Furthermore, this technology is expected to be incorporated into billions of devices in the coming years (184).

The spectrum using this technology is divided into 40 of 2 MHz channels grouped into 37 connection-oriented channels and 3 channels for the primary advertisement (185), admitting two different forms of communication: Unicast (Peer-Peer) Connection and Broadcast Connection. In the first mode (see Figure 14.3), two BLE hosts that are initially disconnected enter a discovery state in which the device that wishes to be discovered becomes the Advertiser and the host that wishes to connect becomes the Initiator or Scanner. Once the Initiator sends the connection request, and it is accepted by the Advertiser, the communication is established, and the Initiator becomes the Master, while the advertiser becomes the slave. In the second mode (see Figure 14.4), the roles of Scanner and Advertiser change to Observer and Broadcaster (the host that sends the packets), respectively. In this communication mode, the messages sent are unidirectional, with a one-to-many link. (183, 186).

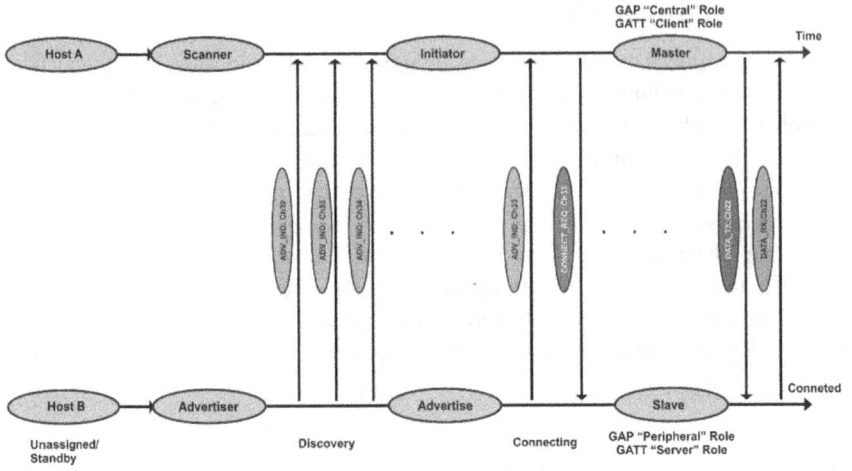

FIGURE 14.3 BLE link layer roles and states for unicast (peer-peer) connection.

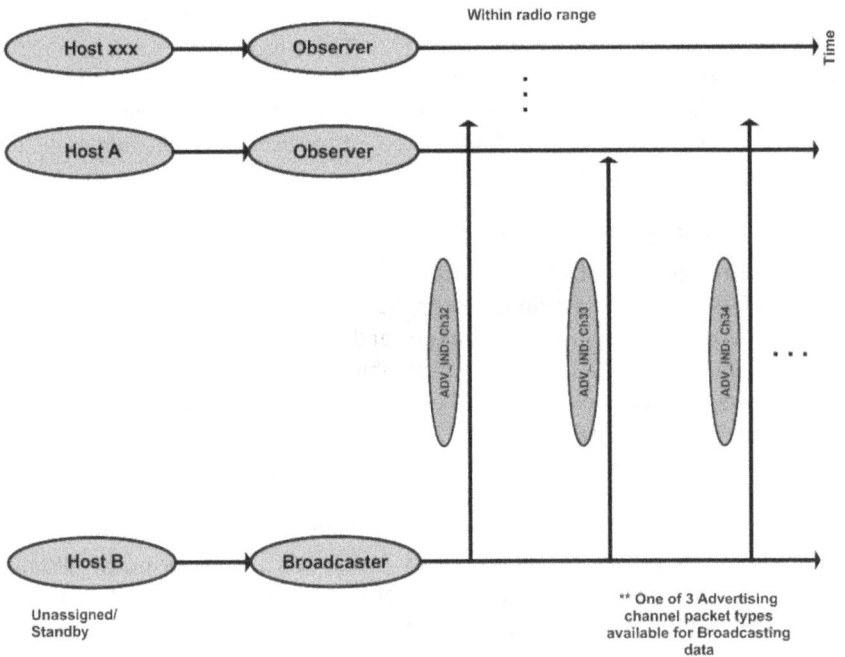

FIGURE 14.4 BLE link layer roles and states for broadcast connection.

14.4.1.2 Radio-Frequency Identification (RFID)

It is a technology that allows the collection of wireless data automatically by sending information in the UHF band from 860 MHz to 960 MHz. This technology is widely used in the manufacturing and consumer industries through the manufacture of identification cards, passports and credit cards that do not require contact (187). Combining biosensors with this technology has allowed the exploration of new fields such as environmental monitoring, medical care and food quality monitoring (188–189). Recently, UHF RFID tags have greater detection and computation capabilities, thanks to lower energy needs, greater computational power, and reduced device size (190).

Devices that implement this technology can be divided into two classes: active devices and passive devices. The former requires being permanently connected to a power source. If this energy supply comes from an integrated battery, its useful life is drastically reduced. This is impractical for many applications. On the other hand, passive devices do not require batteries or maintenance. These obtain their energy from the radio frequency signal generated by the reading device. This same signal is used as the response channel of the passive device by modulating the backscattered signal. (187, 191).

14.4.1.3   Near-Field Communication (NFC)

Near Field Communication (NFC) is a technology that allows communication between two devices a few centimeters apart and is widely used in applications such as smart homes, payments, agriculture, the environment, and healthcare. This technology is part of the set of radio frequency identification systems (RFID) that operates in the 13.56 MHz band, both in the role of reader (active device) and in the role of tag (passive device) (192–195).

NFC implements three modes of operation: Reader/Writer Mode, Peer to Peer (P2P) Mode, and Card Emulation Mode. In the first mode, the device acts as a tag reader, in this way, through an application, the device can write and read the information stored on a tag. In the second mode, NFC devices can exchange information bi-directionally. Lastly, card emulation mode allows the NFC-enabled device to act as a contactless smart card (196–197).

14.4.1.4   Wi-Fi

WiFi stands for the name of the industry consortium in charge of coordinating and implementing the IEEE802.11 wireless Ethernet standards. This standard operates in the 2.4GHz or 5GHz free band with a transmission speed of 1Mbps to 600Mbps. The network administrator can select different channels to reduce possible interference. In the event of a failure in recognizing a packet, the transmission speed is lowered while the recognition of the successive packets is executed and then it recovers its original speed (198–199).

Due to the advancement of IoT technology, the standard allows detecting, monitoring, controlling and managing the different devices connected locally or in the cloud, and articulating with technologies that facilitate data processing such as artificial intelligence (AI), Big Data and Machine Learning (200). This advance allows varied applications in the field of health, agriculture, industrial processes, consumer electronics, and wearable devices in general (201–204). However, low security and stability, and high energy consumption, are some of the disadvantages that must be considered when selecting this technology (205).

14.4.1.5 ZigBee

IEEE 802.15.4 Zigbee is a standard that defines a set of communication protocols that allow establishing and providing control functionalities for networks such as home networking, body area networks, wireless sensor network (WSN) etc. ZigBee operates in the 868 MHz, 915 MHz and 2.4 GHz frequency bands with a maximum data rate of 250K bits per second (206–207).

Among the advantages that ZigBee technology provides compared to the IEEE802.11 standard are: greater security, greater stability, expandability (currently, the actual size of the WiFi network is generally not more than 16 devices), low power consumption, a distributed system and ad hoc network capabilities; however, it exhibits a considerable degree of signal interference and a weak diffraction capacity (205, 208).

14.4.2 SMARTPHONE-BASED *VS* SMARTPHONE-FREE

14.4.2.1 Smartphones and Tablets

Point-of-care (POC) devices have achieved a remarkable advance in miniaturization, precision, reliability and ease of use, through the integration of portable biosensors with modern phones, tablets and other smart terminals in different areas such as care of health (209–210), environmental monitoring (211) and food evaluation (212).

Due to the inclusion of sophisticated cameras, high resolution displays, new communication protocols, integrated circuits with greater processing power and the addition of new sensors, smartphones can perform one or more of the following functions: (i) act as detectors, (ii) as a processing unit, (iii) as a display and control interface, or (iv) as a storage unit (213–214). This can significantly simplify the electronic design, reduce the volume and lower the cost of the system by replacing some hardware modules such as buttons, signal conditioner and display, by the development of applications on the smartphone (215).

Another significant contribution, as a result of this integration, is the ease of adoption and appropriation of technology focused on smart mobile devices due to the high contact of the population with this type of technology. To date, two thirds of the world's population is familiar with smartphones, and more than half with the use of the Internet (216).

14.4.2.2 Garments, Accessories and Custom Electronics

Although most of the publications on wearable devices for health care are supported by the use of smartphones, there is a large group of elements and accessories with embedded hardware resources, such as smart bracelets, smart glasses, smart health clothing, health patch and other accessories, aimed at measuring different physiological parameters in real time without affecting people's daily activities. These devices require powerful microcontrollers with very low consumption, screens and buttons, among others, that allow the visualization of the signals captured and processed (217–218). Figure 14.5 shows an example of some of the elements required in the implementation of this type of technology.

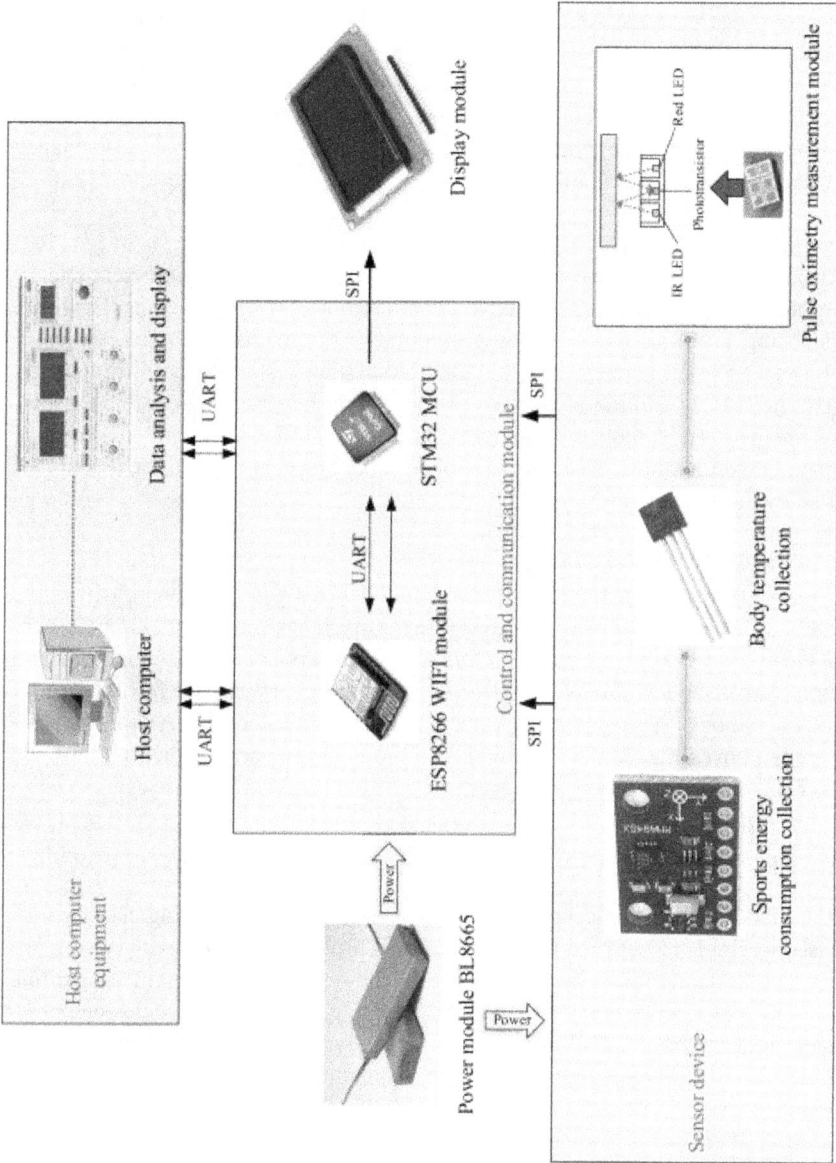

FIGURE 14.5 The hardware structure of the health bracelet terminal. Reprinted from ref. (217) with permission.

14.4.3   Power Sources Type

14.4.3.1   Batteries

Despite the limited lifespan and potential dangers to human health, batteries remain the primary solution for wearable systems and sensor networks (219). Its main challenges are to decrease the weight and size and increase the durability of the cycle. Lithium-ion batteries (LIB) stand out for their higher energy storage capacity, low discharge rate, low weight, and no memory effect. However, the high demand for high-efficiency batteries, Li-air and Li-S batteries are considered the most promising batteries due to their higher energy density and higher electrochemical performance compared to current LIBs (220).

14.4.3.2   Fuel Cells

The fuel cell takes advantage of hydrogen and acts as a clean and efficient energy converter, being an excellent alternative for internal combustion engines. However, despite being a promising technology, its durability remains the main challenge. Generally, a fuel cell is integrated into a hybrid system that consists of three parts: the cell, the battery and the balance of plant (221). For its use in wearable devices, this technology is adapted for its operation in microbial fuel cells with conventional textile processes, it allows the emergence of portable bioenergy collection techniques powered by human sweat (221).

14.4.3.3   Energy Harvesters/Free Batteries

The energy harvesting technology for wearable devices generates energy from the human body's movements while ensuring flexibility. These collectors allow to control of electronics at any time and therefore have broad application potential as health monitors, soft robotics, flexible solar cells, electrical coatings, and flexible energy collectors. Among these devices, portable energy harvesters, apart from achieving energy conversion, detect movements to monitor the parts of the human body (222–223).

14.5   CHALLENGES, CONCLUSIONS AND FUTURE OUTLOOK

In this chapter, we summarized the main principles in developing nanomaterials-based wearable biosensors, from the sampling process to the readout systems. We believe that these biosensors potentially impact future applications, mainly regarding the development of electronic skins (e-skins), offering the possibility to detect metabolites in sweat samples, in real-time, at low concentrations and in a non-invasive way. It is also due to the outstanding features of the nanomaterials that meet the mechanical requirements and improve the sensitivity of the e-skins.

However, there are still some challenges that need to be overcome in the development of wearable biosensors based on nanomaterials. One of them concerns the real-time detection of protein biomarkers in sweat and ISF samples. The use of antibodies and aptamers as bioreceptors is already a standard approach in the literature, but how to guarantee the stability of these bioreceptors against changes in temperature, humidity and ionic strength in a real application of wearable nano-biosensors is still a challenge.

The nanomaterials-based wearable biosensors are at the embryonic development stage and far from going to the market. The significant challenge primarily stems from the microfabrication on a large scale to the heterogeneity that can occur from device to device. The integration of nanomaterials, keeping their properties, in the standard microfabrication technologies and flexible substrates were not totally achieved yet. Besides that, strategies to maintain sensor stability and improve shelf life are still under development.

In addition, we showed that the development of wearable nano-biosensors, goes far beyond the development of the sensor itself, but that the sampling processes (based on microfluidic devices) and the readout system are fundamental parts of this new technology. Wireless communication capability, especially with a smartphone, will become increasingly crucial in wearable sensors. Furthermore, the source of energy for these sensors is still a key issue. We showed the importance of self-powered wearable platforms based on the energy harvest process and biofuels, which will be more prevalent soon. More than that, wearable sensors like RFID are already a reality, and even more, they will be applied in biosensing for healthcare applications. Even though the development of wearable nano-biosensors is still at an early stage, this represents an excellent opportunity for advancing research and development in this area. The many efforts from scientists in the interdisciplinary fields of sensing, materials, chemistry, biology and electronic engineering will push for the realization and future technologies for the nanomaterials-based wearable biosensors for healthcare.

ACKNOWLEDGMENTS

Leyla Karadurmus is grateful for a fellowship from the Scientific and Technological Research Council of Turkey (TUBITAK).

REFERENCES

1. Yetisen AK, Martinez-Hurtado JL, Ünal B, Khademhosseini A, Butt H. 2018. Wearables in medicine. *Advanced Materials*. 30(33): 1706910
2. Kim J, Campbell AS, Ávila BEF, Wang J. 2019. Wearable biosensors for healthcare monitoring. *Nature Biotechnology*. 37(4): 389–406
3. Sharma A, Badea M, Tiwari S, Marty JL. 2021. Wearable biosensors: An alternative and practical approach in healthcare and disease monitoring. *Molecules*. 26(3): 748
4. Chen G, Zheng J, Liu L, Xu L. 2019. Application of microfluidics in wearable devices. *Small Methods*. 3(12): 1900688
5. Heikenfeld J, Jajack A, Rogers J, Gutruf P, Tian L, Pan T, Wang J. 2018. Wearable sensors: Modalities, challenges, and prospects. *Lab on a Chip*. 18(2): 217–248
6. Basra G, Basra S, Parupudi S. 2011. Symptoms and signs of acute alcoholic hepatitis. *World Journal of Hepatology*. 3(5): 118
7. Sohail U, Satapathy SK. 2012. Diagnosis and management of alcoholic hepatitis. *Clinics in Liver Disease*. 16(4): 717–736
8. Sandahl TD. 2014. Alcoholic hepatitis. *Danish Medical Journal*. 61: B4755

9. Lee H, Choi TK, Lee YB, Cho HR, Ghaffari R, Wang L, Choi HJ, Chung TD, Lu N, Hyeon T, Choi SH, Kim DH. 2016. A graphene-based electrochemical device with thermoresponsive microneedles for diabetes monitoring and therapy. *Nature Nanotechnology*. 11(6): 566–572

10. Campbell AS, Kim J, Wang J. 2018. Wearable electrochemical alcohol biosensors. *Current opinion in electrochemistry*. 10: 126–135

11. Biscay J, Findlay E, Dennany L. 2021. Electrochemical monitoring of alcohol in sweat. *Talanta*. 224: 121815

12. Selvam AP, Muthukumar S, Kamakoti V, Prasad S. 2016. A wearable biochemical sensor for monitoring alcohol consumption lifestyle through Ethyl glucuronide (EtG) detection in human sweat. *Scientific Reports*. 6(1): 1–11

13. Wang Y, Fridberg DJ, Leeman RF, Cook RL, Porges EC. 2019. Wrist-worn alcohol biosensors: Strengths, limitations, and future directions. *Alcohol*. 81: 83–92

14. Davis-Martin RE, Alessi SM, Boudreaux ED. 2021. Alcohol use disorder in the age of technology: A review of wearable biosensors in alcohol use disorder treatment. *Frontiers in Psychiatry*. 12: 246

15. Mohan AV, Windmiller JR, Mishra RK, Wang J. 2017. Continuous minimally-invasive alcohol monitoring using microneedle sensor arrays. *Biosensors and Bioelectronics*. 91: 574–579

16. Swift RM, Martin CS, Swette L, LaConti A, Kackley N. 1992. Studies on a wearable, electronic, transdermal alcohol sensor. *Alcoholism: Clinical and Experimental Research*. 16(4): 721–725

17. Lien YHH, Shapiro JI. 2007. Hyponatremia: Clinical diagnosis and management. *The American Journal of Medicine*. 120(8): 653–658

18. Korzelius CA. 2013. *CME Information Diagnosis, Evaluation, and Treatment of Hyponatremia: Expert Panel Recommendations Program Overview*. https://www.amjmed.com/article/S0002-9343(13)00605-0/pdf

19. Buffington MA, Abreo K. 2016. Hyponatremia: A review. *Journal of Intensive Care Medicine*. 31(4): 223–236

20. Upadhyay A, Jaber BL, Madias NE. 2006. Incidence and prevalence of hyponatremia. *The American Journal of Medicine*. 119(7): 30–35

21. Terkeltaub R. 2010. Update on gout: New therapeutic strategies and options. *Nature Reviews Rheumatology*. 6(1): 30–38

22. Major TJ, Dalbeth N, Stahl EA, Merriman TR. 2018. An update on the genetics of hyperuricaemia and gout. *Nature Reviews Rheumatology*. 14(6): 341–353

23. Cronstein BN, Terkeltaub R. 2006. The inflammatory process of gout and its treatment. *Arthritis Research & Therapy*. 8(1): 1–7

24. Terkeltaub RA. 2003. Gout. *New England Journal of Medicine*. 349(17): 1647–1655

25. Smith E, Hoy D, Cross M, Merriman TR, Vos T, Buchbinder R, March L. 2014. The global burden of gout: Estimates from the global burden of disease 2010 study. *Annals of the Rheumatic Diseases*. 73(8): 1470–1476

26. Levine RJ, Conn HO. 1967. Tyrosine metabolism in patients with liver disease. *The Journal of Clinical Investigation*. 46(12): 2012–2020

27. Russo PA, Mitchell GA, Tanguay RM. 2001. Tyrosinemia: A review. *Pediatric and Developmental Pathology*. 4(3): 212–221

28. Diabetes. n.d. Accessed May 16, 2021. www.who.int/news-room/fact-sheets/detail/diabetes

29. Fagherazzi G, Ravaud P. 2018. Iconography: Digital diabetes: Perspectives for diabetes prevention, management and research. *Diabetes & Metabolism*. 45

30. Polonsky WH, Fisher L. 2013. Self-monitoring of blood glucose in noninsulin-using type 2 diabetic patients: Right answer, but wrong question: Self-monitoring of blood glucose can be clinically valuable for noninsulin users. *Diabetes Care*. 36(1): 179–182

31. Whiting DR, Guariguata L, Weil C, Shaw J. 2011. IDF diabetes atlas: Global estimates of the prevalence of diabetes for 2011 and 2030. *Diabetes Research and Clinical Practice*. 94(3): 311–321

32. Zhou B, Lu Y, Hajifathalian K, Bentham J, Di Cesare M, Danaei G, Bixby H, et al. 2016. Worldwide trends in diabetes since 1980: A pooled analysis of 751 population-based studies with 4·4 million participants. *The Lancet*. 387(10027): 1513–1530

33. Chen S, Wu N, Lin S, Duan J, Xu Z, Pan Y, Zhou J. 2020. Hierarchical elastomer tuned self-powered pressure sensor for wearable multifunctional cardiovascular electronics. *Nano Energy*. 70: 104460

34. Panganiban EB, Paglinawan AC, Chung WY, Paa GLS. 2021. ECG diagnostic support system (EDSS): A deep learning neural network based classification system for detecting ECG abnormal rhythms from a low-powered wearable biosensors. *Sensing and Bio-Sensing Research*. 31: 100398.

35. Ramasamy S, Balan A. 2018. Wearable sensors for ECG measurement: A review. *Sensor Review*. 38(4): 412–419.

36. Vlasov DV, Makukha VK, Tikhonov IV. 2017. Development of wearable sensor for cardiographic monitoring. In *2017 18th International Conference of Young Specialists on Micro/Nanotechnologies and Electron Devices (EDM)*, pp. 619–621. IEEE.

37. Jiang C, Faroqi L, Palaniappan L, Dunn J. 2019. Estimating personal resting heart rate from wearable biosensor data. In *2019 IEEE EMBS International Conference on Biomedical & Health Informatics*, pp. 1–4. IEEE

38. Rodriguez-Labra JI, Kosik CJ, Maddipatla D, Narakathu BB, Atashbar MZ. 2021. Development of a PPG sensor array as a wearable device for monitoring cardiovascular metrics. *IEEE Sensors Journal*. 21(23): 26320–26327

39. Okano T, Izumi S, Katsuura T, Kawaguchi H, Yoshimoto M. 2019. Multimodal cardiovascular information monitor using piezoelectric transducers for wearable healthcare. *Journal of Signal Processing Systems*. 91(9): 1053–1062

40. Coelln R, Dawe RJ, Leurgans SE, Curran TA., Truty T, Yu L, Buchman AS. 2019. Quantitative mobility metrics from a wearable sensor predict incident parkinsonism in older adults. *Parkinsonism & Related Disorders*. 65: 190–196

41. Bu LL, Yang K, Xiong WX, Liu FT, Anderson B, Wang Y, Wang J. 2016. Toward precision medicine in Parkinson's disease. *Annals of Translational Medicine*. 4(2)

42. Coelln R, Shulman LM. 2016. Clinical subtypes and genetic heterogeneity: Of lumping and splitting in Parkinson disease. *Current Opinion in Neurology*. 29(6): 727–734

43. Madrid-Navarro CJ, Escamilla-Sevilla F, Mínguez-Castellanos A, Campos M, Ruiz-Abellán F, Madrid JA, Rol MA. 2018. Multidimensional circadian monitoring by wearable biosensors in Parkinson's disease. *Frontiers in Neurology*. 9: 1–14

44. Saif N, Yan P, Niotis K, Scheyer O, Rahman A, Berkowitz M, Isaacson RS. 2020. Feasibility of using a wearable biosensor device in patients at risk for Alzheimer's disease dementia. *The Journal of Prevention of Alzheimer's Disease*. 7(2): 1–8

45. Masters CL, Bateman R, Blennow K. 2015. Alzheimer's disease. *Nature Reviews Disease Primers*. 1(1): 15056

46. Shahid R, Zafar MZ, Ali Z, Erum A. 2018. Schizophrenia: An overview. *Clinical Practice*. 15(5): 847–851

47. Weinberger DR. 2017. The neurodevelopmental origins of schizophrenia in the penumbra of genomic medicine. *World Psychiatry*. 16(3): 225

48. Kahn RS, Sommer IE, Murray RM. 2015. Insel TR. *Nature Reviews Disease Primers*. 1: 15067

49. Bradford DW, Stroup TS. 2004. Schizophrenia, drug therapy, and monitoring. *The New England Journal of Medicine*. 350(4): 415–416

50. Kilic T, Brunner V, Audoly L, Carrara S. 2016. Smart e-Patch for drugs monitoring in schizophrenia. In *2016 IEEE International Conference on Electronics, Circuits and Systems*, pp. 57–60. IEEE

51. Tron T, Resheff YS, Bazhmin M, Peled A, Weinshall D. 2017. Real-time schizophrenia monitoring using wearable motion sensitive devices. In *International Conference on Wireless Mobile Communication and Healthcare*, pp. 242–249. Springer

52. Hammond KB, Turcios NL, Gibson LE. 1994. Clinical evaluation of the macroduct sweat collection system and conductivity analyzer in the diagnosis of cystic fibrosis. *The Journal of Pediatrics*. 124(2): 255–260

53. Farrell PM, White TB, Ren CL, Hempstead SE, Accurso F, Derichs N, Sosnay PR. 2017. Diagnosis of cystic fibrosis: Consensus guidelines from the cystic fibrosis foundation. *The Journal of Pediatrics*. 181: 4–15

54. De Boeck K, Vermeulen F, Dupont L. 2017. The diagnosis of cystic fibrosis. *La Presse Médicale*. 46(6): e97–e108

55. Stern RC. 1997. The diagnosis of cystic fibrosis. *New England Journal of Medicine*. 336(7): 487–491

56. Chung M, Fortunato G, RadaAlessiMohan N. 2019. Wearable flexible sweat sensors for healthcare monitoring: A review. *Journal of the Royal Society Interface*. 16(159): 20190217

57. Liu C, Xu T, Wang D, Zhang X. 2020. The role of sampling in wearable sweat sensors. *Talanta*. 212: 120801

58. Gao W, Nyein HY, Shahpar Z, Fahad HM, Chen K, Emaminejad S, Javey A. 2016. Wearable microsensor array for multiplexed heavy metal monitoring of body fluids. *Acs Sensors*. 1(7): 866–874

59. Ghaffari R, Rogers JA, Ray TR. 2021. Recent progress, challenges, and opportunities for wearable biochemical sensors for sweat analysis. *Sensors and Actuators B: Chemical*. 332: 129447

60. Yu M, Li YT, Hu Y, Tang L, Yang F, Lv WL, Zhang GJ. 2021. Gold nanostructure-programmed flexible electrochemical biosensor for detection of glucose and lactate in sweat. *Journal of Electroanalytical Chemistry*. 882: 115029

61. Nyein HYY, Bariya M, Tran B, Ahn CH, Brown BJ, Ji W, Javey A. 2021. A wearable patch for continuous analysis of thermoregulatory sweat at rest. *Nature Communications*. 12(1): 1–13

62. Zhang Z, Azizi M, Lee M, Davidowsky P, Lawrence P, Abbaspourrad A. 2019. A versatile, cost-effective, and flexible wearable biosensor for in situ and ex situ sweat analysis, and personalized nutrition assessment. *Lab on a Chip*. 19(20): 3448–3460

63. Mohan AV, Rajendran V, Mishra RK, Jayaraman M. 2020. Recent advances and perspectives in sweat based wearable electrochemical sensors. *TrAC Trends in Analytical Chemistry*. 131: 116024

64. Sempionatto JR, Brazaca LC, García-Carmona L, Bolat G, Campbell AS, Martin A, Wang J. 2019. Eyeglasses-based tear biosensing system: Non-invasive detection of alcohol, vitamins and glucose. *Biosensors and Bioelectronics*. 137: 161–170

65. Yu L, Yang Z, An M. 2019. Lab on the eye: A review of tear-based wearable devices for medical use and health management. *Bioscience Trends*. 13(4): 308–313

66. Kim S, Jeon HJ, Park S, Lee DY, Chung E. 2020. Tear glucose measurement by reflectance spectrum of a nanoparticle embedded contact lens. *Scientific Reports*. 10(1): 1–8

67. Kudo H, Arakawa T, Mitsubayashi K. 2014. Status of soft contact lens biosensor development for tear sugar monitoring: A review. *Electronics and Communications in Japan*. 97(12): 52–56

68. Chu MK, Iguchi S, Miyajima K, Arakawa T, Kudo H, Mitsubayashi K. 2011. Development of a soft contact-lens biosensor for in-vivo tear glucose monitoring. In *5th*

European Conference of the International Federation for Medical and Biological Engineering, pp. 1007–1010. Springer

69. Garcia-Carmona L, Martin A, Sempionatto JR, Moreto JR, Gonzalez MC, Wang J, Escarpa A. 2019. Pacifier biosensor: Toward noninvasive saliva biomarker monitoring. *Analytical Chemistry*. 91(21): 13883–13891

70. Hong W, Lee WG. 2020. Wearable sensors for continuous oral cavity and dietary monitoring toward personalized healthcare and digital medicine. *Analyst*. 145(24): 7796–7808

71. Bandodkar AJ, Jeerapan I, Wang J. 2016. Wearable chemical sensors: Present challenges and future prospects. *Acs Sensors*. 1(5): 464–482

72. Yoon JH, Kim SM, Park HJ, Kim YK, Oh DX, Cho HW, Choi BG. 2020. Highly self-healable and flexible cable-type pH sensors for real-time monitoring of human fluids. *Biosensors and Bioelectronics*. 150: 111946

73. Arakawa T, Tomoto K, Nitta H, Toma K, Takeuchi S, Sekita T, Mitsubayashi K. 2020. A wearable cellulose acetate-coated mouthguard biosensor for in vivo salivary glucose measurement. *Analytical Chemistry*. 92(18): 12201–12207

74. Kim J, Imani S, Araujo WR, Warchall J, Valdés-Ramírez G, Paixão TR, Wang J. 2015. Wearable salivary uric acid mouthguard biosensor with integrated wireless electronics. *Biosensors and Bioelectronics*. 74: 1061–1068

75. Rashee A, Iranmanes E, Li W, Fen X, Andrenk AS, Wan K. 2017. Experimental study of human body effect on temperature sensor integrated RFID tag. In *2017 IEEE International Conference on RFID Technology & Application*, pp. 243–247. IEEE

76. Zhang H, Zhang R, Li Q, Yin L, Chen S, Wang T, Chen X. 2012. A GPRS-based wearable electronic thermometric alarm system. In *2012 5th International Conference on BioMedical Engineering and Informatics*, pp. 804–807. IEEE

77. Anuar H, Leow PL. 2019. Non-invasive core body temperature sensor for continuous monitoring. In *2019 IEEE International Conference on Sensors and Nanotechnology*, pp. 1–4. IEEE

78. Chen W, Dols S, Oetomo SB, Feijs L. 2010. Monitoring body temperature of newborn infants at neonatal intensive care units using wearable sensors. In *Proceedings of the Fifth International Conference on Body Area Networks*, pp. 188–194. ACM

79. Haines W, Momenroodaki P, Berry E, Fromandi M, Popovic Z. 2017. Wireless system for continuous monitoring of core body temperature. In *2017 IEEE MTT-S International Microwave Symposium*, pp. 541–543. IEEE

80. Caccami MC, Mulla MYS, Di Natale C, Marrocco G. 2017. Wireless monitoring of breath by means of a graphene oxide-based radiofrequency identification wearable sensor. In *2017 11th European Conference on Antennas and Propagation*, pp. 3394–3396. IEEE

81. Yang CM, Yang TL, Wu CC, Hung SH, Liao MH, Su MJ, Hsieh HC. 2014. Textile-based capacitive sensor for a wireless wearable breath monitoring system. In *2014 IEEE International Conference on Consumer Electronics*, pp. 232–233. IEEE

82. Jiang T, Deng L, Qiu W, Liang J, Wu Y, Shao Z, Lin L. 2020. Wearable breath monitoring via a hot-film/calorimetric airflow sensing system. *Biosensors and Bioelectronics*. 163: 112288

83. Ling TY, Wah LH, McBride JW, Chong HM, Pu SH. 2019. Nanocrystalline graphite humidity sensors for wearable breath monitoring applications. In *2019 IEEE International Conference on Sensors and Nanotechnology*, pp. 1–4. IEEE

84. Maier D, Laubender E, Basavanna A, Schumann S, Güder F, Urban GA, Dincer C. 2019. Toward continuous monitoring of breath biochemistry: A paper-based wearable sensor for real-time hydrogen peroxide measurement in simulated breath. *ACS Sensors*. 4(11): 2945–2951

85. He J, Wang C, Xu C, Duan S. 2017. Human motion monitoring platform based on positional relationship and inertial features. In *International Conference on Intelligent and Interactive Systems and Applications*, pp. 373–379. Springer

86. Wang S. 2017. A survey on wearable human motion state monitoring method using wearable. In *2017 IEEE 3rd Information Technology and Mechatronics Engineering Conference*, pp. 1235–1242. IEEE

87. Gong X, He H. 2020. Signal extraction and monitoring of motion loads based on wearable online device. *Computer Communications*. 154: 138–147

88. Qiao M, Zhang J, Cao Y, Wang Q, Ren X, Ai L, Zuo Y. 2017. A wearable motion monitoring fiber sensor based on graphene. In *Asia Communications and Photonics Conference*, pp. Su2C-3. Optical Society of America

89. Kang TH, Merritt CR, Grant E, Pourdeyhimi B, Nagle HT. 2007. Nonwoven fabric active electrodes for biopotential measurement during normal daily activity. *IEEE Transactions on Biomedical Engineering*. 55(1): 188–195

90. Xiaoxiang Z. 2020. Wearable health monitoring system based on human motion state recognition. *Computer Communications*. 150: 62–71

91. Wang C, Li X, Hu H, Zhang L, Huang Z, Lin M, Xu S. 2018. Monitoring of the central blood pressure waveform via a conformal ultrasonic device. *Nature Biomedical Engineering*. 2(9): 687–695

92. Arakawa T. 2018. Recent research and developing trends of wearable sensors for detecting blood pressure. *Sensors*. 18(9): 2772

93. Nilghaz A, Liu X, Ma L, Huang Q, Lu X. 2019. Development of fabric-based microfluidic devices by wax printing. *Cellulose*. 26(5): 3589–3599

94. Malon RS, Chua KY, Wicaksono DH, Córcoles EP. 2014. Cotton fabric-based electrochemical device for lactate measurement in saliva. *Analyst*. 139(12): 3009–3016

95. Bagherbaigi S, Córcoles EP, Wicaksono DH. 2014. Cotton fabric as an immobilization matrix for low-cost and quick colorimetric enzyme-linked immunosorbent assay. *Analytical Methods*. 6(18): 7175–7180

96. Nilghaz A, Wicaksono DH, Gustiono D, Majid FAA, Supriyanto E, Kadir MRA. 2012. Flexible microfluidic cloth-based analytical devices using a low-cost wax patterning technique. *Lab on a Chip*. 12(1): 209–218

97. Li S, Ma Z, Cao Z, Pan L, Shi Y. 2020. Advanced wearable microfluidic sensors for healthcare monitoring. *Small*. 16(9): 1903822

98. Huang J, Liu Y, Tang Z, Shao X, Zhang C. 2020. A polymer-based microfluidic sensor for biochemical detection. *IEEE S Jiangensors Journal*. 20(12): 6270–6276

99. Gray BL, Chung D. 2018. Wearable microfluidic and electronic frameworks for biomedical applications. In *ECS Meeting Abstracts*, p. 1534. IOP Publishing

100. Lim HR, Kim HS, Qazi R, Kwon YT, Jeong JW, Yeo WH. 2020. Advanced soft materials, sensor integrations, and applications of wearable flexible hybrid electronics in healthcare, energy, and environment. *Advanced Materials*. 32(15): 1901924

101. Liu L, Jiao Z, Zhang J, Wang Y, Zhang C, Meng X, Ren L. 2020. Bioinspired, superhydrophobic, and paper-based strain sensors for wearable and underwater applications. *ACS Applied Materials & Interfaces*. 13(1): 1967–1978

102. Martinez AW, Phillips ST, Whitesides GM. 2008. Three-dimensional microfluidic devices fabricated in layered paper and tape. *Proceedings of the National Academy of Sciences*. 105(50): 19606–19611

103. Abe K, Kaori K, Koji S, Daniel C. 2010. Inkjet-printed paperfluidic immuno-chemical sensing device. *Analytical and Bioanalytical Chemistry*. 398(2): 885–893

104. Ion M, Dinulescu S, Firtat B, Savin M, Ionescu ON, Moldovan C. 2021. Design and fabrication of a new wearable pressure sensor for blood pressure monitoring. *Sensors*. 21(6): 2075

105. Vinoth R, Nakagawa T, Mathiyarasu J, Mohan AV. 2021. Fully printed wearable microfluidic devices for high-throughput sweat sampling and multiplexed electrochemical analysis. *ACS Sensors.* 6(3): 1174–1186
106. Yang B, Kong J, Fang X. 2019. Bandage-like wearable flexible microfluidic recombinase polymerase amplification sensor for the rapid visual detection of nucleic acids. *Talanta.* 204: 685–692
107. He Z, Elbaz A, Gao B, Zhang J, Su E, Gu Z. 2018. Disposable Morpho menelaus based flexible microfluidic and electronic sensor for the diagnosis of neurodegenerative disease. *Advanced Healthcare Materials.* 7(5): 1701306
108. Gao Y, Ota H, Schaler E W, Chen K, Zhao A, Gao W, Javey A. 2017. Wearable microfluidic diaphragm pressure sensor for health and tactile touch monitoring. *Advanced Materials.* 29(39): 1701985
109. Nightingale AM, Leong CL, Burnish RA, Hassan SU, Zhang Y, Clough GF, Niu X. 2019. Monitoring biomolecule concentrations in tissue using a wearable droplet microfluidic-based sensor. *Nature Communications.* 10(1): 1–12
110. Agaoglu S, Diep P, Martini AmjadiM, Samudhyatha KT, Baday M, Araci IE. 2018. Ultra-sensitive microfluidic wearable strain sensor for intraocular pressure monitoring. *Lab on a Chip.* 18(22): 3471–3483
111. Ardalan S, Hosseinifard M, Vosough M, Golmohammadi H. 2020. Towards smart personalized perspiration analysis: An IoT-integrated cellulose-based microfluidic wearable patch for smartphone fluorimetric multi-sensing of sweat biomarkers. *Biosensors and Bioelectronics.* 168: 112450
112. Xiao J, Liu Y, Su L, Zhao D, Zhao L, Zhang X. 2019. Microfluidic chip-based wearable colorimetric sensor for simple and facile detection of sweat glucose. *Analytical Chemistry.* 91(23): 14803–14807
113. Vaquer A, Barón E, de la Rica R. 2020. Wearable analytical platform with enzyme-modulated dynamic range for the simultaneous colorimetric detection of sweat volume and sweat biomarkers. *ACS Sensors.* 6(1): 130–136
114. Li M, Wang L, Liu R, Li J, Zhang Q, Shi G, Wang H. 2021. A highly integrated sensing paper for wearable electrochemical sweat analysis. *Biosensors and Bioelectronics.* 174: 112828
115. Cao Q, Liang B, Tu T, Wei J, Fang L, Ye X. 2019. Three-dimensional paper-based microfluidic electrochemical integrated devices (3D-PMED) for wearable electrochemical glucose detection. *RSC Advances.* 9(10): 5674–5681
116. Liang B, Cao Q, Mao X, Pan W, Tu T, Fang L, Ye X. 2021. An integrated paper-based microfluidic device for real-time sweat potassium monitoring. *IEEE Sensors Journal.* 21(8): 9642–9648
117. Yang Y, Song Y, Bo X, Min J, Pak OS, Zhu L, Gao W. 2020. A laser-engraved wearable sensor for sensitive detection of uric acid and tyrosine in sweat. *Nature Biotechnology.* 38(2): 217–224
118. Kassal P, Sigurnjak M, Steinberg IM. 2019. based ion-selective optodes for continuous sensing: Reversible potassium ion monitoring. *Talanta.* 193: 51–55
119. Castro LF, Freitas SV, Duarte LC, Souza JAC, Paixão TR, Coltro WK. 2019. Salivary diagnostics on paper microfluidic devices and their use as wearable sensors for glucose monitoring. *Analytical and Bioanalytical Chemistry.* 411(19): 4919–4928
120. Vu CC, Kim SJ, Kim J. 2021. Flexible wearable sensors—an update in view of touch-sensing. *Science and Technology of Advanced Materials.* 22(1): 26–36
121. Windmiller JR, Wang J. 2013. Wearable electrochemical sensors and biosensors: A review. *Electroanalysis.* 25(1): 29–46
122. Jiang X, Lillehoj PB. 2020. Microneedle-based skin patch for blood-free rapid diagnostic testing. *Microsystems & Nanoengineering.* 6(1): 1–11

123. Wang B, Facchetti A. 2019. Mechanically flexible conductors for stretchable and wearable e-skin and e-textile devices. *Advanced Materials*. 31(28): 1–53

124. Stoppa M, Chiolerio A. 2014. Wearable electronics and smart textiles: A critical review. *Sensors*. 14(7): 11957–11992

125. Xu K, Lu Y, Takei K. 2019. Multifunctional skin-inspired flexible sensor systems for wearable electronics. *Advanced Materials Technologies*. 4(3): 1–25

126. Kwon YT, Kim YS, Kwon S, Mahmood M, Lim HR, Park SW, Yeo WH. 2020. All-printed nanomembrane wireless bioelectronics using a biocompatible solderable graphene for multimodal human-machine interfaces. *Nature Communications*. 11(1): 1–11

127. Kim K, Park YG, Hyun BG, Choi M, Park JU. 2019. Recent advances in transparent electronics with stretchable forms. *Advanced Materials*. 31(20): 1–20

128. Nguyen EP, De Carvalho Castro Silva C, Merkoçi A. 2020. Recent advancement in biomedical applications on the surface of two-dimensional materials: From biosensing to tissue engineering. *Nanoscale*. 12(37): 19043–19067

129. Lee H, Choi TK, Lee YB, Cho HR, Ghaffari R, Wang L, Choi HJ, Chung TD, Lu N, Hyeon T, Choi SH, Kim DH. 2016. A graphene-based electrochemical device with thermoresponsive microneedles for diabetes monitoring and therapy. *Nature Nanotechnology*. 11(6): 566–572

130. Wang L, Wang K, Lou Z, Jiang K, Shen G. 2018. Plant-based modular building blocks for "green" electronic skins. *Advanced Functional Materials*. 28(51): 1804510

131. Yetisen AK, Qu H, Manbachi A, Butt H, Dokmeci MR, Hinestroza JP, Skorobogatiy M, Khademhosseini A, Yun SH. 2016. Nanotechnology in textiles. *ACS Nano*. 10(3): 3042–3068

132. Yang T, Xie D, Li Z, Zhu H. 2017. Recent advances in wearable tactile sensors: Materials, sensing mechanisms, and device performance. *Materials Science and Engineering R: Reports*. 115: 1–37

133. Seshadri DR, Li RT, Voos JE, Rowbottom JR, Alfes CM, Zorman, CA, Drummond CK. 2019. Wearable sensors for monitoring the physiological and biochemical profile of the athlete. *NPJ Digital Medicine*. 2(1): 1–16

134. Zheng XQ, Cheng HY. 2019. Flexible and stretchable metal oxide gas sensors for healthcare. *Science China Technological Sciences*. 62(2): 209–223

135. Hu J, Meng H, Li G, Ibekwe SI. 2012. A review of stimuli-responsive polymers for smart textile applications. *Smart Materials and Structures*. 21(5): 053001

136. Kim J, Lee J, Son D, Choi MK, Kim DH. 2016. Deformable devices with integrated functional nanomaterials for wearable electronics. *Nano Convergence*. 3(1): 1–13

137. Lee W, Yun H, Song JK, Sunwoo SH, Kim DH 2021. Nanoscale materials and deformable device designs for bioinspired and biointegrated electronics. *Accounts of Materials Research*. 2(4): 266–281

138. Grayson ACR, Shawgo RS, Johnson AM, Flynn NT, Li Y, Cima MJ, Langer R. 2004. A BioMEMS review: MEMS technology for physiologically integrated devices. *Proceedings of the IEEE*. 92(1): 6–21

139. Qi Y, McAlpine MC. 2010. Nanotechnology-enabled flexible and biocompatible energy harvesting. *Energy & Environmental Science*. 3(9): 1275–1285

140. Matsumoto K, Ueno K, Hirotani J, Ohno Y, Omachi H. 2020. Fabrication of carbon nanotube thin films for flexible transistors by using a cross-linked amine polymer. *Chemistry—A European Journal*. 26(28): 6118–6121

141. Kim T, Park J, Sohn J, Cho D, Jeon S. 2016. Bioinspired, highly stretchable, and conductive dry adhesives based on 1D-2D hybrid carbon nanocomposites for all-in-one ECG electrodes. *ACS Nano*. 10(4): 4770–4778

142. Park S, Kim H, Kim JH, Yeo WH. 2020. Advanced nanomaterials, printing processes, and applications for flexible hybrid electronics. *Materials*. 13(16): 3587

143. Singh R, Singh E, Nalwa HS. 2017. Inkjet printed nanomaterial based flexible radio frequency identification (RFID) tag sensors for the internet of nano things. *RSC Advances*. 7(77): 48597–48630

144. Dervisevic M, Alba M, Prieto-Simon B, Voelcker NH. 2020. Skin in the diagnostics game: Wearable biosensor nano-and microsystems for medical diagnostics. *Nano Today*. 30: 100828

145. Tung TT, Tripathi KM, Kim T, Krebsz M, Pasinszki T, Losic D. 2019. Carbon nanomaterial sensors for cancer and disease diagnosis. *Carbon Nanomaterials for Bioimaging, Bioanalysis and Therapy, John Wiley and Sons Ltd*. 167–202

146. Gao W, Emaminejad S, Nyein HYY, Challa S, Chen K, Peck A, Javey A. 2016. Fully integrated wearable sensor arrays for multiplexed in situ perspiration analysis. *Nature*. 529(7587): 509–514

147. Yao S, Swetha P, Zhu Y. 2018. Nanomaterial-enabled wearable sensors for healthcare. *Advanced Healthcare Materials*. 7(1): 1700889

148. Choi SJ, Kim ID. 2018. Recent developments in 2D nanomaterials for chemiresistive-type gas sensors. *Electronic Materials Letters*. 14(3): 221–260

149. Davaji B, Cho HD, Malakoutian M, Lee JK, Panin G, Kang TW, Lee CH. 2017. A patterned single layer graphene resistance temperature sensor. *Scientific Reports*. 7(1): 1–10

150. Parlak O, Keene ST, Marais A, Curto VF, Salleo A. 2018. Molecularly selective nanoporous membrane-based wearable organic electrochemical device for noninvasive cortisol sensing. *Science Advances*. 4(7): eaar2904

151. Trung TQ, Lee NE. 2016. Flexible and stretchable physical sensor integrated platforms for wearable human-activity monitoringand personal healthcare. *Advanced Materials*. 28(22): 4338–4372

152. Souri H, Banerjee H, Jusufi A, Radacsi N, Stokes AA, Park I, Sitti M, Amjadi M. 2020. Wearable and stretchable strain sensors: Materials, sensing mechanisms, and applications. *Advanced Intelligent Systems*. 2(8): 2000039

153. Wang L, Lou Z, Jiang K, Shen G. 2019. Bio-multifunctional smart wearable sensors for medical devices. *Advanced Intelligent Systems*. 1(5): 1900040

154. Amjadi M, Kyung KU, Park I, Sitti M. 2016. Stretchable, skin-mountable, and wearable strain sensors and their potential applications: A review. *Advanced Functional Materials*. 26(11): 1678–1698

155. Jung H, Park C, Lee H, Hong S, Kim H, Cho SJ. 2019. Nano-cracked strain sensor with high sensitivity and linearity by controlling the crack arrangement. *Sensors*. 19(12): 2834

156. Yao S, Ren P, Song R, Liu Y, Huang Q, Dong J, O'Connor BT, Zhu Y. 2020. Nanomaterial-enabled flexible and stretchable sensing systems: Processing, integration, and applications. *Advanced Materials*. 32(15): 1902343

157. Xin M, Li J, Ma Z, Pan L, Shi Y. 2020. MXenes and their applications in wearable sensors. *Frontiers in Chemistry*. 8: 297

158. Barry RC, Lin Y, Wang J, Liu G, Timchalk CA. 2009. Nanotechnology-based electrochemical sensors for biomonitoring chemical exposures. *Journal of Exposure Science & Environmental Epidemiology*. 19(1): 1–18

159. Manickam P, Pasha SK, Snipes SA, Bhansali S. 2017. A reusable electrochemical biosensor for monitoring of small molecules (cortisol) using molecularly imprinted polymers. *Journal of the Electrochemical Society*. 164(2): 54–59

160. Fan R, Andrew TL. 2020. Perspective—challenges in developing wearable electrochemical sensors for longitudinal health monitoring. *Journal of the Electrochemical Society*. 167(3): 037542

161. Valera E, Jankelow A, Lim J, Kindratenko V, Ganguli A, White K, Kumar J, Bashir R. 2021. COVID-19 point-of-care diagnostics: Present and future. *ACS Nano*. 15(5): 7899–7906

162. Mannoor MS, Tao H, Clayton JD, Sengupta A, Kaplan DL, Naik RR, McAlpine MC. 2012. Graphene-based wireless bacteria detection on tooth enamel. *Nature Communications*. 3(1): 1–9

163. Li X, Dunn J, Salins D, Zhou G, Zhou W, Schüssler-Fiorenza Rose SM, Snyder MP. 2017. Digital health: Tracking physiomes and activity using wearable biosensors reveals useful health-related information. *PLoS Biology*. 15(1): e2001402

164. Hwang HS, Jeong JW, Kim YA, Chang M. 2020. Carbon nanomaterials as versatile platforms for biosensing applications. *Micromachines*. 11(9): 814

165. Tiwari JN, Vij V, Kemp KC, Kim KS. 2016. Engineered carbon-nanomaterial-based electrochemical sensors for biomolecules. *ACS Nano*. 10(1): 46–80

166. Zhu Z, Garcia-Gancedo L, Flewitt AJ, Xie H, Moussy F, Milne WI. 2012. A critical review of glucose biosensors based on carbon nanomaterials: Carbon nanotubes and graphene. *In Sensors (Switzerland)*. 12(5): 5996–6022

167. Tabish TA, Abbas A, Narayan RJ. 2021. Graphene nanocomposites for transdermal biosensing. *Wiley Interdisciplinary Reviews: Nanomedicine and Nanobiotechnology*. 13(4): e1699

168. Takei K, Honda W, Harada S, Arie T, Akita S. 2015. Toward flexible and wearable human-interactive health-monitoring devices. *Advanced Healthcare Materials*. 4(4): 487–500

169. Son D, Lee J, Qiao S, Ghaffari R, Kim J, Lee JE, Song C, Kim SJ, Lee DJ, Jun SW, Yang S. Park M, Shin J, Do K, Lee M, Kang K, Hwang CS, Lu N, Hyeon T, Kim DH. 2014. Multifunctional wearable devices for diagnosis and therapy of movement disorders. *Nature Nanotechnology*. 9(5): 397–404

170. Lee SP, Ha G, Wright DE, Ma Y, Sen-Gupta E, Haubrich NR, Branche PC, Li W, Huppert GL, Johnson M, Mutlu HB, Li K, Sheth N, Wright JA, Huang Y, Mansour M, Rogers JA, Ghaffari R. 2018. Highly flexible, wearable, and disposable cardiac biosensors for remote and ambulatory monitoring. *Npj Digital Medicine*. 1(1): 2

171. Mohankumar P, Ajayan J, Mohanraj T, Yasodharan R. 2021. Recent developments in biosensors for healthcare and biomedical applications: A review. *Measurement: Journal of the International Measurement Confederation*. 167: 108293

172. Ray T, Choi J, Reeder J, Lee SP, Aranyosi AJ, Ghaffari R, Rogers JA. 2019. Soft, skin-interfaced wearable systems for sports science and analytics. *Current Opinion in Biomedical Engineering*. 9: 47–56

173. Ponmozhi J, Frias C, Marques T, Frazão O. 2012. Smart sensors/actuators for biomedical applications. *Measurement*. 45(7): 1675–1688

174. Hsien B, Yeh W, Ai HW. 2019. Development and applications of bioluminescent and chemiluminescent reporters and biosensors. *Annual Review of Analytical Chemistry*. 12: 1–22

175. Zhu M, Yi Z, Yang B, Lee C. 2021. Making use of nanoenergy from human—Nanogenerator and self-powered sensor enabled sustainable wireless IoT sensory systems. *In Nano Today*. 36: 101016

176. Ali H, Tariq UU, Hardy J, Zhai X, Lu L, Zheng Y, Antonopoulos N. 2021. A survey on system level energy optimisation for MPSoCs in IoT and consumer electronics. *Computer Science Review*. 41: 100416

177. Kashani MH, Madanipour M, Nikravan M, Asghari P, Mahdipour E. 2021. A systematic review of IoT in healthcare: Applications, techniques, and trends. *Journal of Network and Computer Applications*. 103164

178. Jin X, Liu C, Xu T, Su L, Zhang X. 2020. Artificial intelligence biosensors: Challenges and prospects. *Biosensors and Bioelectronics*. 112412

179. Salim A, Lim S. 2019. Recent advances in noninvasive flexible and wearable wireless biosensors. *Biosensors and Bioelectronics*. 141: 111422

180. Pallavi S, Narayanan, VA. 2019. An overview of practical attacks on BLE based IOT devices and their security. *2019 5th International Conference on Advanced Computing and Communication Systems*. 694–698

181. Rajamohanan D, Hariharan B, Menon KU. 2019. Survey on smart health management using BLE and BLE beacons. In *2019 9th International Symposium on Embedded Computing and System Design*, pp. 1–5. IEEE

182. Zhang Y, Weng J, Ling Z, Pearson B, Fu X. 2020. *BLESS: A BLE Application Security Scanning Framework. Proceedings*, pp. 636–645. Springer

183. Shan G, Im SY, Roh BH. 2016. Optimal AdvInterval for BLE scanning in different number of BLE devices environment. In *2016 IEEE Conference on Computer Communications Workshops*, pp. 1031–1032. IEEE

184. Dian FJ, Yousefi A, Lim S. 2018. A practical study on bluetooth low energy (BLE) throughput. In *2018 IEEE 9th Annual Information Technology, Electronics and Mobile Communication Conference*, pp. 768–771. IEEE

185. Baert M, Camerlynck P, Crombez P, Hoebeke J. 2019. A BLE-based multi-gateway network infrastructure with handover support for mobile BLE peripherals. In *2019 IEEE 16th International Conference on Mobile Ad Hoc and Sensor Systems*, pp. 91–99. IEEE

186. Microchip Technology Inc. 2020. *Bluetooth® Low Energy Discovery Process—Developer Help*. https://microchipdeveloper.com/wireless:ble-introduction

187. Zhi-yuan Z, He R, Jie T. 2010. A method for optimizing the position of passive UHF RFID tags. *2010 IEEE International Conference on RFID-Technology and Applications*. 92–95

188. Kapucu K, Dehollain C. 2014. A passive UHF RFID system with a low-power capacitive sensor interface. In *2014 IEEE RFID Technology and Applications Conference*, pp. 301–305. IEEE

189. Periyasamy M, Karthikeyan S, Mahendran G. 2020. Electromagnetic immunity testing of radio frequency identification devices (RFID) in healthcare environment—a selected review. *Materials Today: Proceedings*. DOI:10.1016/j.matpr.2020.09.372

190. Bevacqua MT, Bellizzi GG, Merenda M. 2019. Field focusing for energy harvesting applications in smart RFID tag. In *2019 IEEE International Conference on RFID Technology and Applications*, pp. 263–266. IEEE

191. Chen YL, Liu D, Wang S, Li YF, Zhang XS. 2019. Self-powered smart active RFID tag integrated with wearable hybrid nanogenerator. *Nano Energy*. 64: 103911

192. Boada M, Lazaro A, Villarino R, Girbau D. 2018. Battery-less soil moisture measurement system based on a NFC device with energy harvesting capability. *IEEE Sensors Journal*. 18(13): 5541–5549

193. Daskalakis SN, Goussetis G, Georgiadis A. 2019. NFC hybrid harvester for battery-free agricultural sensor nodes. In *2019 IEEE International Conference on RFID Technology and Applications (RFID-TA)*, pp. 22–25. IEEE

194. Steinberg MD, Kimbriel CS, d'Hont LS. 2019. Autonomous near-field communication (NFC) sensors for long-term preventive care of fine art objects. *Sensors and Actuators A: Physical*. 285: 456–467

195. Zhou C, Zhou T, Bai W. 2018. The key study of the integration between smartphone NFC technology and ERP system. *2018 3rd IEEE International Conference on Cloud Computing and Big Data Analysis, ICCCBDA*. 500–505

196. Chishti MS, King CT, Banerjee A. 2021. Exploring half-duplex communication of NFC read/write mode for secure multi-factor authentication. *IEEE Access*. 9: 6344–6357

197. Ukalkar GV, Halgaonkar PS. 2017. Cloud based NFC health card system. In *2017 International Conference on Intelligent Computing and Control Systems*, pp. 436–441. IEEE

198. Kawahara Y, Sugiyama T. 2020. IEEE802.11 priority control by using even-odd selection backoff slot in a home environment. In *2020 International Conference on Information and Communication Technology Convergence*, pp. 48–50. IEEE

199. Walrand J, Parekh S. 2010. Communication networks: A concise introduction. *Synthesis Lectures on Communication Networks*. 3(1): 1–192

200. Gunasagaran R, Kamarudin LM, Zakaria A. 2018. Embedded device free passive (EDfP) system: Effect of WiFi protocols. In *2018 IEEE Student Conference on Research and Development*, pp. 1–5. IEEE

201. Lu Y, Shi Z, Liu Q. 2019. Smartphone-based biosensors for portable food evaluation. *Current Opinion in Food Science*. 28: 74–81

202. Shalannanda W, Zakia I, Fahmi F, Sutanto E. 2020. Implementation of the hardware module of IoT-based infant incubator monitoring system. In *2020 14th International Conference on Telecommunication Systems, Services, and Applications*, pp. 1–5. IEEE

203. Wang W, Chen Y, Wang L, Zhang Q. 2017. Sampleless Wi-Fi: Bringing low power to Wi-Fi communications. *IEEE/ACM Transactions on Networking*. 25(3): 1663–1672

204. Li F, Valero M, Shahriar H, Khan RA, Ahamed SI. 2021. Wi-COVID: A COVID-19 symptom detection and patient monitoring framework using WiFi. *Smart Health*. 19: 100147

205. Pan G, He J, Wu Q, Fang R, Cao J, Liao D. 2018. Automatic stabilization of Zigbee network. In *2018 International Conference on Artificial Intelligence and Big Data*, pp. 224–227. IEEE

206. Abd Rahman AN, Habaebi MH, Ismail M. 2014. Android-based P2P file sharing over ZigBee radios. In *2014 5th International Conference on Intelligent Systems, Modelling and Simulation*, pp. 591–596. IEEE

207. Elarabi T, Deep V, Rai CK. 2015. Design and simulation of state-of-art ZigBee transmitter for IoT wireless devices. In *2015 IEEE International Symposium on Signal Processing and Information Technology*, pp. 297–300. IEEE

208. Xiaoman L, Xia L. 2016. Design of a ZigBee wireless sensor network node for aquaculture monitoring. In *2016 2nd IEEE International Conference on Computer and Communications*, pp. 2179–2182. IEEE

209. Bai Y, Guo Q, Xiao J, Zheng M, Zhang D, Yang J. 2021. An inkjet-printed smartphone-supported electrochemical biosensor system for reagentless point-of-care analyte detection. *Sensors and Actuators B: Chemical*. 346: 130447

210. Reda A, El-Safty SA, Selim MM, Shenashen MA. 2021. Optical glucose biosensor built-in disposable strips and wearable electronic devices. *Biosensors and Bioelectronics*. 185: 113237

211. Lu Y, Jiang K, Chen D, Shen G. 2019. Wearable sweat monitoring system with integrated micro-supercapacitors. *Nano Energy*. 58: 624–632

212. Zhao F, He J, Li X, Bai Y, Ying Y, Ping J. 2020. Smart plant-wearable biosensor for in-situ pesticide analysis. *Biosensors and Bioelectronics*. 170: 112636

213. Purohit B, Kumar A, Mahato K, Chandra P. 2020. Smartphone-assisted personalized diagnostic devices and wearable sensors. *Current Opinion in Biomedical Engineering*. 13: 42–50

214. Zhang D, Liu Q. 2016. Biosensors and bioelectronics on smartphone for portable biochemical detection. *Biosensors & Bioelectronics*. 75: 273–284

215. Liao J, Chang F, Han X, Ge C, Lin S. 2020. Wireless water quality monitoring and spatial mapping with disposable whole-copper electrochemical sensors and a smartphone. *Sensors and Actuators, B: Chemical*. 306: 127557

216. TNW. 2021. *Digital Trends 2021: Every Single Stat Marketers Need to Know*. https://thenextweb.com/news/insights-global-state-of-digital-social-media-2021

217. Hu J, Wang J, Xie H. 2020. Wearable bracelets with variable sampling frequency for measuring multiple physiological parameter of human. *Computer Communications*. 161: 257–265

218. Wang Y, Fridberg DJ, Shortell DD, Leeman RF, Barnett NP, Cook RL, Porges EC. 2021. Wrist-worn alcohol biosensors: Applications and usability in behavioral research. *Alcohol*. 92: 25–34
219. He T, Guo X, Lee C. 2020. Flourishing energy harvesters for future body sensor network: From single to multiple energy sources. *Iscience*. 101934
220. Wang L, Hu J, Yu Y, Huang K, Hu Y. 2020. Lithium-air, lithium-sulfur, and sodium-ion, which secondary battery category is more environmentally friendly and promising based on footprint family indicators? *Journal of Cleaner Production*. 276: 124244
221. Vichard L, Steiner NY, Zerhouni N, Hissel D. 2021. Hybrid fuel cell system degradation modeling methods: A comprehensive review. *Journal of Power Sources*. 506: 230071
222. Kim J, Byun S, Lee S, Ryu J, Cho S, Oh C, Kim H, No K, Ryu S, Lee YM, Hong S. 2020. Cost-effective and strongly integrated fabric-based wearable piezoelectric energy harvester. *Nano Energy*. 75: 104992
223. Ni QQ, Guan X, Zhu Y, Dong Y, Xia H. 2020. Nanofiber-based wearable energy harvesters in different body motions. *Composites Science and Technology*. 200: 108478

Index

For Product Safety Concerns and Information please contact our EU
representative GPSR@taylorandfrancis.com
Taylor & Francis Verlag GmbH, Kaufingerstraße 24, 80331 München, Germany